Sand Sheff

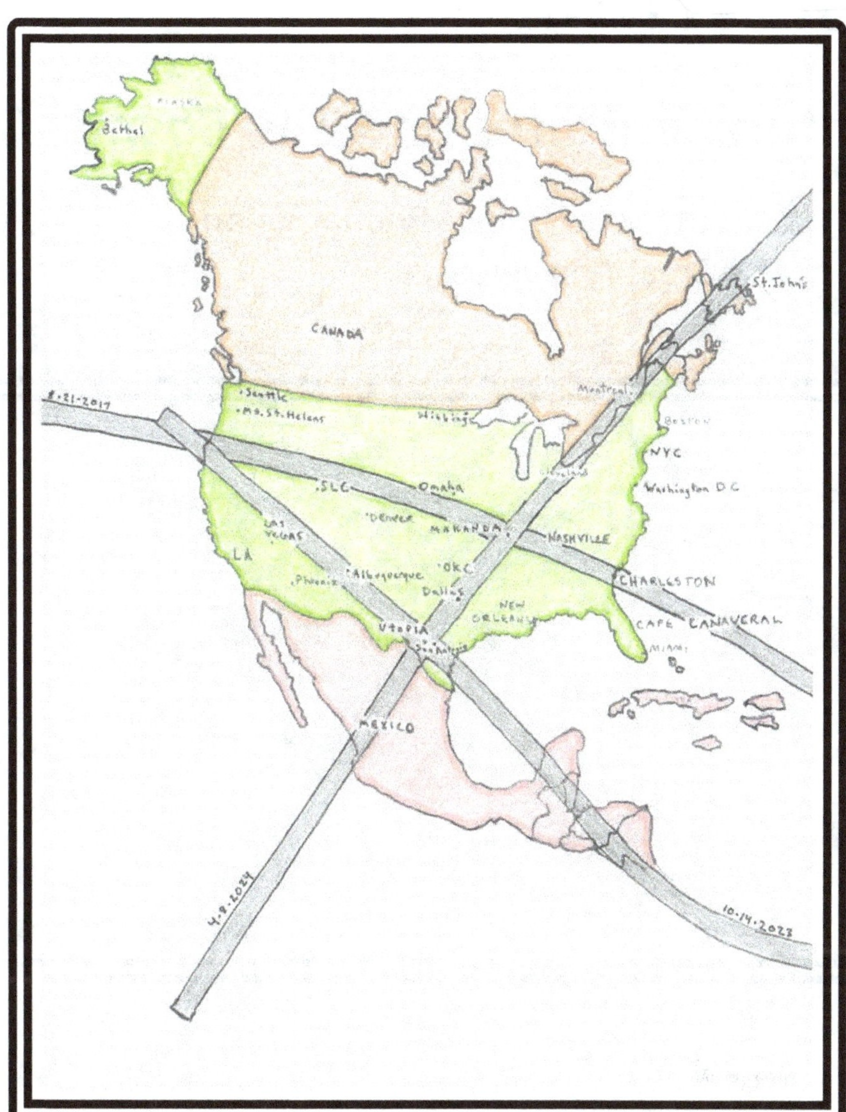

Eclipse Witness

Supernatural Patterns and the Cosmic Clock

Evidence of God in the Solar Eclipse Phenomenon
In the Time of the Great Anniversary

Eclipse Witness

Supernatural Patterns and The Cosmic Clock

Evidence of God in the Solar Eclipse Phenomenon
In the Time of The Great Anniversary

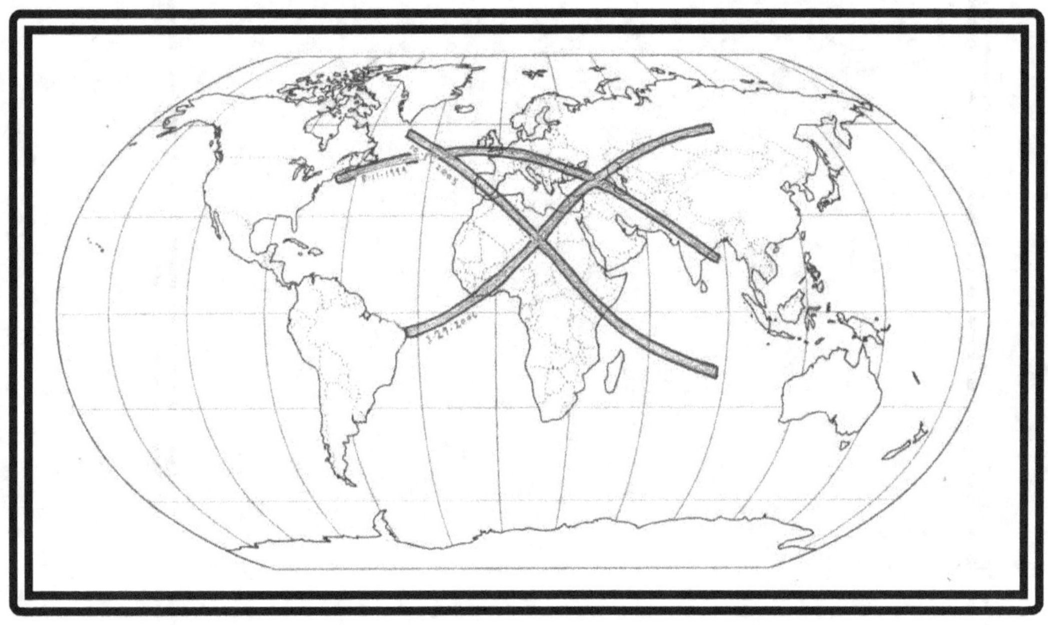

by Sand Sheff

with maps created and illustrated by the Author

Eclipse Witness Table of Contents

Definitions from the cover as intended in Eclipse Witness:

Supernatural Pattern: a hieroglyphic physical expression of Divine perspective involving historical, cosmological and mathematical alignment. Resembles poetry, art or songs.

Cosmic Clock: The created order of Time for the purpose of story. The interaction of the solar, lunar and eclipse calendars.

The Great Anniversary: The 2,000[th] Anniversary of the Crucifixion and Resurrection of Jesus Christ: the solar year anniversary occurring sometime between 2027 and 2034.

Some Quotes to start the book:

In the beginning, God Created the Heavens and The Earth.
Genesis 1:1

"Synchronicity is an ever-present reality for those who have eyes to see."
Carl Jung

"Look Mankind, I say to you all, in essence you are living in a clock. The clock keeps perfect time, to an accuracy of one second/day! How could such a clock in the heavens come to be without there being some being, who, with perception and understanding, who, with a plan and with the power, could form that clock?"

-Norman Bloom, homeless ice cream man and famous antagonist of scientists in the 1970's: quoted by Carl Sagan in *God and Norman Bloom*

"There will be signs in the sun, moon and stars. On the earth, nations will be in anguish and perplexity at the roaring and tossing of the sea. People will faint from terror, apprehensive of what is coming on the world, for the heavenly bodies will be shaken. At that time, they will see the Son of Man coming in a cloud with power and great glory. When these things begin to take place, stand up and lift up your heads, because your redemption is drawing near."

-Jesus / in Luke 21: 25-28

For my family…

Introduction: Does this image look peculiar to you?

This is a drawing of the paths of four total solar eclipses that occur between 2017 and 2034.

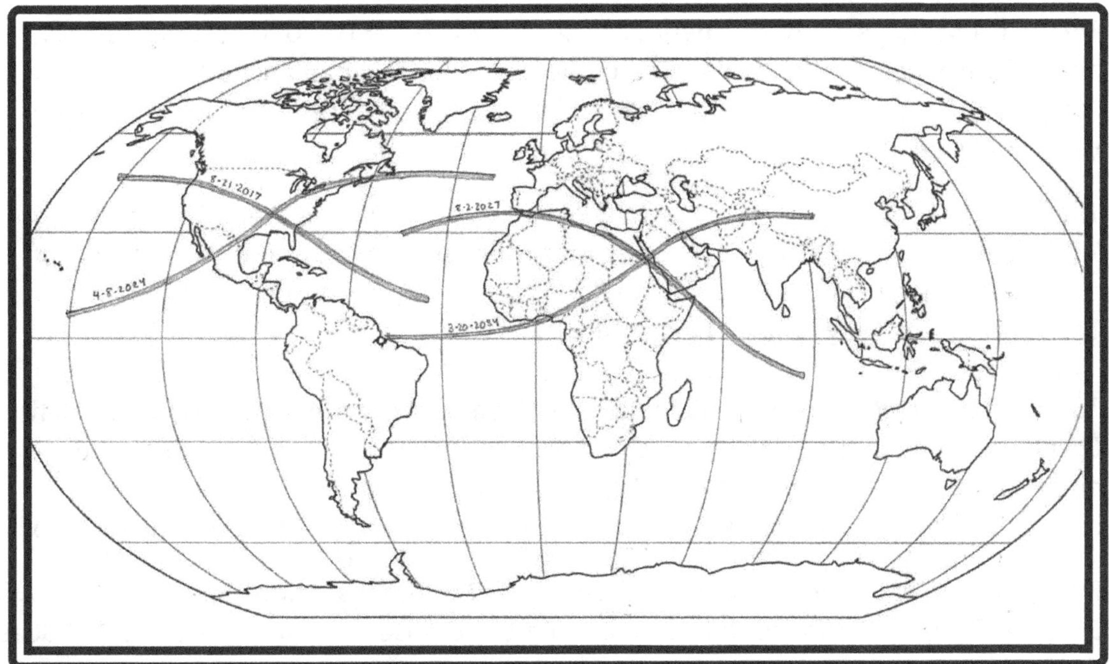

They represent lines of shadow, each between 70 and 150 miles wide, cast on earth by the moon as it passes directly in front of the sun.

Does anything in the image strike you as peculiar? Perhaps by now you've heard about these eclipses. Perhaps not. Of course, most people notice that the four eclipses in the image create two balanced crosses, or x's. Though that pattern is obvious, one might hesitate before calling it peculiar. After all, maybe solar eclipses do that sort of thing all the time? Curiosity about these crosses started me on this work. The work ended up being about a lot more than just them.

These eclipses certainly don't belong to me. Neither do they belong to "science". They are the work of a Master Artist. That acknowledgment opens up a big supernatural can of questions. Why would The Artist do this? Is that Artist saying something to us? I offer historical and geographical associations which might inform one to draw their own conclusions.

This book presents evidence that these two crosses are indeed peculiar. It also asserts that the eclipse phenomenon is likely the most peculiar reality in all of nature, aside from perhaps the existence of intelligent life. While eclipses have been studied by science for centuries, I believe there is much to be learned by looking at solar eclipses in a different light.

Consider again the two crosses at the top of this page. One cross is over the western hemisphere, centered on North America. The other is over the eastern hemisphere, centered on the Middle East. I witnessed the first one in Wyoming USA in August of 2017, just after publishing a children's book about eclipses, *Eclipse Miracle*.

After that experience, I began researching the next USA eclipse, which comes in 2024. I noticed that online maps of these two USA eclipses formed a nicely shaped x. I felt an urge to figure out if such crosses were commonplace. As I researched, I noted certain peculiarities. Here's a few:

- The first set of eclipses, over America Aug 21 2017 and Apr 08 2024 occur 2422 days apart. Both happen on a Monday. The paths "intersect" over an area of southern Illinois, Tennessee, Kentucky and Missouri that's been known as "Little Egypt" for almost 200 years. They are the first cross-country US eclipses in 99 years.

- The second set of eclipses, over the Middle East Aug 2 2027 and Mar 20 2034 also occur 2422 days apart. They also both happen on a Monday. The paths intersect over Egypt and the Red Sea, near the Valley of the Kings, the Necropolis of the Egyptian Pharaohs. They are the first total eclipses in Egypt for 122 years.

- To spell it out plainly, the apex of one cross is Egypt and the other Little Egypt. Ancient Egypt was home to the first distinct empire in human history. America is the last distinct empire. Both were culturally influential, complicated, and scientifically advanced societies with a 400-year history of famous slavery. There is a pyramid on the dollar bill.

- From the date of the first American Eclipse in 2017 to the first Mideast Eclipse in 2027 is 3633 days. From the date of the second American Eclipse in 2024 to the second Mideast Eclipse in 2034 is 3633 days. A cycle was at work of 1211, 2422 and 3633 days

- Between the two American Eclipses, at the 1211-day mark, a total eclipse occurred in the southern western hemisphere, on Dec 14 2020, precisely at the moment the Electoral College voted in the President and Vice-President of the United States.

- 1211 days turns out to be exactly 7 eclipse seasons. 2422 days is 7 eclipse years. These types of seasons and years are almost unknown to the general public.

- The two sets of x's or crosses are strikingly balanced. As it turns out, the horizontal angle of both sets of intersections are close to 66 degrees.

- Two intriguing Ring of Fire Eclipses also intersect the general areas of the crosses in each hemisphere, one in America in 2023 and one in Eurasia in 2030. The American Ring of Fire creates the distinctive A pattern seen on the cover of this book.

I wanted first to determine if these crosses were peculiar. Why? Because pattern recognition and peculiarity of appearance are two of our only ways to determine if something has significance. I set out to find a way to consider that peculiarity. I did this by grouping eclipses based on time frames and cycles.

Pattern recognition is at the center of human intelligence. Determining what patterns are considered meaningful is at the forefront of social behavior, science and education.

After quite a bit of study, I concluded that I was dealing with supernatural arrangements of an extraordinary natural phenomenon. I determined this in the same way that one determines a series of notes played in a certain arrangement are clearly music. A Beethoven sonata is distinct from the sounds of a busy city, but only so by the acknowledgment and intuition of the listener. It is perspective, no doubt, but one informed by experience and intuition.

But to demonstrate patterns and peculiarities? We must examine a large set of historical eclipses for comparison. It took lots of trial and error to find a workable mechanism and presentation for such a complicated and slippery phenomenon. The book you hold is what I came up with. It's not perfect, but it's also the only work of its kind.

Fortunately, we are not attempting to chronicle thousands of eclipses in this work. Complete solar eclipses are rare enough that, even over the course of a century and a half, we are only dealing with about 200 eclipses. I also map eclipses in the US and Mideast since 1492, so there's some more there too. There are many ways in which eclipses relate to one another and to history, so that took some organizing. Fortunately, solar eclipses tend to organize themselves quite remarkably- in ways I believe you'll find interesting.

After three years of considering these extraordinary events, I am convinced there are great stories in these peculiar patterns. I now call them God Songs. They connect peculiar places at peculiar times with cosmic shadow. They do it for a reason, it seems to me, even if that reason was just to tell a story only once.

The Story: A Story is told by a solar eclipse. The subjects are earth's two great companions and the people which behold them, who are both audience and participants. One subject, the sun, is the source of light and necessary for life. The other, the moon, is a dead orb which can only reflect light.

One day, the great light appears to be swallowed or hidden by darkness, causing distress among the animal kingdom (and until recently, human kingdoms as well). For a few minutes reality is inside out. It gets dark in the day. You can look at the sun, which has become transformed into a black hole crowned with ethereal light. Does darkness win?

No. Eventually the light returns. The good guy wins. We live on, but with a message or omen lingering in our minds. In a universe of repeating patterns, we are given a cosmic story template written in the sky. From that basic pattern, eclipses go on to more varied levels of symbolism.

Past eclipses perhaps said something in their moment of time to those who experienced them. As it turns out, these patterns of "shadows past" can be very confidently recreated by scientists and cartographers. Eclipses follow mathematically precise repeatable equations, though they show up in great geographical diversity. These equations can be applied reliably to history, with remarkable precision. Though these patterns have been mapped by science for 130 years, they've only become widely available at this hour.

Simultaneously, there is proof of a "Cosmic Clock" that beats with miraculous regularity and is defined by the co-ordination of provable cycles. This work intends to showcase some of the attributes of this clock in a manner worthy of its dignity.

From an initial attempt to determine the peculiarity of two eclipse crosses, this veritable encyclopedia was assembled. Perhaps this work will lead you to draw similar conclusions to the ones I did. Perhaps it leads you to draw radically different conclusions, or perhaps none at all. I am content with any consideration the work elicits, even negative.

These eclipses were made by God. He created the mechanism and arranged this reality at the beginning of Time. He has that power and he's clearly got a reason. I readily admit I do not have Him all figured out. But clearly, these events and patterns speak to the power of their Author. I have a long-held spiritual belief that all reality was created by this same Author. I call Him God and believe in the truth of Jesus, and these beliefs inform my life.

I hope this work will be considered by people who don't believe everything I do. I encourage everyone to flip through the pages and consider eclipses for themselves. The shadows have no denomination or prejudice. They appear all over the world to all peoples.

By combining historical associations with cosmic events, this book offers perspectives that might challenge what we've been taught about reality. Eclipses have been segregated from the human history which accompanied them. To me, that seems inappropriate. The phenomenon of the solar eclipse seems obviously created with human perspective in mind. Some may not feel eclipses should be viewed from historical perspectives. That's fine. A more modern "scientific" perspective is readily available from many other sources.

Once, solar eclipses were considered omens. They were seen as messages from God or the "gods". Was that instinct absurd? Western civilization's inheritors have told us for hundreds of years that it's foolish to ascribe supernatural meaning to solar eclipses. Why? Usually we're told that since smart people have diagrammed the mechanics of the eclipse phenomenon, there is no need to offer the phenomenon supernatural context. But is the capability of describing the mechanics of any natural phenomenon sufficient grounds for dismissing the possibility of supernatural origins? Can't visible natural phenomenon have supernatural origin or intent? The solar eclipse reality goes to the heart of this vital question. How much do we want to know?

Are omens even possible? What constitutes an omen? Is there a timeline an eclipse must follow in order to be proved ominous? Or should the idea be dismissed altogether as ridiculous superstition? Only a careful study might reveal insight about the matter, and even if truth is found, it can still only manifest from within an individual, and can always be denied by another.

I try not to say any particular eclipse "means" one thing or another, even ones that could be easily portrayed as ominous. They speak for themselves. The real meaning rests in the fact that this undeniably strange phenomenon exists, and that the phenomenon contains the astonishing varieties of peculiarities that it does. Eclipses have attributes that resemble art, poetry, music, and story. They are aesthetic phenomenon just as much as scientific. Still, thankfully, science and mathematics lend great weight and depth to their consideration. Like art or music or stories, they invite meaning from the observer.

I have always been grateful for the science of astronomy but I am not an approved "expert" in that field, or any other. I do my best to keep my facts straight. That's very important to me.

"I'll know my song well before I start singing."
Bob Dylan-*Hard rain's Gonna Fall*

We clearly live in a kind of "tyranny of experts" at this time in history, just as we live in a singular story machine. The process of having one's work "approved" by institutions or committees or editors has become ingrained in our consciousness. After generations of this ideal, we are left in a place where intellectual blind spots are huge, where withering digital criticism is regularly used as a silencer of unpopular ideas, and where entire fields of science and art and theology are manipulated by various entities to control and direct society.

The sun is the same apparent size as the moon in the sky. The refusal to teach this most basic aspect of our reality in our educational system is glaring proof of the type of blind spots endemic to both education and science.

Sun and moon are the same apparent size, though of course slightly shifting within a few percentage points of exactness. Virtually no children are taught this fact. I learned it at age 32, from a late-night TV public access astronomy lecture. That our physical reality is a Constructed Design was the starting premise of the founders of modern science, men such as Newton, Kepler and Kelvin. This science of design is now ruthlessly attacked and denigrated by high priests of a hijacked scientific method. Lately these priests have perhaps over-played their hand, and appear more and more defensive. It's alright if we can't make them happy.

Maybe I look at eclipses like this because I'm a musician. I map relations of solar eclipses like notes or musical arrangements. Just doing that occupies a lot of space in this book. There isn't much room left for opinions. Certainly, I can't help but show a level of disdain for historical events like genocide, war, and the slave trade, as they appear in the history of eclipses. But even then, deeds of empire tend to speak for themselves. Just as pieces of music don't do well being overly dissected in search of meaning, so solar eclipses don't need many conjectures. Still, like music and story, eclipses are created for consideration. I'm not guaranteeing that opinions don't slip out. I'm only human.

Most of these eclipses have already been scientifically mapped and re-mapped as individual eclipses many times in the last couple of centuries. This work does not replace those maps, which are more appropriate for detailed scientific consideration. Instead, this work presents artistic renderings of the eclipses in intelligible relation to one another. It's perhaps the first time they've been organized in this fashion. Not bragging. Just saying.

This is not astrology, numerology or fortune-telling. We're not investigating whether Uranus was in Aquarius last month or trying to read the future in my shoe size. This is a true history of actual alignments in three celestial bodies without whom none of us could live. That alignment leaves distinct, confirmable shadow paths at unusual intervals. They create the only possible means of a hieroglyphic message system between God and His earth-bound creation.

The cosmic clock, which involves the interaction of several different calendars based on positions and rotations of earth, sun and moon, has left evidence of its presence through repetitive eclipse dates. I didn't invent these calendars, and I don't manipulate them to fit my "scheme". Their reality has long been established by science and yet remains absent from most educational settings. Why? Good question.

There's no "alternative science" here. I believe the earth is quite round, the sun is really big and the moon is not made of green cheese. I use accepted astronomical and historical dates, distances, motions and calendars. I present common knowledge that's not commonly shared. By presenting these associations, this work can't help but point out blind spots in our modern worldview. That doesn't make this book "wrong", or science "wrong" or make your teachers "wrong". It just shows the obvious. We all have blind spots. Perhaps these eclipses provide evidence or inspiration or provoke impressions. Perhaps they'll just be eclipse maps.

Total solar eclipses are the rarest and grandest spectacle in creation. The vast majority of humans who ever live on our planet never see one. We are living in an extraordinary time where great eclipses travel over significant populations and imperial landmarks.

I find it impossible to call this phenomenon coincidence. That's not to say someone *can't* say it's coincidence. Of course they can. One can also say "so what?". That's the beauty of personal opinion, something we all get plenty of these days. However, I would like to suggest that pattern recognition has long been a matter of human survival. Who decides which patterns are meaningful and which are meaningless? Throughout history, "experts" or "priests" have often decided which patterns are appropriate for the masses to recognize. Were they always right?

I believe the evidence presented in this work is as close to proof of intelligent design as can be reasonably expected. But is it undeniable? No, of course not. We've all heard plenty of sincere people deny things that seemed undeniable. That's part of living in a free world.

These eclipses are songs to me. I sure didn't write them, but I've come to love them.
I'm trying to play them for you the best I can. I don't play them perfect. I don't know what every note or line means. I was given an opportunity to hear them and I want others to hear them too. It's just like I always enjoyed playing music for others and I'm not sure why.

God wrote beautiful, mysterious messages with eclipses. Have a listen for yourself.

Here's the Cross over North America: 2017 and 2024

The first American total eclipse (upper left to lower right) occurred on August 21 2017. It was the first total solar eclipse in the USA in 38 years. It traveled the entire USA and touched no other nation. The second total eclipse of the pair occurs April 8 2024. That path travels

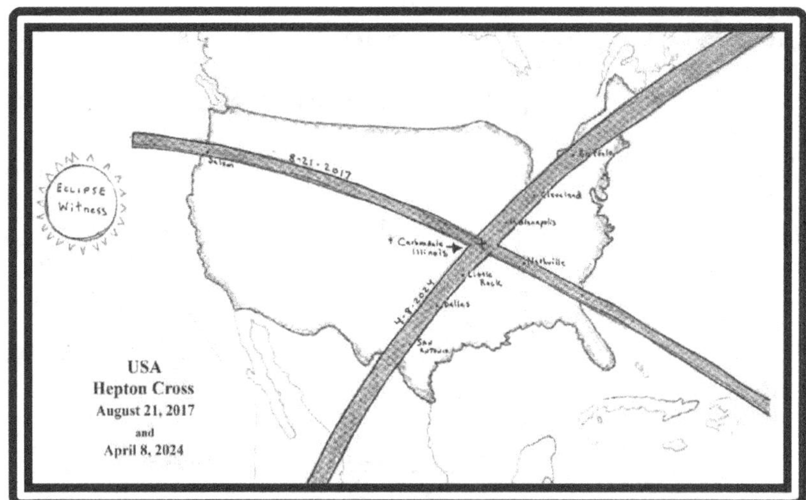

south to north to make the impression of X seen on the map. It crosses Mexico, the USA and Canada. Both occur on Monday. The individual eclipses contain extraordinary peculiarities which are detailed in their own chapter.

The Second Cross-Over the Mideast- 2027 and 2034

Each of these eclipses begin 10 years minus 19 days (3633 days) from their American counterpart. The first eclipse travels upper left to lower right on August 2 2027, hugging the north coast of Africa before crossing Arabia. The second eclipse in the pair is on the Spring Equinox, March 20 2034. It travels across Africa, Arabia, Iran, Pakistan, and ends in Tibet.

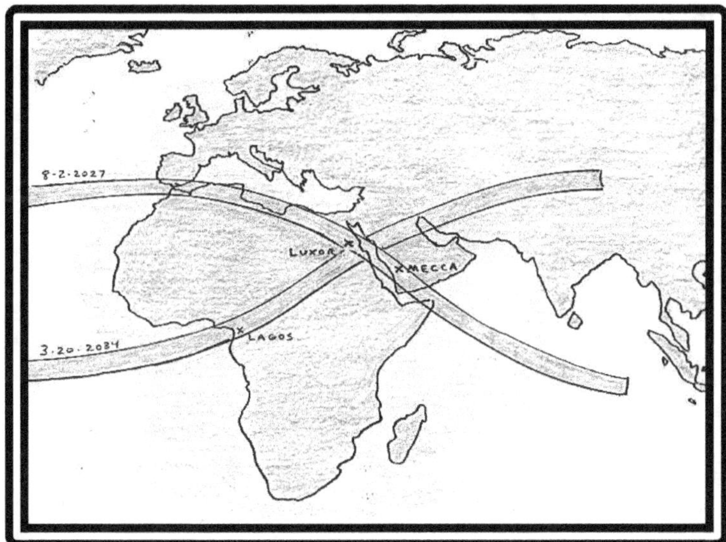

The cross has its apex on the map in southeastern Egypt and the Red Sea.
Both pairs on this page are spaced 2422 days apart. All the eclipses occur on Monday.
Perhaps it seems obvious to state, but Monday is "Moon" day. Are these symmetrical dates and geography unusual? What's with the "x marks the spot" appearance?

God Signs His Eclipse Book

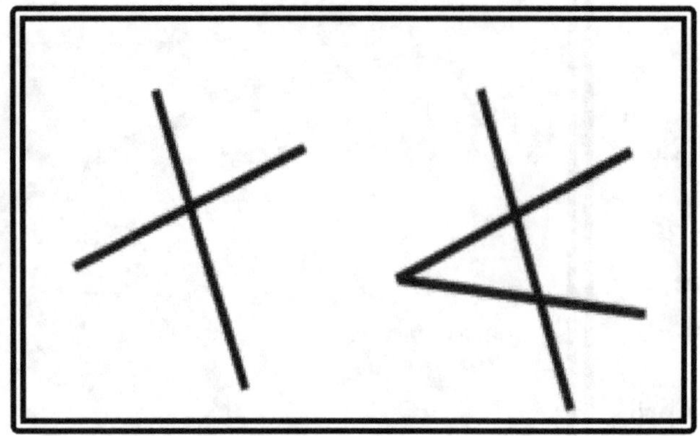

Paleo-Hebrew Tav   Paleo-Hebrew Aleph

These are ancient fundamental letters. The first and the last. Paleo-Hebrew Writing is not to be confused with modern Hebrew, which arose after Babylonian Exile. It is the original written word of the earliest Bible -including the Torah, most of the Psalms and Prophets.

The Aleph is the first letter of the alphabet, the source of our modern letter A. It's on the right, Hebrew being read from right to left.

Tav is on the left. It's the last letter (the 22nd) of the Hebrew alphabet. It is an X, or cross. This character x is essentially the same in Egyptian Hieroglyph, Phoenician, and Paleo-Hebrew.

Here on these pages, we see God has signed His book called history.

First, below, the Turkish Aleph takes shape. Composed of three eclipse paths from 1999, 2005 and 2006. Note the triangle over the heart of Western Civilization, the Mediterranean Sea.

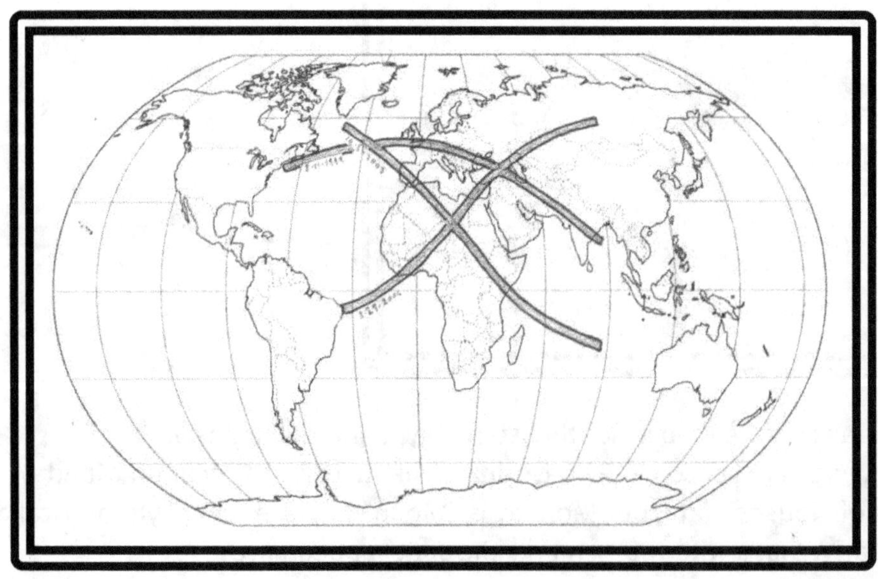

Next, the same Aleph moves over the USA, with three eclipses 2017, 2023 and 2024

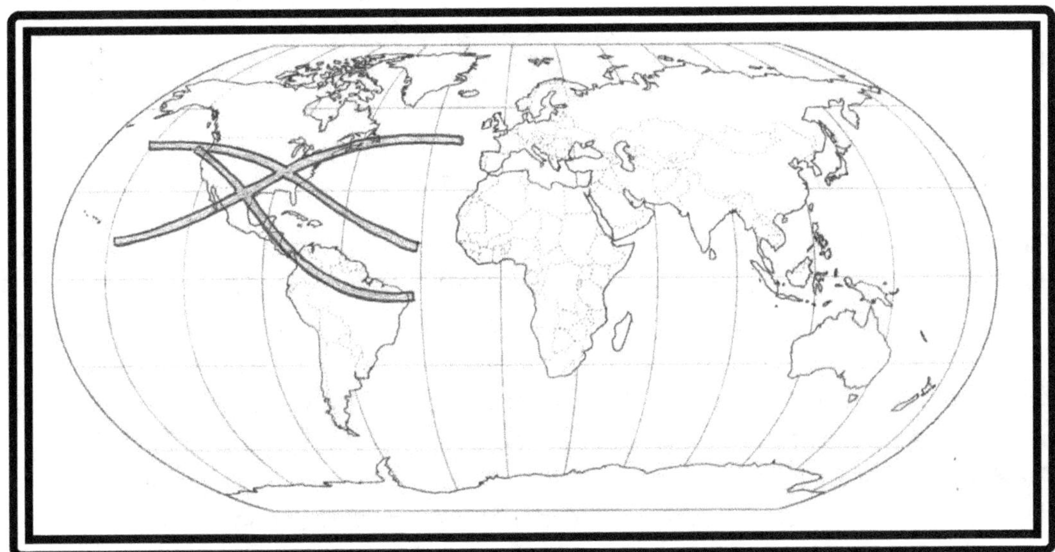

One of the most amazing things about this Aleph is how tight the top of the A is. This is not an easy cosmic task, and as far as I can tell, such a tightly constructed Aleph hieroglyph is not seen in the last 200 years of worldwide eclipses.

Below, again, the two Total Tavs. The one on the left is part of the above USA Aleph. Occurring in 2017 and 2024. It has apex over what is called Little Egypt (southern Illinois). The one on the Right is from the eclipses 2027 and 2034. It has an apex over Egypt.

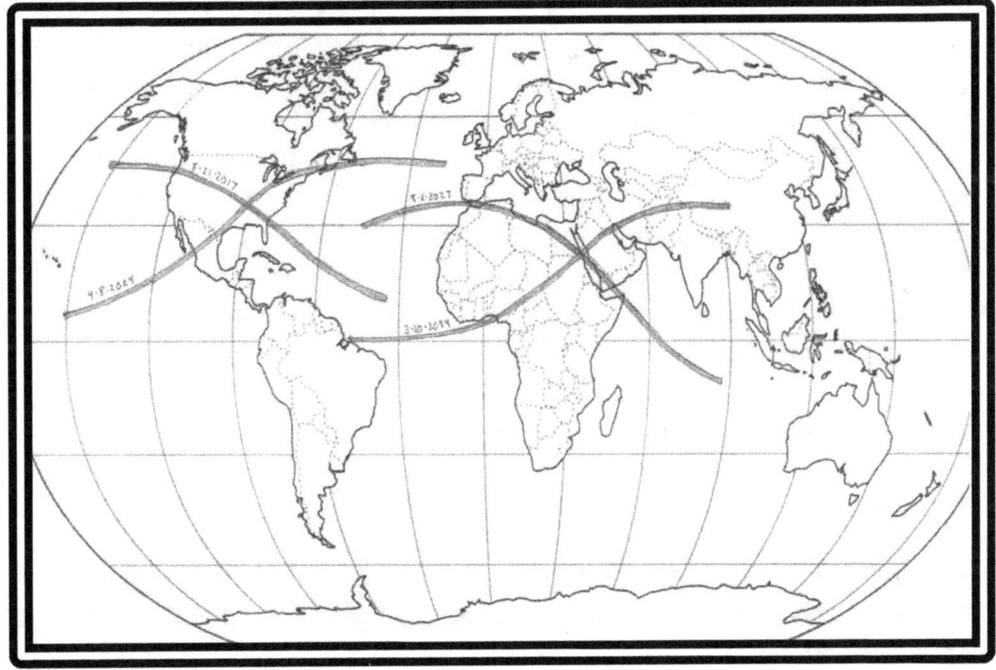

God is writing, "I AM the Alpha and the Omega" on the face of earth.

A Quick Glance at Twelve Clearly Odd Eclipses

There are dozens of unusual eclipses in this book. But here's 12 for starters. Remember that the phenomenon itself is a result of perhaps the most astounding coincidence in all of natural reality, the matching apparent size of the sun and moon. There were only 66 total solar eclipses on earth from 1900 to 2000 and slightly more ring of fires. Any particular place on earth has about a one in 400 chance of a total eclipse any given year, and some places can potentially go a thousand years without a complete solar eclipse.

1. **Titanic Eclipse April 17 1912.** Two days after Titanic sank, a total eclipse crosses Europe. 90 % at Titanic port of Southampton. Total in Paris and Russia. Two years before WWI.
2. **WWI Eclipse August 21 1914.** Total eclipse across the Ottoman Empire, Europe and Russia in the opening days of WW1. The same day the British lost their first casualty.
3. **Jerusalem/Babylon August 21 1933.** Exactly 19 years after WWI eclipse, a ring of fire across Jerusalem and Babylon (Baghdad) as Nazis rise to power in Germany.
4. **America August 21 2017.** Same Aug 21 date as last two eclipses in this list. America is divided almost perfectly geographically in half just as political and social divisions escalate. 7 months after Trump's first full day in office. Hebrew year 5777.
5. **Atom Bomb Eclipse July 9 1945.** One week before the first atom bomb is detonated in New Mexico on July 16 1945, a total solar eclipse travels from the USA to the USSR.
6. **JFK Premonition Eclipse Nov 22 1919.** When JFK is two, a ring of fire eclipse occurs 44 years to the day before his death. Eclipse begins at sunrise in Texas, the state where he would be murdered. It crosses town of Stonewall, where his successor, a ten-year old Lyndon Johnson lives. Eclipse also crosses Cuba, where JFK's fate was so entwined.
7. **Trail of Tears Eclipse Nov 30 1834.** A total solar eclipse traces the path of the forced march of Cherokees from Georgia to modern Oklahoma while the Cherokees were on it.
8. **Mount Saint Helens/ Microsoft Feb 26 1979.** Total Eclipse over Mount Saint Helens and Greater Seattle area seven weeks after Microsoft relocates to Seattle/Bellevue and one year, 2 months and 22 days before Mount Saint Helens erupts.
9. **Vietnam June 20 and Dec 14 1955.** In the first year of the Vietnam War, two solar eclipses over Vietnam. Part of an eclipse swarm from 1944 to 1965 of six eclipses. Bangkok Thailand has 4 complete eclipses in 10 years.
10. **Wuhan Eclipse July 22 2009.** Totality in Wuhan China 10 years before Covid-19. That same day, Windows 7 is released to manufacturing. Part of a three-eclipse cluster in 17 months with apex near Chongqing.
11. **Mogadishu Sunrise September 11 1988.** Mogadishu, Somalia is the only city to experience a complete eclipse in a 10,000-mile eclipse path. (9-11 is also ancient Egyptian New Year). 13 years before 9-11 attacks and three years before the "Blackhawk Down" Battle of Mogadishu.

Shahrud Tav Iran Feb 25 1952 and June 30 1954. Two total solar eclipse paths intersect over city of Shahrud bracketing the August 1953 western backed coup which installed Shah Reza Pahlavi. The 1954 eclipse travels from the USA at sunrise.

Maximum Expression and the Eclipse

Time is a place where stories occur. Within the story of time, we find a phenomenon that can be called *maximum expression*. Maximum expression is a fundamental property and signature of existence that occurs in personalities, nature, ideas, civilizations, and empires.

Maximum Expression: *A supreme and peculiar characteristic or attainment, a peaking of story, identity, purpose, essence, growth, personality, meaning or resonance.*

Natural example of maximum expression:

1. A certain tiger finds a maximum expression of his adult personality at a great fight in his 7th year, when he defeats a pack of hyenas to keep a kill. He also had maximum expression of playful kitten-ness and old age. Each day had a maximum expression.
2. That tiger, being particularly brave and powerful, was also a kind of maximum expression of his species. A peak of tiger-ness.
3. The tiger species itself is a maximum expression of solitary carnivorous reality.
4. The large carnivore is a maximum expression of the order of mammals.
5. The mammal is a maximum expression of extraordinary biological construction.
6. Biological creation is a maximum expression of design.
7. Creation is a maximum expression of God.

Important: Notice that it's not THE maximum expression. It's A maximum expression. It's not a competition. It's the diversity of expressions that is crucial, all feeding into an essential relationship with the Creator. You are surely the maximum expression of you-ness. Maybe your high school days were the maximum expression of your youth. Maybe your dog is a maximum expression of dog-ness. Life is itself a maximum expression of creation. Within these diverse existences are myriad maximum expression.

Is there any doubt that last night's thunderstorm or the Grand Canyon or Martin Luther King or Beethoven or Marilyn Monroe or Joan of Arc or the Milky Way or Sitting Bull or a Hawaiian volcano or the guitar solo in Hotel California or redwood trees or children at play are maximum expressions? This flowering of existence, this persistence of personality, provides deep and fervent mixtures of pure reality and tenacious symbolism. It's a dance of duty and desire. Maximum expression is unapologetic divine essence and purpose. You know it when you see it.

The solar eclipse and Maximum Expression:

1. The sun/moon alignment is a maximum expression of cosmic peculiarity. The two most important cosmic partners of earth are the same apparent size in the sky.
2. The peculiar cycles and orbital realities of the sun, moon and earth allow for complete eclipses to occur at predictable yet diverse locations. The cycles themselves are maximum expressions of an almost unreal cosmic clock.
3. By creating a pinpoint of shadow, called an umbra, the alignment of sun and moon draw an artistic curved line on earth, which when illustrated over years, give maximum

expression of God's potential to write hieroglyphic messages without human mediation. These hieroglyphs, when all their associations are considered, appear to offer a maximum expression of a kind of divine poetry and song.

4. The hieroglyphs often point to maximum expressions of empire, creating aesthetic representations of imperial energy exchanges in the implicate and explicate order. They also occasionally draw attention to great personalities, events and historical crimes.
5. These hieroglyphs only became easily accessible with the climatic maximum expression of history: the advent of the universal technological super machine.
6. Our maximum expression of history occurs alongside the 2000th anniversary of the crucifixion and resurrection of Jesus Christ. The idea of millennium is of great importance to the last chapters of the Bible. Simultaneously, the Cosmic Clock which creates eclipses has re-aligned to the exact position of his Jesus' life. Jesus, of course, is the maximum expression of human and divine storytelling.
7. Eclipse cycle peculiarities seem to reach their maximum expression in the Hepton Cycle, composed of seven eclipses seven eclipse seasons and seven eclipse years apart. Septiform (seven-based) reality is fundamental to the Bible and ancient wisdom schools.
8. The maximum expression of the Hepton Cycle appears to occur with eclipses over the Mideast, Eurasia, Africa and the USA between 1999 and 2052, with particular emphasis between 2017 and 2034.

Okay, I'll be honest. I believe we stand very near the end of a story. Is that really such a strange opinion? Personally, I feel it's not something to get upset about. Nail-biting won't help. If the doctor says you may have only 12 years left, are we going to freak out? What if it's only three years? Then again, maybe 29 years? Then the doctor admits they aren't really sure, but it sure looks pretty terminal. I don't have dates for the "end of the world". I am not sure any of these eclipses point to the exact "end of the world", an idea which is itself subjective.

They do seem to indicate "judgment". Take that as you will. I believe these signs in the sun and moon point to a season. Does it matter if we have twenty years or two weeks? Shouldn't we live our lives in maximum expressions of dignity, decency and faith either way? My hunch is that we might have some years left, but I'm not placing bets, or dates. At least for me, I feel it's time to toughen up, get a sense of service and a sense of humor. If there was ever a time to act sane, smart, responsible, kind and strong, it's probably right now.

A net is closing on free will and human maximum expression. That is the clearest clue to our predicament. How much longer can God tolerate story without free will and maximum expression? It's unclear, but probably not long. The direction of this story seems untenable. We face a spiritual enemy opposed to free will. It despises creation. It is historically committed to the degradation of innocence and dignity, constantly inspiring various destructions and deceptions, and is also greatly encouraged when people don't believe it exists. Once confined to manipulating the wills of men and institutions, this adversary has now gained access to a universal machine that thinks. That machine, which we stare at most of the day, clearly wants universal control, total deception and access to the human soul, at which point all bets are off. Read between the riddles. It's nothing to get scared of. You are of the eternal spirit. Why worry about spooky shows on a temporary plane? The eternal promises strength and purpose. Be sane and patient. Keep going to work. Do something nice for others. Step back from illusion.

God made the Universe. Let's see what he's up to. He made solar eclipses, after all. If that doesn't give us an idea of how powerful he is, I don't know what will. They are not illusion. Solar eclipses are *the* maximum expression of peculiarity in cosmic natural reality. They are unmistakably improbable. They are at once beautiful, essential and symbolic.

Once, people believed them to be signs. When did we stop believing that? It's an important question. Solar eclipses are not a scientific phenomenon, as if there was such a thing. Solar eclipses are natural phenomenon which bear clear markings of supernatural authorship. There is nothing like it in the known universe: a host star perfectly eclipsed for sentient souls.

Solar eclipses show up no matter what. You cannot cancel them or deepfake them away. No matter how absurd this story gets, they will appear on planet earth to rich and poor alike.

"And there shall be signs in the sun, and in the moon, and in the stars; and upon the earth distress of nations"- Luke 24:25

Our Creator makes brilliant art. Considering just the physical properties alone, eclipses are unsurpassed genius. Once mapped, these hieroglyphs become complex and sublime God Songs. Drawn with the tip of a cosmic pen, the paths are poetry- marking human empire and implicate energies. Sometimes these shadow lines seem a thundering accusation, other times a sorrowful song. They are deeply profound in a way unlike anything else in nature. They belong to God.

One thing's for sure, it's not astrology. I don't believe the sun and moon are sentient entities, much less gods. I am guilty, however, of sometimes drawing smiley faces on them.

The sun and moon have, besides their physical relationship to earth, an interactive role; one stated at the beginning of Genesis.

"let them be for signs and seasons, and for days and years; and let them be for lights in the firmament of the heavens to give light on the earth"; and it was so. Then God made two great lights: the greater light to rule the day, and the lesser light to rule the night. He made the stars also. ... And God saw that it was good- Genesis 1:14

If that wasn't enough, the sun is the same size as the moon in the sky.

Welcome to the final hoedown, the big art opening, the cosmic symphony coming to a world near you. The stage is set, and the actors introduced. A broken world looks up. We approach the 2000th anniversary of the crucifixion and resurrection of Jesus Christ and find signs in the sun and moon, casting peculiar shadows on both the first and the final empires of man.

Sand- July 22 2023

Chapter Two:

Eclipse Miracle

The Basics: What is a Solar Eclipse and Why Should we Care?

Total Solar Eclipse—Photo from NASA

What is a Solar Eclipse?

A solar eclipse happens from our perspective when the moon goes in front of the sun.

Of course, any object can eclipse any other object depending on distance and perspective. I can eclipse a cat from your perspective if I walk in front of it. You can also eclipse a hundred-foot tree with your thumb if the tree is far enough away. You can do the same with a mountain, or a million galaxies for that matter. The action of eclipse is an action of perspective. There is a distinctive difference in these analogies and the reality of the solar eclipse. Sticking my thumb out may eclipse a tree, but it does not cast a perfect small tree-like shadow. Also, the tree is not an enormous cosmic light furnace nor my thumb an orbiting satellite.

The alignment of sun and moon to our perspective, and the shadow cast by their alignment are what make solar eclipses peculiar. The sun is 400 times bigger than the moon and also 400 times further away. During a complete solar eclipse, these two objects line up and the shadow cast on earth is a small, perfect representation of that spherical union. When the Sun is in total eclipse, it's the only time when we can stare at the sun without damaging our eyes. It is also the only time that the Corona of the sun is visible.

Lunar eclipses are very different. In lunar eclipses, the moon enters the earth's shadow. That is a darkening of the lunar surface, not an obscuring of the surface, as what happens in a solar eclipse.

One of the oddest elements of the total solar eclipse is the nature of that darkest part of the shadow which reaches the earth. In our daily lives, we are accustomed to shadows that are either larger than the object casting the shadow, or roughly the same size. But the moon's umbral shadow during a total solar eclipse is anywhere from ½ mile to 150 miles wide, far smaller than the diameter of the moon. You can experiment with this yourself. Try to artificially cast a shadow that is smaller than the object you are shining the light on. Try with a flashlight and a golf ball. It doesn't work, at least visibly.

That's because there are different parts to a shadow. The center point of a shadow, the umbra, is the darkest part of a shadow. During a total eclipse, the moon creates a pinpoint shadow on earth as it goes before the sun. The Moon's umbra is only 380,000 km long, just long enough for the tip to touch the Earth but not large enough to cover the entire Earth. This peculiar design creates rarity.

An Umbral shadow is like the Tip of a Cosmic pen-

Image courtesy of NASA

A larger area of partial shadow extends thousands of miles out from the narrow umbral path.

Solar eclipses only happen at a new moon. That's when the Moon is closest to the plane of the Earth's orbit. If the Moon were in a perfectly circular orbit and in the same orbital plane as Earth, there would be total solar eclipses every new moon. But because the Moon's orbit is tilted at about 5 degrees to Earth's orbit, its shadow usually misses Earth. This creates both geographic variation and rarity.

The moon varies in distance every month from earth from about 222,000 miles to 253,000 miles away. Difference in distance changes the nature of eclipses. The moon is closest to the earth for total eclipses and farther away for ring of fires

Solar (and lunar) eclipses happen during eclipse "seasons". Eclipse seasons occur every year separated by 173-day spans (though you'll often see 177 day intervals for actual eclipses). Two eclipse seasons make up an eclipse (Draconic) year of 346.62 days. Some kind of solar eclipse will happen during each of these seasons. Like lunar years, draconic years rotate through our solar year. The beginning of the lunar and draconic dates happen differently on our common calendar every day. As of this writing, for instance, the Chinese New Year was on Jan 22 and will be different each year. The Draconic calendar is similarly odd, but I assume it is new year's is probably near April 17 here in 2023.

There are rarely more than two complete eclipses per year. There are 4 types of solar eclipses:

Partial solar eclipses are just what they sound like. The moon only partly goes in front of the sun. Partial solar eclipses comprise about 35% of all eclipses. Some of these can be impressive, but many can easily occur without anyone visually noticing. This book does not chronicle partial eclipses except occasionally as they relate to dates. Complete eclipses create a large area of partiality as well. Most online resources map this entire area. I avoid mapping partial areas as I find them distracting.

Photo courtesy NASA

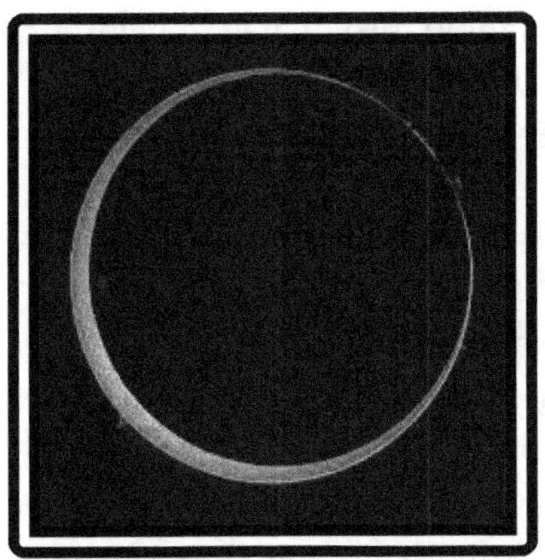

Annular (Ring of Fire) Eclipses (left) are a "complete" eclipse. The moon goes in front of the sun completely. However, the moon is slightly farther away from earth than it is for total eclipses, so the disc of the moon leaves a sunny Ring of Fire around its dark face. They comprise about 33% of solar eclipses.
Photo courtesy NASA

Hybrid Eclipses are part "ring of fire" and part total. A path of annularity (ring of fire) will be mixed with a brief total eclipse in a very thin line. Thus, during a Hybrid, the apparent size of the moon is very closely aligned with the apparent diameter of the sun. It will be very close to a 100% match. The May 9 1948 solar eclipse (classified as a ring of fire by some) that divided North and South Korea was 99.9999% total. The Exmouth Australia Hybrid 420 eclipse of Apr 20 2023 had an incredible 100.00% coverage. The Hybrid eclipse that passed near San Francisco on April 28 1930 had a path of totality only 1/3 of a mile wide, not much more than a big city block, and totality would have lasted a few seconds. Hybrids are the rarest complete eclipse, comprising less than 5% of all solar eclipses.

Total eclipses have been called "the grandest spectacle in all of nature". On many levels they are, aside from supernovas, the rarest possible cosmic phenomenon visible from earth. Far less humans (by orders of magnitudes)have seen a total eclipse than have seen a comet. Total eclipses require alignment between the centers of the Sun and Moon and a lucky placement of the observer. Total solar eclipses occur on average about once every 18 months somewhere on earth. But they appear very rarely at any given place on Earth, on average perhaps once every 375 to 400 years, depending on who is calculating.

Photo courtesy NASA

Annular eclipses are nearly as rare as totals but tend to have a slightly wider path of "annularity". So it may be that any given spot on earth averages a complete eclipse of some kind every 150 years or so. Averages are rarely realities. The vast majority of people live and die without ever seeing a complete eclipse, be it total, hybrid, or ring of fire. Many people have the opportunity to witness a partial solar eclipse at some point, depending on weather.

Are Solar Eclipses Rare?

Good question. It depends on what you define as rare. Some information sources insist that solar eclipses are common. For instance, *Space.com* says this on its website:

"It's a popular misconception that solar eclipses are rare, but, in fact, they happen on average once every eighteen months". -Space.com

Hmm. Well, there we go. But what is rarity? Is a moving thin line shadow turning day to night for a few minutes on 1/400th of a planet's surface once every 18 months rare? Depends on your perspective, I suppose. Such an event seems to me quite rare. Solar eclipses certainly don't happen a lot in any one place. Indeed, complete (especially total) solar eclipses can go a *very* long time without returning to any specific location.

The entire continental USA (all 48 states) went 38 years without a total eclipse in its boundaries between the eclipses of 1979 and 2017. Those two eclipses combined brought totality to much less than 10% of the USA's land mass. From the birth of the USA in 1776 to 2017, there were only 20 total solar eclipses within the boundaries of the continental USA. That means that a total solar eclipse occurred in the USA .0002% of the days. All told, less than 2 hours in nearly 250 years was spent with somewhere in totality. Seems pretty rare. But then, I'm no expert, like space.com!

Rome hasn't had a total eclipse since 1386.
Toronto has not had a Total Solar Eclipse since 1142.
Sacramento has not had one since 1084.
Jerusalem hasn't had a Total solar eclipse since 993 AD and won't have another one until the 2300's.

Computing all those endless possibilities is not really the issue. The reality is this: the earth is a rather big place, about 200 million square miles. The path of a complete eclipse is thin, usually between 50 and 150 miles wide. That path can go for maybe ten thousand miles. The path of a total eclipse can thus roughly cover an average of about 500,000 to 700,000 square miles. There is only one total eclipse on average every 18 months, and it only takes a few hours to traverse its path. It brings totality to any spot for 7 minutes or less. Any one spot has about a one in 400 chance of seeing one any year.

Here is that image again.

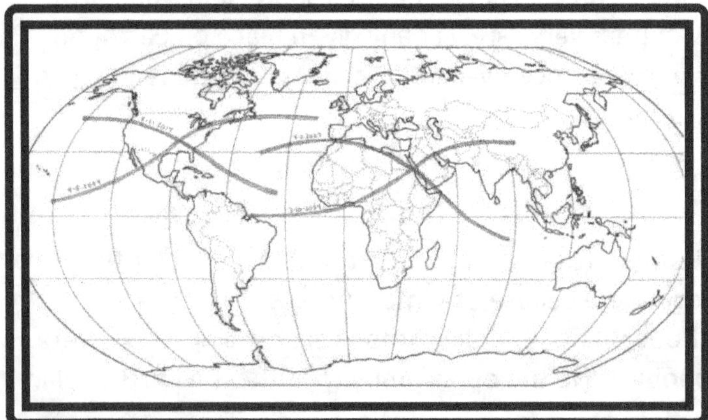

Is truth served by saying they are not rare? Are "popular misconceptions" needing to be corrected? What's almost as peculiar as solar eclipses themselves are repeated assertions that they *aren't* exceptionally peculiar. It's odd to live in such a miraculous world where deliberate efforts are made to downplay how amazingly bizarre our situation is. Perhaps it is precisely the rarity of the complete solar eclipse that allows for its non-consideration, for lack of a better word. They actually occur infrequently enough that they can be glossed over, and indeed, not heartily studied-even by many astronomers!

The Sun is the Same Size as the Moon in the Sky

The formation of intelligent life required a great deal of favorable conditions. The cosmic, molecular and biological synchronicities necessary for existence provide a continually growing body of evidence for an intelligent and active Designer. Nevertheless, such evidence is generally dismissed as coincidence by scientific orthodoxy, and likely will be until the end of the story.

While the science of Intelligent Design is generally focused on the extraordinary constructions of the things necessary for life, a much-neglected piece of evidence for Intelligent Design is to be found outside of conditions linked to necessity. Aesthetic (meaning artistic) evidence is placed directly in view of all the world, the symmetry of the two visible spheres in the sky.

Apparent Size: The sun and moon are both .5-degree angular diameter in earth's 360-degree sky. That's about the size of a pea held at arm's length. Sun and Moon often seem to appear larger than that, particularly when they rise and set, but that psychological illusion is not the subject of this work. The sun is essentially the same apparent size as the moon in the sky.

The orbits of the Moon around the Earth, and the Earth around the Sun are not perfectly circular but slightly elliptical, so the apparent angular diameters vary a bit. Still, the similarity in size remains very close. Sun and Moon average 98% similarity in size. So, on average, if you use a pea, 98% similarity would be the difference between a pea that was 12 millimeters wide and one that was 11.76 millimeters held at arm's length. The difference between the apparent size of those peas would be .24 millimeters, equal to about twice the width of a human hair perceived at a three-foot distance. If you considered basketballs held at 80-foot distances. 98 % similarity would mean one basketball was 25.6 centimeters wide and the other 25.1 centimeters 80 feet away. Obviously, there could be little perceptible difference...and perception is what this phenomenon is all about.

Such is the difference between average distances. The average, as we all know, is rarely the reality. As the moon's orbit swings between 225 and 253 thousand miles away, it can regularly line up nearly perfectly with the sun so that the percentage similarity shifts to 100%. The recent hybrid eclipse in Exmouth, Australia in 2023 featured an astounding 100.00% coverage. When the moon is closer to earth and the apparent size of the Moon appears slightly larger than the sun, from 101% up to 104%, it allows for the perfection of the total solar eclipse phenomenon.

Because the Moon's orbit seems to be moving away from earth (apparently 1.5 inches a year), astronomers tend to say that in the distant past, the moon appeared much larger than the Sun. Likewise, in the distant future, the Moon would appear smaller. This is a unique time. Apparently, intelligent life appeared upon earth while the Sun/Moon diameter is essentially identical.

Facts and figures:

Actual sizes: The Sun is 865,370 miles.
The diameter of the Moon is 2,159 miles.
There is about a 400.5 to 1 size ratio between sun and moon. That doesn't change.

Actual distances: The Sun is an "average" of 93 million miles from earth, but actual distance can vary between 91.5 million miles (147 million km) and 94.5 million miles (152 million km). Earth's orbit is also an ellipse. Over the course of the year, earth moves closest to the Sun (called perihelion), usually around January 3rd. It reaches its farthest position from the Sun (called aphelion) around July 4th.

The Moon's orbit around the Earth is also elliptical and varies between 225,000 thousand miles (called perigee) and 252,000 miles (called apogee) over the course of 27 days. On average, the Moon is 238,000 miles away. In other words, two elliptical orbits are at work with earth, sun and moon, constantly changing the perceived angular diameters of sun and moon by very slight degrees. Generally, even the most detail-oriented astronomers just simply say the sun is 400 times bigger than moon and 400 times further away. For one instance, when the Sun is 92 million miles away and the moon is 230 thousand, the Moon is exactly 400 times closer than the Sun.

Comprehending the Actual Sizes and Distances Involved

To comprehend the scale of this cosmic reality is not easy. The vast distances involved are impossible to simulate in a published medium. You just can't draw a picture of an object that is 400 times bigger and 400 times further away than something and then put it on a page. It must be done with words.

We'll substitute a one-foot wide (30cm) basketball for planet earth.
The Moon will then be a softball, about 4 inches (10 cm) wide. At that size ratio, the moon will be 34 feet away from Earth, a lot farther than many people realize.
Given that model ratio of the earth as a basketball and the moon as a softball 34 feet away, how big and far away is the sun? Well, the Sun would be a sphere the size of a 15-story building (133 feet in each direction) located about 2 ½ miles away from the basketball.
This gives some idea of the scale which we are dealing with.

It's tough to comprehend this. During a total solar eclipse, sun and moon line up perfectly. They line up so perfectly that sunlight often actually ripples through valleys of the lunar surface, creating what are known as "Bailey's Beads". It's all quite impossible. But here we are, thinking about it! Peculiar, don't you think?

What were we taught?

Below is a quote defining solar eclipses from *Kids Britannica,* a popular online educational resource. It is by no means unique. Indeed, it's the common description of solar eclipses in both children's and adult educational material, and it certainly sums up the explanations I remember from my youth:

In a solar eclipse the Moon passes between the Sun and Earth. This prevents the Sun's light from reaching Earth. As the Moon passes in front of the Sun, the Moon's shadow sweeps across Earth. The sky gradually grows darker. --Kids Britannica. com

That's it. While nothing about this explanation is untrue, it fails to mention a rather important fact. Namely, the sun is the same size as the moon in sky! The alignment that allows for the most incredible qualities of the phenomenon, such as the appearance of the corona, is ignored. As it turns out, generations of scientists, even eclipse specialists, find a way to downplay that truth.

Certain astonishing facts of our existence are either hidden from us by clever obfuscation or denigrated as "coincidence." Whether it's intentional or not is beside the point. What might this mean for the way we have viewed reality, the way we perceive our setting, history and each other?

How about the moon? Is it peculiar?

The Moon is Big relatively: Earth's moon is by far the most significant satellite around any rocky planet in the solar system. Mercury and Venus have no moons. Mars' moons are tiny odd-shaped rocks, barely 15 miles across. All great moons in the solar system outside of our moon belong to the Gas giants. But the earth's moon is still comparable to those. The largest satellite in the solar system, Jupiter's Ganymede, is only 1000 miles wider than our moon. Pluto and Charon at the edge of the solar system are really a dual dwarf planet system, being as they are only 12,000 miles apart. Frozen Pluto is smaller than the moon and Charon is a mere 700 miles across.

It makes the tides and more: The moon is the primary force creating our ocean tides, stirring the water and cleaning the seas. Life would likely be impossible without the moon. Also, the moon stabilizes the earth's rotation, making for a more consistent climate and acts as an asteroid shield.

I've Just Seen a face: The moon shows only one face to the earth, because it rotates on its axis at the exact same rate which it revolves around the earth. This phenomenon is called tidal locking.

We have bright moonlight! Who else has that?: We have the brightest moonlight known in the Cosmos. Our bright moon gives nighttime a distinct and changeable character with phases that also allow for the establishment of chronological time by creating observable months. and establishes circadian body rhythms in animals and humans. It encourages diverse nocturnal animals.

How about the Sun?

Perfect Sphere: The Sun is the closest thing to a perfect sphere yet observed in nature. It has less than a six mile difference in diameter between north and south axis, despite being 865,000 miles wide.

We are in the perfect place: The huge fusion furnace of the sun provides the source of all life. The Sun holds us in a favorable "Goldilocks Zone"; not too close, not too far...just right!
The sun creates proper photosynthesis for plants and proper UV light.

How about the Earth? Are we Peculiar?

We have life: Earth is the only known planet where life exists. Some type of life is found in virtually every place on the surface of earth.

Nice Rotation: Earth's rotational speed is just right for life. If the earth rotated on its axis slower, life would die from overexposure to either heat or cold. Faster rotation would cause environmental chaos.

Right distance from the Moon too: The earth is just the right distance from the moon. If the moon were much closer, huge tides would overflow onto the landscapes. If it were much further away, we would not have our beneficial cleansing tides.

Cool Water: The earth is the only planet known to contain large bodies of liquid water. 70% of the earth is water. Water stabilizes the earth's temperature and provides the chemical foundation for life.

Good gases too: The mixture of gases in our atmosphere is perfect for life. If this chemistry was changed even very slightly, the atmosphere would become poisonous or volatile.

Good size and nice Magnetic Field: Earth is a good-sized planet for life, with excellent gravitational conditions. The earth also has a powerful electro-magnetic shield that protects us from radiation, and an ozone layer to protect us from UV.

We're here: There are intelligent beings on Earth capable of considering such phenomenon. Love, memory, worship, self-sacrifice, music, dance and dreams are peculiar to Earth.

An order that "educates" us to not reckon such peculiarities as the gift of a Creator is peculiar.

How about the Universe? Is it Peculiar?

A wealth of information exists about the many atomic, chemical and gravitational variables finely tuned to allow for any matter at all to properly exist as we know it. Suffice it to say that even being here is practically impossible. Despite this, for nearly 200 years an institutional dogma has taught that everything popped out of nothing for no particular reason at all.

Is life Peculiar? Of course. True Peculiarity is relational.

Existence is odd. Our environment is peculiarly suited to existence. Our ability to intelligently consider this environment and existence is perhaps the strangest reality of all. When you have such a "preponderance of peculiarities", it is reasonable to consider them as evidence of meaning, function and authorship. Odds against intelligent life's coincidental development are so great as to be incalculable. Some say billions to one, some trillions to one. No one has any clue because no one's ever been able to make life happen without previous life already existing. As absurd as these tremendous numbers and conjectures are, many maintain the possibility for coincidence.

But at some point, possibility of coincidence must become impossible. When is that Point?

Some peculiarities seem to demand assessment. When does a phenomenon cease to have the possibility for coincidence? This is not a little question. Can one always say something is coincidence? If your missing wallet turns up in my pocket, should you first turn to the law of coincidence to explain it? We were born into a world that has instituted a coincidental worldview. A kind of "law of coincidence" is tacitly enforced among educational institutions. Any talk of design is dismissed, and its advocates "canceled" and denigrated.

If we pretend life is coincidence, society will reflect that belief. Origin stories have significance. They form the basis of belief systems which lead directly to actions. We will believe *something*, after all.

Tyrants once at least felt obligated to offer lip service to religious faith. They now substitute a kind of evolutionary theology to justify their deeds. Networks of entrenched "experts" assume the right to give meaning and take away meaning, at their discretion. They exert power to censor pattern recognition that makes them uncomfortable.

In other words: They've got their story, and they're sticking to it! I sincerely bless them and wish them well. I have tried to ignore God a few times in my life too. It's not very fun.

If life is a coincidence, then there is no active God. If there is no active God, there is no moral responsibility. If there is no moral responsibility, there can be no organic social order. Without moral law, social order exists only as it is imposed by intimidation and brute force. Consideration of the difference between a designed and a coincidental worldview is not just philosophical exercise.

The possibility of coincidence evaporates aesthetically in the face of a total solar eclipse. Math turns into music when the sun/moon alignment is really listened to. After all the improbable fine-tuned conditions for existence are met on a single beautiful diverse planet, it so happens that the sun is the same size as the moon in the sky when you look from earth. It's clearly an aesthetic touch that reverberates with deep symbolism about perspective and story, light and darkness.

In any normal story, this great truth would be marveled at the world over, especially since we figured out scientifically just how precise this alignment is. But there is yet more evidence of authorship.

The Mind-Blowing Reality of the Draco-Metonic Calendar.

Let's approach perhaps the greatest peculiarity in the shared reality we call Time. The Metonic, Draco-Metonic and the Saros Calendars.

The Metonic Calendar is 19 solar years. The Saros is 19 eclipse Years.

The Metonic Cycle is an intersection of the lunar and solar calendars described 2500 years ago by Meton of Athens. It involves the alignment of 19 Solar Years with 235 lunar months.

A Solar year is 365.2422 days. So 365.24 x 19 = 6,939.60 days
A Lunar month is 29.5302 days. So 29.53 x 235 = 6,939.68 days
This is usually rounded to 6940 days.

The difference between 19 solar years and 235 Lunar months is only about two hours. In other words, these cosmic patterns align with each other nearly perfectly every 19 solar years.

19 years ago was a lot like today. What does that mean? It means that if there is a new moon tonight, there will be a new moon 19 years from now on this same date, with two hours lag time. If there is a full moon tonight, or a crescent, or a quarter moon, or any phase, that exact same phase occurs again in 19 years. Likewise, 19 years ago on this date, there was also the same phase of the moon. After about 240 years, the date finally shifts ahead in a very long and hard to fathom cycle.

It is not a mathematical or gravitational necessity that the two primary astronomical calendars of humanity should align in such an accessible time frame as every 19 years. They could just as easily have done so for the first time after 214 years or every 2,378 years given different orbital realities.

The 19-year Metonic is a convenient synchronicity still utilized for calculating movable feasts such as Easter. This 19-year cycle was known to the Celts, who used it as their "Great Year". It was also used by the Egyptians, Babylonians, Mayan, Chinese and other ancient societies. It is still a fundamental calendar reality for the Bahai Faith. As interesting as it is to have such an alignment, the synchronicity becomes astonishing when we consider the full Draco-Metonic Reality. But first we should learn about the "eclipse year", the year most of us have never heard of.

The Draconic Year. The Dragon Year.

There is a type of year hardly anyone knows about except astronomers and alchemists. But it *is* a year, nevertheless. The Eclipse Year is vital for deepening the peculiarities presented in this book. The Eclipse Year is also called the Draconic Year. Draconic means "dragon", as in, the dragon that eats the sun, probably in reference to Chinese folk belief. China has maintained the oldest continuous astronomical studies on earth.

The eclipse year is 346.62 days. It's the return of the moon to the same eclipse node in the sky as the sun. An eclipse season is 173 days, where the moon is in the exact opposite node. The eclipse month, or Draconic month, is 27.212220 days, the time it takes for the moon to return to the node of the eclipse plane. The eclipse month involves a different cosmic perspective than a lunar month or a solar year. The Eclipse month is an alignment irrelevant to both solar and lunar calendars. It is difficult and possibly misleading for me to try and portray the cosmic draconic positions (with their vast scale and movements) in a graph illustration. So I don't try. Check online.

Trust that there is truly an ecliptic plane which sun and moon periodically cross. The moon has its "month" of crossing that ecliptic plane.

Draco Metonic Alignment:
This is where it gets amazing. Remember that 19-year spot of the Metonic where:

19 Solar Years   (19 x 365.242 days) is 6,939.60 days....and...
235 lunar months (235 x   29.53 days) is 6,939.68 days...well...it turns out:

255 Draconic months (255 x 27.212 days) is 6,939.11 days!

Only hours over the course of almost two decades separate three completely different arrangements of time at the 19-year mark. The result: These calendars are close enough that eclipses **often repeat on the same days of the year** every 19 years for up to a century.

For instance: there was a solar eclipse on March 20 2015. There is another eclipse 19 years later on March 20 2034, and still another 19 years later on March 20 2053. So, of fifty complete eclipses on planet earth in those 38 years, three are on March 20. Repeating eclipses on specific calendar dates. How could that be? How weird is that?

This seems an almost impossible co-ordination. This truth is likely only known to the best astronomers (and perhaps alchemists), and almost totally untaught in schools of the world, be they astronomical or theological. The Draco-Metonic Cycle would seem to be *the* maximum expression of calendar peculiarity. Calendar synchronism of this magnitude almost inevitably points to a Clockmaker, and perhaps it is this fact which has kept it from the institutions of higher learning.

It's ok if you don't get it all at once. I didn't. By the end of this book, it should be second-nature.

So concludes the summary of the Draco-Metonic Calendar, which gets its own chapter later.
Again, the Draco-Metonic is the alignment of 3 Calendars at precisely 19 solar years.
The other great alignment, the Saros Cycle, is the alignment of 3 calendars at 19 eclipse years.

The Saros Cycle—The Eclipse Personalities

The Saros Cycle is the last piece of this Cosmic peculiarity puzzle. Its incredible reality allows for the science of eclipse prediction. Importantly, the Saros Cycle helps define eclipses as individual personalities. These Saros personalities repeatedly grace the earth with unique geometric shadow patterns returning on a clockwork. Probably more than any other cosmic phenomenon, their cycle most resembles a lifespan. Composed of around 70 "events", they are born and mature and die. They also get their own section later. But I will summarize the alignment.

Remember the lunar month of 29.53 days? That provides the phases of the moon.
Remember the Draconic month of 27.21 days? That is the return of the moon to the eclipse plane.
It turns out they align almost perfectly at 19 eclipse (Draconic) which is 18 years, 11 days and 8 hours.
As it happens, 19 eclipse years is 19 solar years minus one *lunar* year. Boy, that's something.

See if you can keep this straight. It took me a while to take in, but it's not because the numbers are nonsense. It's because the synchronistic truth they convey is difficult to believe.

See, there's also an "anomalistic" month, based on the closeness of the moon to the earth. The lunar orbit is elliptical. The time it takes for it to return from closest-to-closest point, called perigee, or from farthest-to-farthest point, called apogee, is the anomalistic month. It is 27.5545 days. It turns out that 239 anomalistic months is also 19 eclipse years.

What is the Saros alignment:
A Saros Cycle is a period of exactly:
242 draconic months: 6585 days (6584.82)
223 lunar months: 6585 days (6585.19)
239 anomalistic months 6585 days (6585.45)

This alignment is how we can predict future eclipses and determine the paths of past eclipses.

One Saros period after an eclipse, the Sun, Earth and Moon return to about the same relative geometry, a near straight line, and a very similar eclipse will occur but at a different place on earth, landing about 1/3 of the way around the earth since the last time this geometry occurred.
A Saros Cycle is exactly 19 eclipse years (or 18.999%).

This alignment of calendars makes eclipse prediction possible. When this reality is brought up at all in textbooks, it is almost universally called coincidence. There seem to be a lot of coincidences that pile up around existence and eclipses. There is no scientific "reason" for any of these alignments of different cosmic timekeeping, just as there is no scientific "reason" for the sun and moon being the same size in the sky. It's not due to any scientific laws, nor do they break any laws. It all just is.

The Recap—here's our years:
Solar Year: 365.24 days
Lunar Year: 354.36 days
Eclipse Year: 346.62 days

Here's our months:
Lunar (synodic) Month: 29.53 days
Draconic (eclipse) Month: 27.21 days
Anomalistic (moon distance) Month: 27.55 days

Here's 19 Year Metonic Alignment (same phase of moon every 19 years)
19 solar years: 6939.60 days
235 lunar months: 6939.68 days

Here's 19 Year Draco-Metonic Alignment (same phase of moon and repeating eclipse)
19 solar years: 6939.60 days
235 lunar months: 6939.68 days
255 Draconic months 6939.11 days

Here's Saros Alignment (repeating eclipse personality -19 eclipse years):

242 draconic months: 6585 days (6584.82)
223 lunar months: 6585 days (6585.19)
239 anomalistic months 6585 days (6585.45)

Summary of peculiarities outlined so far:

- It's peculiar that you are reading this, and exist at all, endowed with acute observational abilities. It's peculiar that I care about this phenomenon, or your ability to understand it.

- It is peculiar that life exists on earth, made possible only through an extraordinary combination of peculiar circumstances, including balanced astronomical and chemical conditions.

- It is peculiar that we have a large Moon directly facilitating the existence of life on earth, which provides significant light at night, and conceptions of the passage and division of time.

- It is peculiar that our life-giving light source, the Sun, is perfectly situated in relation to our moon to create perfect eclipses, the sun being 400 times bigger and 400 times further away. Our two great cosmic partners in this journey are the same apparent size. The sun's light is peculiar as well, creating a pinpoint of shadow known as an umbra during a total eclipse.

- It is peculiar that this arrangement is distinguished by certain repeatable variables that make the phenomenon of complete eclipses rare but also regular. The tilt of the Earth at 5 degrees, the variation in the physical distance from the Earth to Moon during the Moon's orbit, and the orbital distinctions all allow for the phenomenon of eclipses to not occur every month in the same places, but instead appear fairly infrequently and at varied geographic locations. Complete eclipses occur on average about once every nine months, but the earth can go as long as nearly 2 years without a complete solar eclipse appearing anywhere. Most human beings go their entire life without ever viewing a complete solar eclipse. It is peculiar that we have a Saros Cycle every 19 Eclipse years. The alignment of three cosmic phenomenon—the Lunar (synodic) Month, the Draconic (eclipse) Year, and the Anomalistic (perigee to perigee) Month- allow for the creation of unique eclipse personalities. This regularity creates the ability to both predict future eclipses and map out past eclipses following mathematical formulas, thus creating the template for what I term God Song hieroglyphs.

- It is peculiar that we have a Metonic Calendar. That is, that the two different primary realities of human calendar time—solar year and lunar month-- should align at a convenient time, 19 solar years and 235 lunar months.

- It is peculiar that we have a Draco-Metonic Calendar reality, where the eclipse year also aligns with that same 19 year solar Metonic. 19 solar years equals 235 lunar months equals 255 Draconic months. This alignment allows for eclipses to often repeat on the same day of the year every 19 years, sometimes over the course of a century.

- It is peculiar how much of this reality remains unconsidered and untaught. Children are not even taught that the sun is the same size as the moon in the sky.

There you have it. Just preliminary peculiarities. Each one of these is worthy of a book by a better author than me. But I'm doing it anyway.

Technical Information and Acknowledgments

The Maps:

The paths have been mapped over the course of a few centuries by great scientific minds. I am not one of those minds, and am grateful for their work. This book presents artistic presentations of eclipse paths. They are very close to the true paths, but I have widened the areas of totality or annularity for visibility. I do this for very simple reasons. The area of pure totality in an eclipse can easily be less than seventy miles. That is $1/340^{th}$ of a 24,000-mile globe. Try drawing that on a map just a few inches wide. It is a very thin line, to say the least. Presenting it that way is both difficult and feels slightly absurd. It's not a line. It's a shadow. To best represent this shadow, I include areas of "almost" totality or annularity. The shaded paths in this book are experiencing profound eclipse, but only the very center of the shaded area would be in direct totality or annularity.

The mapping I present is not perfect, but all flat maps have difficulties presenting global phenomenon. I insist on creating hand drawn maps because I am convinced that I was supposed to present them in this primitive analog fashion. I think God might be sick of computers. I pray there's none in heaven.

I base my drawings on four different scientific sources-the earliest from 1887 and the latest from NASA. I then try and convey their reality as best as I can. Since I don't use computers directly for the drawing process, there are some rough spots. I get shaky. There's some funky curves. See if you can spot the "whiteout" correction fluid. It's a little late in this game for fixing all the details. So please forgive me if Orlando is slightly out of line in the eclipse of 1600. I tried very hard to keep it as close as possible, and accuracy is important to me, but this medium has limitations, as do I. For closer geographical "certainty", please refer to the sources I credit further below.

My maps show the dominant area of most eclipse. For total eclipse, shading will usually be where the eclipse is 95% or more, for annular (ring of fire) eclipses, where it is 85% or more. I believe this method helps the viewer "see" the signature of the eclipse best. I use pencils to shade in. Good old lead No.2's. After trying other formats, ink and pencil seemed the best way to convey the impression of shadow. A hand-drawn eclipse path has a 3-D quality that vanishes in computer drawn maps. I also limit or exclude some political boundaries. The eclipse paths themselves are obtained from several modern and classic sources and then artistically adapted for comparative effect.

Here are the main sources.

- 20th Century eclipse resources owe their primary debt to Theodore Oppolzer's *Canon of Solar Eclipses* (1887), one of the greatest computational works of all time.

- In our day and age, Fred Espenak (NASA's eclipse scientist) and Jean Meeus's book "*Five Millennium Canon of Solar Eclipses*" Volume One and Two are indispensable and should be bought. Their maps are available online at NASA, but it's not the same as the physical book which allows for quick comparisons of large time frames and patterns.

- The tremendous website "Time and Date.com" is the most comprehensive source for detailed eclipse maps online. Their site allows the viewer to locate individual towns and cities on maps from 1900 to 2100. You can learn all sorts of information there, including the percentage of coverage at any given location, and the exact time the eclipse occurred. Their website also has a very easy to use date calculator.

- "Solar-Eclipse.info", the site of Andreas Moller, provides easy to read and clear guides for historical eclipses going far back in time. His site allows you to search national maps.

- Christopher Grant of the Mathematics Department of BYU did the vital calculation that determined the Great 391 Year Cosmic Clock (first hinted at by H. Grattan Guiness but missing the draconic calculation). He also did an extended Saros calculation.

- There are several resources for the upcoming American Eclipses. My favorite is Dan McGlaun's "eclipse2024.org".

- Dean Coombs at 1260d.com has one of the deepest considerations of Biblical calendar reality.

- George Van Den Bergh named many of the cycles mentioned here, including the Octon and Hepton. He also numbered the Saros Characters. His ideas remain in spotty places on the internet. Much of his greatest work has almost completely "disappeared" from modern scholarship and the web. Part of his unusual story is presented in the personalities section. His great work *Universe in Space and Time* is occasionally available at used bookstores.

- I find the best books on the general topic of God in human history are the non-fiction works of CS Lewis, especially *Mere Christianity.* The late Chuck Missler is another good resource.

- Lewis Mumford's *Myth of the Machine: The Pentagon of Power* provides the essential big picture study of the progressive trajectory of Empire and scientific priesthood.

- *The Approaching End of the Age* and *Creation Centered in Christ* by 19th Century Irish evangelist and astronomer H. Grattan Guiness are deep dives into the cosmic calendars of God.

- *The Holographic Universe* by the late Michael Talbot is a great overview of cutting-edge physics confronting the many unusual characteristics of the implicate and explicate order.

- Of course, The Bible is the great book and clearly Living Word.

I couldn't do any of this without publicly available maps. I don't expect any of my sources to necessarily sympathize with the direction of this work, but I am still grateful for their efforts. Clearly there's a lot of ways to love eclipses. Some people love them with a heart for a "purer science" than me. God Bless them! It's all good.

Historical and astronomical information in this work is "common knowledge". I don't own it and neither does anyone else. It is available in the Columbia History of the World, Britannica Encyclopedia and other sources. I credit all specific quotes and details. Mostly just basic history and geography.

All illustrations besides my own are from NASA, or other public domain sources. Publicly available NASA maps are the work of Fred Espenak and Jean Meuss, and constitute a true public asset.

Who I am

People want to know who is telling them a story. I understand that. So I'll say where I come from.

I am currently the author of a huge book which I am very much looking forward to finishing. I'm also a husband, father and manager of a small hay farm. I do outdoor work and play old-time style country music for tour buses. I have a small town bluegrass band.

My name is Sand, and that is my birth name. My parents were Norman and Norma.
For a while growing up, we lived on a place called Normal Street. I kid you not.

Norman was from working-poor Jewish stock (so I'm half) and was raised in Bedford-Stuyvesant, Brooklyn USA. His father William was from Ukraine. He fought in the Battle of the Marnes in WWI for the USA and was injured for life by mustard gas. His father before him had been born a serf.

His wife, my grandmother Lily, was from Austria.

My mother Norma was from working-poor Protestant American stock. She was born in Oklahoma and raised in the Dust Bowl, mostly in Northern Arizona. Her bloodline goes back to the American revolution on both sides, the Russells and the Waltmans. All our people fought for the Union in the civil war. They were westward moving people. Mom's grandpa was a free-will Baptist minister who built 5 churches. He had been saved following a tornado experience working as a roughneck in Ripley Oklahoma. My mother Norma was a teacher and a poet and much more.

I was conceived in New Zealand while my parents had briefly emigrated there. I was in Mom's belly when we took the ship back to the USA. I was born in Flagstaff Arizona Sunday Evening June 4th 1967, just as the Six Day War began in Israel. The Beatles *Seargeant Pepper's* was released that week. At two months old, my family moved to Tahlequah, Oklahoma which is the capital of the Cherokees, at the end of the Trail of Tears. I learned guitar there at 11 and was there until I was 14.

My youngest memory: I was five. My brother and sister took riding lessons at a place out in the country near Tahlequah. Someone decided to stick me on the back of an unsaddled and untied horse so I could see what it felt like. The horse startled and ran across a field with me at a dead run. I see the horse ears and my hands on the mane in my mind's eye to this day. I held on for dear life and leaned forward. I had never been on a horse before. They chased me down in a pick-up truck across the field. I rode the horse to a standstill and they got me off. My sister remembers this. It actually happened.

I left Tahlequah at 14 with my Mom and attended a private high school on scholarship in Tulsa. I attended Colorado College on scholarship until graduating in 1989. After college, I worked for 7 years off and on as a dude ranch cowboy and ranch hand in Utah. I also lived and worked in several small towns in Colorado. At 28, I moved to Nashville to make it in Country Music and had a band.

I found out that the sun is the same size as the moon in the sky when I was 32, watching a late-night public access channel. I left Nashville at 33. I became a believer in the Way of Jesus 3 days after I left Nashville in the Flat Top Mountains of Colorado on July 14 2000. I watched the mountain Mesa Verde burn about ten days later. A few weeks after that I Joined a wildland firefighting crew in Oregon and at the end of the season wrote a book called *"What the Fire Said"* about story and fire and technology.

Since then, I've lived in the Four Corners mostly, aside from a three-year stint in Texas. I have done mostly outdoor labor and music most of my adult life. I have made lots of original music albums and do lots of gigs where I play lots of classic country and rock requests.

In 2017 I completed a children's book called *Eclipse Miracle: The Sun is the Same Size as the Moon in the Sky* and I traveled along the 2017 eclipse path with my book that summer. I began work on *Eclipse Witness* in early 2020.

But enough about me! Back to the God Songs, the hieroglyphs, the Cosmic Timekeeping of the Real Author, the Author of the Real.

Chapter 3

USA Complete
Eclipses from 1492 to 2012

The first gathering of all USA complete eclipses described and illustrated in this fashion.

We search for similar overland Hepton crosses as they appear in our current history (2017 to 2024) to determine if our current cross is peculiar. I assert the Hepton Cycle is the maximum Expression of eclipse cycles. It is explored in the Hepton Section of this book. I also assert that the two Hepton Crosses (USA and Mideast between 2017 and 2034) are the maximum modern expression of the Hepton.

In this section, you'll find only two Hepton Crosses in the 500-year time frame, one in the 1500's and one in the 1700's. Neither of them is composed of true totals. Each has one hybrid.

The maps in this chapter only depict the USA. Obviously, the eclipses extend further. I sometimes describe where the extended path reaches, but not always.
You will see clumps of eclipses, what I call swarms, often followed by long periods of inactivity. These swarms are especially pronounced around the most interesting times in USA history.
Certain Saros Personalities such as 120 and 145 return often, playing important roles.

You will notice an unusual cycle called the 300/44 pairing which is a 300 year minus 44-day period which can (but not always) bring similar eclipses to the same general geographic regions.

Certain places (i.e., cities) receive more eclipses than others. This may have significance to the idea of certain geographic or spiritual energy "hot spots". I will give a round-up of cities in a moment.

I highlight certain anniversaries if I feel they have significance, both solar year anniversaries and eclipse year anniversaries. When an anniversary is highlighted, it signifies a true re-alignment of earth, sun and moon, and not just a random date.

I do not include partial eclipses.
These maps are artistic hand-drawn representations of the eclipse paths based on several sources. They are not perfectly precise but very close I believe. Any locations mentioned are all verified, however, and are based on NASA alignments and common knowledge history. I highly suggest learning about the history of the USA if you haven't already.

History In Eclipses; A Brief Perspective

I will suggest certain possibilities in eclipse history.
- Imperial energies appear to be highlighted. The deeds and movements of empires appear in the paths, appearing to build bridges between geographic imperial powers. The pace of peculiarity seems to pick up towards the later history.
- Anniversaries and unusual dates can appear. Extremely important events are often marked in advance in unusual and profound ways.
- Certain maximum expression personalities and times are reflected.
- There is an underlying story of numbers, mathematics and calendars that is clearly present, yet again, difficult to fully pin down.
- Eclipses are the only way for hieroglyphs to be written on earth. War, conquest and slavery are one of the main subjects. Maximum expression of story and empire seems the overall theme.

America is very special. We are clearly a maximum expression, not only of empire but of music and a uniquely creative spirit, seen in both its original inhabitants and in its settlers.

What we know as the USA only began to take shape in the early 1600's. The Mayflower in the north in 1620 and Jamestown to the South in 1607. The first slaves arrived there in 1619. Manhat-tan Island is settled in 1624. Energies struggled in the USA, mirrored in a flesh and blood battle for the new world. Exchanges occurred between peoples and cultures and faiths.

List of selected US Cities and how many eclipses since 1600, including 2023 and 2024.

Albuquerque: 3 eclipses- of which 1 is total eclipse

Anchorage: 2 eclipses (1 total)

Atlanta: 8 eclipses (2 total)

Austin: 7 eclipses (1 total)

Boston: 5 eclipses (2 total)

Charleston: 5 eclipses- 3 total

Chicago: 3 eclipses (1 total)

Cincinnati: 1 eclipse (0 total)

Dallas: 5 eclipses (3 total)

Denver: 3 eclipse (3 total)

Houston: 5 eclipse (0 total)

Indianapolis: 2 eclipses (1 total)

Los Angeles: 6 eclipses (3 total)

Minneapolis: 1 eclipse (1 total)

New Orleans: 7 eclipses (3 total)

New York City: 3 eclipses (1 total)

Oklahoma City: 2 eclipses (2 total)

Salem: 2 eclipses (2 total)

San Antonio: 8 eclipses (1 total)

San Francisco: 1 eclipse (0 total)

Washington DC: 5 eclipses (0 total)

Interesting eclipses Just Before Columbus Voyage

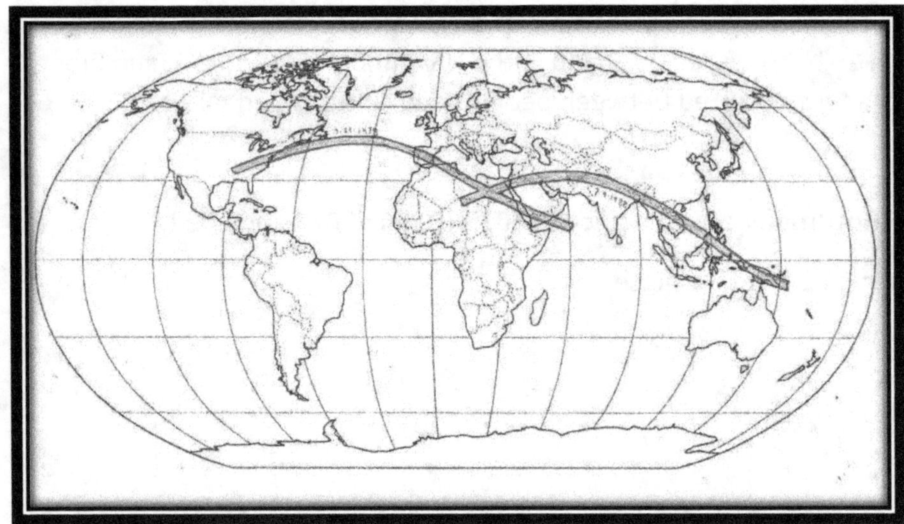

1478 July 29 Total (Saros 127)
Very close to exactly 14 years before Columbus set sail from Spain, a total eclipse from America to Spain to Egypt. This is the last total eclipse over what will be Washington DC. Still has not seen totality again. Also future New York City, Cape Cod, Memphis. Totality in Spain, Egypt and Arabia.

1488 July 9 Total Egypt/ Sinai/ Arabia/ Babylon/ Persia/Himalaya/ India/ Southeast Asia (Saros 118)
Geographic power spots eclipsed 4 years before Columbus.

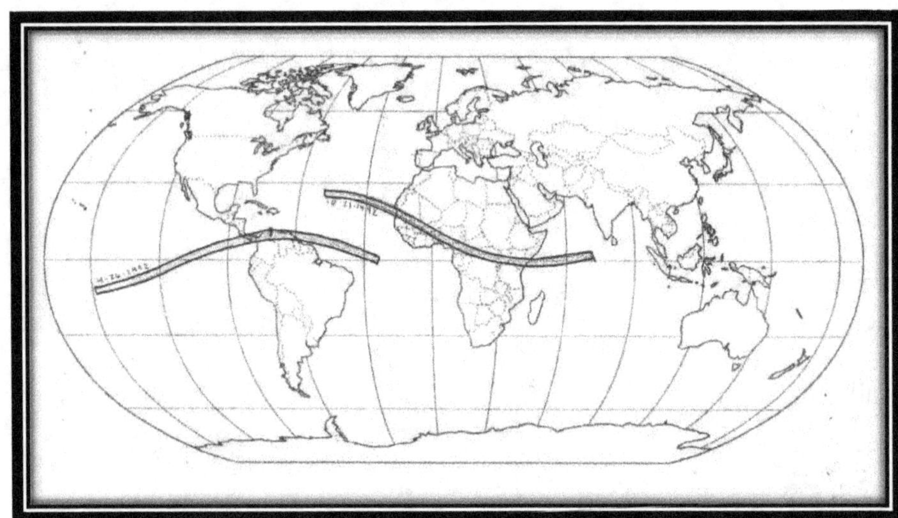

1492. The Americas, the Atlantic, and Africa

1492 Apr 26 Ring of Fire Pan- American Isthmus (Saros 120) The Saros Cycle that perhaps most highlights the American Eclipse experience (120) has its first complete eclipse. The isthmus of the Americas is covered 94 days before Columbus sets sail.

1492 Oct 21 Ring of Fire Africa (Saros 125) Nine days after Columbus touches ground in North America, Africa has a ring of fire across the heart of the continent.

Simply put, the three continents most affected by the Imperial expansion of 1492 all experience eclipse in 1492, and the pattern is simple and symmetrical.

I. 1496 to 1506- Mayan Apocalypse to Columbus Death

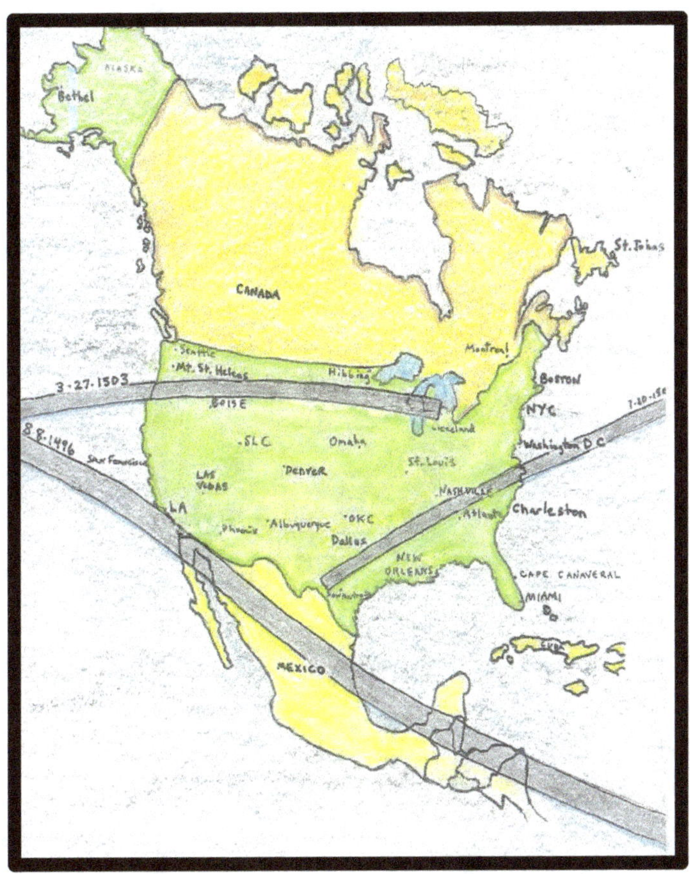

1496 Aug 8 Mayan Apocalypse Total Eclipse- (Saros 127) Path from extreme Southern California across Aztec and Mayan empires. For the star-gazing astronomers of the Maya, this solar eclipse foretold doom from the recently arrived Spanish Conquistadors. Within a generation of the eclipse, perhaps 3/4 of the natives succumbed to war and pestilence.

Saros 127: Saros 127's last eclipse prior to this -featured on last page- went from the USA to Spain to Arabia through Egypt. 1496 is the last appearance of Saros 127 in the USA. Saros 127 is the last eclipse of the current Monday Hepton Cycle, bisecting Australia and NZ July 13 2037.

300/44 Relation: The eclipse is in a 300/44 pairing with the July 4 1796 (20th anniversary of founding of USA) Pacific eclipse, which was the only solar eclipse to occur on July 4th in seven centuries.

1503 Mar 27- Total Eclipse- (Saros 121) Path travels across northern Great Plains

1506 Jul 20-Total Eclipse- (Saros 118) Two months after death of Columbus on May 20 1506. Twenty years before first African Slaves brought to the Americas by Spanish in 1526. Exactly 463 years before moon landing of 7-20-1969. Through Texas across what becomes Jamestown, Virginia.

Saros 118 : Saros 118 will appear in many interesting eclipses, including 1777 USA Revolution Eclipse, 1831 Lincoln Birthday/ Nat Turner Ring of Fire and Hitler Birthday/Death Day 1939 and 1957.

Summary: Total eclipses outlining rough border of the USA. 14 years before next USA eclipse.

II—The Only Two USA Eclipses in a 50-year span happen in 1520 and 1531

1520 Oct 11 Ring of Fire (Saros 116) ---On the eve of the 28th anniversary of Columbus landing in the New World. Path crosses extreme Northeast, including what becomes New York and Boston.
Saros 116: Saros 116 reappears in Lincoln Birthyear/death day eclipse of 4-14-1809.

1531 March 18 Total Eclipse (Saros 112)—Only US eclipse near Spring Equinox in survey. A razor thin hybrid with only seconds of totality. Crosses Little Egypt Illinois and directly over future Oklahoma City.

Summary: These are the only solar eclipses between 1506 and 1557. It was 14 years without an eclipse up until 1520 and then a full 26-year span after 1531 before next eclipse in 1557. In 1526, the first African slaves are brought to the western hemisphere by the Spanish to Brazil. The first 100 brought to North America rebelled and disappeared among the natives. Millions of natives die during this time from war and disease. No overland cross in eclipses.

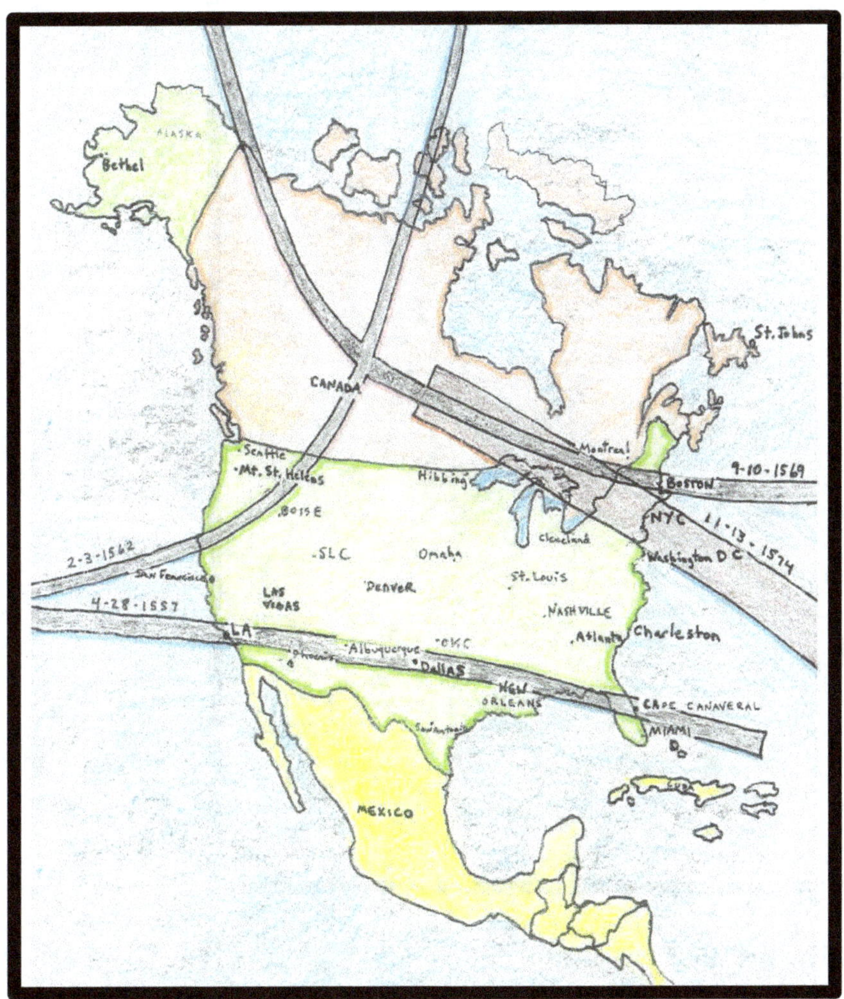

1557 April 28 Total Eclipse (Saros 121) 15 years after first Spanish visit Alta California. Path across modern LA, Santa Fe, Dallas, and Cape Canaveral. Saros 121's last appearance in the USA.

1562 Feb 3 Total Eclipse (Saros 133) Path over northwest.

1569 Sept 10 Total Eclipse (Saros 137) Path over extreme northeast. 300 years and 44 day pairing with USA eclipse of Aug 7 1869 (that is a long eclipse cycle I will point out a few times).
Saros 137 has series of profound eclipses: Titanic eclipse of 1912 , San Francisco 1930, Korea 1948, Columbus Death day 1966, USA 1984, Canadian Judgment 2021.

1574 Nov 13 Ring of Fire (Saros 116) Ring of Fire over what becomes New York City 50 years before the Dutch buy it for trinkets in 1624. Saros 116 reappears on Christmas Day 1628 with Ring of Fire over Charleston and Atlanta

Section 3 Summary: Once again, no real crosses overland of future boundaries of USA. An Octon Total Cross is visible over south-central Canada 1562 to 1569.

IV. California Hepton Cross 1578 and 1585

1578 Sep 1 Total Eclipse (Saros 118) Path travels length of future State north to south.

1585 April 29 Hybrid Eclipse (Saros 112) Path travels across the Northwest.
Saros 112: This is Saros 112's last appearance in the USA.

300/44 Pairings: Eclipses are both paired with the 1878 and 1885 Hepton Buffalo Cross.

While they could be organized as part of a 28-year six eclipse grouping beginning in 1557, I separated them because they are a distinct Hepton 2422 day cross. One of only eight Hepton Crosses in the survey, it is the only one over the US in the 16th Century. It occurs over California in the decades following the first Spanish explorations of that future state. This California Cross is not a True Hepton Total Cross like the 2017/2024 as the 1585 eclipse is a Hybrid, featuring only a few seconds of totality.

Apex appears to be Mount Shasta/ modern Redding area.

The 1500's end with only 10 complete eclipses having occurred over the contiguous USA, the lowest number of eclipses occurring during any of the centuries in the survey.

V. Cape Canaveral to Khartoum 1600

1600 July 10 Total Eclipse (Saros 120)

Eclipse begins at sunrise offshore of western Florida and crosses what will be Cape Canaveral/ Cape Kennedy. It goes on to Africa, including historic Egypt/Ethiopia, where it marks the first line of a Hepton Cross of 1600 to 1607 (see Middle East Maps).
300/44 Pairing: Paired with USA to Luxor 1900 total eclipse, the first eclipse of the 20th Century.
Saros 120: One of the greatest Saros of them all. No less than 10 profound eclipses, many in USA.

In what becomes the USA, this is the first complete solar eclipse in 15 years and the last eclipse for 18 years. It crosses a relatively tiny patch of ground with great significance. The Apollo Eleven Mission will leave from Cape Canaveral 369 years and 6 days later.

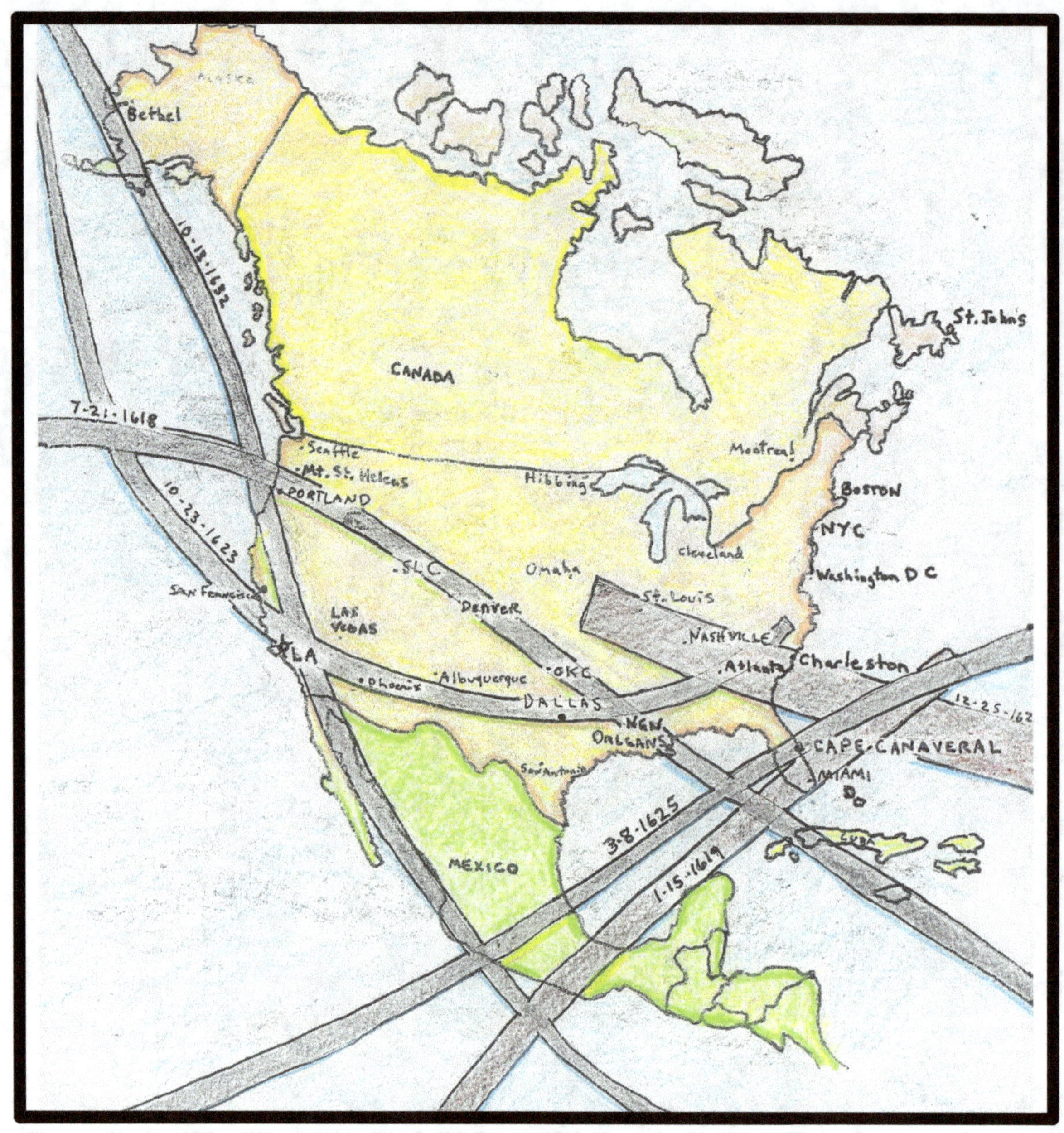

VI. USA Eclipse Swarm 1618 to 1632

(For 300/44 Pairing Comparison, look at 1918 to 1932 Swarm)

1618 July 21Total Eclipse (Saros 120)
1619 Jan 15 Ring of Fire (Saros 125)
1623 Oct 23 Total Eclipse (Saros 137)
1625 Mar 8 Total Eclipse (Saros 114)
1628 Dec 25 Ring of Fire (Saros 116)
1632 Oct 13 Total Eclipse (Saros 118)

VI. First USA Eclipse Swarm 1618 to 1632

1618 July 21 Total Eclipse (Saros 120) First complete solar eclipse in USA in 18 years and one of only a handful of true cross-country eclipses in 500 years, though it does not quite fully reach East Coast. It occurs as the settlement of the future USA has begun in earnest. 1618 is known as the beginning of "The Great Migration". It is one year before the first importation of African slaves by the British in 1619. Smallpox moves from New England to Virginia, most notably killing Chief Powhatan, father of Pocahontas, in April of 1618. The European population of the Jamestown area will climb from 400 to 4500 between 1618 to 1623.**Hot Spots:** Denver, Oklahoma City, New Orleans
300/44: Paired with true USA Cross Country Eclipse of June 8, 1918
Saros 120: The Great Saros 120 was also the last eclipse in this survey (1600 July 10)

1619 Jan 15 Ring of Fire (Saros 125) The only time Martin Luther King's actual birthday appears in an American Eclipse, six months before the arrival of the first African slaves into USA and 310 years before his birth in 1929.

1623 Oct 23 Total Eclipse (Saros 137) The eclipse nearly crosses the country but ends just shy of the east Coast in North Carolina. Over modern LA, Phoenix, Dallas, Atlanta
300/44: Paired with 9-10-1923 Japan to LA Eclipse.

1625 Mar 8 Total Eclipse (Saros 114) Cape Canaveral's second total eclipse in 25 years; 344 years before Moon Launch. Florida only.
300/44: Paired with Jan 24 1925 NYC/ Roaring 20's/Bob Dylan Eclipse

1628 Dec 25 Christmas Ring of Fire (Saros 116) Christmas Monday Eclipse begins at sunrise near modern Columbia, Missouri and ends at sunset over Africa after creating a near approximation of northern boundary of future US slave states. The only complete US eclipse to occur on Christmas. Crosses 11 modern states. A very long eclipse lasting 12 minutes. **Saros 116** died completely on 7-22-1971. Memphis, New Madrid/Makanda, Lincoln Birth Home, Nashville, Shiloh, Savannah, Charleston (USA's biggest slave port).

1632 Oct 13 Total Eclipse (Saros 118) California only. Bisects Silicon Valley and covers modern LA for its second total eclipse in 9 years.
300/44: Paired with Aug 31 1932 USA New England Total Eclipse

Notes: In these 14 years there were 6 solar eclipses. Several crosses occur in grouping, but none are between totals over the US except 1618 and 1623, with apex at modern-day Shreveport, Louisiana. Apex of 1628/1623 annular-total cross is north of Atlanta. This swarm corresponds (through 300/44 pairing) to many eclipses in the 1918 to 1932 USA swarm. Four of the six have corresponding eclipses. This swarm coincides with the "Great Migration" to the New World from Britain. In 1618 there were only hundreds of settlers in what would become the USA. By 1640, there would be nearly 25,000. It would be 38 years after this swarm before another solar eclipse in the USA in 1670, the longest stretch between complete eclipses in US history. By then, there would be more than 100,000 colonists of European heritage in America and the US would have another extraordinary eclipse swarm. The 1618 and 1625 eclipses are a Hepton pair. The 1625 and 1632 eclipses are an Octon pair. The same Hepton/ Octon configurations occur in 1918 to 1925 and 1925 to 1932 eclipses.

VII. Eclipse Swarm Two—1670 to 1684

1670 Apr 19 Total Eclipse (Saros 133)
1672 Aug 22 Total Eclipse (Saros 120)
1673 Feb 16 Ring of Fire (Saros 125)
1679 Apr 10 Total Eclipse (Saros 114)
1683 Jan 27 Ring of Fire (Saros 116)
1684 July 12 Hybrid Eclipse (Saros 131)

VII. Eclipse Swarm Two—1670 to 1684

1670 April 19 Total Eclipse (Saros 133) An extraordinary eclipse touches perhaps the smallest amount of US land in the survey, the very northwest tip of Washington's Olympic Peninsula, bringing totality to Mount Olympus (named 108 years later).
300/44: paired with Mar 7 1970 USA "End of 1960's" eclipse.

1672 Aug 22 Total Eclipse (Saros 120) Only Maine.
300/44: Paired with July 10 1972 "You're So Vain" Nova Scotia eclipse.

1673 Feb 16 Ring of Fire (Saros 125)
San Antonio/Utopia Texas, Jackson, Birmingham, Atlanta, Charlotte. Ends almost precisely at Kitty Hawk, which will be home of the first airplane flight in 1903.

1679 Apr 10 Total Eclipse (Saros 114)
Hot Spots: LA, Las Vegas, Zion Utah, Wounded Knee, Hibbing (home of Bob Dylan)
300/44: paired with Feb 26 1979 Microsoft/ Mount Saint Helens Eclipse
Saros: This is Saros 114's last appearance in the USA. It dies completely Sep 12 1931.

1683 Jan 27—Ring of Fire (Saros 116) Mostly Texas...San Antonio, Utopia, Corpus Christi, Houston, extreme south Florida.

1684 July 12 Hybrid Eclipse (Saros 131) 23 seconds of totality in narrow band...
300/44: paired with Washington DC/Atlanta/New Orleans 5-30-1984 Hybrid/Annular

Notes: 38 years (Two 19 year Metonics) after the last complete eclipse in the USA, another swarm of six complete solar eclipses occur in 14 years as America becomes increasingly settled. European and slave Populations in America double between 1670 and 1690. Both the slave trade and wars against the natives increase. Two crosses occur over Canada. The 1672 and 1679 are a Hepton Pair meeting in northern Quebec. Central Texas sees three eclipses, with modern San Antonio near the apex of a 10 year minus 19 day Cross.

The two 17th century eclipse swarms occupy only a small part of the USA's chronological history but contain 1/6th of its eclipses.

VIII. 17th Century Stand Alone Ring of Fire 1694

1694 June 26 Ring of Fire (Saros 122) Midland TX, Tahlequah OK (modern capital of Cherokee Nation) Crosses apex of 2017-2024 Cross at Makanda Illinois. Also crosses Lincoln Birthplace in Kentucky, Nashville, almost entire state of West Virginia in one of its only eclipses, and exits after passing over future Washington DC, ending at sunset over Africa just before Slave port of Porto Novo in Benin. **300/44**: paired with great May 10 1994 USA ring of Fire.

Saros 122: This is the last time Saros 122 will appear in the future US as a complete eclipse. It was responsible for the notable Christmas Dec 25 2000 USA partial eclipse.

Notes: Ten years minus 19 days (9 solar years plus one eclipse year) after last USA eclipse in 1674 and 19 years before the next one, this eclipse stands alone in time. It touches 14 future US states and symbolically divides the future nation very close to the coming Mason-Dixon Line. By now the US European and African American slave population has reached nearly 250,000 and the exponential direction of migration, the slave trade and native conquest are firmly established.

IX. Standing Rock Cross Swarm 1713 to 1724

1713 Dec 17 Hybrid Eclipse (Saros 137)
1717 Oct 4 Hybrid Eclipse (Saros 139)
1722 Dec 8 Ring of Fire (Saros 118)
1724 May 22 Total Eclipse (Saros 133)

1713 Dec 17 Hybrid Eclipse (Saros 137) 56 seconds of totality in certain spots. Path from sunrise in southwest Utah. Crosses Los Alamos and Santa Fe. Exits Texas. Crosses Columbus landing site, San Salvador, in the Bahamas 233 eclipse years after event.

1717 Oct 4 Hybrid Eclipse (Saros 139) Third hybrid eclipse in USA out of last four eclipses, the only time that happens in this survey. First eclipse of Standing Rock Hepton Cross. Crosses 11 states.
Saros 139: Saros 139's first US appearance. It reappears in the USA on Apr 08 2024, as the opposite position of another Hepton Cross. Will become most the powerful Saros of the 21st and 22nd century.
300/44: Paired with August 21 2017 USA Great Eclipse.

1722 Dec 8 Ring of Fire (Saros 118) Only Maine.

May 22 Total Eclipse (Saros 133) Tremendous solar eclipse (one of the widest paths in entire survey) completes the Standing Rock Hepton Cross.
Hot Spots: LA and Las Vegas in centerline, Scrapes Salt Lake City, Covers Pine Ridge and Wounded Knee, South Dakota
300/44: Paired with April 8 2024 Great North American Eclipse.

In the space of ten and ½ years, there are four eclipses. The Standing Rock Cross is the best example of a Hepton Cross in the contiguous USA outside of the 2017/2024 American event, the primary subject of our survey. In the amazing reality of the eclipse cycles, 300 years minus 44 days separate the two great crosses of the Standing Rock Cross and The American Judgment Cross.

Saros 139 appears in both events at opposite ends of the Cross. In the 1717 Cross, Saros 139 is the first eclipse, traveling northwest to southeast. In 2024, Saros 139 will be the second eclipse of the cross, traveling southwest to northeast. The apex of the Standing Rock Cross is in the modern Standing Rock Sioux Indian reservation which straddles the border of North and South Dakota.

Epicenter of apex is about 50 miles from the geographical center of North America, most recently determined to be in town of Center, North Dakota, which oddly enough received its name well prior to the discovery. Two other towns in North Dakota also claim this title.

As unique as the Standing Rock Cross is, it must be noted that the 1717 eclipse is a hybrid eclipse, and therefore not a true Hepton Total Cross as we see in the 2017/2024 Tav. Hybrids typically only provide a few seconds of totality and can waiver in and out of totality altogether. By 1724, the European and Slave population of the USA had grown to more than half a million.

X. Florida Duo 1737 and 1752

1737 Mar 1 Ring of Fire (Saros 116) Covers same area as Texas Cross of 2023/2024. Centerline at Corpus Christi. Occurs at sunrise over Texas. almost exactly 302 (301.99) eclipse years before Oct 14 2023 ring of fire eclipse, which also crosses Texas and also featuring a centerline exit at Corpus Christi. San Antonio's third Ring of Fire in 64 years. Cape Canaveral/ Kennedy again.

1752 May 13 Total Eclipse (Saros 124) Eclipse travels from Mexico City through Miami. In the 44 years between the 1724 Standing Rock Cross and the first eclipse of the Revolution (1768), these are the only two eclipses. They intersect (but do not cross) over South Florida. During this time frame, Florida is a contested area with battles between British, Spanish and native tribes.

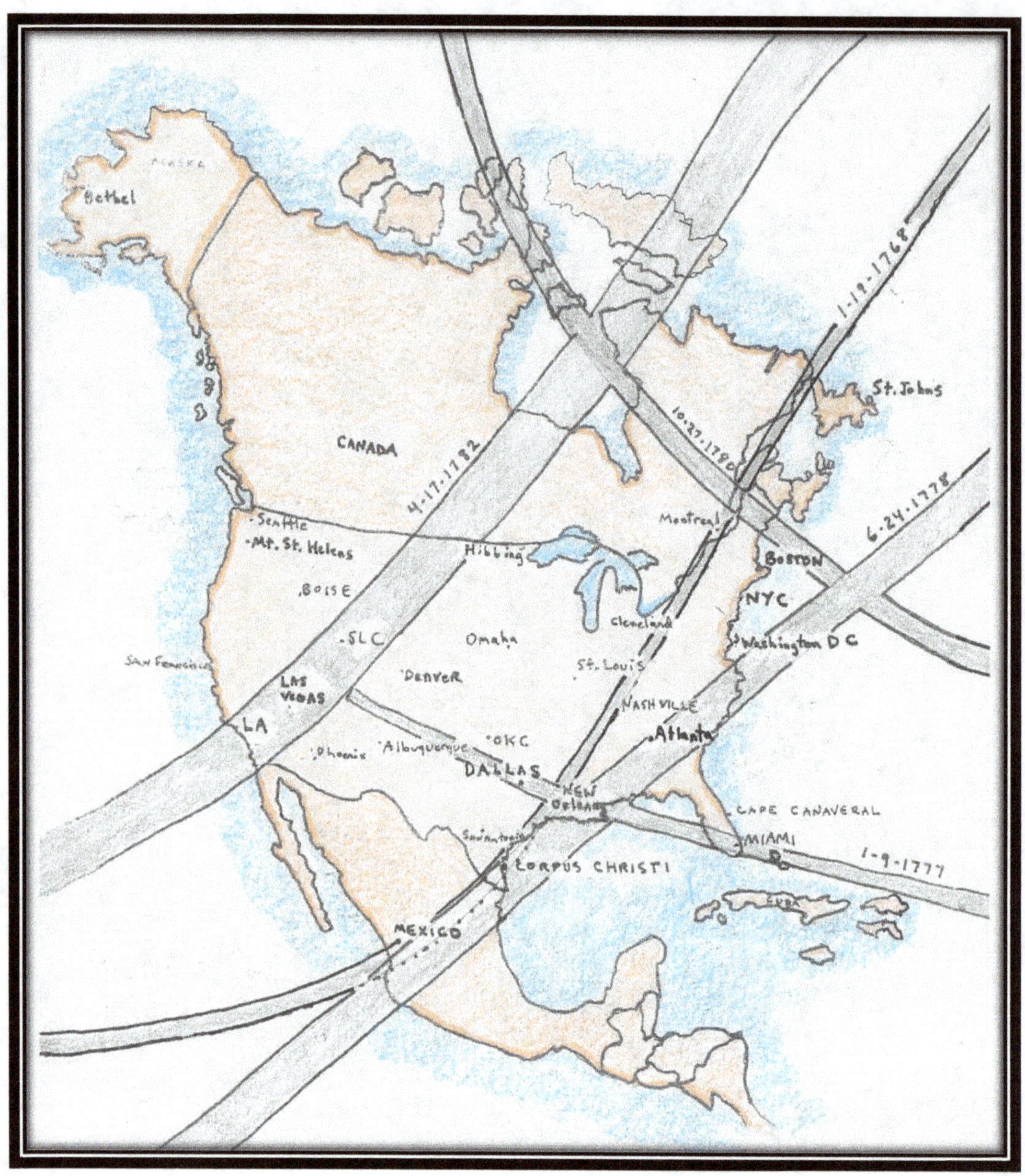

XI. US Revolution Eclipse Swarm 1768 to 1782

1768 Jan 19 Hybrid Eclipse (Saros 137)
1777 Jan 9 Ring of Fire (Saros 118)
1778 June 24 Total Eclipse (Saros 133)
1780 Oct 27 Total Eclipse (Saros 120)
1782 Apr 12 Ring of Fire (Saros 135)

1768 Jan 19 Hybrid Eclipse (Saros 137) 13 seconds of totality four miles wide spottily shoots across the USA, loosely along the western side of the 13 colonies. 1768 is widely considered the beginning of the revolution as tensions begin to boil over between Colonists and the British.

1777 Jan 9 Ring of Fire (Saros 118) 190 days after the Declaration of Independence, a ring of fire … Begins over Mesa Verde, Colorado (last home of the Anasazi). Thin eclipse path travels over Dallas, New Orleans and Miami. Also Columbus Landing site in the Bahamas and ends at La Coruna, Spain (The Crown?) **Saros 118:** Saros 118 returns for Lincoln's 19th Birthday Feb 12 1831 eclipse over USA.

1778 Jun 24 Total Eclipse (Saros 133) An important total eclipse noted by historians and troops at the time. New Orleans receives its second eclipse in a year and a half. Totality in Atlanta and Virginia. Occurs four days before the Battle of Monmouth, a strategic moral victory for the revolutionary Army.

1780 Oct 27 Total Eclipse (Saros 120) … Eclipse only in Maine. Three weeks after major defeat of British at Battle of Kings Mountain. Very close to one lunar year before final defeat of British at Yorktown on 10-19-1781

1782 April 12 Ring of Fire (Saros 135) Six months after British defeat at Yorktown, as British Forces evacuate the colonies, a great ring of fire goes across the still wild west. Los Angeles (founded 7 months before on 9-4-1781), future Las Vegas and Salt Lake City all in the path.

The American revolution is neatly outlined by 5 eclipses in 14 years. The 1777 and 1778 eclipses were both noted by George Washington and witnessed by troops and scientist historians. By the end of the Revolutionary War, the US population will be at nearly 3 million. It will be 24 years after the 1782 eclipse before another solar eclipse occurs in the USA in 1806. By then, the territory of the new nation will have expanded dramatically with the Louisiana purchase and the population will double again.

Of interest, there also happened to be a total eclipse on July 4, 1796 (the 20-year anniversary of the signing of the Declaration of Independence). That eclipse travels across Empty Pacific Ocean, touches virtually no land except beginning on a tiny sliver of beach at sunrise on the Philippines and ends off the west coast of Mexico without touching land. It was the only solar eclipse on July 4th date for hundreds of years past or future on planet earth.

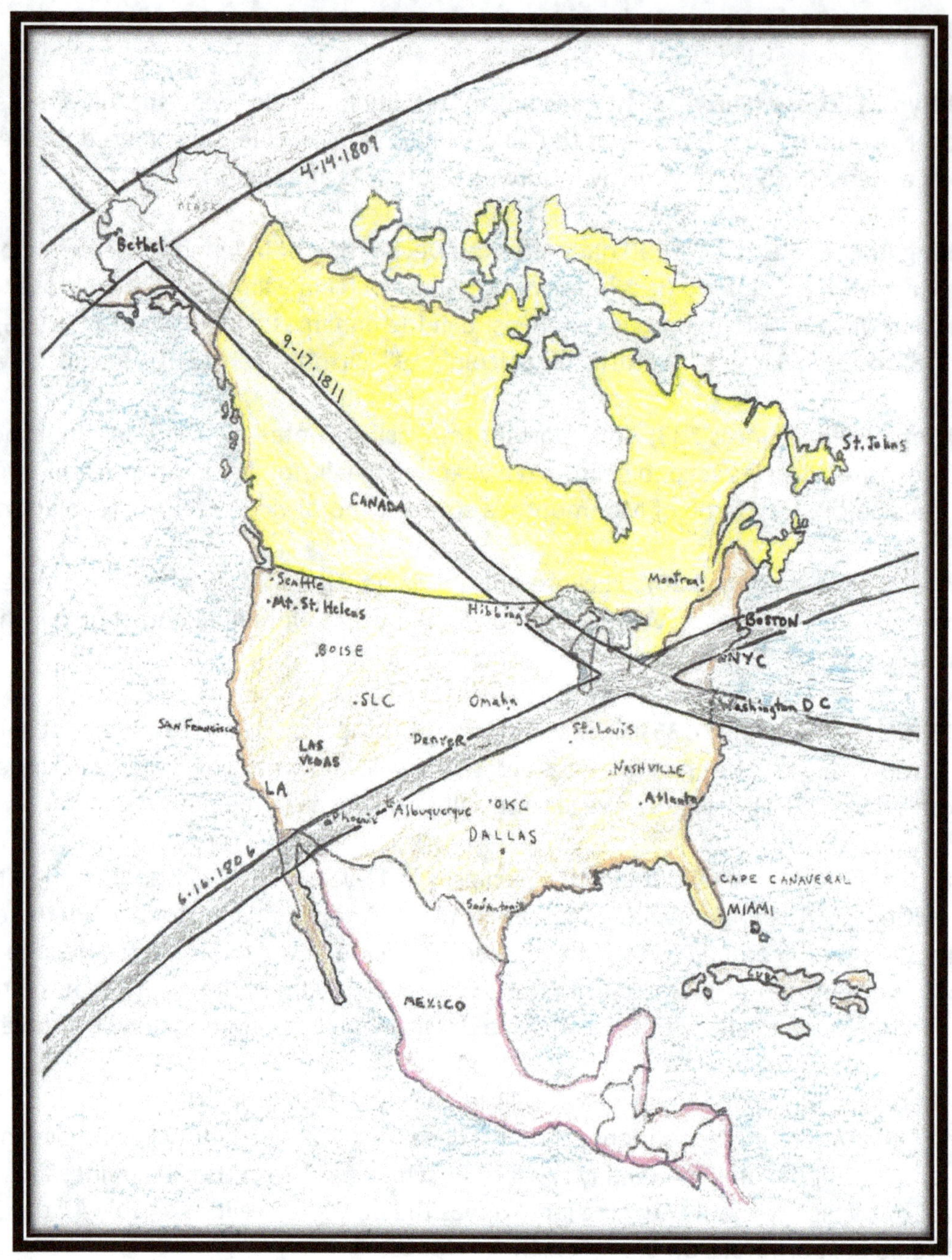

XII: Lincoln Birth/Bethel and Cleveland Cross 1806 to 1811

    1806 June 16 Total Eclipse (Saros 124)
    1809 April 14 Ring of Fire (Saros 116)
    1811 Sept 17 Ring of Fire (Saros 141)

XII. Lincoln Birth/-Bethel and Cleveland Cross 1806 to 1811

1806 June 16 Cross Country Total Eclipse (Saros 124) A truly great American eclipse, almost fully Cross-country two and ½ years before Abe Lincoln's birth and 62 eclipse years exactly before his death. The only total cross-country eclipse to travel southwest to northeast, directly across American heartland. It begins in the desert of Arizona and travels the length of the 30-year-old nation. Phoenix, Albuquerque, Kansas City, Springfield, south Chicago, Boston
Saros 124: Notably appears for Civil War Seattle eclipse Jul 18 1860 and WWI eclipse Aug 21 1914.

1809 April 14 Lincoln Ring of Fire (Saros 116) Only eclipse included that doesn't appear over contiguous USA. Passes over what will come to be named Bethel ("House of God"in Hebrew), Alaska on the same day upon which Lincoln will be shot 56 years later. Eclipse occurs 2 months and 2 days after he is born on Feb 12 1809. Very slight partial coverage over his birthplace in Kentucky.
Saros 116: has its last complete eclipse 2 eclipses later in 1845 and dies completely on 7-22-1971, which is 2 years and 2 days after the moon landing, exactly 38 solar years before Wuhan Eclipse and exactly 112 eclipse years after Lincoln's assassination on April 14/15 1865.

1811 Sep 17 Ring of Fire (Saros 141) The great Saros 141 crosses the 1809 eclipse in modern Bethel. Crosses Canada and then bulk of Northeast US (12 states). Hibbing MN, Michigan, Columbus Ohio and Washington DC, Jamestown.
Saros 141: 141 is one of the greatest Saros of the last thousand years, producing the longest eclipses. Returns in Farewell Abraham Oct 19 1865, 188 days after the murder of Lincoln. Also creates JFK Premonition in 1919, 1955 Southeast Asia and 1992 LA Sunset.

Abraham Lincoln's life and death bear an unusual relation to solar eclipses. The midpoint of these 1806 and 1811 eclipses is Feb 4 1812, just 8 days before the President's birth on Feb 12 1809. Springfield Illinois, (Lincoln's most famous hometown) receives totality in the first eclipse and Washington DC, where he served and died as President is in the last. The middle eclipse bears the date of his death 2 months and 2 days after his birth. These are just the beginning of eclipse peculiarities surrounding the martyred President. See section on Lincoln and Kennedy.

The 1806 and 1811 eclipses cross in Cleveland, which was founded in 1796. It is not a Hepton Cross, and one of the eclipses is annular, but it's still probably the most pronounced US cross until our current ones.

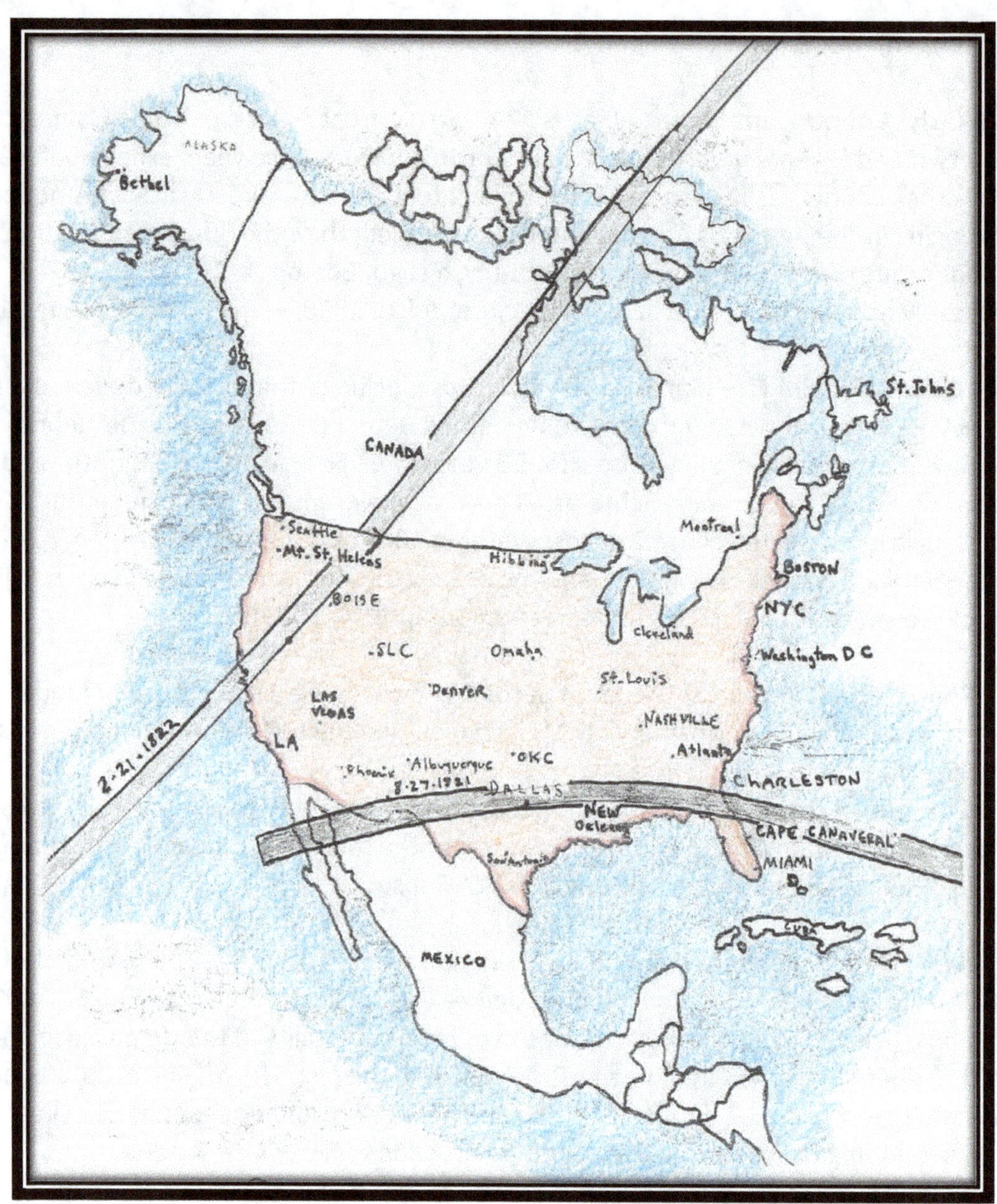

XIII.   Two Ring of Fires in 178 days- 1821 and 1822

1821 Aug 27 Ring of Fire (Saros 132) An "almost" Cross Country Eclipse through the Deep South.
Hot Spots: El Paso, Midland, Dallas, Shreveport, Jackson, Columbus, Savannah

1822 Feb 21 Ring of Fire/ Hybrid (Saros 137) Occurs over far west, including Mount Shasta

These two eclipses occur 178 days apart. This is the shortest period between American eclipses in this 531-year survey. 177 days is typically the shortest time possible between complete solar eclipses on earth at any time and are called "semester series", beginning days of opposite eclipse seasons.

XIV. Trail of Tears Octon 1831 to 1838

1831 Feb 12 Lincoln Birthday/ Nat Turner Ring of Fire (Saros 118)
1834 Nov 30 Trail of Tears Total Eclipse (Saros 120)
1838 Sep 18 Washington DC Ring of Fire (Saros 122)

1831 Feb 12 Lincoln Birthday/ Nat Turner Ring of Fire (Saros 118) On Abraham Lincoln's 19th birthday, this eclipse across the southeastern USA inspires the young (and mystical religious) Nat Turner to begin preparations for the largest slave rebellion in US history, which he leads on August 21, 1831 (an important Metonic date—see great eclipses of 1914, 1933, and 2017). Simultaneously, forced relocation of the Southeastern Native Tribes of Cherokee, Creek, Choctaw, Seminole and Osage is in full swing after Indian Removal Act of 1830 (famously opposed by Davy Crockett). It becomes known as The Trail of Tears. Lincoln could have seen this eclipse on his birthday, as least as a deep partial.

In all, 9 slave states experienced the eclipse, probably the most geographically possible. Houston (5 years before Texas revolution), Atlanta, Richmond, Jamestown.
Saros 118: returns to US for Buffalo Eclipse of Mar 16 1885 and finishes a thousand years of complete eclipses with successive eclipses on Hitler Birthday and Death day in Apr 19 1939 and Apr 30 1957.

1834 Nov 30 Trail of Tears Total Eclipse (Saros 120) One of the most extraordinary eclipse paths in history traces the Trail of Tears from Georgia to Oklahoma at the same time as the forced marches were proceeding. Atlanta, Charleston, Columbus GA, Tahlequah OK (modern capital of Cherokees) first settled beginning in 1832. This is Tahlequah's only total eclipse in survey. In path again in 2045.
Saros 120: Saros 120 continues its run up to the present day as one of the most clearly ominous eclipses of the Saros. It ends its history of complete eclipses in 2033 connecting the Eastern and Western Hemispheres simultaneously across the Bering Strait, within 4 days of the 2,000[th] anniversary of the death of Jesus Christ.

1838 Sep 18 Washington DC Ring of Fire (Saros 122) In the most deadly year of the Trail of Tears, a wide ring of fire across 13 northern states. Many of the great northern cities. Hibbing, Detroit, Cleveland, Columbus OH, Pittsburgh, Philadelphia, New York City, Baltimore. Washington DC, where the Cherokees were betrayed, is the centerline of the Ring of fire.

How to view such a piece of time art? History marked with shadows. It sings in a cosmic record the hand and song of God. The eclipse paths are eternal art. The subject matter? God only knows (and He does). It certainly seems like a story song. The worst kinds of human treachery and bondage were being lived out in a broken world under the banner of the freest nation in human history. God sees everything.

Between the 1831 and 1838 eclipses, there are exactly 7 years, 7 months, and 7 days (an Octon—exactly 8 eclipse years). The 1834 Trail of Tears eclipse is situated at the precise mid-point on the calendar (at 8 eclipse seasons either way).

It is also at the midpoint of the map geographically. One of the greatest God Songs I've seen.

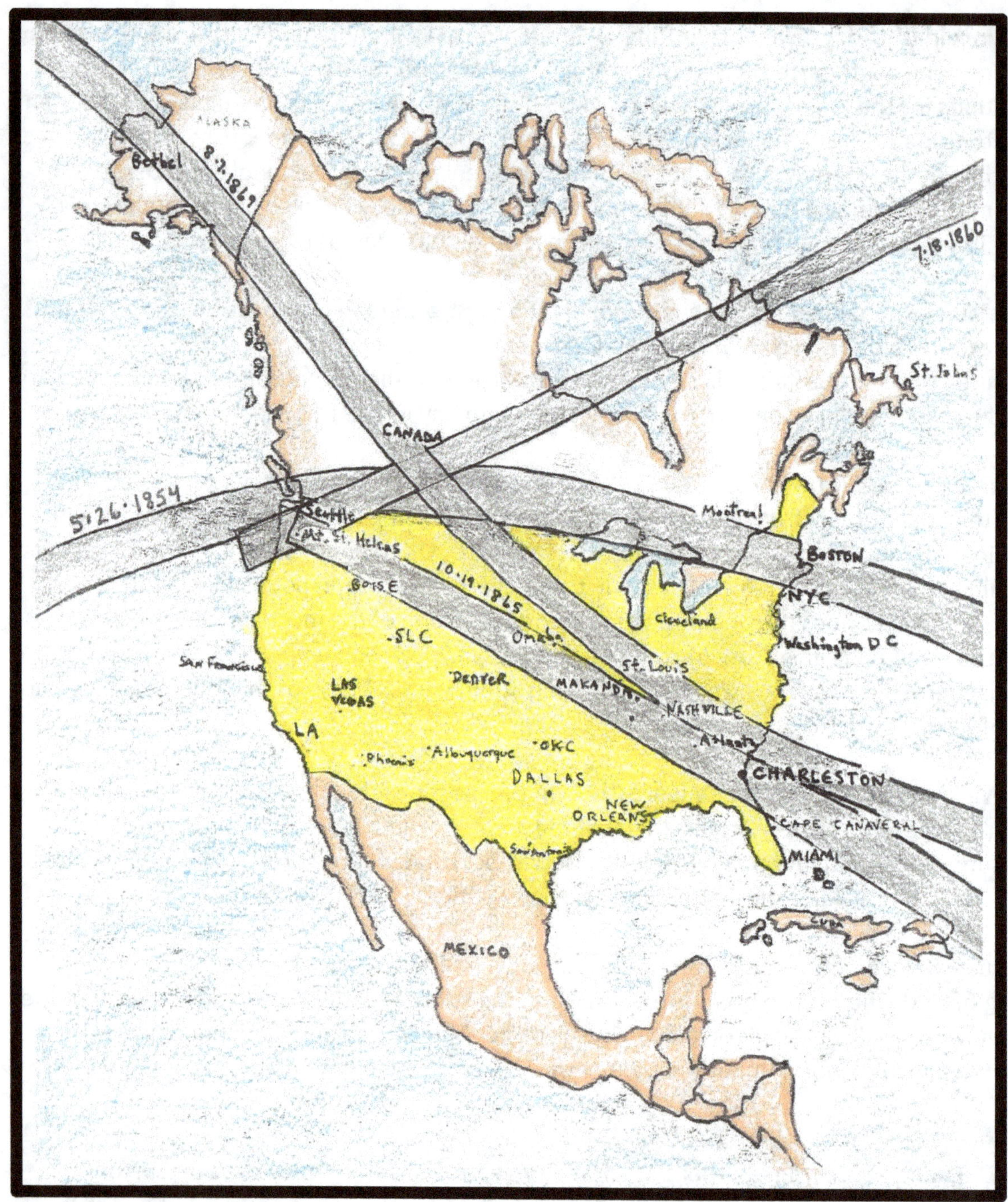

XV. Civil War 1854 to 1869

1854 May 26 Northern Border Ring of Fire (Saros 135)
1860 Jul 18 Total Eclipse Seattle (Saros 124)
1865 Oct 19 Farewell Abraham Ring of Fire (Saros 141)
1869 Aug 7 Golden Spike Total Eclipse (Saros 143)

1854 May 26 Northern Border Ring of Fire (Saros 135) There's never been another eclipse that so perfectly straddles the US/ Canada border. Washington to Maine, all 12 states of the northern border. Eclipse happens 4 days before the passing of the Kansas-Nebraska Act on May 30, 1854, which opened up new territories to slavery, prompting violent clashes which accelerated the race to Civil War. Abraham Lincoln opposed the Act and gave one of his most famous anti-slavery speeches in October 1854, resurrecting his political career and leading him eventually to the White House.
From that speech:

" Little by little, but steadily as man's march to the grave, we have been giving up the old for the new faith. Nearly eighty years ago we began by declaring that all men are created equal; but now from that beginning we have run down to the other declaration, that for some men to enslave others is a 'sacred right of self-government.' These principles cannot stand together. They are as opposite as God and Mammon; and whoever holds to the one must despise the other."— Abraham Lincoln

1860 Jul 18 Total Eclipse Seattle (Saros 124) Three and ½ months before Lincoln is elected in November, a total eclipse over Seattle, founded 9 years before. 120 years before Mount Saint Helens eruption. First of two St. Helens eclipses in a row. it has four in 119 years between 1860 and 1979. The secession of the southern states began five months after this eclipse on December 20 1860.

1865 Oct 19 Farewell Abraham Ring of Fire (Saros 141) 188 days after the assassination of Lincoln on Passover and Good Friday 1865, the Cross-country eclipse starts over Mount Saint Helens at sunrise and crosses the entire USA to exit at the notorious slave port of Charleston, SC. A total of 20 states and territories witnessed the eclipse, the most crossed by any eclipse in this 500-year survey. At 9 minutes 27 seconds, it would be among the longest eclipses in American history. It would cross Lincoln's birthplace and boyhood home in Kentucky. It also crosses the apex of the 2017/2024 eclipses at Makanda, Illinois, as it follows an almost identical path as the August 21 2017 USA eclipse. The eclipse manages to cross the Atlantic and end at sunset on the Western coast of Africa. Seattle, Portland, Billings, Lincoln NE, Kansas City, Columbia, St. Louis, Nashville, Atlanta, Charleston.

Saros 141: Saros 141 create 11-22-1919 JFK death day eclipse. Saros 141 delivers the longest eclipses in the last thousand years, including ominous ones on 11-11-1901 over Egypt, Dec 14 1955 over Vietnam and Southeast Asia—the longest in 1000 years at 12 minutes 9 seconds just as war began. Also comes for the LA ring of Fire Jan 4 1992 three months before the Rodney King riots. Its next eclipse is Australia Day Ring of Fire South America to Spain on 1-26-2028.

1869 Aug 7 Golden Spike Total Eclipse (Saros 143) A total eclipse across the heart of the nation including Lincoln's adopted hometown of Springfield, Illinois. 3 months after "the golden spike" at Promontory Utah completes the American transcontinental railroad. Within 15 years of this eclipse nearly every buffalo in America was dead and the final free native tribes of the western United States would be brutally subjugated, and the survivors herded onto reservations.

XVI. Far West—Buffalo Eclipse Swarm 1878 to 1889

1878 Jul 19 Buffalo Cross Total Eclipse (Saros 135)
1880 Jan 11 Salt Lake City Total Eclipse (Saros 139)
1885 Mar 16 Buffalo Cross Ring of Fire (Saros 118)
1889 Jan 1 New Years Day Total Eclipse (Saros 120)

1878 Jul 19 Buffalo Cross Total Eclipse (Saros 135) First half of a Hepton Cross makes a swooping pass from Alaska to the Gulf of Mexico. Denver and Dallas are in the path of Totality.

1880 Jan 11 Salt Lake City Total Eclipse (Saros 139) The great Saros 139 ends at sunset just past Salt Lake City, after giving the capital of Mormonism its only total eclipse in US history. The 1880 census showed 20,768 residents of Salt Lake City. Only 18 months after the 1878 eclipse. one of only three times American total eclipses occur so close together in survey. The eclipse stops just shy of interacting with the other paths.

1885 Mar 16 Buffalo Cross Ring of Fire (Saros 118) Mount Shasta in centerline as well as Boise, Idaho. Second half of a Hepton with the 1878 eclipse.

1889 Jan 1 Total Eclipse (Saros 120) New Years Day, USA. Of course, the amazing Saros 120 gives an ominous New Year's Eclipse. Passes just north of San Francisco. Reno, Idaho Falls, Billings MT, almost the entire Yellowstone park. Less than two years before Wounded Knee Massacre.

Saros 120: Saros 120 returns to the USA for NYC Eclipse of Jan 24 1925 then Mount Saint Helens/Microsoft Eclipse on Feb 26 1979. It brings the almost unreal Bering Strait eclipse of 3-30-2033, which links the eastern and western hemispheres. Saros 120 had the last eclipse on earth before Columbus came to the Americas, over the isthmus of Panama connecting North and South America on Apr 26 1492.

There's never been such a systematic extermination of so grand an animal as the Buffalo, primarily carried out between 1860 and 1885. This unprecedented catastrophe accompanied the conquest of the Plains and Intermountain Tribes who depended on the Buffalo to survive. From perhaps 60 million Buffalo in 1850, only a few hundred remained by 1885. The vast majority were killed from 1870 to 1880 for hides and often left to rot.

The solar eclipse swarm of 1878 to 1889 draws a unique and dramatic pattern across the land.

XVII. 20th Century's First Total Eclipse 1900 May 28 (Saros 126)

May 28 1900 The first eclipse of the 20th Century on planet earth, either lunar or solar, connects the imperial characters of the USA and Egypt. 17 years (and one day) before the birth of John F. Kennedy. The next solar eclipse on earth will occur on November 22 1900, 63 years to the day before his murder. JFK's birth and death days run side by side throughout eclipse history (see "personalities in eclipse" section). Eclipse begins in the Pacific and ends just before Luxor, Egypt with 95% coverage as the sun sets. Mazatlan, New Orleans, Columbus GA, Jamestown, Spain and Portugal, Algiers, Tripoli (which is in several 20th century eclipses), Luxor Egypt at sunset.

300/44: paired with 1600 Jul 10 Cape Canaveral to Khartoum

Saros 126: Saros 126 makes its first appearance in the USA with only its second eclipse as a Total. It returns again on its next cycle as the great June 8 1918 USA cross-country eclipse. It also brings the June 30 1954 USA to Iran Shahrud Eclipse Total.

Next up for **Saros 126**? Aug 12 2026 in Spain. Hasta Luego, amigo.

XVIII: Temple of Sun and Moon-Mexico City/Cape Canaveral 1908 June 28

Jun 28 1908 Ring of Fire (Saros 135) Eclipse begins just offshore of Mexico, crosses both great metropolises of the two great civilizations of ancient Mexico, Teotihuacan and Tenochtitlan—including the Temple of the Sun and Moon. The path of Annularity then crosses Cape Canaveral 61 years before the liftoff of Apollo 11 on Jul 16 1969 on its way to the Moon. Path ends in Ghana, Africa.
The eclipse occurs 60 days before the birth of Lyndon Johnson on August 27 1908. 66% coverage where he would be born in Stonewall Texas. In 11 years, an eclipse will go directly over his home. He would be the president who would nurture the Apollo moon program (inherited from Kennedy) before his term ended six months exactly before the moon landing.

XIX. Great American Eclipse Swarm 1918 to 1932

1918 June 8 Cross Country American Century Total Eclipse (Saros 126) One of the greatest and most witnessed eclipses in US history. Ten years minus 10 days from the Temple of Sun/Moon-Cape Canaveral Eclipse. It happens as the nation is simultaneously embroiled in World War One and facing the opening months of the Spanish Flu Pandemic. The path is an inverse reflection of the Aug 21 2017 eclipse. Eclipse over Cape Canaveral, its second eclipse in ten years. The beginning of the American Century, reckoned here as from 1918 to 2017. Mount Saint Helens (third in 58 years--next up 1979), Denver, OKC, Cape Canaveral **300/44 Pairing:** 1618 Jul 21 US Eclipse
Hepton: paired with Roaring 20's 1925 Jan 24 NYC eclipse.

1919 Nov 22 JFK Premonition Ring of Fire (Saros 141) One of the incredible John F. Kennedy eclipses. JFK is a two-year-old in Boston when it occurs. This Ring of Fire begins 44 years to the day before his murder on Nov 22 1963. Path begins at sunrise in Texas, in the state JFK where will be murdered, passing over the boyhood home of a ten-year-old Lyndon Johnson, the man who will replace him.

Path crosses entire island of Cuba, the only time this happens in our 500-year survey. Cuba would be forever associated with Kennedy's Presidency. Path ends in Africa. Sunrise at Midland Texas (sometimes home of the Bushes), San Antonio, Austin, Houston, Havana. Ends at sunset in ancient city of Timbuktu, Mali in Africa.

300/44 Pairing: 1619 Jan 15 USA to Africa 310 years before Martin L. King birth.

Saros 141: Almost the longest eclipse in US history at 11 minutes 37 seconds.

1923 Sep 10 Earthquake Companions Japan to LA Total Eclipse (Saros 143) Eclipse begins in the northern islands of Japan 9 days after worst disaster in Japanese history, the Great Tokyo/Yokohama Earthquake and Fire of Sep 1 1923. Bisects southern metropolis of Los Angeles and gives totality to San Diego. One week before the birth of singer Hank Williams in Mount Olive, Alabama (58% coverage) on 1923 Sep 17.

300/44 pairing: coupled with 1623 Oct 23 eclipse, which also had totality in LA.

Saros 143: Last time in USA was 1869 Total. Returns for Boston single City in 1959

1925 Jan 24 Roaring 20's/ NYC/ Bob Dylan Total Eclipse (Saros 120): The great Saros 120 brings totality to New York City at the peak of its new maximum expression. The path begins at sunrise over Bob Dylan's Birthplace and Home (Duluth and Hibbing) goes over Woodstock and New York City 36 years to the day before a 19-year-old Dylan shows up from Minnesota in New York on Jan 24 1961. Eclipse is within months of birth of Robert Kennedy, Malcolm X, George HW Bush and an astonishing variety of famous people. Path crosses to England.

300/44 pairing: paired with 1625 Mar 8 eclipse over Cape Canaveral.

Saros 120: What more can you say about Saros 120? Trail of Tears 1834, Death of the Buffalo New Year's 1889, Iron Curtain 1961, Mount Saint Helens 1979, North Pole Equinox 2015 and the Great Bering Strait Jesus Anniversary of 2033.

1930 Apr 28 Bizarre San Francisco Hybrid (Saros 137) the pinpoint of totality was less than a half mile wide and lasted but a few seconds as it passed apparently just north of downtown San Francisco in this Hybrid eclipse. 36 years and two days before founding of Church of Satan by Anton LaVey in San Francisco in April 30 1966 (itself the 21st anniversary of Hitler's death).

Saros 137: Korea in 1948, US in 1984, Pandemic Solstice 2021

1932 Aug 31 Northeastern US Total Eclipse (Saros 124) small area of US totality includes New Hampshire and Vermont. Kennebunkport Maine (spiritual center of the Bush family) Cape Cod (spiritual center of the Kennedy's) in totality as well as totality bisecting birthplace of both George HW Bush and the Kennedy' s in Boston. JFK was 15 years old, RFK and Bush were both 7 years old.

Summary of perhaps the greatest US swarm ever? Six eclipses in fourteen years with endless associations and implications. The next eclipse occurs in the USA 7 years, 7 months and 7 days later- (Apr 7 1940) in an Octon Cycle with the 1932 final eclipse of swarm. That will be the longest period of time between eclipses in the years 1918 to 1984. The 20th Century was a rich time for US eclipses.

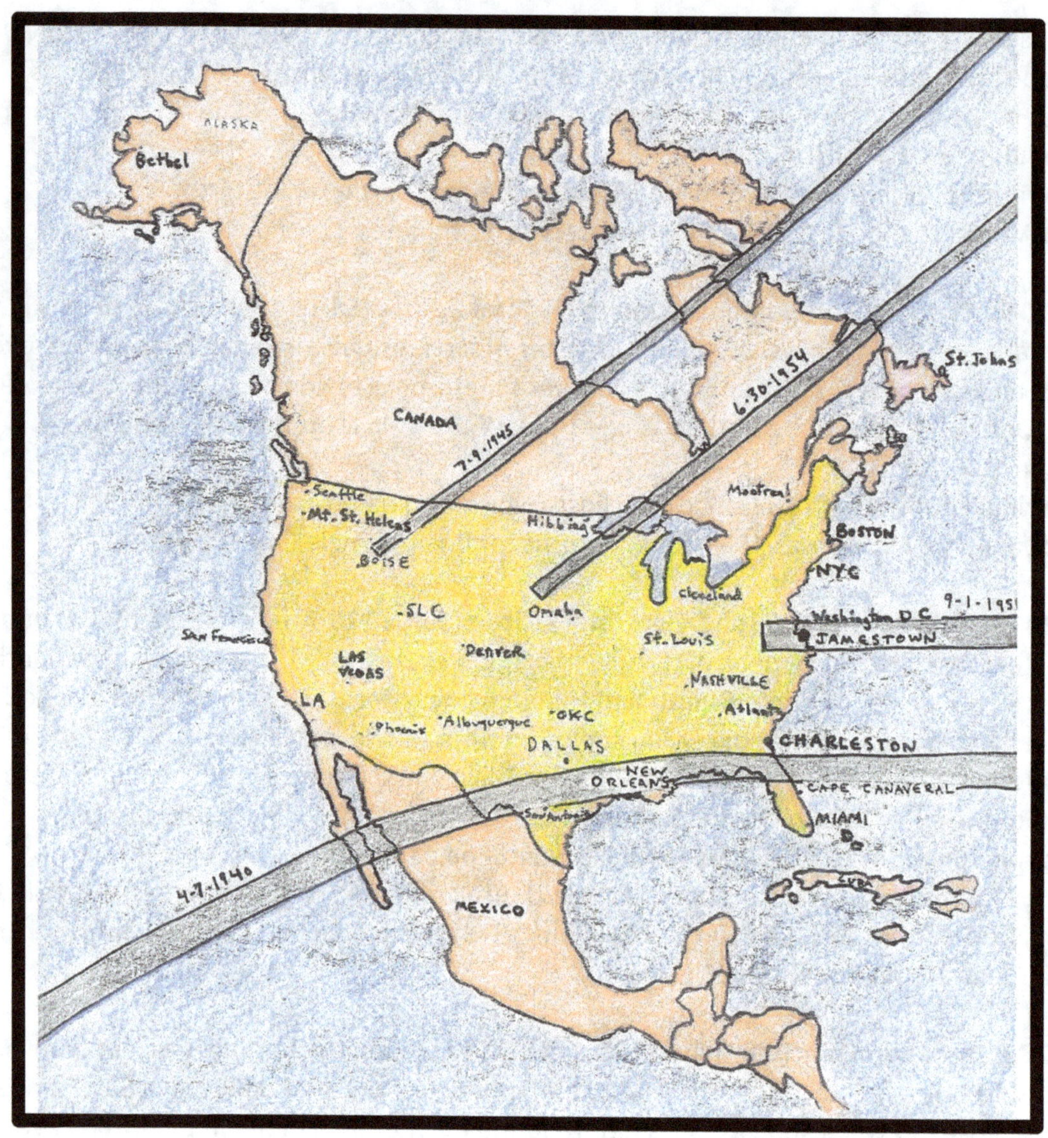

XX. USA 1940 to 1954 –Emanations

1940 Apr 7 Ring of Fire Deep South (Saros 128)
1945 Jul 9 Atom Bomb Total Eclipse US to USSR (Saros 145)
1951 Sep 1 Jamestown to Angola Ring of Fire (Saros 134)
1954 June 30 USA Heartland to Shahrud Iran Total eclipse (Saros 126)

1940 Apr 7 Ring of Fire Deep South (Saros 128) Eclipse runs along Gulf Coast 64 years and one day before upcoming 2024 Apr 8 eclipse. Those are the closest eclipses to Easter in the survey, both 8 days after Easter, apparently as close as you can get a solar eclipse to Easter. 20 months to the day (609 days) before Pearl Harbor. As with Pearl Harbor, also on a Sunday. San Antonio, Houston, New Orleans. **Saros 128:** Saros 128's first appearance in USA. Returns for two consecutive eclipses in the US in 1994 and 2012. Reappears over Athens and Istanbul in 2030.

1945 July 9 Atom Bomb Total Eclipse US to USSR (Saros 145) July 9 1945 is the day White Sand's Missile Range in New Mexico was established by the USA to create a place for "the bomb". There was an eclipse that day from the USA to the USSR. One week later, on July 16th the world's first atomic bomb was detonated. Hindu scripture ran through the mind of Atomic Bomb father Robert Oppenheimer as he watched: *"Now I am become Death, the destroyer of worlds."*
The bomb was used on humanity 27 days later in Hiroshima and then in Nagasaki, Japan. This total solar eclipse travels from the USA to USSR one week before the detonation. Path of Totality in Idaho and Montana (Butte), Canada, Greenland, Scandinavia and then bisecting Russia. 98% totality in St. Petersburg and 95% in Moscow. Path ends at western border of China. 60% coverage at nuclear test site at White Sands. 70% at Los Alamos Laboratory. **Saros 145:** returns to the US in 1963 and 2017. Crosses USSR in 1981 and is the incredible last eclipse of the 20th century on 8-11-1999, which crosses Hitler's Birthplace and Ninevah. Scheduled to cross Beijing, Pyongyang and Tokyo on Sep 1 2035.

1951 Sep 1 Jamestown to Angola Ring of Fire (Saros 134) One of many 20th century eclipses with Slave Trade associations. This Ring of Fire eclipse happens 332 years after first slaves brought to English colonial Jamestown on August 20 1619. Path begins in Virginia with annularity in Jamestown and colonial Williamsburg. It crosses the Atlantic to the Slave Coast, including Angola, home to the 20 or so Africans brought on that first slave ship to the British colony. Eclipse crosses Africa and ends in Madagascar almost precisely on capital of Anananarivo. One of only a couple of true "single city" eclipses in our survey that begin on eastern seaboard. Another happens five years later in Boston. **Saros 134:** Aug 21 1933 Jerusalem and Oct 14 2023 USA are some notable eclipses of Saros 134

Jun 30 1954 USA Heartland to Shahrud Iran Total eclipse (Saros 126) begins in Nebraska and ends in Iran, intersecting the path of a total eclipse that occurred 2-25-1952 over the city of Shahrud (The Shahrud Cross) one year after the Shah was re-installed in August 1953 by a US and British backed coup. The path brings totality in Saint Paul, 97% coverage in Hibbing Minnesota for a 13-year-old Bob Dylan, bisects Canada, crosses the Atlantic, dives through Scandinavia into the Iron Curtain of the Eastern USSR (totality in Kiev). Intersects with last total eclipse over city of Shahrud, Iran. Path then goes over Afghanistan, and Pakistan (totality in Quetta) before ending in India.
Saros 126--USA cross country Jun 8 1918. Spanish judgment Aug 12 2026)

Summary of Eclipses: No crosses. The eclipses seem oddly spaced from each other, 3 of them look like emanations from the nation itself, all beginning at sunrise one right after another between 1945 and 1954. In the next section, the 1959 Eclipse does the same. These occur as the USA enters its most powerful phase of mystery empire building. Look at the Southeast Asia Eclipse Swarm (1948 to 1965) for historical and astronomical reference.

XXI. The 1960's Hepton and JFK Boston Eclipse

1959 Oct 2 Total Eclipse JFK Boston Sunrise Eclipse (Saros 143) The fourth emanation in a row. As if it weren't enough that JFK's birthday and death day appear side by side in the eclipse Metonic during his lifetime, a single city total eclipse occurs over his hometown of Boston while he is running for President. Beginning at sunrise on border of Massachusetts, Vermont and New York, the eclipse crosses the Atlantic and travels to Africa. bringing totality to Senegal and Mali (both of whose Presidents visited Kennedy in the White House, a first for either of those countries). Boston experienced a total solar eclipse for the first time since 1806 (it won't have another until 2200). Kennedy gives a speech that night on the campaign trail in Rochester, New York. Boston is the only major city in America to witness the eclipse. Kennedy is elected President the following year on Nov. 8th, the day after a transit of Mercury was visible across the Sun's face from the western hemisphere. (See personalities section). **Saros 143.** Saros 143 includes 1869 USA total and WWII fall Equinox Eclipse (9-21-1941) over USSR and China. It brings totality to Colombia on Columbus Day Oct 12 1977 and then returns for its last total eclipse on Oct 24 1995--over Vietnam and Southeast Asia.

1963 July 20 Total Eclipse Beginning of the Sixties (Saros 145) 6 years to the day before Apollo 11 astronauts stepped on the moon. 39 days before the famous March on Washington that featured both Martin Luther King (he would die at 39 years old). March also presented a young Bob Dylan, whose song, "Blowin' in the Wind", was climbing the pop charts at the time (the Peter, Paul and Mary version) at number 14. It would eventually reach number 2 on August 14. The Eclipse occurs 125 days before the assassination of John F. Kennedy. It occurs 18 years (minus 17 days) after the twin atomic bombs were dropped by the US on Japan. The eclipse begins in Japan (the island of Hokkaido), bisects new state of Alaska before crossing Canada. It touches the USA only in Maine, with totality in Bangor. Four months after the eclipse, John F. Kennedy would be shot dead in Dallas Texas (35 % coverage). His hometown Boston saw 95 % coverage. 97% totality at Bush family haven of Kennebunkport. Washington DC had 77% coverage. A total lunar eclipse followed JFK assassination 38 days later.

Probably the beginnings of the USA 1960's, as far as cultural and solar eclipse cosmic timekeeping associations go. The first Beatles record released in America happened two days later. It was "Please Please Me". The decade culturally will end 7 eclipse years later with the March 7 1970 USA eclipse.
Saros 145 which appeared at the Atom Bomb eclipse of 1945, Liverpool 1927
Beijing/Pyongyang/Tokyo on Sep 2 2035
Hepton: linked with Saros 139 2422 days apart.. Saros 145 and 139 return for the USA in 2017/2024.

1970 Mar 7 Total Eclipse Eastern Seaboard (Saros 139) Seven eclipse years after the 1963 USA event, the southeastern seaboard of the USA experiences coastal totality. The seaboard was mostly cloudy though Florida remained fair. 93% totality at Cape Canaveral 234 days after men shot from there landed on the moon. Totality at Charleston South Carolina, the greatest slave port in North America (it receives totality again in 2017). 95% totality in Washington D.C., 96% in New York and 97% in Boston. The end of the sixties. The Beatles announced their break-up a month later. 3 months after the infamous Altamont Rolling Stones Concert. 2 months before the Kent State shootings.
Saros 139: returns to America in 2024 Apr 8 with similar path about 1,000 miles to the west.
Hepton: Hepton Cross a few hundred miles offshore of Boston/NYC. Linked with the 1963 eclipse.
300/44: paired with 1670 Apr 19 eclipse which would touch the least amount of USA in this 531-year survey, only the extreme northwestern edge of the Olympic Peninsula, including Mount Olympus.

Summary: Most cultural historians would likely agree that the era we refer to as "the sixties" fits neatly between the two Hepton eclipses of 1963 and 1970. Late 1963 was a turning point in American history with the rise of the counterculture movement, the March on Washington and the murder of JFK. 1970 saw the Hippie Dream collapse almost immediately as the calendar turned. Also, there was probably no more famous politician in the history of the 20th century than John F. Kennedy, so it seems quite remarkable that a solar eclipse in 1959 would pass *only* over his hometown of Boston as he runs for President. All three of these eclipses saw more than 95% totality in Boston and Cape Cod. They also saw similar coverage at Bush family center of Kennebunkport, Maine. We see peculiarity in honest consideration of such eclipses, a peculiarity that points towards meaning, towards story. It knocks at the doors of our perception of reality with songs, questions and suggestions.
None of it means anything without humility.

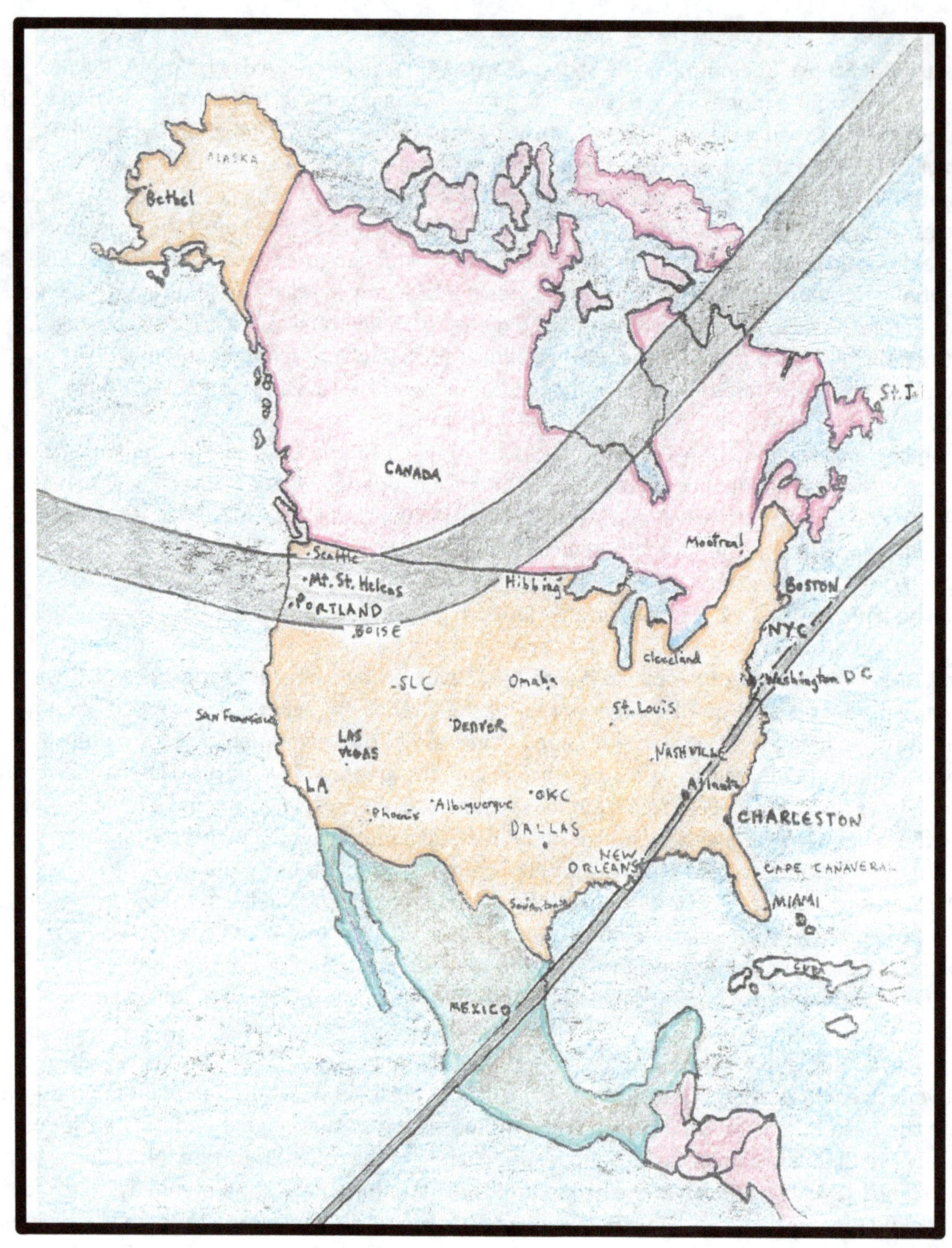

XXII. Brave New World-Mount Saint Helen's and 1984

1979 Feb 26 Mt. St. Helens/Microsoft Total Eclipse (Saros 120)
1984 May 30 Atlanta/ New Orleans/ DC Ring of Fire (Saros 137)

1979 Feb 26 Mt. St. Helens/Microsoft Total Eclipse (Saros 120) The great harbinger of 1979. Jimmy Carter is in the White House. During the coldest winter in a century, eight months before the hostage crisis in Iran, ten months before Soviets invade Afghanistan. America is set to rise again with a new kind of world dominance, defined by the universal machine that thinks. The eclipse occurs seven weeks after Bill Gates and Paul Allen move Microsoft Corporation to Gate's hometown of Bellevue, Washington, 99.6% Totality in Bellevue. The beginning of the technological era, the final era of the American Empire and the creation of the last world Empire. The number one song in America was "Do you think I'm Sexy" by Rod Stewart. The number one non-news show was "Three's Company".

Salem, Oregon is in Totality (it will again be in totality for Aug.21, 2017 USA eclipse, the only major city that has that distinction). Portland has totality. Crosses 5 northwestern states before bisecting Canada (Totality in Winnipeg).

Totality over Mount Saint Helens (122.2 degrees longitude) 388 days before she awoke with an earthquake on Spring Equinox of March 20 1980. Saint Helens explodes to the due north in the direction of Bellevue Washington (97 miles away --122.2 degrees longitude) on May 18 1980 in the largest volcanic eruption in modern North American History.. First eclipse over Saint Helens since Farewell Abraham on Oct 19 1865.
Saros Cycle 120: Again! Saros 120! The last time Saros 120 will touch contiguous USA. It returns to Alaska US for Bering Strait/ 2,000th anniversary of Jesus Crucifixion Eclipse on 3-30-2033.

1984 May 30 Atlanta/ New Orleans/ DC Ring of Fire-Almost Total (Saros 137) An interesting annular eclipse that is about as close as you can get to a hybrid/total. At its Maximum, the sun is 99.77% covered (near Christchurch, Virginia). Atlanta is 99.67% covered. What does this mean? It means that the alignment of sun and moon is almost perfect over the path. The moon is almost exactly 400 times closer to the earth than the sun. They almost perfectly fit in perspective. May 30 1984 is the JFK Birthday Metonic. His actual birthday is May 29 but that's how the Metonic works, sliding around on the hours around a date. He would have still been celebrating his 67th Birthday 12 hours before if he had not been murdered. Ring of Fire begins in Mexico and crosses eastern USA before heading to Africa. Near perfect coverage in Atlanta at Martin Luther King's Home exactly 5900 days after his assassination (and almost exactly 17 eclipse years).

95% coverage in Washington DC, where the Reagan/Bush Presidency was firmly in power. 90% in JFK's birthplace of Brookline, Massachusetts, 80% in Dallas Texas, where he was murdered. 93% in New York City. Direct path in the Deep South includes New Orleans (the last of its astonishing 7 eclipses in 500 years, birthplace of Lee Harvey Oswald and site of the Jim Garrison JFK conspiracy trial),

The eclipse crosses the ocean and enters Africa over Casablanca ("white house") and Marrakesh (Land of God) and ends in desert of the Sahara in Algeria. The last year of JFK birthday/Death Day Metonic conjunction which began with May 28/ Nov 22 in 1900. The next eclipse on earth after this amazing annular occurs on Nov 22 1984 (the 21st anniversary of his murder) over the site of his famous PT-109 heroics in the Solomon Islands of New Guinea.
Saros 137: Saros 137 brought the Titanic Eclipse of 1912, the San Francisco Hybrid of 1930, the Korean War eclipse of 1948, 137 returns for the 2020 Father's Day Solstice Pandemic Ring of Fire.

XIII. LA and Cross-Country Rings of Fire 1992 and 1994

1992 Jan 4 California Sunset (Saros 141)
1994 May 10 Cross Country Ring of Fire (Saros 128)

LA and Cross-Country Rings of Fire 1992 and 1994

1992 Jan 4 California LA Metropolis Sunset (Saros 141) The Saros responsible for bringing the longest duration eclipses of the last thousand years ends its millennium with a path over open Pacific ocean until ending at the metropolises of Los Angeles and San Diego/Tijuana at sunset. San Diego had a good view. Clouds obscured the eclipse in Los Angeles.
It had been nine days since the Soviet Union officially dissolved. George HW Bush will be President for one more year. It is the 500th year since Columbus in 1492 sailed the ocean blue. Southern California's Michael Jackson had the number one song with "Black or White". Nirvana's "Smells like Teen Spirit" was number 13.

Martin Luther King would have been 63 solar years plus one lunar year old at the time of eclipse if he hadn't been murdered. The worst riots in California history break out in Los Angeles 114 days after the eclipse (April 29-also a Metonic date) following the acquittal of police officers involved in the beating of motorist Rodney King. 64 people died and more than 2300 were injured in the violence.

Located at the very eastern extension of the eclipse path was Landers, California. On June 28, 1992, 60 days after the riots and 174 days after the eclipse, an Earthquake in the Mojave desert near Landers registered 7.3 on the Richter Scale. It was the largest earthquake to hit California in 40 years but damage was limited because of its remote location. 3 people were killed and 400 injured in the quake.
Single City: Southern California was the only land touched on a 10,000-mile path across the ocean. It was one of the longest eclipses of the last millennium at 11 minutes and 41 seconds near its mid-ocean maximum. Note 11:41 number peculiarity with Saros 141.
Saros 141: One of the greats. 1865 Farewell Abraham, 1919 JFK Premonition Texas Eclipse. Longest eclipse of the millennium (12 minutes and 9 seconds) in 1955 over Southeast Asia. Its final eclipse to last over 10 minutes occurs January 26 2028 -part of the upcoming Spanish Swarm.

1994 May 10 Cross Country Ring of Fire (Saros 128) 856 days after the Southern California Sunset ring of Fire. a cross-country annular eclipse starts in the Pacific, bisects Baja and the Sonoran Desert before crossing the heartland of the United States and exiting over the Bush family haven of Kennebunkport, Maine. Serial killer John Wayne Gacy is executed just after midnight on this day in Joliet Illinois, which borders total annularity. Barrack Obama is working in Chicago with 89% coverage.
Annularity in many cities, including El Paso, Amarillo, Lubbock, Tulsa, Joplin, Detroit, Springfield, Cleveland, Buffalo, Albany, and much of Maine. The USA ring of fire crosses the Atlantic and ends just after going over Casablanca ("white house") Morocco.

George W. Bush wins the Texas State Governorship 6 months after eclipse. Path goes by his childhood home of Midland with 87% coverage. His ranch at Crawford has 80% coverage. Austin has 77%. Bush center Kennebunkport, Maine is in path of full Annularity. George W. Bush birthplace of New Haven Connecticut sees 85% and Washington DC sees 80%. 6 years later he would be running for President.

This is the last complete eclipse in the USA for 18 years and 11 days until May 20 2012 Columbus Death Day Ring of Fire, which is also Saros 128. Remarkably, that eclipse *ends at Midland*. This 18 year stretch between eclipses (1994 to 2012) is the first time since 1854 that the US goes more than ten years without an eclipse.

2012 May 20 Hanoi Sunrise/Midland Sunset Ring of Fire (Saros 128)

I have this one eclipse sort of sitting by itself. It was 2012. It had been eighteen years since the USA had a complete eclipse, way back in 1994. A lot went on in those 18 years, to say the least. For one thing we went from landlines to facephones. I went from digging ditches to ...well, digging ditches and liking it better. But that's not all that happened between 1994 and 2012.
This eclipse doesn't neatly fit in with the next trio of eclipses that show up five years later in 2017. They are of different times. This 2012 one, its different, like the end of an era. My family and I watched this eclipse riding horses in the high desert, The pinon and juniper trees make little crescents on the ground out of the eclipsed sunlight. We were a little too north for annularity, but it was still a deep partial (maybe 80%) It was the first solar eclipse I'd seen since I was a kid, when I had watched the Feb 26 1979 in Oklahoma, also a deep partial.

Yeah, this 2012 eclipse sits by itself. Not only because it's a nice memory, but because it travels from sunrise in Hanoi Vietnam to sunset in Midland Texas. It travels that route on Columbus's death day.

2012 May 20 Hanoi to Four Corners Ring of Fire/ Midland Sunset (Saros 128)

Spanning the gulf between the East and the West. Saros 128 last visited the US in 1994. It now returns to the USA with a ring of fire ending at the boyhood home of George W. Bush three years after he left the office of President. Path begins at sunrise just east of Hanoi, Vietnam.

China gets its 3rd eclipse in 4 years, Annularity in Taipei, Taiwan and Tokyo, Japan. Path crosses Pacific and enters Northern California.

It crosses Mount Shasta, Reno, and Zion National Park in Utah. Much of the Navajo Indian reservation (including 3 of 4 of their sacred corner mountains-- San Francisco Peaks, Navajo Mtn, and Mount Taylor in New Mexico). All of the Hopi Mesas are in annularity as well as many other Pueblo tribes. Annularity over the Four Corners cross for the first true time in this 500-year survey. A ring of fire will perfectly cross it again in 2023, 12 eclipse years later. Chaco Canyon, Santa Fe, Los Alamos and Albuquerque are in both paths. Goes just north of White Sands Atom Bomb Site (87%) ...Annularity over Roswell, which is almost certainly home to real live space aliens (or so I've heard)!!

The path enters West Texas and reaches George W. Bush boyhood home of Midland at sunset, stopping shy of Abilene. 2012 is 10 years since US forces occupied the Middle East under his Presidency and 21 years since they did so under his father's Presidency.

Columbus died in Spain on 5-20-1506, exactly 506 years prior to the eclipse. Tradition claims his last words were "*Father, into thy hands I commit my Spirit*", also the last words of Jesus Christ. 2012 is the 520th anniversary of his discovery of America...thus, the 506 and 520 numeric are in peculiar juxtaposition in 5-20-1506 and 5-20-2012 506 years later. Numbers are just funny, sometimes.

The 500th anniversary of the Mayan/Aztec Apocalypse Eclipse of 1496 occurred in 1996, the same year there was a Columbus Day partial Eclipse visible from North America on October 12th. These sorts of fractal oddities appear in lots of directions when one considers eclipses.

If you remember, The Mayan calendar also supposedly ended on 12- 21-2012, which is yet another peculiar set of reflective numbers.

On the day of the May 20 2012 eclipse, President Obama hosts the NATO Summit in Chicago, Illinois It was the last NATO summit held in the USA. Chicago has 49% coverage at sunset.

The US eclipses of 2017 to 2024 are considered in the 2017 to 2034 section.

Chapter Four

MEET THE SAROS CYCLE TRIBE

Personalities in the Eclipse Story

MEET THE SAROS CYCLE TRIBE

Personalities in the Eclipse Story

The Saros Eclipse Cycle creates Saros Characters, or personalities. They are born. They mature. They fade. They die. They have characteristics specific only to them. You will meet them in this section. They exist only because of one of the greatest synchronicities in all of visible existence, the co-ordination of three separate cosmic calendar alignments. Perhaps you aren't supposed to know about this. Perhaps it's reserved for the alchemists.

Three cosmic calendars align perfectly every 19 solar years in the Metonic Cycle.
Three cosmic calendars align perfectly every 19 eclipse years in the Saros Cycle.
And yes, the sun is the same size as the moon in the sky.

Each Saros Character is a repeating and slowly evolving curved geometrical pattern. Its existence is defined by the shadow path of a repeating solar eclipse, which returns to the earth at a regular interval and at shifting locations.

Who discovered Saros? Knowledge of the cycle has been around since at least the Babylonians (ca. 700 BC). They determined its existence by observing lunar eclipses repeating on 18-year and 11 day (19 eclipse year) intervals. Other ancient astronomers in both hemispheres were likely aware of the Saros repetition as it pertained to Lunar eclipses.

Determining the solar Saros cycle took much longer (until the early 18th century). Solar eclipses are hard to predict without a measured globe and understanding of earth's rotation and revolution.

Who named it and why? The honor of naming the Saros fell to Edmund Halley, British astronomer and mathematician. He is famous for the comet named after him, the orbit of which he first calculated. Halley predicted and calculated the timing and path of an eclipse over England May 3 1715, the most accurate early prediction of a solar eclipse. It is a great peculiarity that this eclipse would occur in such a specific place as London (Saros 114 with 3 minutes and 33 seconds of totality) while Halley was at the peak of his profession. He was a friend of Sir Isaac Newton. It was the last complete solar eclipse over London in our history.

Halley named the Saros for a Babylonian word that ended up meaning 3600. Saros means "sweep" in Greek. It is said he was mistaken at its meaning.

Who numbered the Saros? The name of each Saros Character is a number. These numbers were given by the Dutch astronomer George Van Den Bergh in the 1950's in his book, *Periodicity and Variations in Solar Eclipses* (good luck trying to find a copy) and has been since adopted worldwide. The peculiar tale of Van Den Bergh (or what little we know) appears in the personalities section.

Because of the Saros Cycle's astonishing precision, eclipses have been mapped (probably quite accurately) backwards about 4,000 years and forwards at least a thousand years. They are given numbers as they are born.

Saros Cycle 0 was born May 23 2955 BC. It had 71 events. It began life as an almost imperceptible partial, matured into a robust ring of fire and faded again as a partial until dying June 29 1675 BC. It had "lived" about 1300 years and traveled the world, in its own fashion, bringing a strange dance of the sun and moon to the inhabitants of earth. Saros Cycle 1 showed up on June 4 2872 BC.

Saros receive their number from the order in which they are born. Saros Characters mature into complete eclipses at different rates, and have different "lifespans". One Saros may have 70 events. Another may have 77, or 69. That is why, when we review the Characters in a minute, they do not appear in the earth's sky in numeric order.

In recent history, Saros Characters bear numbers between 112 and 152. There are about 40 characters, 26 of whom are currently creating complete eclipses. You'll meet them in a minute.

The Facts Again:

Every Saros Cycle takes exactly 19 Draconic, or eclipse, years. That is 19 x 346.62(one eclipse year, or the time it takes for the sun, moon and earth return to the same ecliptic node. This amounts to 18 years, 11 days and 8 hours. At that time, the Sun, the Earth, and the Moon return to approximately the same relative geometry in relation to the orbits, tilts and nodes of the three celestial objects. One Saros period after a given eclipse, another similar eclipse occurs, displaced by 120 degrees in longitude or one third of the earth. This displacement is due to the remaining eight hours in the duration of the Saros cycle.

The eclipses shift sometimes in latitude as well, due to the tilt of the planet. It creates an individualized pattern, as seen in the maps in this section. Only serious mathematicians can accurately predict exactly where the eclipse will land. Thankfully, they have already done this.

Important:
- Saros characters return every 19 eclipse years. That is 19 x 346.62 or 6585.78 days.
- A Saros is also, to within a few hours, exactly 223 Lunar (synodic) months.
  (The math: A lunar month is 29.53 days. 29.53 x 223 equals 6585.19 days)
- A Saros in solar years is 18 years, 11 days and 8 hours, or 19 *solar* years minus one lunar year.
- A Saros is also "coincidentally" almost perfectly equal (to within a day) to 239 anomalistic months (27.55 days). The math: 239 x 27.55 equals 6585.45 days. The anomalistic month is the time between perigee to perigee of the moon. It's when the distance of moon to earth as it orbits returns to about the same. This gives Saros their character, so that they stay consistent (as partial, total or ring of fire) for lengthy periods of time. Over time, they slowly change.

In the Saros Synchronicity, three different calendars align, the lunar, the draconic and the anomalistic. Quite an arrangement. God arranged the Cycle to be both predictable and unique.

So again, a Saros Cycle is all four of these at once: 19 Draconic (eclipse) years, 223 Lunar (synodic) months, 239 Anomalistic months, and 18 solar years minus one lunar year.

The Saros synchronicity is its own unique alignment. It's not to be confused with the 19-year Solar Year Metonic "coincidence" (the 19 Solar Year alignment). We look at that in the next section.

How long do they live? The Saros "lives" between 1200 and 1500 years. It begins its life as a partial eclipse, typically matures into a complete eclipse after a century or two, becoming Annulars, Hybrids, Totals or often, combinations of the three. After centuries, it fades again into a partial before eventually fading. They usually begin life near one of the polar regions and usually die in the opposite polar region. The Solar Saros Cycle has about 40 members at any one time, as it does now.

There are currently 14 partial eclipse personalities and 26 complete eclipse characters.

A note about Lunar Eclipse Saros.

Lunar Eclipses have their own Saros Cycle, repeating at opposite fortnights from the Solar Saros. Lunar eclipses accompany solar eclipses on the opposite phase of the moon. In other words, if there is a solar eclipse on a certain new moon, there will be a lunar eclipse on the full moon either two weeks before or after. That is when sun, moon have aligned into the same cosmic "node".

Partial solar eclipses

We do not focus on or map partial solar eclipses in this work. I will note current partials and then leave them be. There are currently 14 Saros members which only create partial eclipses.

Every Total or Ring of Fire solar eclipse also creates a wide area of partiality over millions of square miles. That area may receive 99% coverage. It might receive .02% coverage. But that is not the same as the moon going perfectly in front of the sun. That alignment only happens in the narrow band of shadow produced by a complete solar eclipse, whether Total, Hybrid or Ring of Fire.

Partial solar eclipses are sometimes not even physically noticeable. 35% of all solar eclipses are partial. These are either immature and growing Saros characters or dying Saros members. By reducing consideration of partials, I believe we are better able to focus on our current cosmic tribe.

SAROS PARTIALS: List of the 14 Saros currently creating Partial Solar Eclipses

There are two types of Saros Partials.
Dying Saros: These are the surviving Saros who had a history, a distinct lifetime as a complete eclipse. Now they are growing old. They will die someday.
Growing Saros: These are the immature Saros who have not yet become a complete eclipse.
There are seven dying and seven growing partial Saros. Let's briefly meet them.

7 Dying Saros still in cycle making partials:

Saros 117: Consists of 71 Events. Born on June 24, 792 AD. Matured into complete eclipse as a ring of fire on September 18, 936. Became a Hybrid on May 25, 1351. Dies Aug 3, 2054
Saros 118: 72 events. Born May 24 803 AD. Matured into complete eclipse as a total on Aug 19 947. Became a Hybrid in 1668, Annular on 1704. Last complete eclipses Apr 19 1939, the last eclipse before WWII-Hitler birthday/ Darwin Death Day. Last complete non-central annular April 30 1957, 12th anniversary of Hitler's death. Saros 118 dies in 2083

Saros 119: 71 events. Born on May 15 850 AD. 2 totals beginning Aug 9 994 AD (one lunar year after last Jerusalem Total in Aug 20 993 AD). Hybrid in 1030. Annular in 1048. Dies in 2112.
Saros 122: 70 events. Born on Apr 17 991 AD. Total begins 1135. 2 hybrids beginning 1189. Annular from 1225 to 1874 (last complete eclipse). Dies 2235.
Saros 123: 70 events. Born on Apr 29 1074. Annular begins 1182. Hybrid begins 1669. Total begins 1723. Last complete total on Oct 23 1957. Dies 2318.
Saros 124: 73 events. Born Mar 6 1049. Became total 1211. one hybrid on 1986. then partial beginning Oct 14 2004. dies in 2347.
Saros 125: 73 events. Born Feb 4 1060. totals beginning 1276. 2 hybrids beginning 1348 (Black Death). Annular begins 1384. Last complete Aug 22 1979. dies in 2358.

* Notice that the list seems to skip Saros 120 and Saros 121. That's because they are still creating complete eclipses. Saros get their rank number from when they are born, not when they die. Saros 120 and 121 were born before Saros 122 to 125, but are currently still spending this portion of their life creating complete eclipses.

7 Growing Saros (currently partial):

Saros 149: 71 events. Born Aug 21 1664. Becomes Total on 2043. Recent partial Mar 9 2007.
Saros 150: 71 events. Born on Aug 24 1729. Becomes Annular on 2126. Recent partial Feb 15 2018
Saros 151: 72 events. Born on Aug 14 1776. Becomes Annular on 2107. Recent partial on Jan 4 2011
Saros 153: 70 events. Born Jul 28 1870. Becomes Annular on 2104. Partial on Oct 23 2014.
Saros 154: 71 events. Born on Jul 19 1917. Recent partial on Sep 11 2007.
Saros 155: 71 events. Born on Jun 17 1928. Recent partial on Aug 11 2018
Saros 156: 69 events. Born on Jul 11 2011.
Note: Saros 152 is creating complete eclipses.

Case Study: SAROS CYCLE 136. An eclipse personality as an example.

We will use Saros 136 as a case study to demonstrate the life cycle of a single Saros Eclipse.

Saros 136 is the eclipse which creates the upcoming Valley of the Kings Eclipse on August 2 2027, the first line in the Mideast Cross which begins this book.

Facts:

Solar Saros 136 contains 71 events or "members", each occurring exactly 19 eclipse years apart.
It was "born" with a partial solar eclipse on June 14 1360 (number peculiarity of 136 and 1360).
It became an annular eclipse in 1504, which it remained through 1594.
It morphed into a hybrid event from November 22 1612 through January 17 1703.
It began total eclipses January 27 1721 and is scheduled to continue doing so through May 13 2496. The series is scheduled to end at member 71 as a partial eclipse on July 30, 2622, with the entire series lasting 1262 years.

The longest eclipse of Saros 136 occurred on June 20, 1955, with a maximum duration of totality at 7 minutes, 7.74 seconds over Southeast Asia. That was the longest total eclipse since the 11th Century. Since then, duration has become shorter with each eclipse, but Saros 136 is still producing the longest total solar eclipses of all Saros. It produced the six longest total solar eclipses of the 20th century. It also produced the longest total eclipse of our current 21st century at 6 min 38.8 sec with the 7-22-2009 Wuhan Eclipse. Saros 136 is scheduled to ultimately produce a total of 44 total eclipses. It produces the most complete total eclipses, 44, of any Saros between the years 1209 and 2718. It brought the greatest magnitude of any eclipse since the year 540 on July 11, 1991 (Hawaii and Central America). Solar Saros 136 contains 15 partial eclipses, 6 annular, 6 hybrid and 44 total eclipses. Below is a snapshot of about 10% of its life's work.

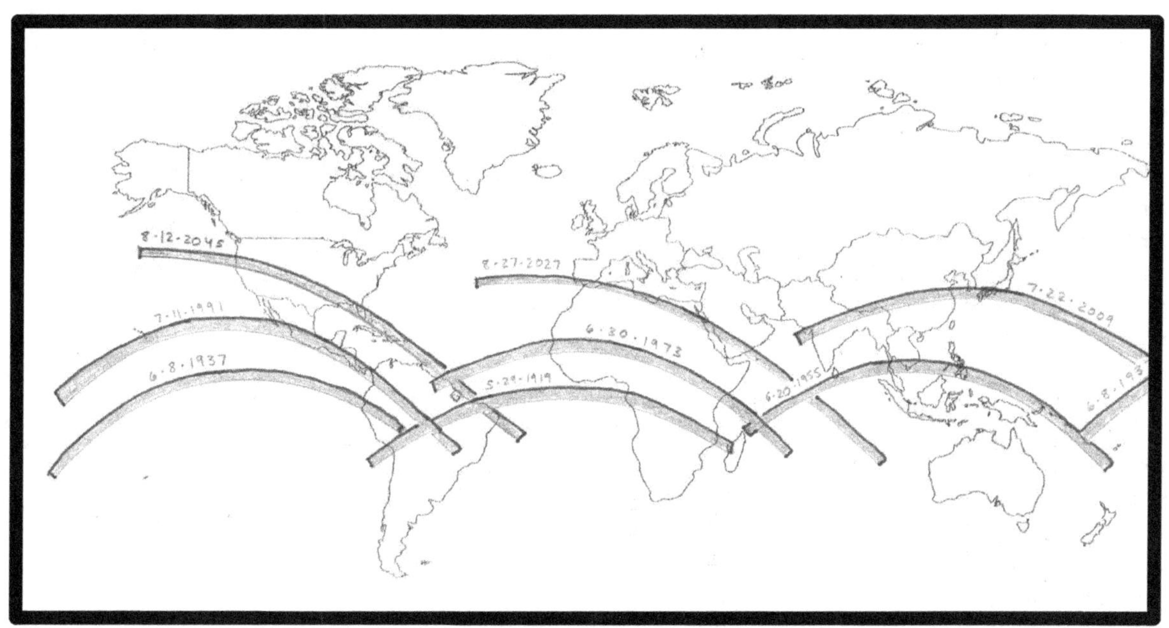

Saros 136 – 1919 to 2045

145 years of Saros 136's shadows. Eight different eclipses. You can see the pattern of the geographical shift where the shadows land over the course of the centuries. That pattern slowly shifts over the centuries. Included in this diagram are some amazing eclipses. The 1955 Southeast Asia Eclipse just as the war in Vietnam began is on the lower right. The Wuhan Totality of 2009 is just above that. Go to the left of the Wuhan Eclipse and you'll find Valley of the Kings Eclipse of 2027 in the center of the map. Go to the bottom of that row and you find Einstein's Eclipse (May 29 1919) that perfectly crosses South America and Africa and proved his Theory of Relativity on John F. Kennedy's 2nd birthday.

Maximum Expression: Perhaps you will notice that eclipse lifespans achieve a maximum expression, like all forms which appear in reality. Duration of eclipse and width of the path are obvious means of determining maturity. Saros 136 achieves maximum duration in the 1955 Southeast Asia Eclipse and maximum width is still increasing, maximum width peaking with the 2027 Valley of the Kings Eclipse. The Chart below has date of eclipse, type and duration of complete eclipses in minutes and seconds. Note Metonic dates advancing 11 days and 8 hours at a time across the solar calendar.

Lifespan of Saros 136
BIRTH: 1360 Jun 14 Partial...
1378 Jun 25 Partial
1396 Jul 05 Partial
1414 Jul 17 Partial
1432 Jul 27 Partial
1450 Aug 07 Partial
1468 Aug 18 Partial
1486 Aug 29 Partial

Becomes Ring of Fire : 1504 Sep 08 Annular ...duration (refering to annularity) 00m 32s

1522 Sep 19 Annular...duration 00m23s

1540 Sep 30 Annular ...duration 00m17s

1558 Oct 11 Annular ...duration 00m12s

1576 Oct 21 Annular ...duration 00m08s

1594 Nov 12 Annular...duration 00m04s

Becomes Hybrid:1612 Nov 22 Hybrid 00m01s...Duration (refers to totality within hybrid)0 m 01s

1630 Dec 04 Hybrid 00m07s

1648 Dec 14 Hybrid 00m14s

1666 Dec 25 Hybrid 00m24s

1685 Jan 05 Hybrid 00m35s

1703 Jan 17 Hybrid 00m50s

Becomes Total Eclipse: 1721 Jan 27 Total (note all the 7's,2's and 1's).Duration of totality 01m07s

1739 Feb 08 Total. Duration ...01m27s

1757 Feb 18 Total 01m51s

1775 Mar 01 Total 02m20s

1793 Mar 12 Total...duration 02m51s

1811 Mar 24 Total...duration 03m27s

1829 Apr 03 Total...duration 04m05s

1847 Apr 15 Total...duration 04m44s

1865 Apr 25 Total duration 05m23s

1883 May 06 Total...duration 05m58s

1901 May 18 Total...duration 06m29s

1919 May 29 Total...duration 06m51s

1937 Jun 08 Total...duration 07m04s

1955 Jun 20 Total ...duration 07m08s

1973 Jun 30 Total...duration 07m04s

1991 Jul 11 Total...duration 06m53s

2009 Jul 22 Total...duration 06m39s

2027 Aug 02 Total...duration 06m23s

2045 Aug 12 Total...duration 06m06s

Saros 136 Continues as a Total eclipse for 24 more eclipses until May 13, 2496, then has 7 partial eclipses. Its last dying eclipse is scheduled for July 30, 2622.

THE CAST OF COMPLETE SAROS CHARACTERS

There are 26 different Saros Cycles currently creating complete solar eclipses on Earth.

Let's meet the 26 complete Saros Personalities in order of appearance using the 2015 Spring Equinox (Saros 120) as an initial milestone. By following every eclipse that happens between Saros 120's Mar 20 2015 North Pole Eclipse and its return again on Mar 30 2033 return, you will meet all current Saros characters in the order they appear on planet earth in that time frame of 19 eclipse years.

Why use Saros 120 as our benchmark? First off, it is the oldest Saros still creating complete eclipses, born as a partial in 993 AD (coincidentally the last year Jerusalem had a total eclipse). Secondly, it has a particularly extraordinary past, appearing at important times in important places. Finally, its last two complete eclipses, in 2015 and 2033, help frame the portrait this work seems to paint.

The Mar 20 2015 eclipse ended directly at the North Pole. The 2033 Mar 30 eclipse connects both hemispheres simultaneously at the Bering strait. It is unclear whether either of these "achievements" have ever occurred before. They are almost impossibly precise in their peculiarities. Also, if the traditional date of Apr 3 33AD for the Crucifixion of Jesus Christ is accepted, then the Mar 30 2033 AD Bering Strait eclipse would be almost exactly 2,000 years since that momentous event. That's why we'll begin with Saros 120. Again, they do not advance in numeric order but in order of appearance.

The Saros, of course, runs two ways: into history and into the future. They can be calculated very precisely due to their extraordinary alignment with our concept of standard calendar time, allowing for both prediction of future eclipses and the transposing of eclipses into history.

A careful look at these eclipses reveals them linked in various fashions with other eclipses in the cycle, For instance: looking at the dates of every other eclipse, we find the days of the year eclipses occur will often be identical or sometimes one or two days apart, occurring at 19 year intervals in the Metonic cycle. These Metonic pairings can be found in eclipses exactly 10 Saros apart. Meaning, identical calendar dates 19 years apart are found in Saros 120 and 130 and again with Saros 140, Saros 135 and 145 share the same dates, etc.. You'll see. Don't get bogged down. Just meet them.

Certain eclipses are paired during much of their lives. Saros 136 and Saros 141 happen 177 days apart century after century, and respectively give the longest Total and Ring of Fire Eclipses. There are scientific reasons for that.

Also, slight wobbles in dates along calendar cycles are to be expected because of both the artificial time zones and very real orbital realities of earth.

This extraordinary reality of the Saros (and its interactions with the Metonic Cycle) involves the brilliant co-ordination of cosmic timekeeping set forth by an omniscient Creator God. We look deeper at the Metonic Cycle after we have met the Saros.

SAROS 120

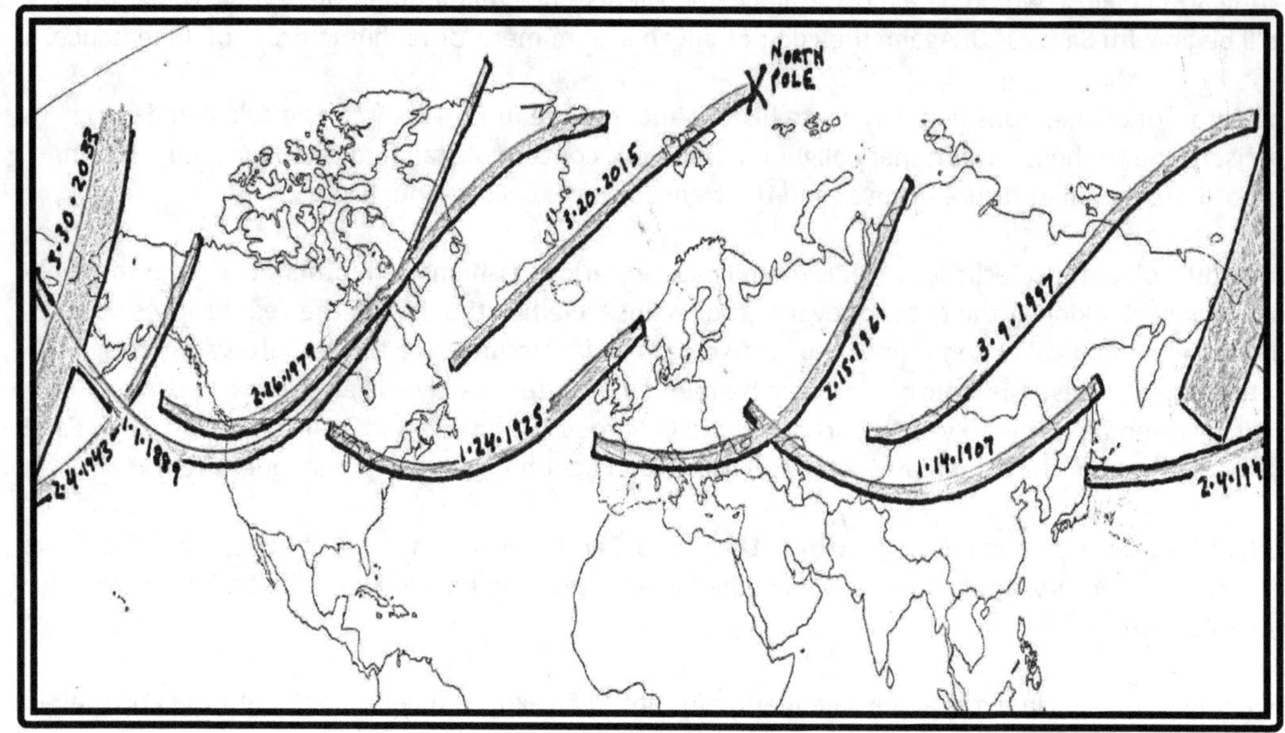

Saros 120 Total Eclipse: A snapshot from 1889 to 2033

Saros 120 has 71 events Born: May 27, 933 AD Matured: August 11, 1059.

1889 Jan 1 New Year's Day USA/Canada
1907 Jan 14 Europe/China/ Russia
1925 Jan 24 Roaring Twenties USA to England
1943 Feb 4 USA/Japan Wartime Eclipse
1961 Feb 15 The Great European Iron Curtain eclipse
1979 Feb 26 Mount Saint Helens/ Microsoft Eclipse
1997 Mar 9 Mongolia/USSR
March 20 2015 North Pole Spring Equinox Total Eclipse. Ended immediately at North Pole.

Bering Strait Eclipse of Mar 30 2033, Links the USA and Russia and International Date Line simultaneously. 4 days before the traditional 2000th anniversary of Crucifixion. The 2033 event is Saros 120's last complete eclipse, ending a run of complete eclipses that began August 11 1059.

All Saros 120 eclipses in this 125-year study period are over Northern Hemisphere Modern Empires.

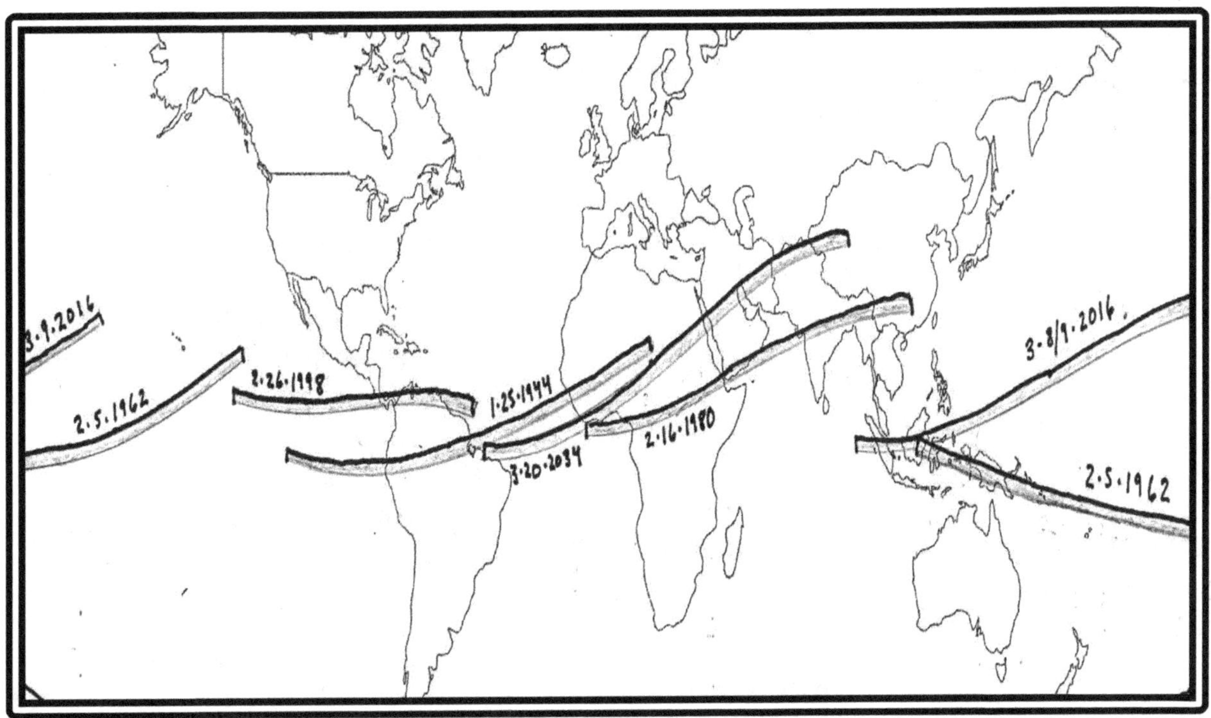

Saros 130 Total Eclipses 1926 to 2034

Saros 130 has 73 events Born: Aug 20 1096. Matured: Apr 5 1475

1926 Jan 14 Africa to Indonesia
1944 Jan 25 South America to Africa
1962 Feb 5 New Guinea/Open Ocean
1980 Feb 16 Africa to India
1998 Feb 26 Pan American Isthmus Eclipse
2016 Mar 9 Indonesia and Pacific 3-8/9 2016

2034 Mar 20 Its next eclipse is the tremendous Spring Equinox Eclipse creating the Mideast Cross, on the 19th Metonic year to the day from Saros 120's North Pole Equinox.

Look at the dates above and you will see them Metonically linked with Saros 120 and with Saros 140...the days will be either identical or one day apart, at 19 year intervals. This extraordinary reality is underappreciated and involves the co-ordination of cosmic timekeeping set forth by God.

Saros 135 - Ring of Fire eclipses 1926 to 2034

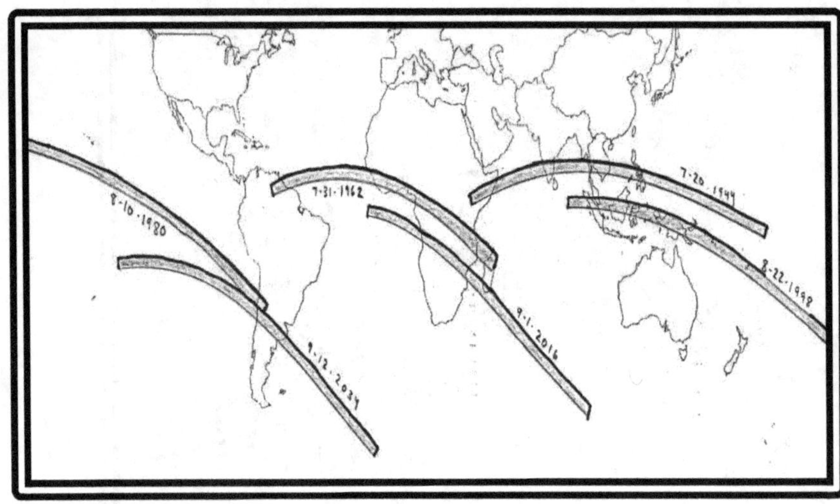

Saros 135 has 71 events. Born Jul 5 1331 matured Oct 21 1511

1926 Jul 9 South Pacific (not pictured)
1944 Jul 20 South Asia to SE Asia
1962 Jul 31 S. America to Africa
1998 Aug 22 Indonesia/Philippines
2016 Sep 1 Africa to ocean.
2034 Sep 12 South America

You will see these same Metonic dates appear again in Saros Cycle 145.

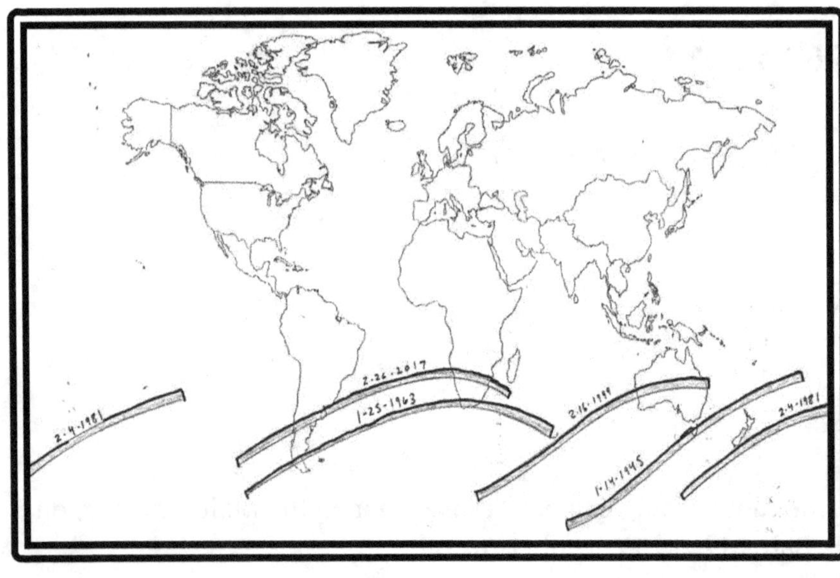

Saros 140 –Ring of Fire 1945 to 2017

Saros 140 has 71 events. born Apr 6 1612 matured Jul 21 1656

1927 Jan 3 Ocean--Australia to Patagonia (not pictured)
1945 Jan 14 Ocean to Hobart Tasmania
1963 Jan 25--Extreme South America to extreme South Africa
1981 Feb 4 South pacific no land.

Grazes southern New Zealand.
1999 Feb 16 Atlantic Ocean to Australia (the second to last eclipse of the 20th Century)
2017 Feb 26--Patagonia South America to Africa.
Next appearance--2035 Mar 9 Wellington New Zealand (single city eclipse not pictured)

This eclipse is currently residing in the Southern hemisphere, preferring mostly open ocean. You will notice the calendar dates are in line with Saros 120 and 130 i.e. Feb 26 and Jan 24/25

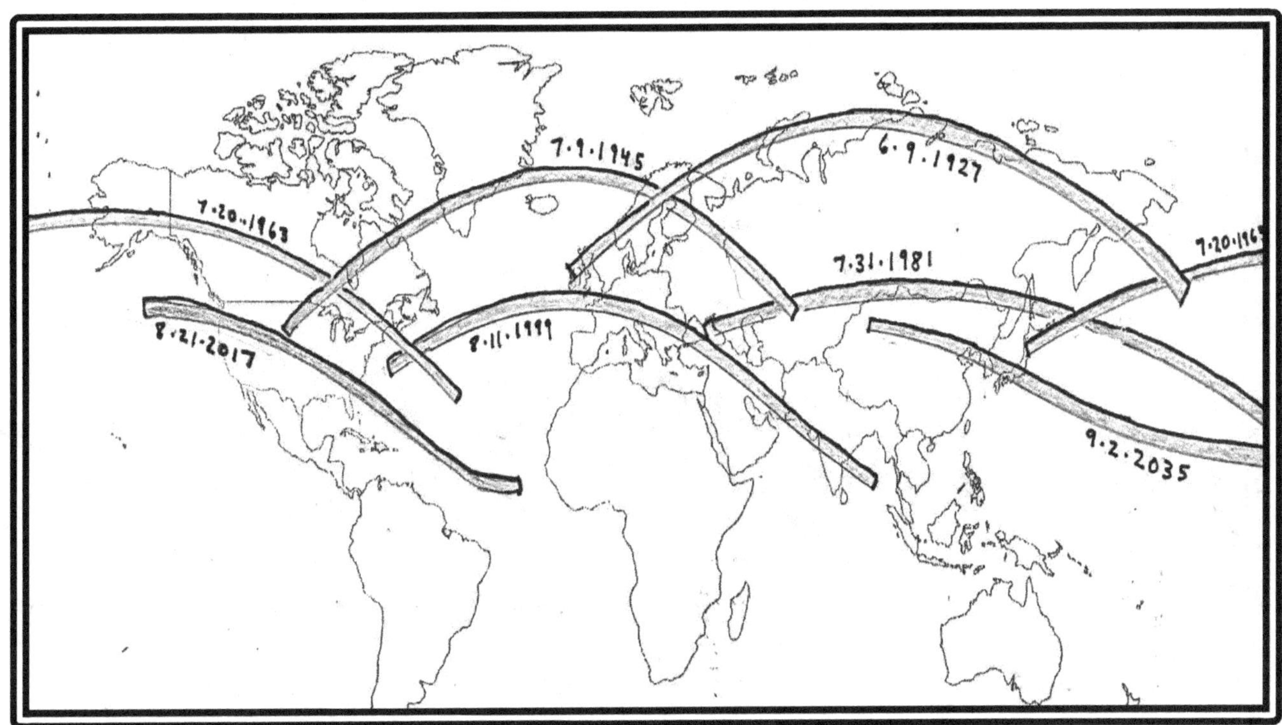

Saros 145 Total Eclipse 1927 to 2035

Saros 145 has 77 events. It is the only current Saros to have 77 events.
Born Jan 4 1639 matured Jun 6 1891 for one ring of fire exactly 77 years before RFK assassination.
One hybrid in 1909. Became Total in 1927
One of the greats.

1927 Jun 29 Liverpool

1945 Jul 9 Atom Bomb Eclipse USA to USSR (one week before first A-Bomb Test)

1963 Jul 20 The 60's Eclipse USSR to USA (Maine)

1981 Jul 31 USSR Judgment

1999 Aug 11 End of the Millennium/ Asia Mideast and Europe

2017 Aug 21 The Great American Cross-Country eclipse.

2035 Sep 2 China/Beijing/ Pyongyang/Tokyo

Too awesome to be dismissed. Dates Metonically linked with Saros 135.

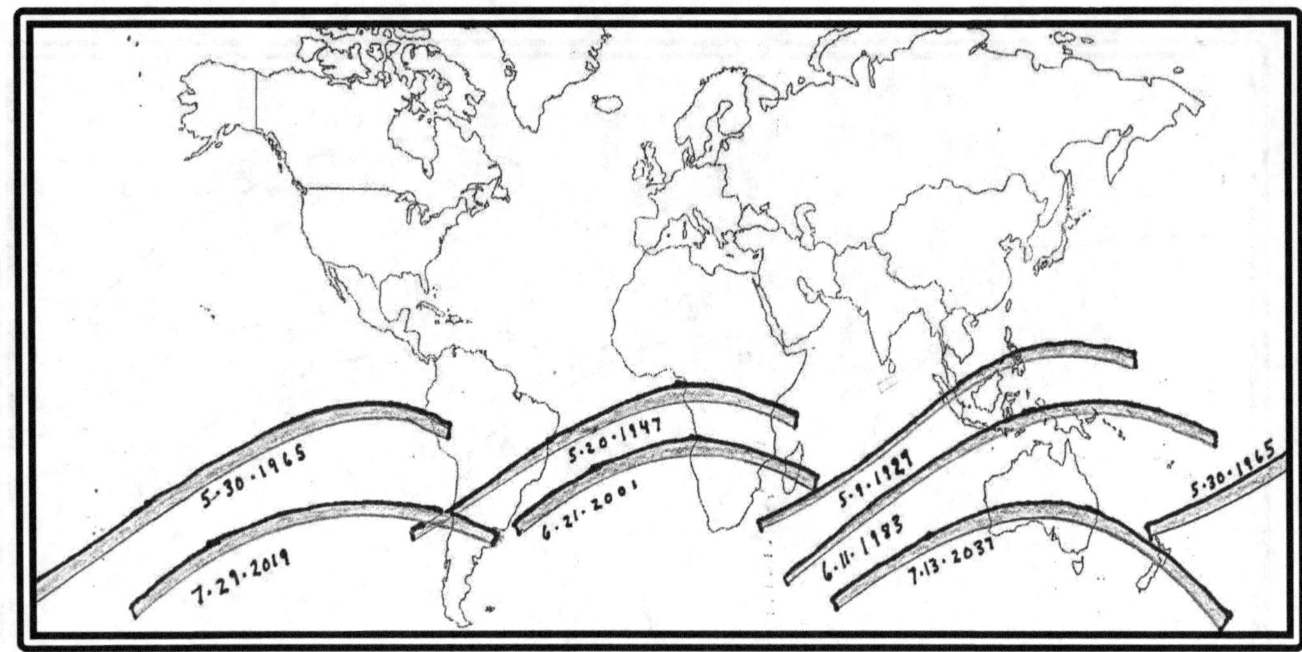

Saros 127 Total Eclipse: Snapshot 1929 to 2037

Saros 127 has 82 events born 10-10-991 matured May 14 1352

1929 May 9 Open Ocean SE Asia

1947 May 20 South America to Africa/ Columbus Death metonic

1965 May 30 Open Ocean/JFK Birthday

1983 Jun 11 Indonesia

2001 Jun 21 South Africa Ring of Fire Summer Solstice. The last eclipse before 9-11

2019 Jul 2 Pope Francis Eclipse (ends at sunset at Buenos Aires)

2037 Jul 13 Australia/New Zealand; the last of the Monday Hepton Judgment series.

Famous ancient history: Saros 127 also created the Aug 8 1496 Mayan/Aztec Apocalypse Eclipse. When the Great Pyramid of Tenochtitlan was dedicated in 1487 the Aztecs recorded that 84,000 people were ritually sacrificed in four days.

Metonically in line with Saros 137

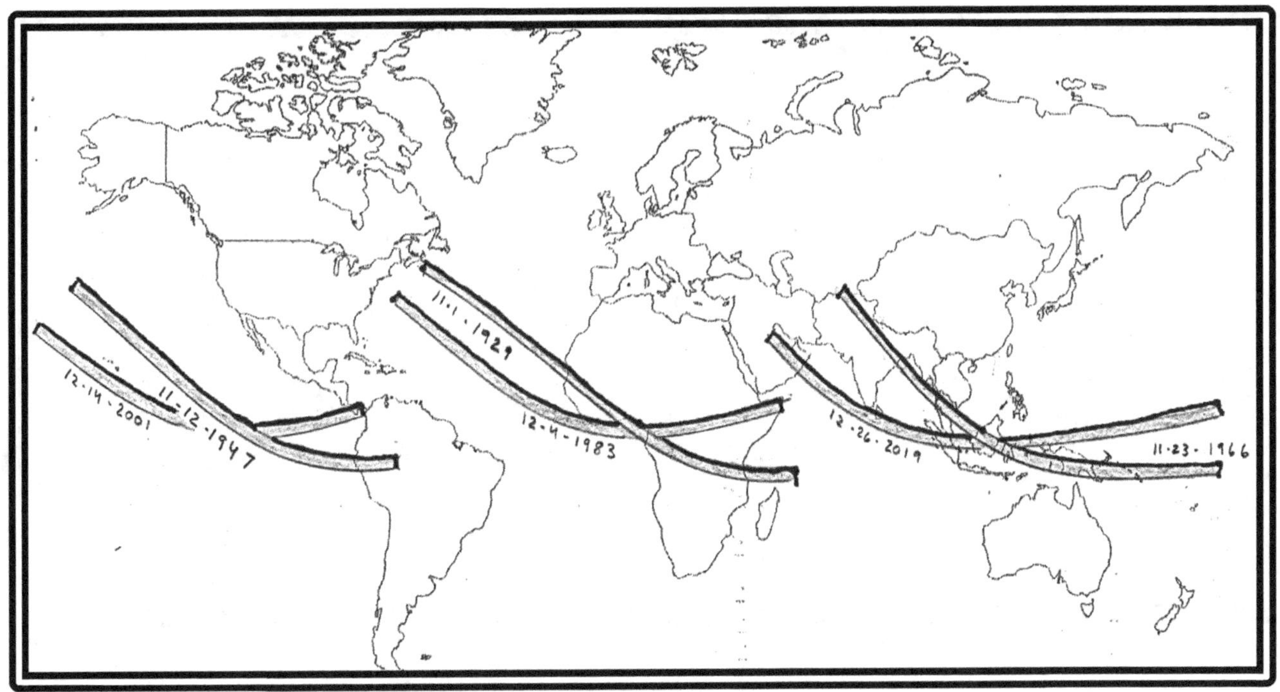

Saros 132 Ring of Fire Snapshot 1929 to 2019

Saros 132 has 71 events born Aug 13 1208 matured Mar 17 1569

1929 Nov 1 Stock Market Crash Eclipse (3 days after Crash)—North American coastal waters to Africa
1947 Nov 12 Pacific Ocean to barely touch South America
1965 Nov 22/23 Southeast Asia/New Guinea PT 109 Eclipse on JFK Death Day Metonic two years after his assassination.
1983 Dec 4 Atlantic across Africa to Horn of Africa
2001 Dec 14 Pacific Ocean to Pan-American Isthmus (Panama Canal)-first eclipse after 9-11.
2019 Dec 26 Tsunami Anniversary Ring of Fire. Travels path of great tsunami backwards from Oman to Sri Lanka to Sumatra on 15th year anniversary of that catastrophe.

Peculiar in its self-crossing pattern every 54 years. Linked Metonically with Saros 142

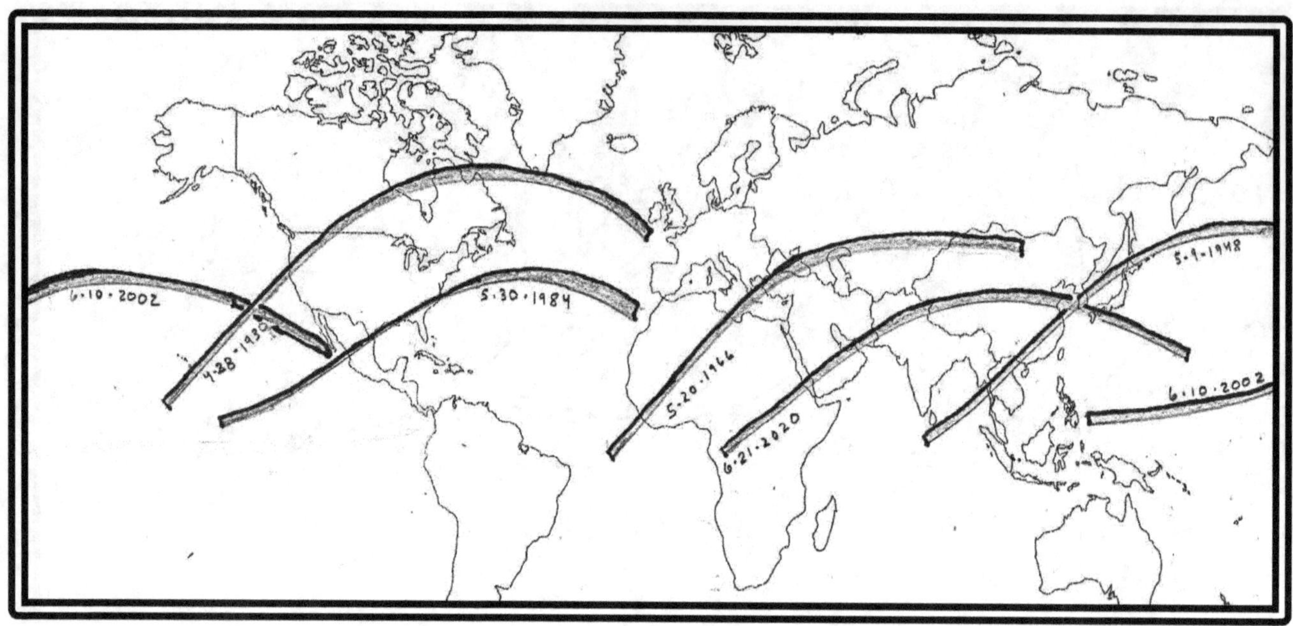

Saros 137 Hybrid/ Ring of Fire Snapshot 1912 to 2020

Saros 137 has 70 events: born May 25 1389 matured Aug 20 1533

1912 Apr 17 (as Hybrid) Titanic Eclipse...3 days after Titanic sank, eclipse across Europe

1930 Apr 28 (as Hybrid) San Francisco...San Francisco's only totality in centuries. It was only 1 kilometer wide and lasted for a few seconds.

1948 May 9 (one of the closest to total ring of fires possible. 99.999% in places) Korea/ SE Asia a few years before Korean War. Splits both Vietnam and Korea in half.

1966 May 20 (very close to total/hybrid --99.9% in places) Africa/ Europe/USSR /China on Columbus Death Day Metonic 560 years after his death...Athens and Istanbul and Tripoli in the path

1984 May 30 (almost hybrid-99.77%) Washington DC/New Orleans/Atlanta on JFK birthday Metonic

2002 Jun 10 Ocean

2020 Jun 21 Summer Solstice Father's Day Pandemic Ring of Fire: Africa to China. 83% in Wuhan.

In Metonic alignment with Saros 127

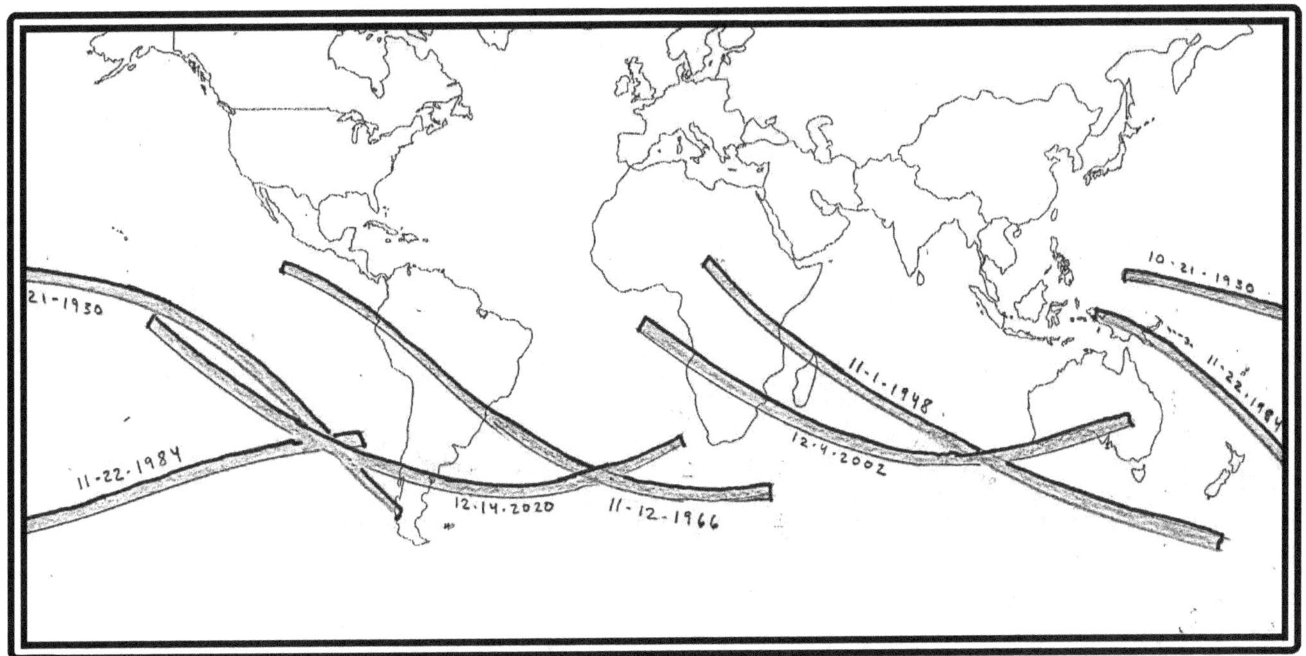

Saros 142 Total Snapshot 1912 to 2020

Saros 142 has 72 events- Born Apr 17 1624 Matured Jul 14 1768

1912 Oct 10 South America/ Ocean

1930 Oct 21 all Ocean

1948 Nov 1 Small part of Africa/ Ocean

1966 Nov 12 South America/ Ocean

1984 Nov 22 Almost open ocean/ New Guinea PT 109---last JFK Death Day Metonic

2002 Dec 4 South Africa/ Indian Ocean

2020 Dec 14 Electoral College Eclipse

Linked Metonically with Saros 132

SAROS 147

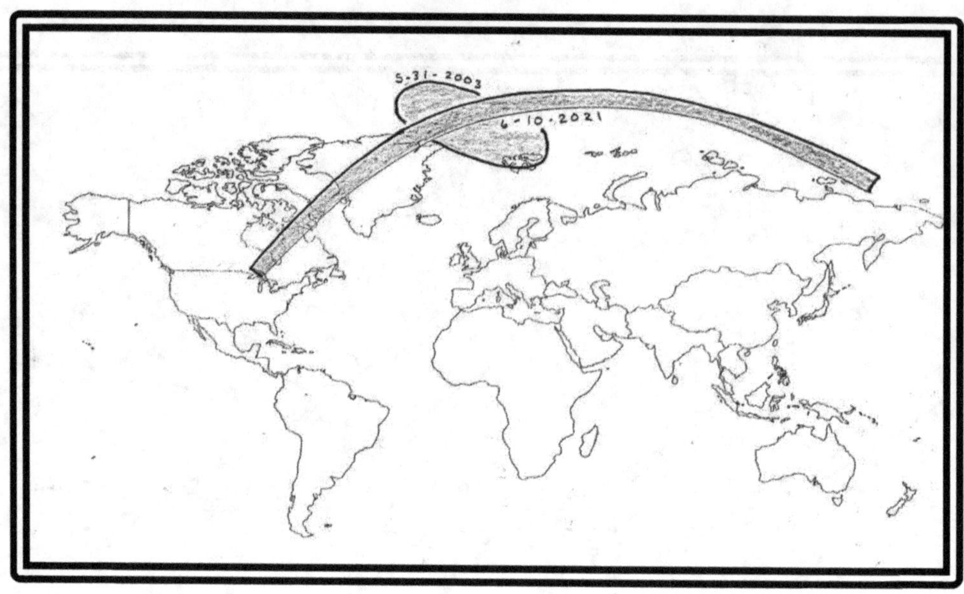

Saros 147 Ring of Fire snapshot 2003 to 2021
Saros 147 has 80 events born Oct 12 1674 and matured May 31-2003

2003 May 31 Northern latitudes
2021 Jun 10 Canadian Judgment Ring of Fire

Saros 147 is just now maturing into a complete eclipse, becoming annular for the first time in 2003.

Saros 152

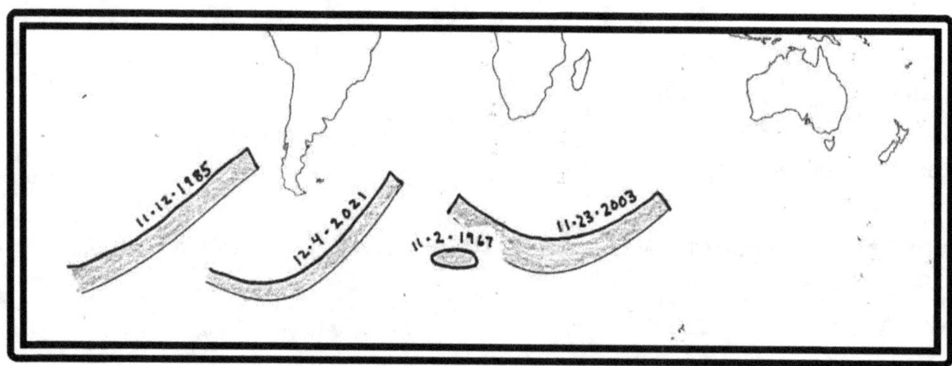

Saros 152 Snapshot 1967 to 2021
Just began maturing in 1967...all these eclipses occur near Antarctica 152
Saros 152 has 70 events and was born Jul 26 1805
1967 Nov 2
1985 Nov 12
2003 Nov 23
2021 Dec 4
Metonically linked with Saros 142 and 132

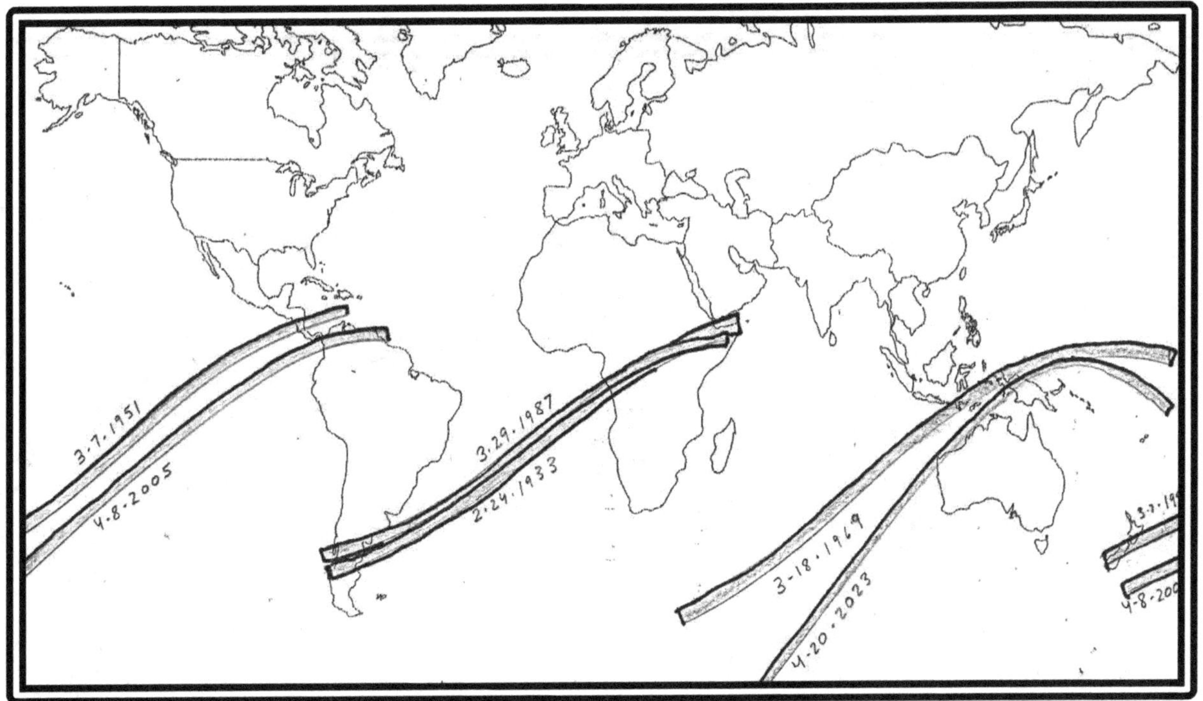

Saros 129 Ring of fire maturing into a Hybrid: Snapshot 1933 to 2023

Saros 129 has 80 events-Born Oct 3 1103 Matured: May 6 1464

1933 Feb 24 South America to Africa ...has an astonishingly similar path to the 3-29-1987 eclipse/ almost identical. Only Saros I found that precisely overlapped itself in such a way.

1951 Mar 7 Isthmus of the Americas

1969 Mar 18 Ocean and Indonesia

1987 Mar 29 South America/ Africa (not only the overlap of this eclipse and 1933 eclipse is unique); This 3-29-1987 eclipse has partner in 1987 with eclipse on 9-23, thus a palindrome of 3-29 and 9-23.

2005 Apr 8 No land but Pan-American Isthmus on 19-year Metonic before 4-8-2024 USA

2023 Apr 20 Australia Hybrid. First time Hitler birthday appears in 500 years (Metonic had been lodged at 4-19 Darwin death Day for a century) ...4-20 is a catch phrase in America and elsewhere...

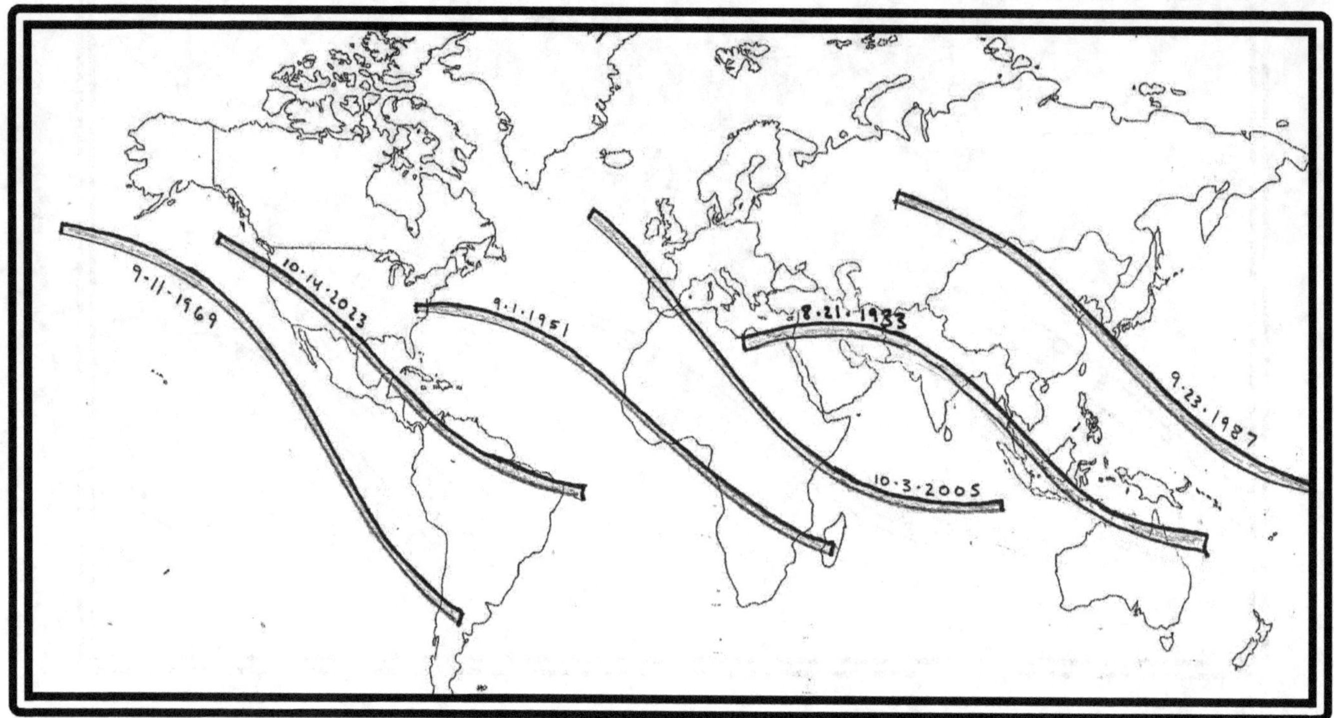

Saros 134 Ring of Fire Snapshot 1933 to 2023

Saros 134 has 71 events Born Jun 22 1248 matured Oct 9 1428

1933 Aug 21 Great Jerusalem Ring of Fire

1951 Sep 1 Jamestown USA to Angola Africa

1969 Sep 11 Ocean and extreme South America. First appearance of 9-11 Metonic in centuries,

1987 Sep 23 China and Ocean. Here's that 9-23 palindrome in 1987, see 3-29-1987 Saros 129

2005 Oct 3 in Spain and Africa

2023 Oct 14 Great American Ring of Fire

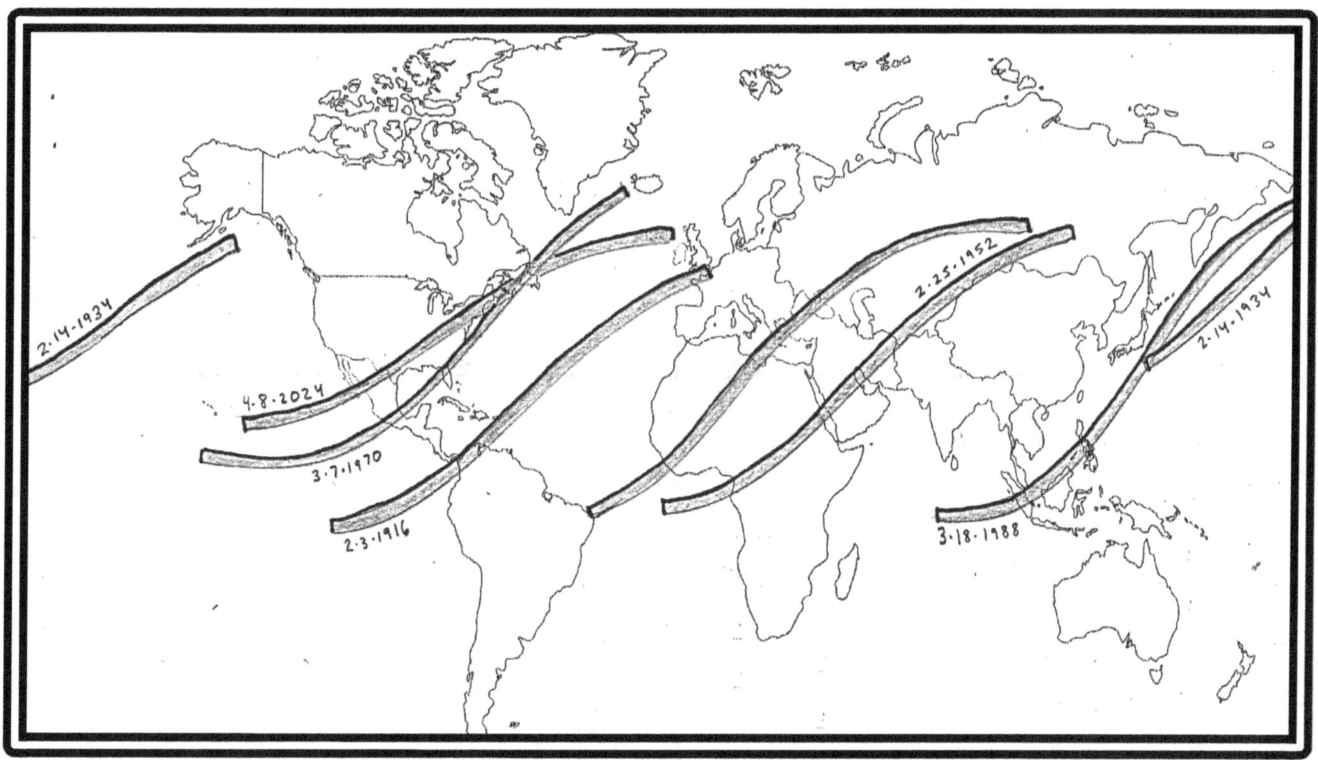

Saros 139 Total Eclipses Snapshot 1916 to 2024

Saros 139 has 71 events. Born May 17 1501. Matured Aug 11 1627.

1916 Feb 3 Pan-American Isthmus to France during WWI

1934 Feb 14 Ocean Valentine's Day Pacific

1952 Feb 25 Iranian Shahrud Eclipse--Africa/ Arabia...first half of the Iran Shahrud Cross. This one is the year *before* the Shah was installed in a coup.

1970 Mar 7 USA end of 60's eclipse, along Eastern seaboard.

1988 Mar 18 Indonesia, Southern Philippines

2006 Mar 29 second line of Great Turkish Cross

2024 Apr 8 First New Moon after Easter; a total eclipse crosses Mexico, USA, and eastern Canada as second line in Hepton Cross. This is the fourth significant cross 139 makes in relation with 145.

SAROS 144

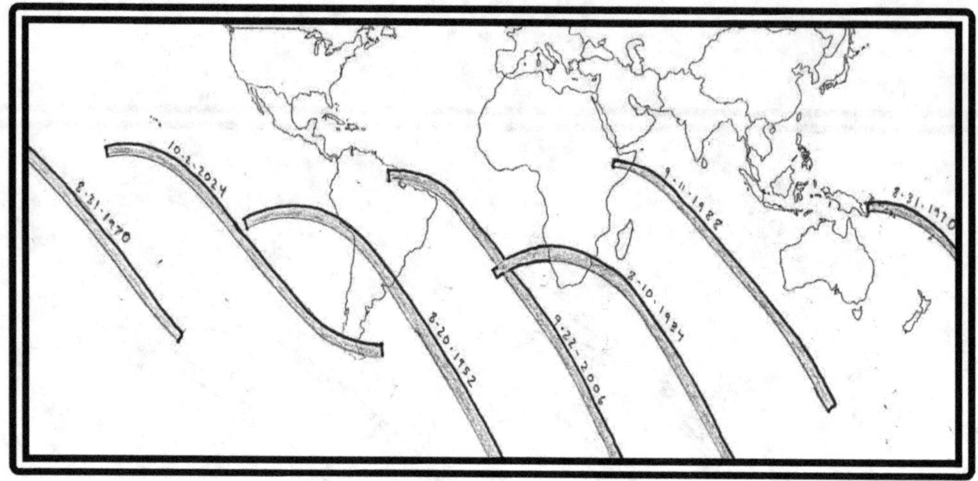

Saros 144 Ring of Fire: Snapshot 1934 to 2024
Saros 144 has 70 events. Born Apr 11 1736 matured Apr 7 1880
1934 Aug 10 South Africa on Temple Destruction Metonic
1952 Aug 20 South America and Ocean
1970 Aug 31 Ocean
1988 Sep 11 Mogadishu Somalia 9-11 Sunrise Single City Eclipse
2006 Sep 22 Fall equinox and Ocean
2024 Oct 2 Chile and Ocean

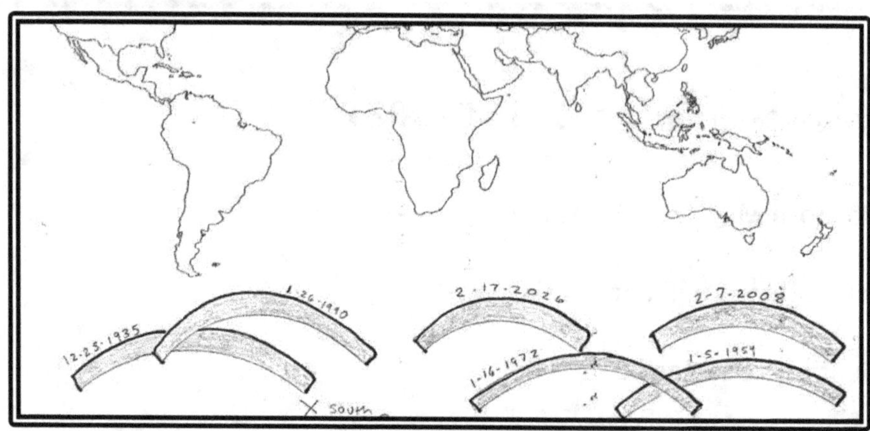

SAROS 121

Saros 121 Ring of Fire snapshot 1935 to 2008
Saros 121 has 71 events born Apr 25 944 matured Jul 10 1070
1935 Dec 25 Christmas in Antarctica
1954 Jan 5
1972 Jan 15/16 Martin Luther King would have been 43.
1990 Jan 26
2008 Feb 7
Saros 121 has its last complete eclipse in 2044 after being lodged near Antarctica for centuries.
Connected Metonically with Saros 131 and Saros 141. Next appearance 2-17-2026

SAROS 126

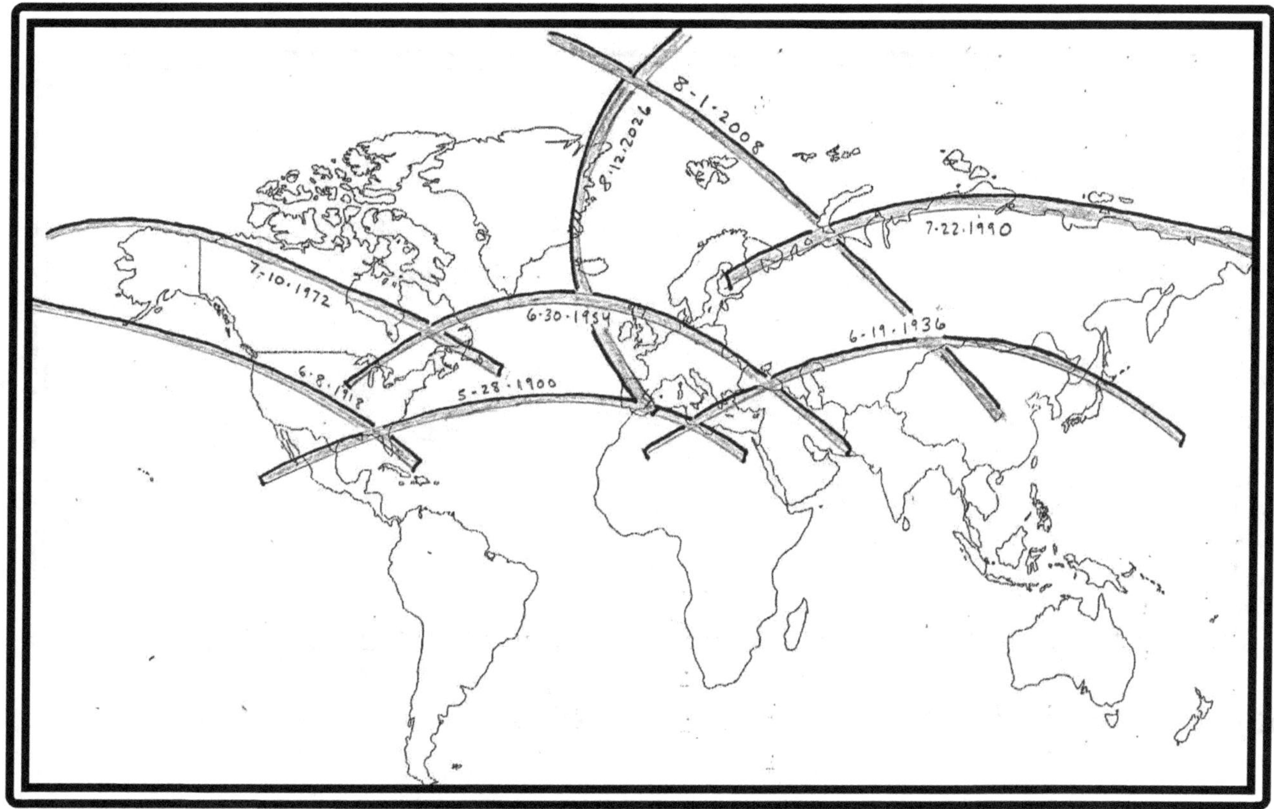

Saros 126 Total eclipse Snapshot 1900 to 2026
Saros 126 has 72 events Born Mar 10 1179 matured Jun 4 1323

1900 May 28 USA to Luxor, Egypt. First eclipse of the 20th Century.

1918 Jun 8 Great USA Century Eclipse, Spanish Flu, WW1 ends 5 months later.

1936 Jun 19 USSR/ Europe

1954 Jun 30 USA to Iran (via Europe)..second half of Shahrud Cross one year after coup in Iran.

1972 Jul 10 "You're So Vain" eclipse---Nova Scotia. The only time a specific solar eclipse has been mentioned in a popular song. Lyric: "you flew up to Nova Scotia to see a total eclipse of the sun". "You're so Vain" reached number one in the USA for 3 weeks in early 1973.

1990 Jul 22 Northern latitudes

2008 Aug 1 Siberia

2026 Aug 12 Spanish Judgment. Diving down from the north over Reykjavik, Iceland and then Spain. Like the previous Saros (121) Saros 126 also has its last complete eclipse in 2044. Saros 120, 121 and 126 all experience their last complete eclipses within 11 years of each other.

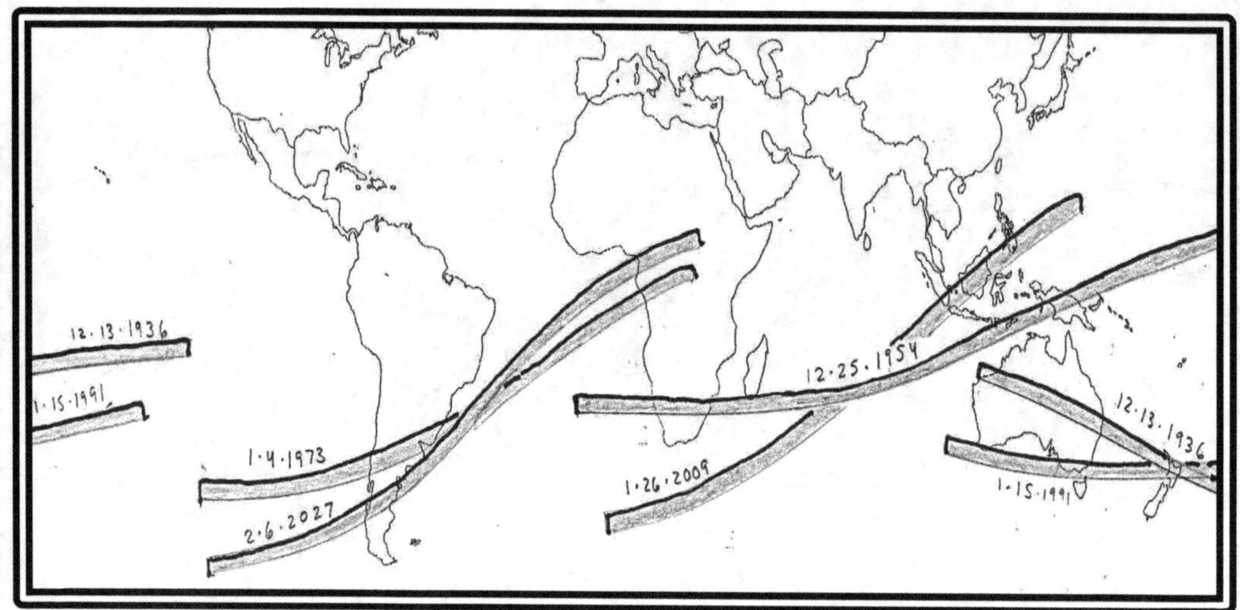

Saros 131 Ring of Fire: Snapshot 1918 to 2027

Saros 131 has 70 events Born Aug 1 1125 matured Mar 27 1522

1918 Dec 3 (not pictured). Three weeks after Armistice of 11-11-1918 that ended WW I. Enters Atacama Desert of South America. Less than 24 hours later, a massive earthquake devastates the city of Atacama, Chile.

1936 Dec 13 New Zealand/ Australia...3 days before birth of Pope Francis.

1954 Dec 25 Christmas in South Africa...one of two Christmas complete Eclipses in 20th Century, the other occurred in Antarctica with Saros 121, 19 years previously.

1973 Jan 4 Eclipse in Chile 9 months before 9-11 coup. Ends in central Africa.

1991 Jan 15 Tasmania/New Zealand. Martin Luther King birthday Metonic on what would have been his 62nd birthday.

2009 Jan 26 Ocean

2027 Feb 6 South America to Africa

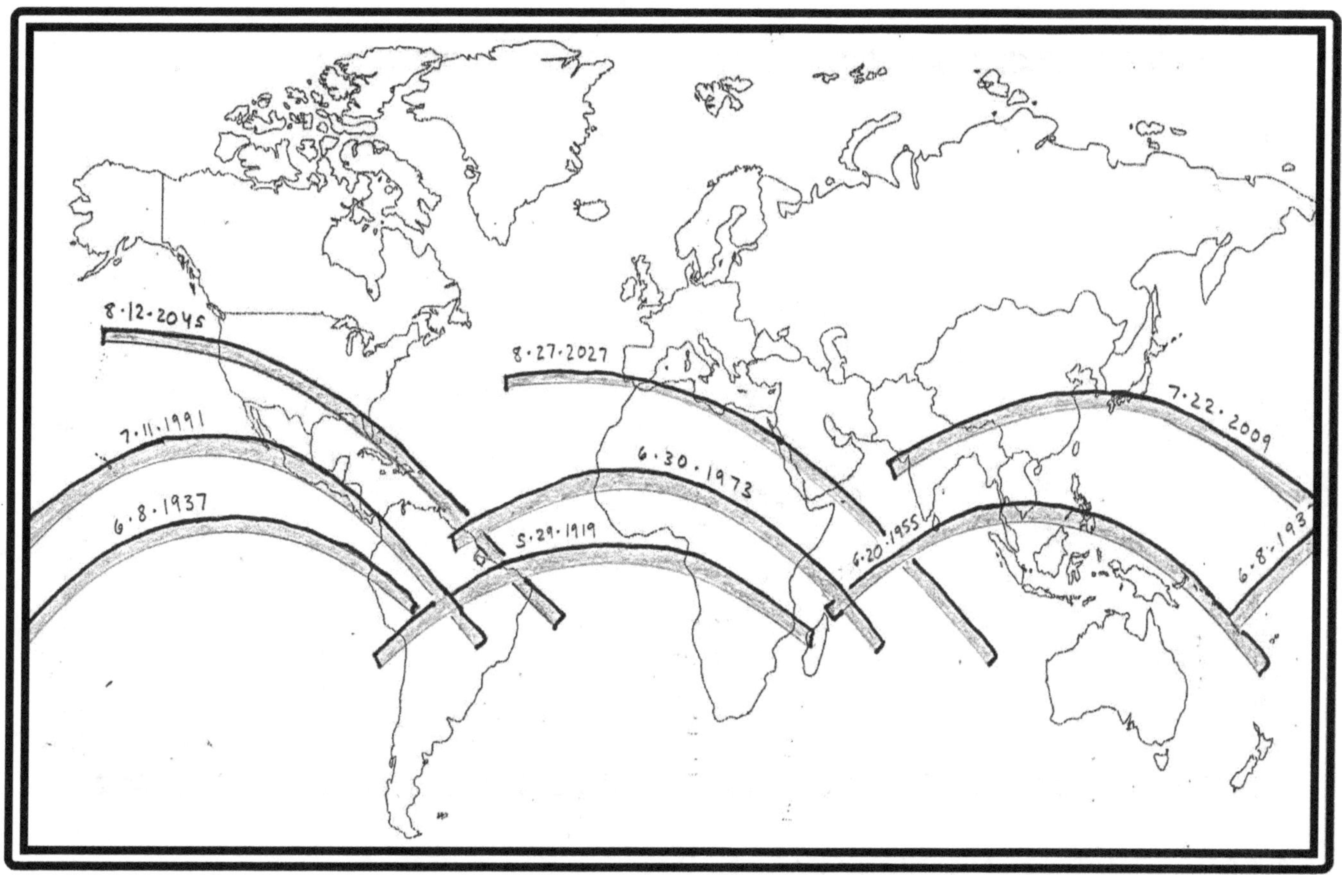

Saros 136 Total: Snapshot 1919 to 2045

The Saros that we earlier used as an example. More info at the beginning of Chapter.

1919 May 29 JFK 2nd Birthday/Einstein's Eclipse

1937 Jun 8 ---Ocean

1955 Jun 20 Vietnam/ Southeast Asia...longest total eclipse in 1,000 years as war begins in Vietnam

1991 Jul 11 Hawaii/ Central America

2009 Jul 22 Wuhan Eclipse, ten years before pandemic. India/China (including Shanghai)/ Japan

2027 Aug 2 Valley of the Kings...Rock of Gibraltar/Tripoli/Luxor/Mecca

2045 Aug 12 Synchronicity Eclipse USA

7 extraordinary eclipses. Saros 136 is Metonically connected with Saros 126. It has maintained a "semester series" with Saros 141 for centuries, which follows with ring of fires every 177 days

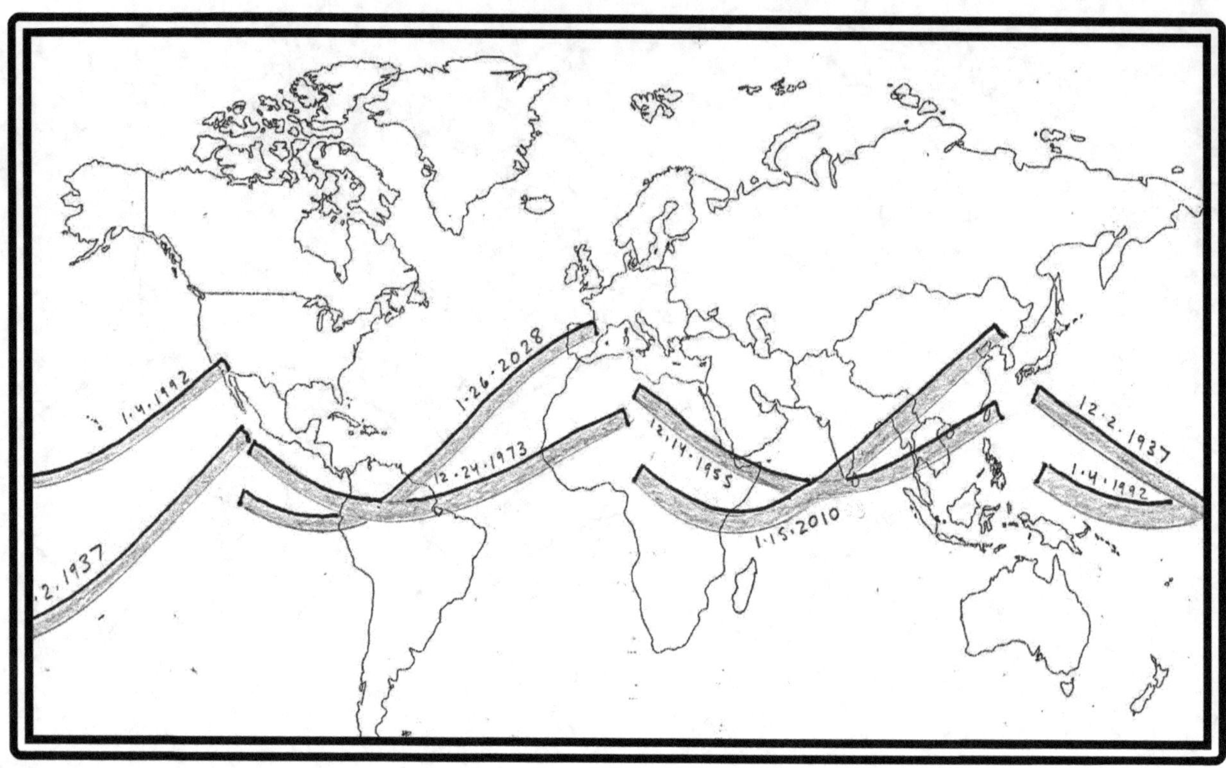

Saros 141 Ring of Fire: Snapshot 1919 to 2028
Saros 141 has 70 events born: May 19 1613 matured Aug 4 1739

1919 Nov 22 JFK Death Premonition Eclipse (not pictured). More at USA and world eclipses---Texas to Africa 44 years before assassination

1937 Dec 2 Ocean

1955 Dec 14 Vietnam/ Southeast Asia -177 days after 1955 Southeast Asia Total Eclipse. Longest Ring of Fire in the last thousand years at more than 12 minutes.

1973 Dec 24 Christmas Eve/South America to Africa, on Anthony Fauci's 33rd Birthday, ho, ho, ho.

1992 Jan 4 LA Sunset Eclipse. Path across ocean to end over Southern California at Sunset. Four months before the LA riots.

2010 Jan 15 Martin Luther King's 81st Birthday --Africa to China

2028 Jan 26 Spanish Ring of Fire. South America to Spain. Ends at sunset on eastern Spanish border.

Saros 141 has a great history, including the farewell Father Abraham USA eclipse of Oct 19 1865. It produced the longest central eclipses of the last thousand years. It shares Metonic with Saros 131.

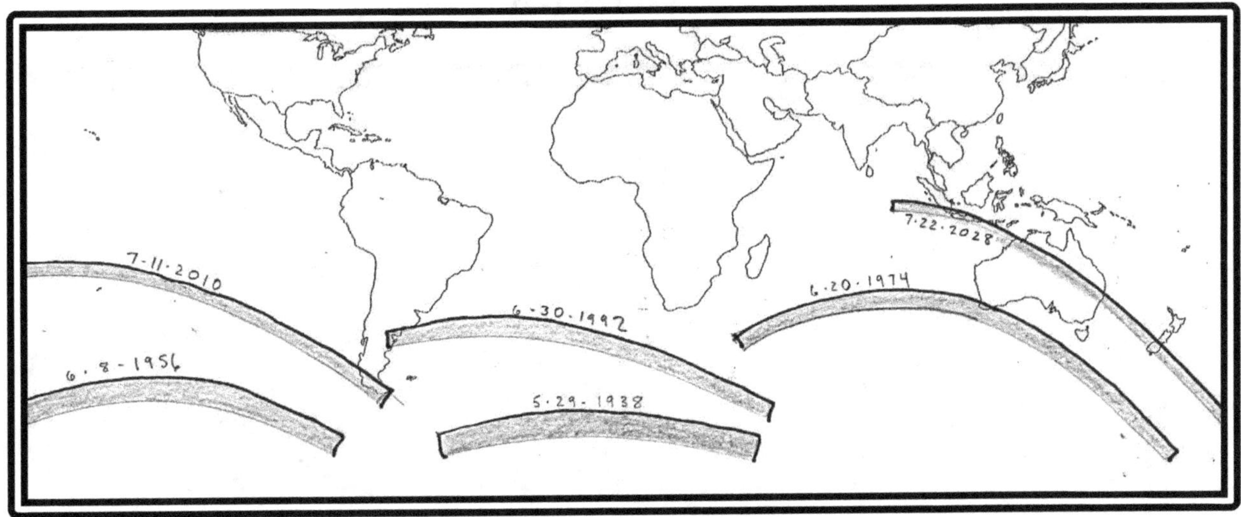

Saros 146 Total Eclipse: Snapshot 1920 to 2028

Saros 146 has 76 events born Sep 19 1541 matured May 29 1938

1920 May 18 Partial (not pictured). The day Pope John Paul II was born. 60 years to the day before Mount Saint Helens eruption.

1938 May 29 JFK's 21st Birthday. Eclipse over wreck of Shackleton's destroyed ship Endurance. First complete eclipse of Saros Cycle 146

1956 Jun 8 Antarctica

1974 Jun 20 Antarctica--scraping Australia.

1992 Jun 30 Montevideo, Uruguay Single City Eclipse

2010 Jul 11 Easter Island and Extreme South Pacific...tiny and mysterious Easter Island receives a Total eclipse exactly one lunar year after Wuhan Totality.

2028 Jul 22 Australia Judgment. Totality at Sydney and Botany Bay

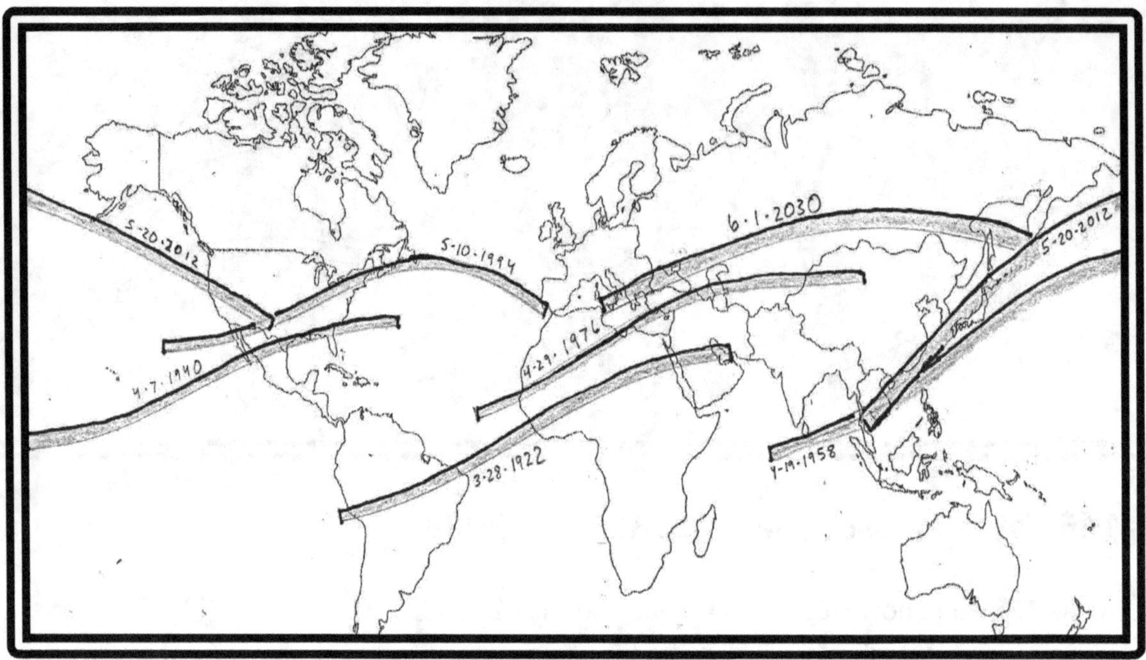

Saros 128 Ring of Fire Snapshot 1922 to 2030

Saros 128 has 73 events Born Aug 29 984 Matured May 16 1417

1922 Mar 28 Egyptian Independence Eclipse/ King Tut -South America to Africa

1940 Apr 7 Mexico/USA Deep South. 84 years before 2024 Apr 8 Great North American Eclipse

1958 Apr 19 Darwin Death Day Metonic Vietnam etc... Part of Southeast Asia Eclipse Swarm

1976 Apr 29 North Africa to China

1994 May 10 USA Ring of Fire

2012 May 20 Columbus Death Day -Hanoi, Vietnam to USA (ends at Midland Sunset)

2030 Jun 1 Athens/Istanbul / Russia/Kazakhstan/ Japan

Metonically aligned with Saros 138

Saros 133

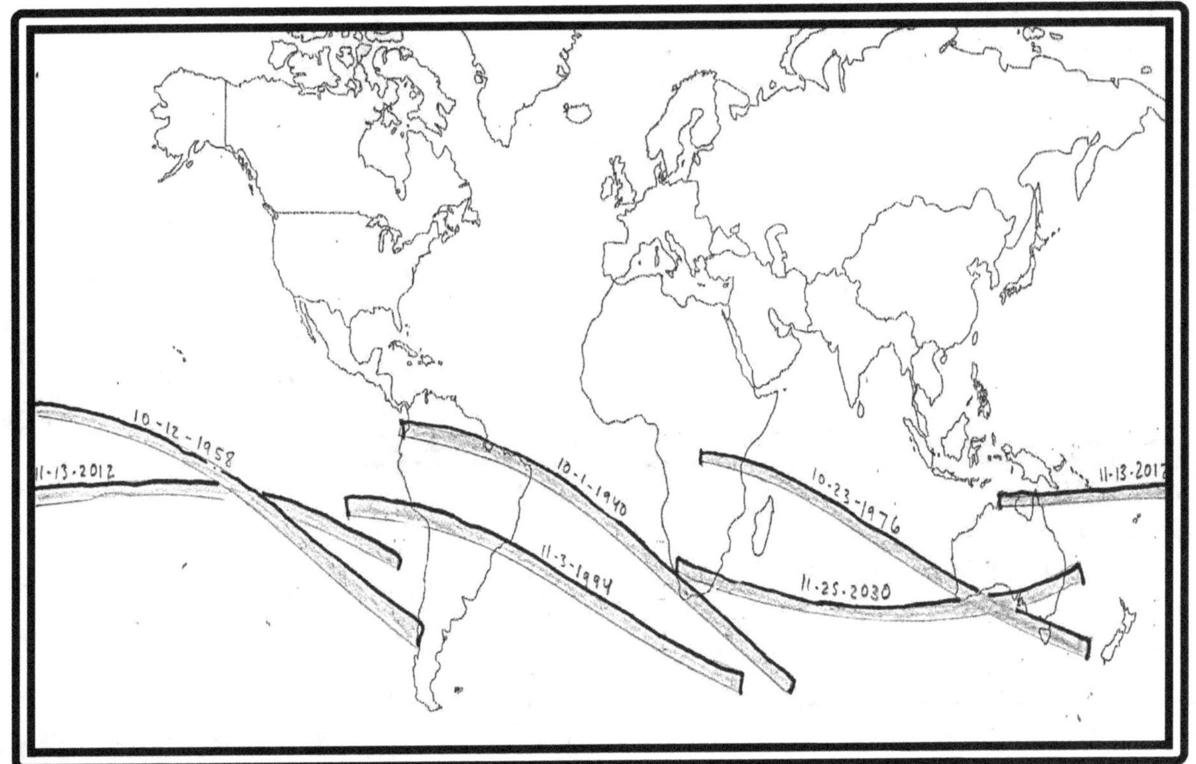

Saros 133 Total eclipse: Snapshot 1922 to 2030

Saros 133 has 73 events: born Jul 13 1219 matured Nov 20 1435

1922 Sep 21 Australia Equinox (not pictured)

1940 Oct 1 South America to Africa-8 days before John Lennon is born.

1958 Oct 12 Columbus Day Pacific Ocean

1976 Oct 23 Australia barely

1994 Nov 3 South America

2012 Nov 13 Ocean/North Australia

2030 Nov 25 Part of Monday Hepton Series---British Empire /South Africa to Australia

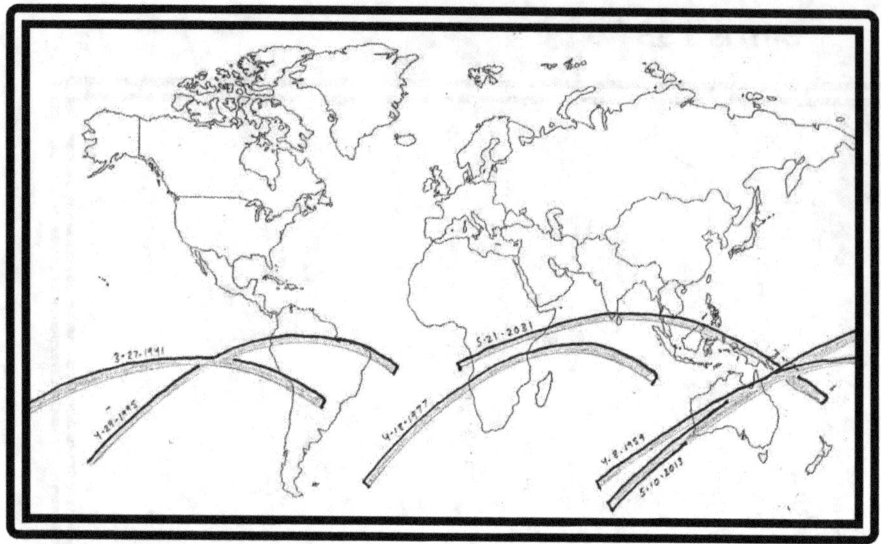

Saros 138 Ring of Fire
1941 to 2031

Saros 138 has 70 events: B. Jun 6 1472. Matured Aug 31 1598.

1941 Mar 27 Ocean to South America

1959 Apr 8 Australia –first appearance of April 8th Metonic, returning in 2024.

1977 Apr 18 South Africa
1995 Apr 29 South America
2013 May 10 tiny bit Australia/South Pacific
2031 May 21 ---Africa/India/ Indonesia...end of Columbus Death Day Metonic

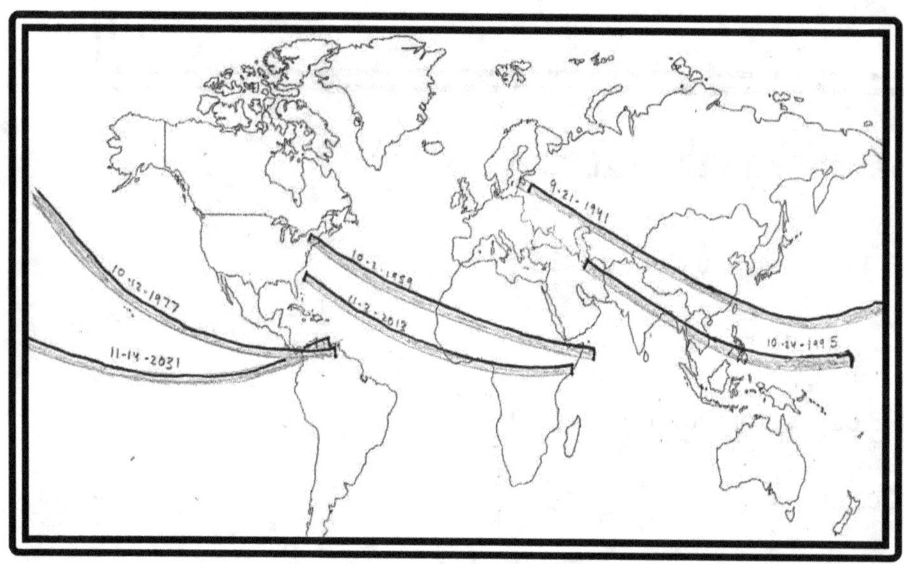

Saros 143 Total Eclipse: Snapshot 1923 to 2031

Saros 143 has 72 events. B. Mar 7 1617 matured Jun 24 1797

1923 Sep 10 Japan Earthquake/ L.A Metropolis

1941 Sep 21 Equinox USSR/China

1959 Oct 2 JFK Election Boston

Single City Sunrise

1977 Oct 12 Columbus Day Metonic North Pacific to Colombia

1995 Oct 24 Iran to Southeast Asia

2013 Nov 3 Hybrid US coastal waters to Africa

2031 Nov 14 Hybrid Pacific to Panama as Totality begins to fade from Saros 143

Saros 148

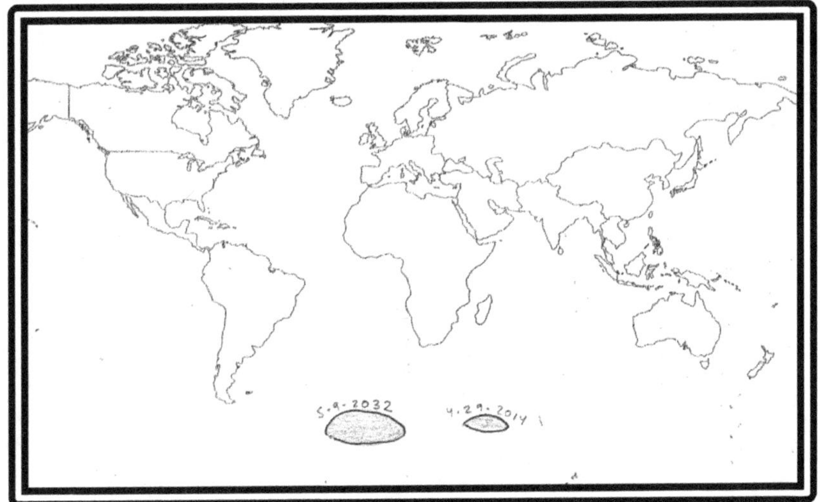

Saros 148 is just becoming an annular. It has 75 events and was born 9-21-1653

2014 Apr 29 first complete eclipse with a tiny spot of annularity in Antarctica
2032 May 9 South Atlantic

In this chapter we have outlined the recent behavior of the 26 Saros which are currently creating complete eclipses. In perusing these last pages, you will have seen the artistic bearings of all complete solar eclipses visible from earth in the last ninety years or so.

Some of them are truly superstar personalities. Others show up with a significant eclipse every now and then. Some linger around the uninhabited oceans and poles. Some are young. Some are old. Surely they all are necessary to create the cosmic mechanics and aesthetic intrigues of the eclipse phenomenon.

Solar eclipses are a largely unexplored territory of creative reality. They are the work of an Artist who lives outside of Time and weaves stories on Time with the shadows of the moon and the human mind. He created the Saros and all these alignments so we could witness this element of His art someday. We are at Saros 152 by the way, for those familiar with the 153 number of fish caught by the disciples in the presence of the resurrected Jesus in John 21:11.

God's designs are thoughtful and fun, like something a great Dad would do for his kids. These personalities interact with each other and with history in many ways. We can only study a few.

Since 1900, we have been considering about 177 complete eclipses. Of these, we will edit down to only the most peculiar eclipses, combinations, and crosses.

After we review the Metonic Cycle, we'll list them all since 1900 and take a closer look at some.

All Glory to God, who set the sun and moon.

Chapter Five

Mideast Eclipse
History 1492 to 2034

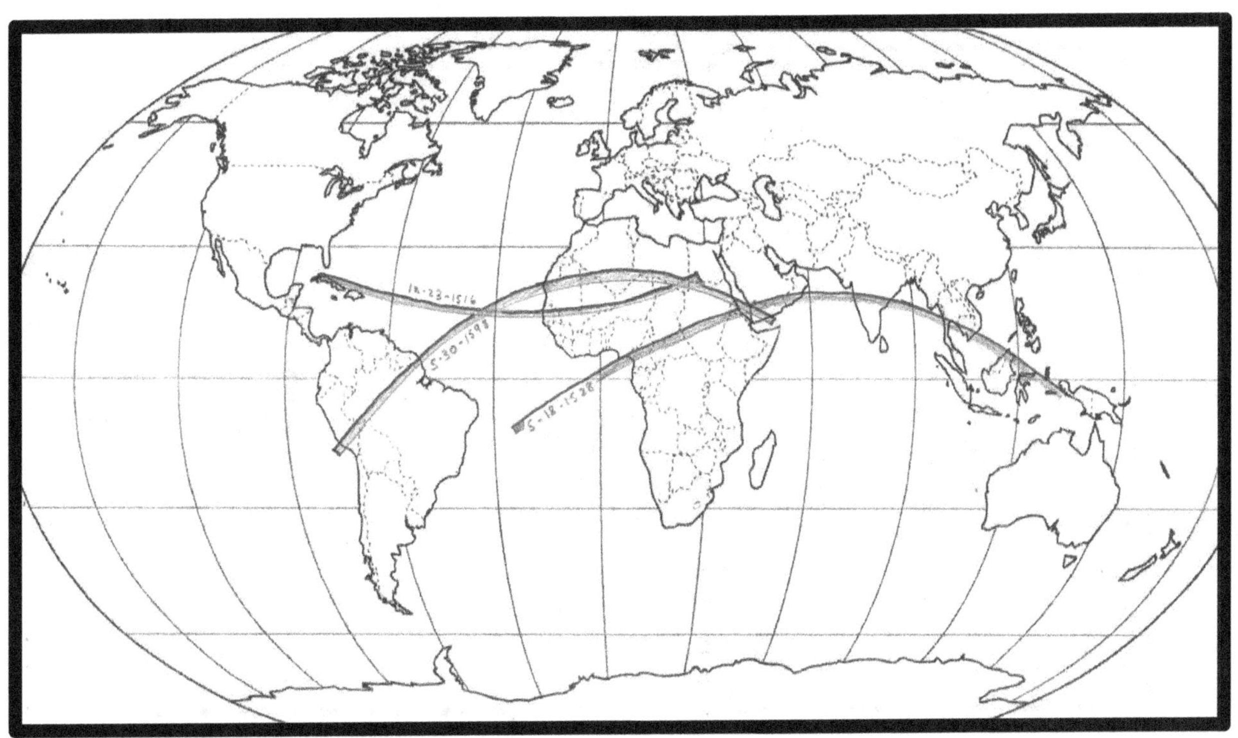

I. 1500 to 1600: The Eclipses Are Barely There Century
1516 Dec 23 (Saros 114) Total Eclipse
1528 May 18 (Saros 120) Hybrid Eclipse
1593 May 30 (Saros 121) Total Eclipse

1516 Dec 23 Mamluk Defeat Eclipse-Total Eclipse (Saros 114) The great Sultanate of the Mamluks, who had ruled Egypt for nearly 300 years, is conquered by the Ottoman Empire between the years 1516 and 1517. The Ottomans will then rule Egypt until 1867, except for a brief period of French Occupation in the early 1800's. **Saros 114** was born July 651 AD, or 19 years after the death of Mohammad. It produced the famous Hailey's Eclipse of May 3 1715, accurately mapped and predicted by Edmund Hailey in London. Saros 114 died September 12, 1931.

1528 May 18 Southern Arabia Hybrid (Saros 120) The great Saros 120 makes its second Hybrid. Its last complete eclipse occurs on 3-30-2033, the 2,000th anniversary of the Lord's Crucifixion.

1593 May 30 Totally in Timbuktu (Saros 121) Path near Timbuktu in Mali the year it became a permanent settlement. Saros 121 is scheduled for last complete eclipse, a ring of fire, in 2044. The two Saros currently leaving the tribe of Complete Saros, 120 and 121, both make an appearance in our first set of eclipses. Between 1488 and 1600, only two eclipses touch the farthest corner of modern Egypt, and only two barely touch southern Arabia. Both regions were absorbed into the Ottoman (based in modern Turkey) Empire.

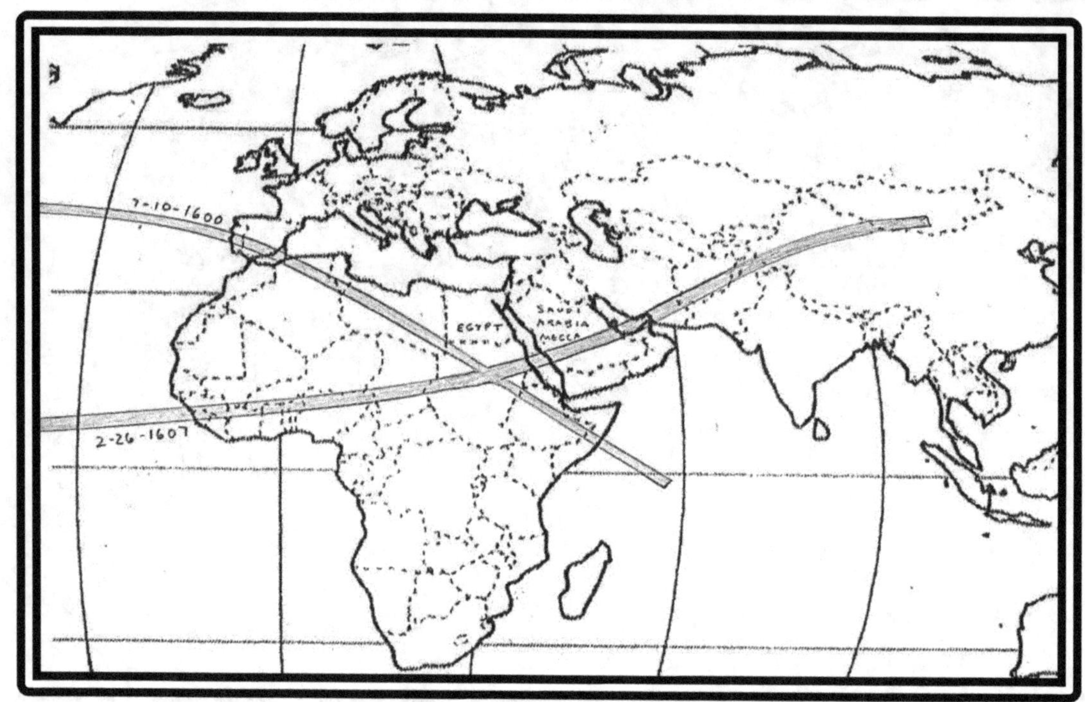

II. Nubian Monday Hepton Khartoum Cross 1600 and 1607

1600 Jul 10 (Saros 120) Total Eclipse
1607 Feb 26 (Saros 114) Total Eclipse

1600 Jul 10 (Saros 120) Total Eclipse Cape Canaveral to Africa This Total eclipse begins at sunrise Monday over the future USA's Cape Canaveral (see US eclipse maps) only crossing Florida, scrapes lower Egypt and bisects the ancient Kingdom of Nubia during the reign of the Dinka.
300/44 Pairing paired with May 28 1900 Monday USA to Egypt first eclipse of the 20th Century.

1607 Feb 26 (Saros 114) Total Eclipse Arabia/Persia: Also on Monday
300/44 Pairing: paired with Jan 14 1907 Russia to Russia Total Eclipse, also on Monday!

These are the same two Saros which appeared in the last two eclipses of 1528 and 1593. They create a true Hepton Total Cross on Monday, the same day of the week as our current Hepton Crosses. Apex of Cross is very near modern-day Khartoum, Sudan.

Note: You may think you are seeing the same eclipses on the next page. But they are 127 years later. They appear almost as an opposite mirror reflection of these eclipses with apex over generally the same area of modern-day northern Sudan.

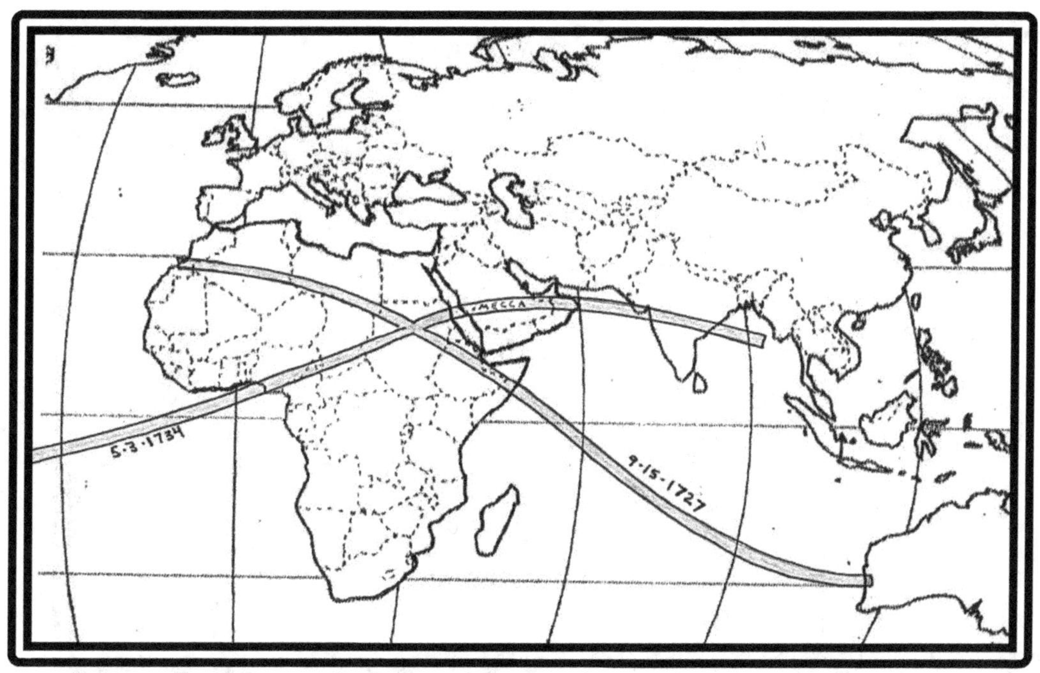

III. Mecca Monday Khartoum Encore Hepton Cross 1727/1734
1727 Sep 15 (Saros 130) Total Eclipse
1734 May 3 (Saros 124) Total Eclipse

1727 Sep 15 (Saros 130) Total Eclipse First eclipse of an extraordinary cross locked in uncanny relationships. Eclipse begins almost directly at Africa's western coast. Ends at Australia's western coast. 1727 which is the exact year of the establishment of the Saudi Kingdom in Arabia. Mecca will be in totality in the next Hepton Eclipse which follows this exactly 7 eclipse years (2422 days) later. Occurs on Monday. **Saros 130:** Saros 130 reappears 307 years later in 2034 as second half of 2027/2034 Hepton Cross. In that eclipse it follows an opposite directional pattern.
300/44 pairing: Paired with Aug 2 2027 Valley of the Kings Eclipse (Mecca)

1734 May 3 (Saros 124) Mecca Total Eclipse This eclipse traverses Africa beginning at the Slave Coast (crossing Lagos Nigeria) in the midst of the Transatlantic Slave Trade, crosses ancient Nubia, Mecca and ends shortly after bisecting India. **Saros 124** This is Saros 124's maximum eclipse, at 5 minutes 36 seconds ; its longest eclipse in its 1200 year life. Returns for eclipses in USA in 1860 and 1878. Also, the 1914 WWI Eclipse, USSR Equinox Judgment 1968. Its Last complete eclipse in 1986.
300/44 Paired with Mar 20 2034 Equinox Great Civilization Eclipse.

These are the only two solar eclipses of the 17th Century in the Egypt/Arabia region. Monday Hepton Crosses directly related to 2027 and 2034 Monday great eclipses via the 300 minus 44 cycle.

This cross, like the last section (1600 and 1607) also is a Hepton Cross with both eclipses occurring on Monday. It is an almost perfect mirror reflection of that cross which occurred 127 years prior. The entire cross is in 300 year minus 44 day relationship with the Mideast Judgment eclipses of 2027 and 2034, which also are on Mondays. The first line of this cross, Saros 130, reappears in the *second* line of the 2027/2034 Cross, which crosses Mecca in its first line. See if you can keep that straight!

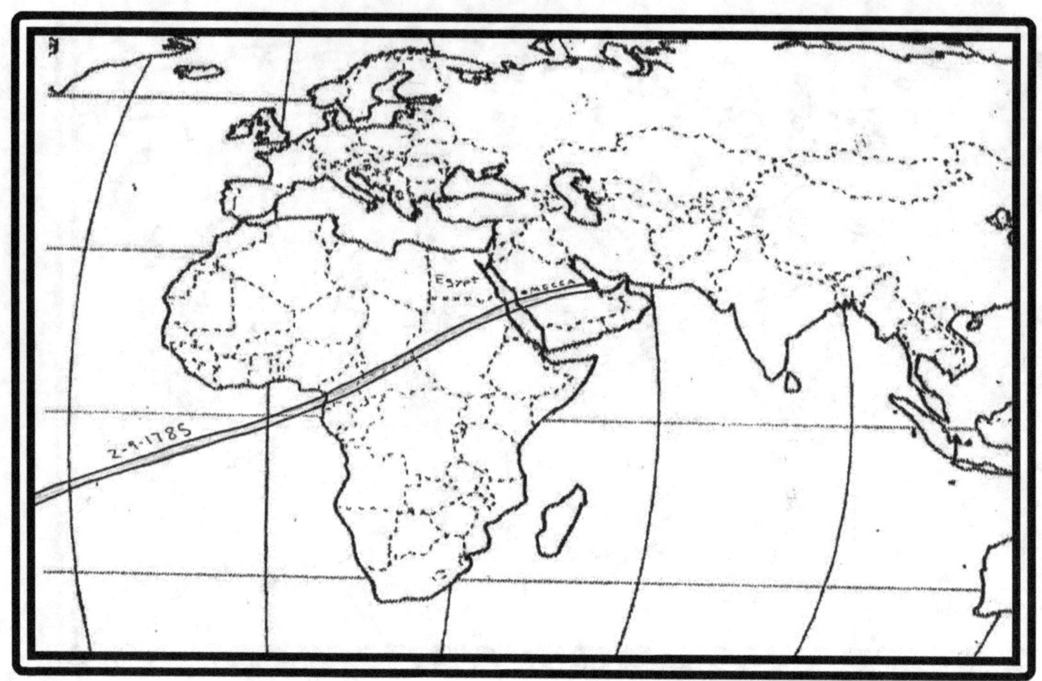

IV. Arabia (Khartoum) Feb 9 1785 Total

1785 Feb 9 Total (Saros 127) Eclipse again enters Africa at the Slave Coast, travels across modern Khartoum region AGAIN (their 5th total eclipse in 185 years...average should be one every 400 years) and narrowly misses bringing totality to Mecca. This is the 8th eclipse out of 9 in a row to come to the region as a total (the other was a hybrid).

Saros 127 was responsible for the so-called Mayan Apocalypse Eclipse of 8-8-1496 and reappears in the last Monday Hepton in Australia and New Zealand on July 13, 2037.

This eclipse stands alone in time, removed by 51 years from the last complete eclipse (1734) in the region and 18 years from the next (1803)

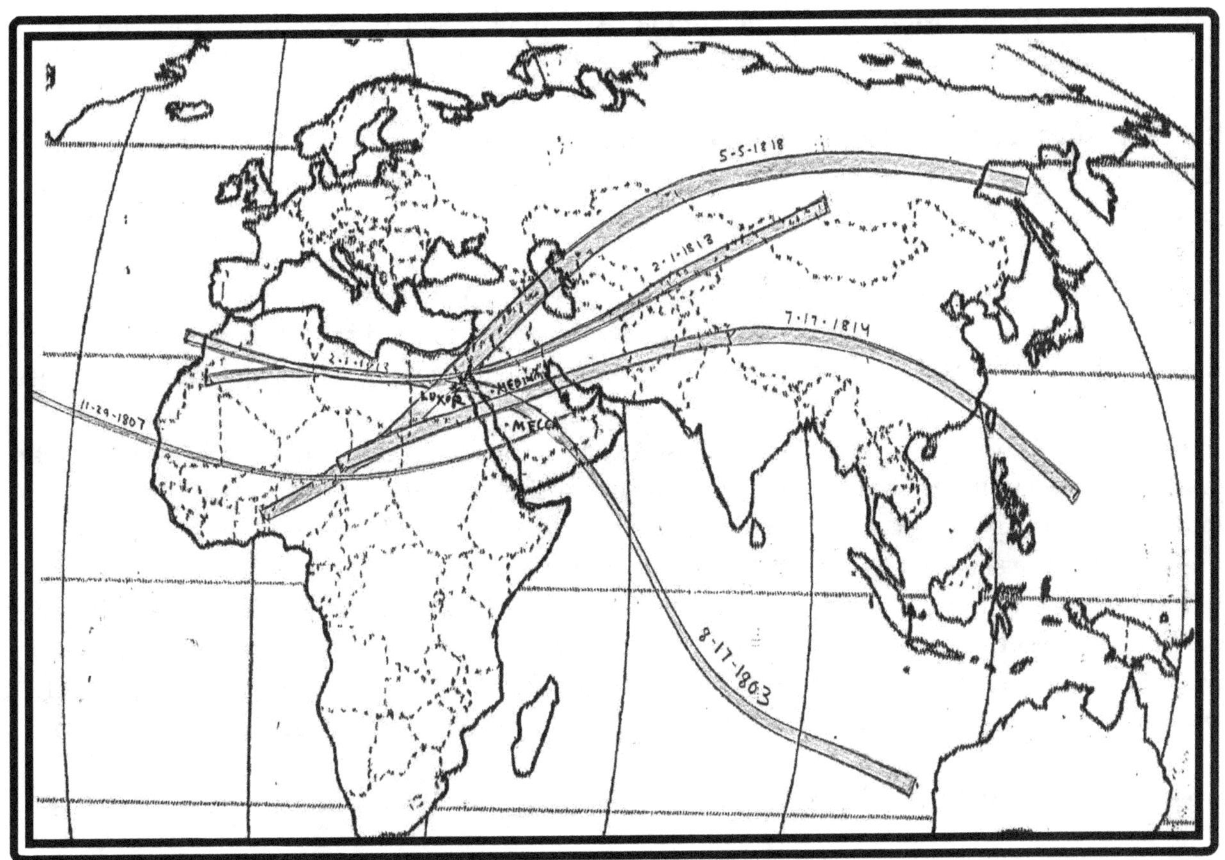

V. Great Mideast Eclipse Swarm 1803 to 1818

1803 Aug 17 Luxor/Medina Ring of Fire (Saros 132)
1807 Nov 29 Arabian Hybrid (Saros 139)
1813 Feb 1 Egypt/Arabia Monday Ring of Fire (Saros 118)
1814 Jul 17 Great Swarm Total (Saros 133)
1818 May 5 Luxor/Sinai/ Ninevah Ring of Fire (Saros 135)

1803 Aug 17 Luxor/Medina Ring of Fire (Saros 132) Two years after the defeat of Napoleon in Egypt, the first eclipse in a 15-year swarm of solar eclipses crosses North Africa, the Valley of the Kings in Egypt and Medina, last home of Prophet Mohammad and second most holy place in Islam.
Saros 132: Saros 132 was the first eclipse after 9-11, on Dec 14 2001. Saros 132 also brought the Tsunami Anniversary Eclipse of Dec 26 2019.

1807 Nov 29 Arabian Hybrid (Saros 139) Hybrid Eclipse crosses Khartoum Region for its sixth eclipse in 207 years. The average should be one every 150 years) Ends at Eastern border of Arabia. Goes south of Mecca. Part of a Sunday Hepton with 1814 Jul 17.
Saros 139 comes to USA Apr 8 2024

1813 Feb 1 Egypt/Arabia Monday Ring of Fire (Saros 118) Ring of fire crosses all of North Africa , Mount Sinai, southern Babylon, Persia, ends in Mongolia

1814 July 17 Great Swarm Total (Saros 133) Starting in Modern Chad, Egypt, Arabia and then southern Babylon, Persia, Modern Afghanistan, Pakistan, Kashmir, China/Taiwan. Part of a Sunday Hepton with 1807 Nov 29 (not a cross).

1818 May 5 Luxor/Sinai/ Nineveh Ring of Fire (Saros 135) A ring of fire on 5-5 (with maximum duration of 5 minutes 5 seconds, honestly) brings central Egypt its third complete eclipse in 15 years. Crosses all of Sinai Peninsula, goes to Jordan, northern Iraq (Nineveh), Kazakhstan and Eastern Russia.

1803 to 1818 brings the greatest Mideast eclipse swarm in our 500-year study. 5 eclipses in 15 years. Arabia sees all five of them. Egypt sees four. It marks a turbulent time in the history of the region. The Egyptian and Ottoman war against French occupiers had ended in 1801 in French defeat. British troops had fought against the French in the Battle of Alexandria in 1801. The British evacuated in 1803 just months before the first eclipse in the swarm. The French had fought among the Pyramids under Napoleon. The Rosetta Stone (which allowed for the science of Egyptology) was found by the French in 1799. The Saud Dynasty was fighting the Ottomans in Arabia in the Wahhabi Wars of 1811 to 1818, finally forced to surrender in September of 1818, four months after the last eclipse in the swarm.

There is also a Sunday Hepton there (Hybrid 1807 and Total 1814), but it doesn't cross. Here it is:

Sunday Hepton Pair in Arabia, Asia and Africa 1807 and 1814

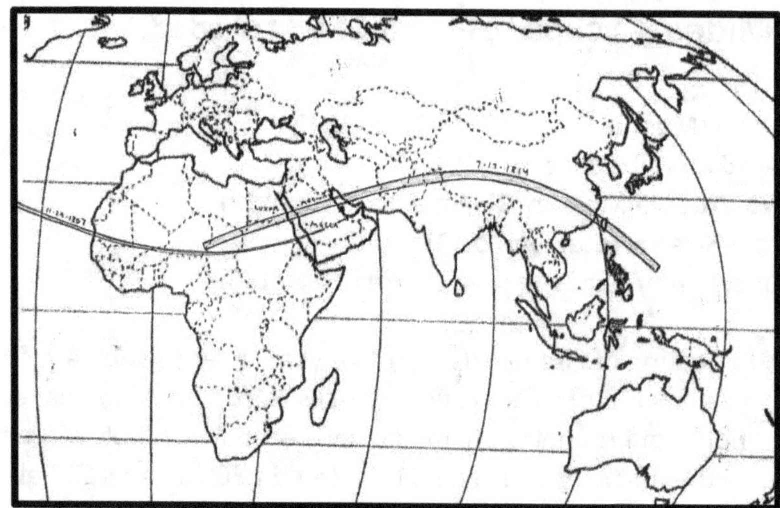

As an exchange of imperial energies rages on the battlegrounds of Egypt and Arabia, the swarm manifests its shadows of the sun and moon. French, British and Ottoman Armies roamed through the deserts of the ancient ruins of the first empire. Before this swarm, there had been only 8 complete eclipses in the region in nearly 350 years. Now there are 5 eclipses in 15 years.

64 years would pass before there was another complete eclipse in the region.

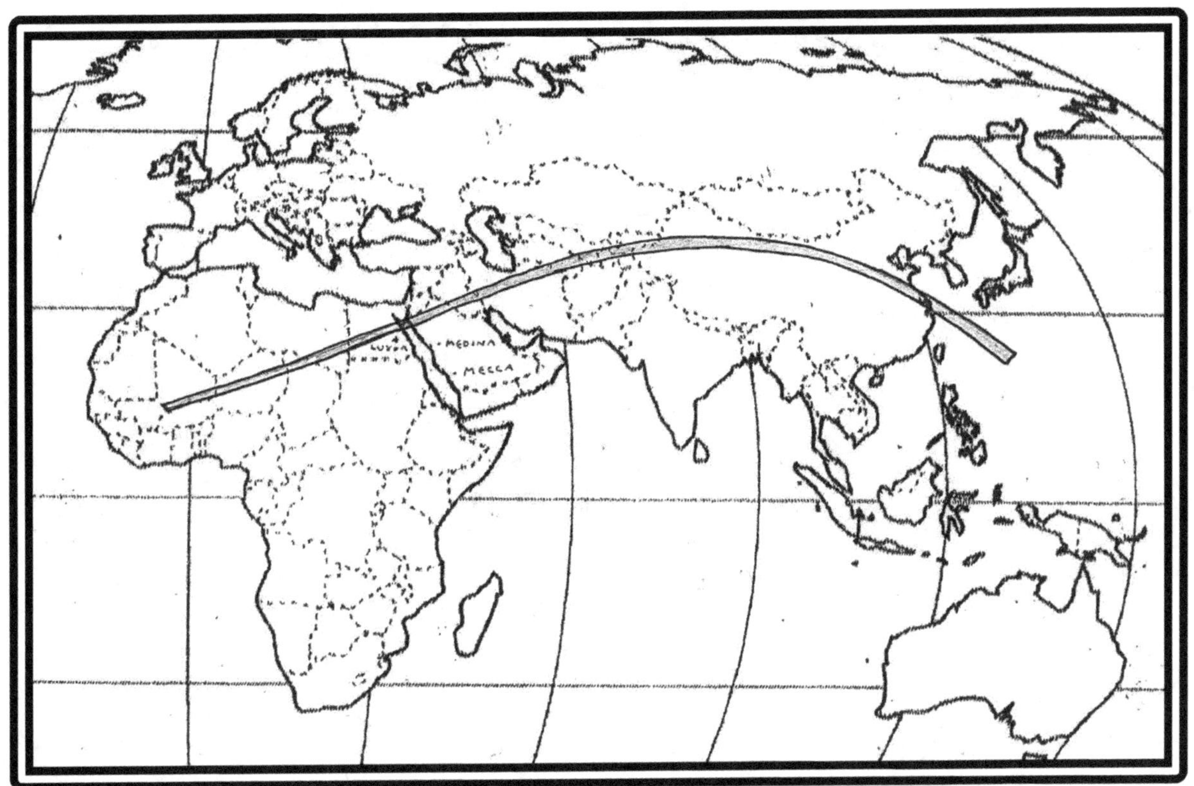

VI. Sinai Total Eclipse 1882

1882 May 17 Great Civilization Total (Saros 126) One of the greatest Saros of them all arrives for the first time in this study, passing over many key sites in the history of empire. This historically isolated total eclipse bisects Egypt in the first year of its Occupation by the British (1882), with totality just 50 miles from Luxor and the Valley of the Kings. The path crosses likely Mount Sinai site at Jabal Al-Lawz in Saudi Arabia. It also crosses ancient Babylon and ancient Persian capitals. Totality in Tehran. Bisects China.

Saros 126: Here, Saros 126 creates its first total eclipse after 540 years of annularity. It will return in spectacular fashion to link the USA to Egypt in its next eclipse 19 eclipse years later, on May 28 1900, the first eclipse of the 20th Century.

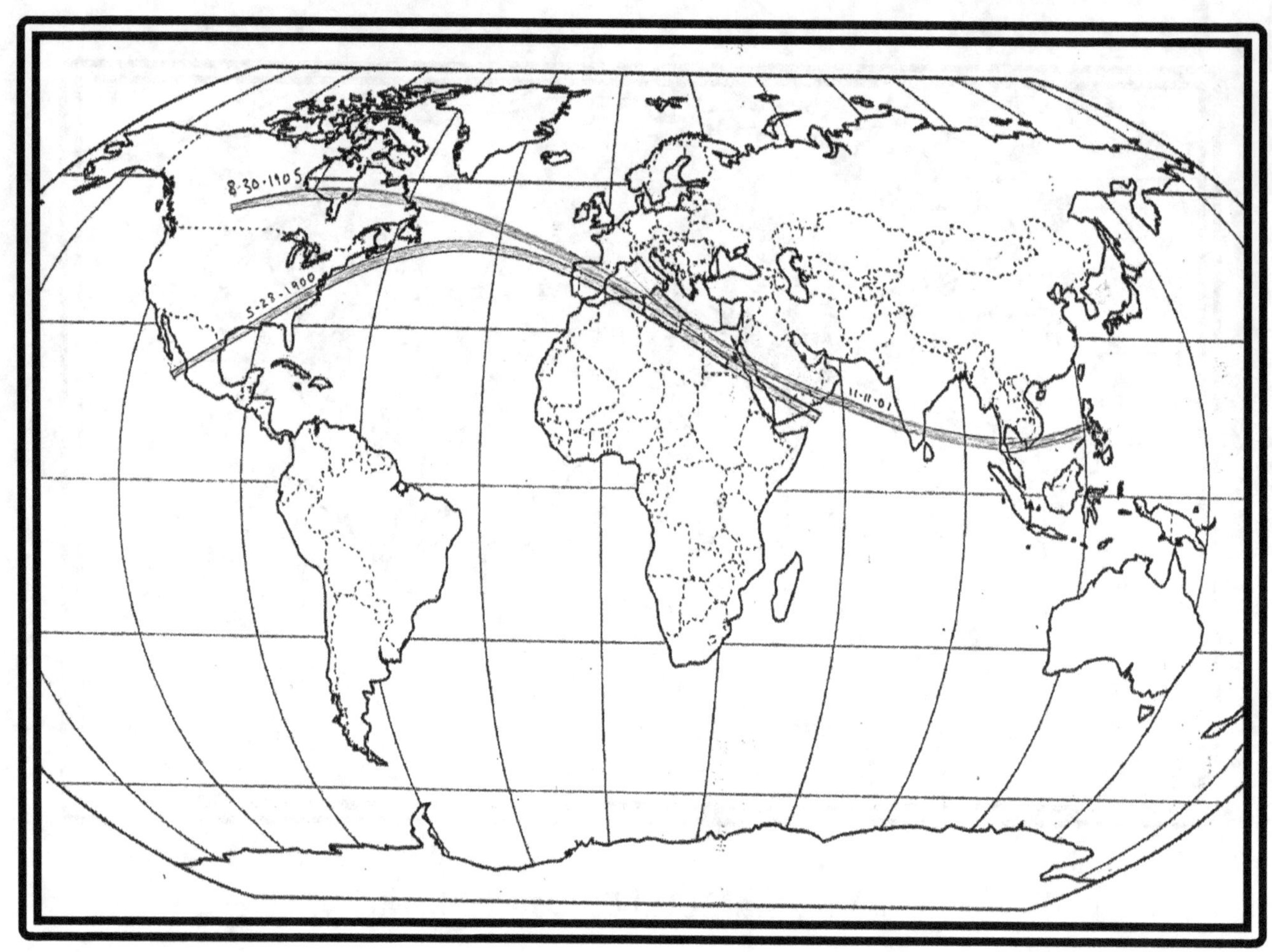

VII. Dawn of 20ᵗʰ Century Egyptian Eclipse Swarm

1900 May 28 USA to Luxor Total First Eclipse of 20ᵗʰ Century (Saros 126)
11-11-1901 Amazing Elevens Ring of Fire (Saros 141)
1905 Aug 30 Luxor and Mecca Total (Saros 143)

*note: the 5-28-1900 eclipse ends just before Luxor in Egypt.
It is difficult to see that on this map because the 1905 eclipse overlaps it in Egypt.*

1900 May 28 USA to Luxor Total First Eclipse of 20th Century (Saros 126) The first eclipse of the 20th Century links the Aztec empire of the New World, the newest empire in the world (USA) and the first empire in the world (Egypt). Eclipse travels from the USA to Egypt, ending at sunset in Luxor. Looked at on a map, this 1900 eclipse would be seen to dead end at the 1882 in the last section. They would meet just to the west of Luxor, the Valley of the Kings, in Egypt. Both eclipses pass within 50 miles of there and created 99% totality.

The 1900 eclipse is exactly 17 years and one day before the birth of John F. Kennedy. Kennedy and his wife Jackie helped bring the Egyptian Temple of Dendur to the United States to be reassembled in New York City. It had been scheduled to be submerged by the Aswan Dam. The next eclipse on Earth after this one occurs on Nov 22 1900, 63 years to the day before JFK's murder. Eclipse also crosses American power spot New Orleans (latitudinally aligned with Pyramids of Giza) and bisects Spain.
300/44 Pairing with Jul 10 1600 Cape Canaveral to Khartoum Eclipse.
Saros 126 returns with Cross country USA 1918 and USA to Iran 1954

1901 Nov 11 Amazing Elevens Ring of Fire (Saros 141) The greatest ring of fire Saros of the last thousand years, Saros 141, begins its first 20th Century run at sunrise on the island of Malta, famous as the site of the Apostle Paul's Biblical shipwreck. Path bisects Egypt just north of the 5-28-1900 eclipse 177 days before, and brings complete annularity to the Pyramids of Giza, Alexandria and both possible Mount Sinai locations. Then it divides Saudi Arabia before encountering India, Vietnam, Cambodia.

Volcano allusions: Eclipse ends at sunset over Mount Pinatubo in the Philippines, site of one of the greatest volcanic explosions of the 20th Century. The next eclipse on earth, May 18 1901, occurs 79 years to the day before another great volcano, Mount Saint Helens.

Eclipse occurs on 11-11-01. Maximum duration of eclipse according to NASA was 11 minutes and 01 seconds. Saros 141 has two 1's in it for good measure. Exactly 17 years before the end of World War One, which happened at 11:11 am on 11-11-1918, which becomes known as Armistice day.
Saros 141 has an extraordinarily history. Farewell Abraham eclipse of 1865, California Sunset 1992, the longest ring of fire in 1,000 years over Southeast Asia in 1955.

1905 Aug 30 Aswan/ Mecca Total (Saros 143) To finish the swarm, Total Eclipse extends from central Canada across Spain (Madrid) and ends up overlapping the path of 1900 Total eclipse almost precisely (including 99% at Luxor and totality at Aswan) before bringing totality to Holy city of Mecca.

Saros 143 has a string of great eclipses, including JFK Boston single city of 10-2-1959

Note: This eclipse trio is difficult to map. The eclipses literally overwhelm and run over the top of each other through the nation of Egypt. Three eclipses in five years (it will have been 4 in 23 years counting the 1882 total) in a small nation that can go centuries without a significant solar eclipse.

Separating the Three Solar Eclipses of 1900 to 1905:

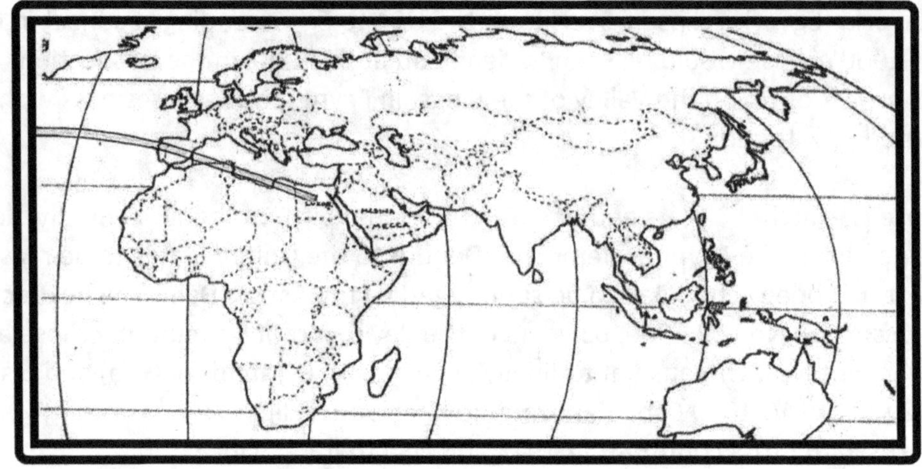

ABOVE: **1900 May 28 Total Eclipse (Saros 126)**
Ends slightly closer to Luxor than depicted (Luxor experienced 99% eclipse at sunset)

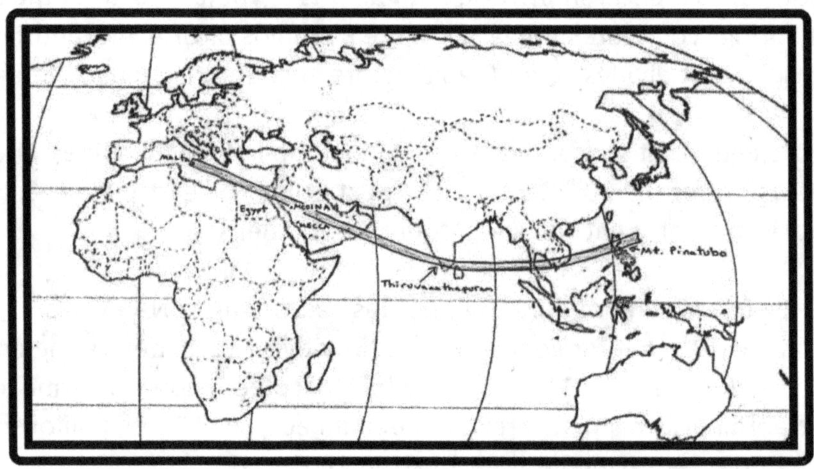

ABOVE:**1901 Nov 11 Ring of Fire Giza Eclipse (Saros 141)**

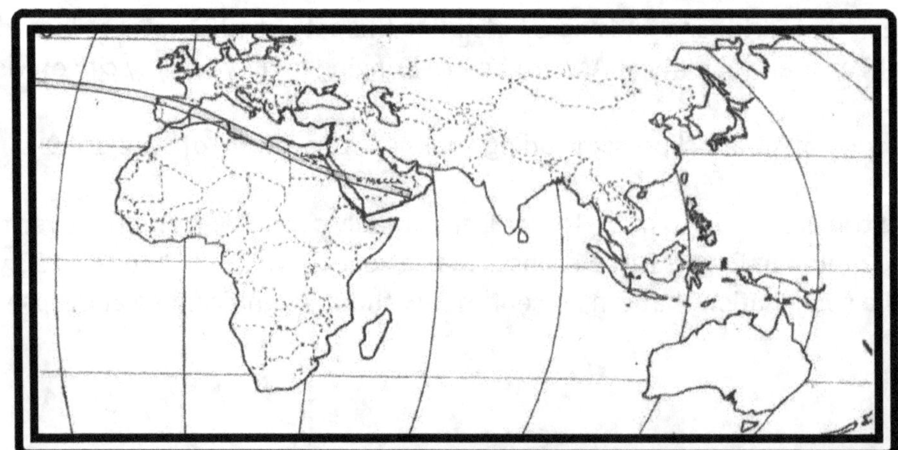

ABOVE: **1905 Aug 30 Total Eclipse** note extraordinary similarity in path of
1900 and 1905 total eclipses. Both brought totality to Spain and to Tripoli, Libya.

 Egyptian Independence/ King Tut Ring of Fire 1922

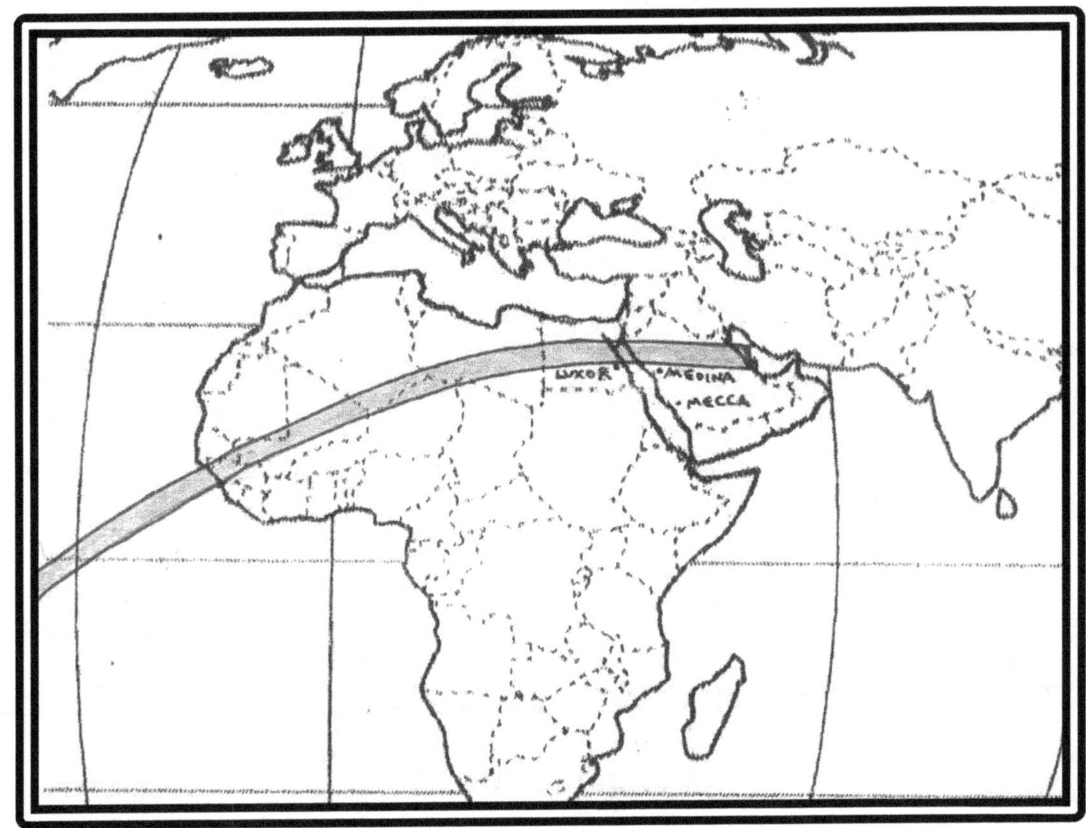

1922 Mar 28 Ring of Fire (Saros 128) Ring of Fire begins in South America, crosses north Africa and bisects Egypt just north of Luxor and across mountain of Jabal Al Lawz (the likely Mount Sinai), then onto end at Eastern border of Arabia.

Occurs one month after Egyptian Independence was declared on Feb 28 1922.

Occurs six months before the discovery and subsequent plunder of the tomb of Pharaoh Tutankhamen by British archaeologists led by Howard Carter on November 4, 1922. King Tut had died at the unusual Metonic age of 19 years old more than 30 centuries before around 1323 BC. His tomb was certainly the greatest single find in the history of archaeology, in terms of material value, especially lots of gold.

It is claimed that the tomb was cursed.

Saros 128 will create back-to-back ring of fires across the USA 1994 and 2012.

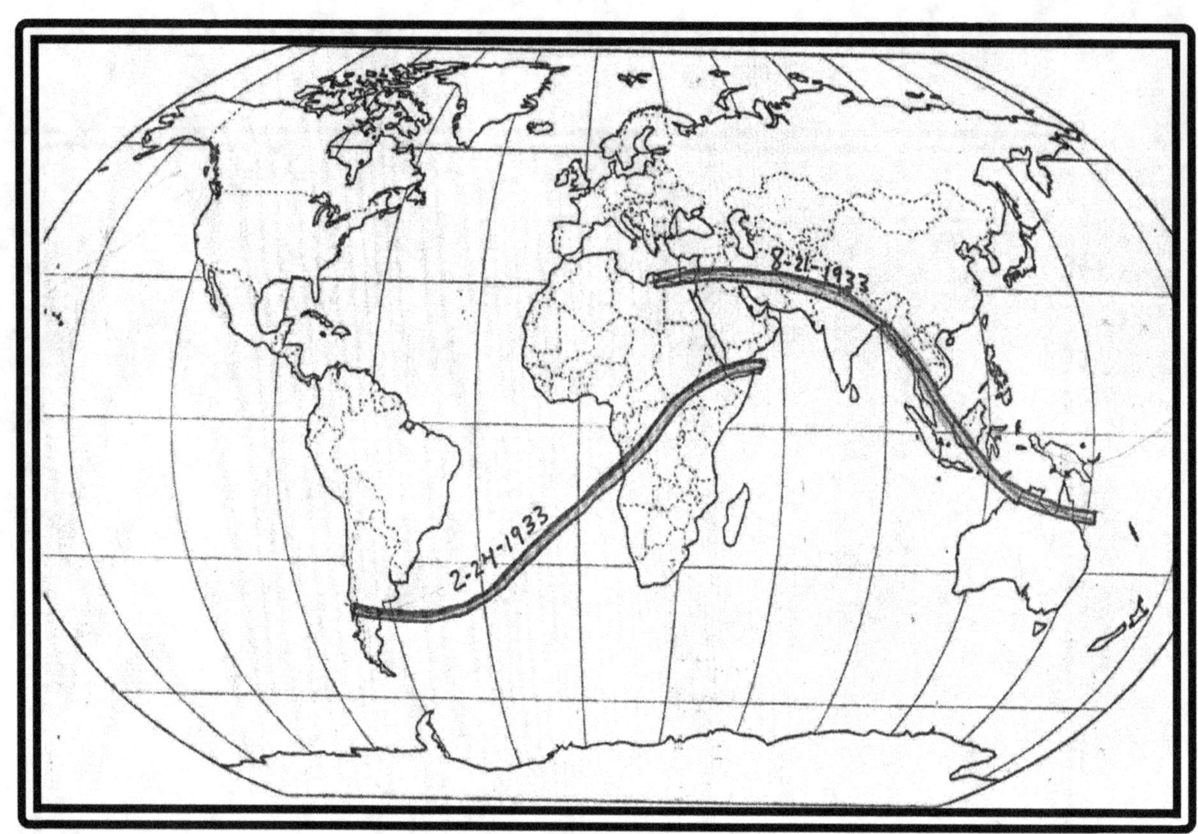

IX Ring of Fires 1933

Pictured are the ring of fires in 1933. These were the only solar eclipses on earth in 1933. The Jerusalem eclipse is the top one. Note the aesthetic pattern created by the framing of the Arabian Peninsula. I do not include the 1933 Feb 24 in the final numbered list of Mideast eclipses, but it is unusual for its perfect symmetric reach of western South America and Eastern Africa Coast.

1933 Aug 21 Jerusalem Ring of Fire (Saros 134) As the Nazis seize control in Germany and at the peak of a worldwide depression, a Ring of Fire chooses the great Metonic date of Monday August 21. Exactly 19 years after WWI eclipse of August 21 1914. Exactly 84 years before USA division eclipse of Aug 21 2017. Path of Annularity across the Mideast--Palestine, Jordan, Iraq, Iran and Afghanistan-- then travels over India and Southeast Asia, total annularity in Jerusalem, Baghdad (Babylon), Amman, Gaza, Alexandria Egypt and the Pyramids of Giza ---path cuts through Iran, Afghanistan, Northern India, Bangladesh, Thailand, Indonesia, and Darwin, Australia before ending at east coast of Australia.

It was the year that Germany went from democracy to dictatorship--the Reichstag fire, the first anti-Jewish laws, the purge of the opposition, the construction of Dachau (the first concentration camp). August 21st 1933 was the start of the 18th Zionist Congress in Prague . In the Hebrew calendar it was 1st day of month of Elul.

Also in August 1933 in Iraq, thousands of Assyrian Christians were killed in the Simele massacre, The "great white spot " of Saturn was seen by astronomers for the first time in 30 years in early August, and stayed for 4 weeks, allowing for the measurement of Saturn's rotation. Mahatma Gandhi was on a hunger strike in India and was taken to the hospital on August 21st, the day of eclipse.

Palestinian leader Yasser Arafat was born 3 days after the eclipse in Cairo (on border of annularity) on August 24, 1933.

This was the first complete eclipse in Jerusalem since 1820 and the last for 300 years. Jerusalem's last total eclipse occurred on Aug 20 993 AD, it also had a ring of fire eclipse on 1133 Aug 2. Note date similarity with Aug 21 1933 and upcoming Aug 2 2027 Valley of the Kings total eclipse.
Saros 134 brings the Great American Ring of Fire in 2023

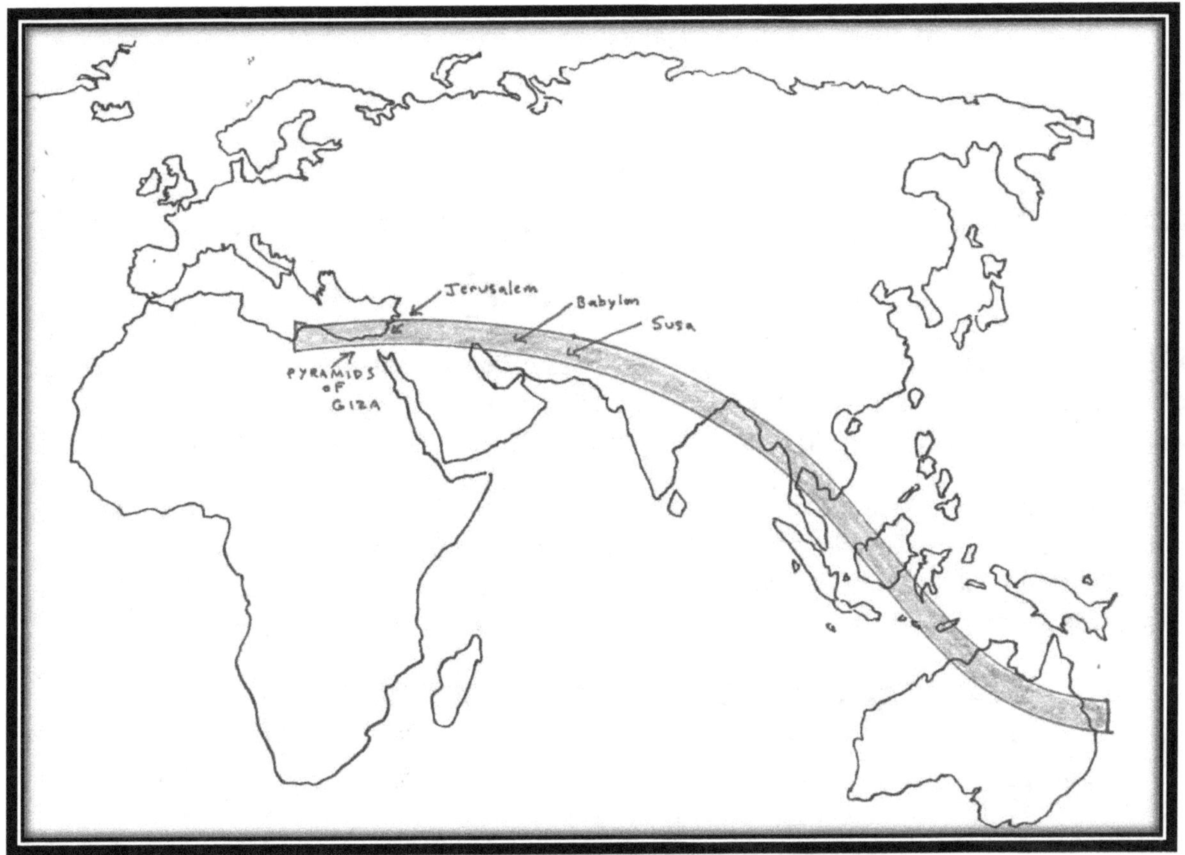

ABOVE: **Jerusalem Ring of Fire 8-21-1933 Closeup**

Above is a closer rough look at the path of the August 21 1933 Jerusalem Ring of Fire. Path passes over the Pyramids of Giza and great city of Alexandria in Egypt, then Jerusalem, Amman, Bagdad (Babylon), and Esfahan, Iran before heading East.

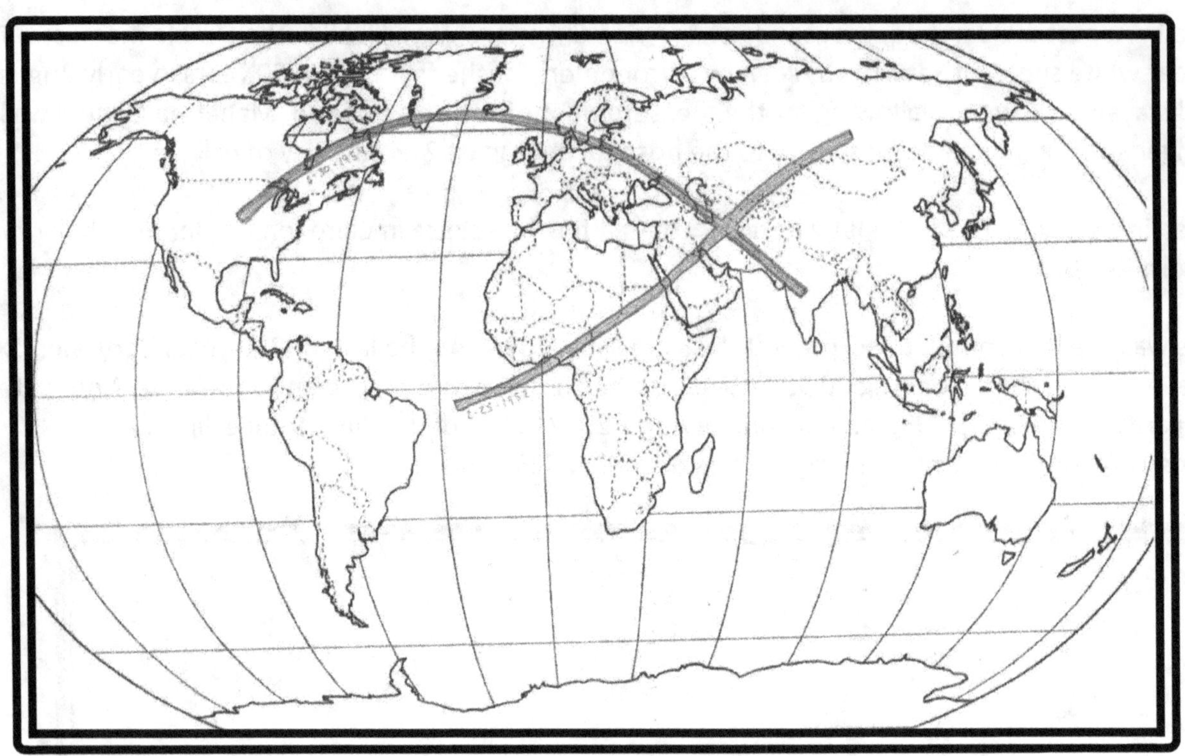

X. Shahrud Iran Cross 1952 and 1954

1952 Feb 25 Total Eclipse (Saros 139) Total eclipse traverses Africa (Totality in Khartoum), then crosses Arabia (totality in Jeddah), crosses Iraq near Basra, and then Iran (totality in Shahrud) ends in Russia. 1952 saw upheaval in Iran, leading to the US backed coup of August 1953 that installed the Shah.

1954 June 30 Total Eclipse USA Ukraine Iran (Saros 126) Path starts in Nebraska, goes north (totality in Saint Paul, bisects Northern Canada, crosses the Atlantic, dives through Scandinavia into the Iron Curtain of the Eastern USSR (totality in Kiev, Ukraine)...intersects with last total eclipse over city of Shahrud, Iran. Goes over Afghanistan, and Pakistan (totality in Quetta) before ending in India.

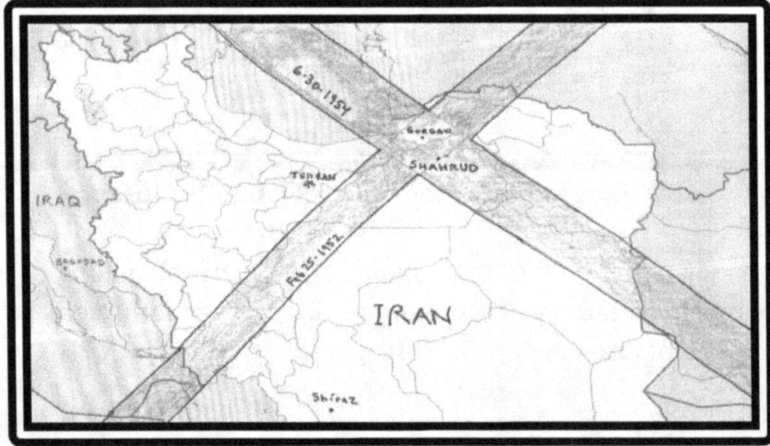

ABOVE: **Up Close Shahrud Cross 1952/1954** A Western backed coup in August 1953 installed Shah Reza Pahlavi over a democratically elected government of Iran. The Coup is nestled between successive total eclipses with intersection over city of Shahrud (literally "city of the Shah"). A good example of Cosmic Timekeeping with broad paint strokes mixed with finer details.

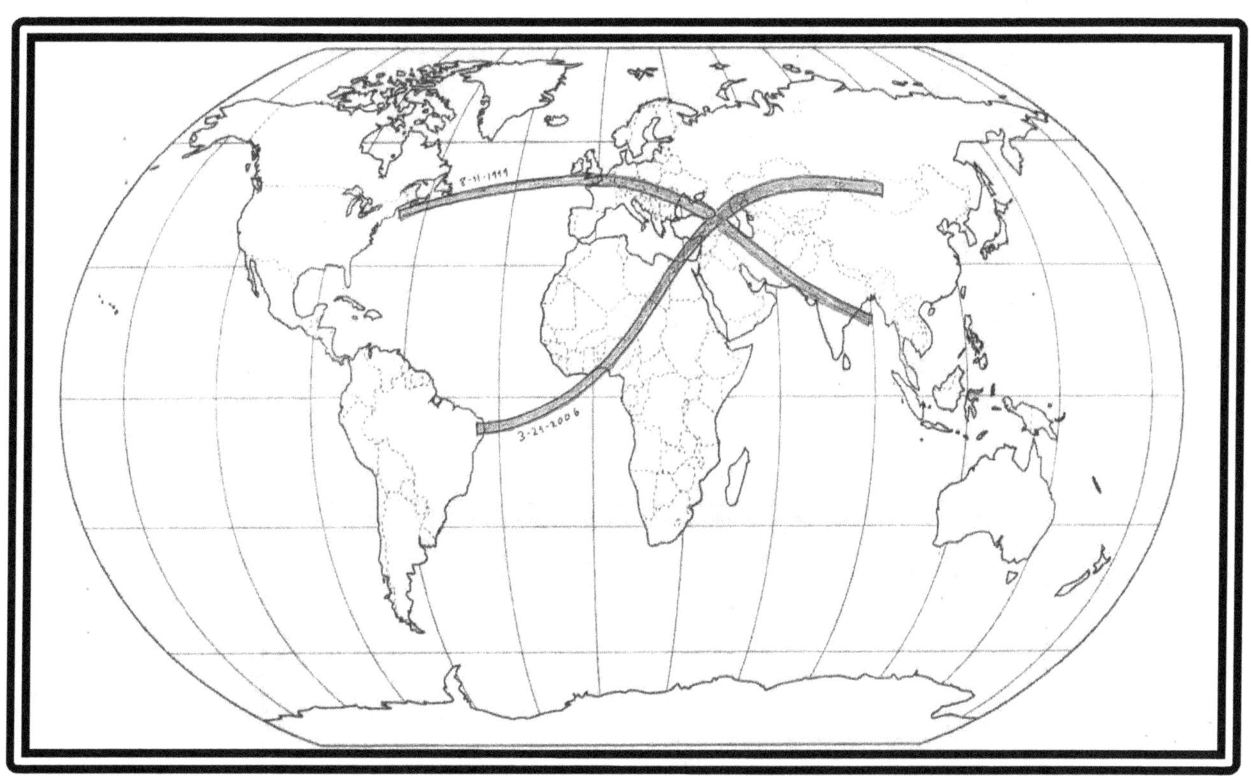

XI. Turkish Hepton Tav/Cross 1999 and 2006

Turkish Hepton Cross 1999 and 2006

1999 Aug 11 End of Millennium Total Eclipse (Saros 145) One of the greatest eclipse paths of all time begins at sunrise just offshore of the USA. Crosses Atlantic to Southern England, bisects France, goes over Germany--Hitler's hometowns of Branau Aum Inn and Munich are in totality for first time in centuries. Mozart's home of Strasbourg, Austria then to Romania, Hungary and across Turkey.

Six days later, on Aug 17 1999, the worst earthquake in Modern Turkish history devastates its western coast and kills perhaps 20,000 people, most in town of Izmit (96% totality). Eclipse continues over Iraq, bringing totality to Nineveh (Mosul) and bisects Iran, brings totality to Karachi, Pakistan. One of the greatest eclipses of the century finishes out the 20th century by crossing the entire nation of India, exiting in centerline of totality at Srikakulum, a sacred city for both Hindus and Buddhists, location of the Arasavalli Sun Temple --the "abode of the Sun god".

One of the most incredible displays of the cosmic art of the Master--our God, who made and placed the sun and moon and stars. He knows the empires as he knows the names of all the stars. The August 11 1999 eclipse has enough levels to be worthy of study for ages. More on this at world eclipses

2006 Mar 29 Total Eclipse (Saros 139) Totality begins directly over city of Natal, Brazil ("Nativity " in Portuguese. Crosses Atlantic to enter Africa at Acrra, Ghana 39 years almost to the day after Martin Luther King's historic visit there in March of 1957. King would be murdered at 39 years of age.

Path continues on across doomed nation of Libya, where it intersects the path of the eclipse of Oct 3 2005, which occurred 6 lunar months before. The path of totality goes over the Mediterranean before entering Turkey. The path heads on to bisect Georgia 20 months before its war with Russia then crosses Kazakhstan, including Totality at Capital of Nur-Sultan, formerly known as Astana. Within a month of 500th anniversary of death of Columbus. May 20 1506)

Close -Up Turkish Cross

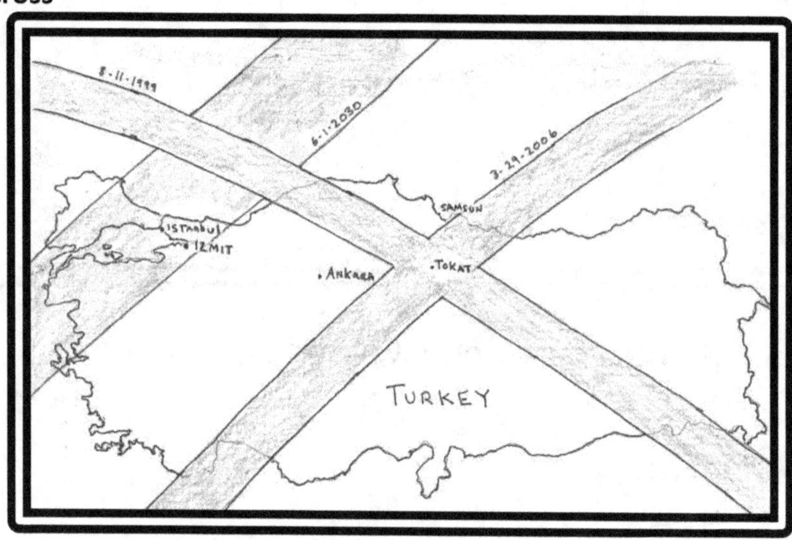

ABOVE: **Turkish Tav/ Cross** The two total eclipses of 1999 and 2006 are spaced 2422 days apart. They are Hepton eclipses exactly 7 eclipse years apart, both occurring on Wednesday. They are a true Hepton Total Cross and are the exact same eclipse personalities (Saros 139 and 145) which bring the USA Hepton Monday Total Cross of 2017/2024.

The intersection of these two great paths occurs over Northern Turkey about 200 miles east of Ankara in the region of Tokat, an ancient city from the time of the Hittites, which was also a Roman stronghold and had a thriving Armenian Christian community until "what happened" during WWI.

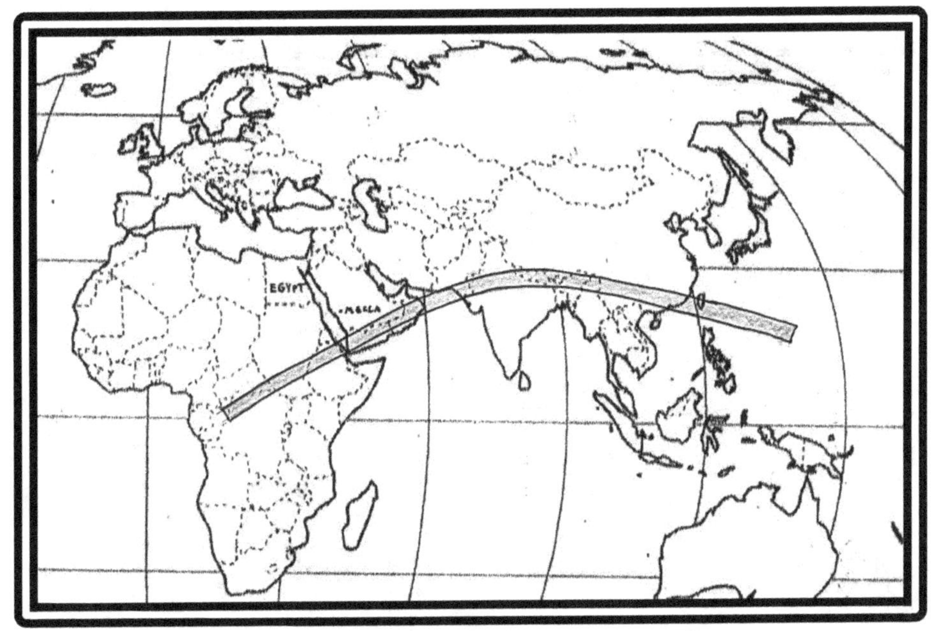

XII. Solstice Pandemic Father's Day Ring of Fire 2020

2020 June 21 Pandemic Solstice Father's Day Ring of Fire (Saros 137)

At the height of the worldwide pandemic, how about a nice Summer Solstice Father's Day Eclipse? Across the continent of Africa across the Red Sea through Yemen, Oman, India, Pakistan, Nepal, Tibet exiting through China and Taiwan. 83% annularity in Wuhan, China, received totality in 2009.

XIII.   Below: The Mideast St. John's Cross 2027 and 2034

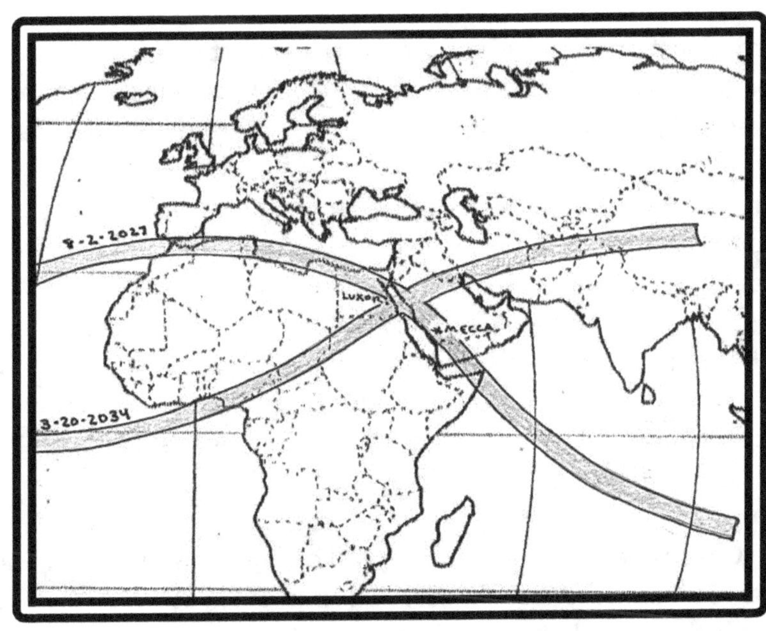

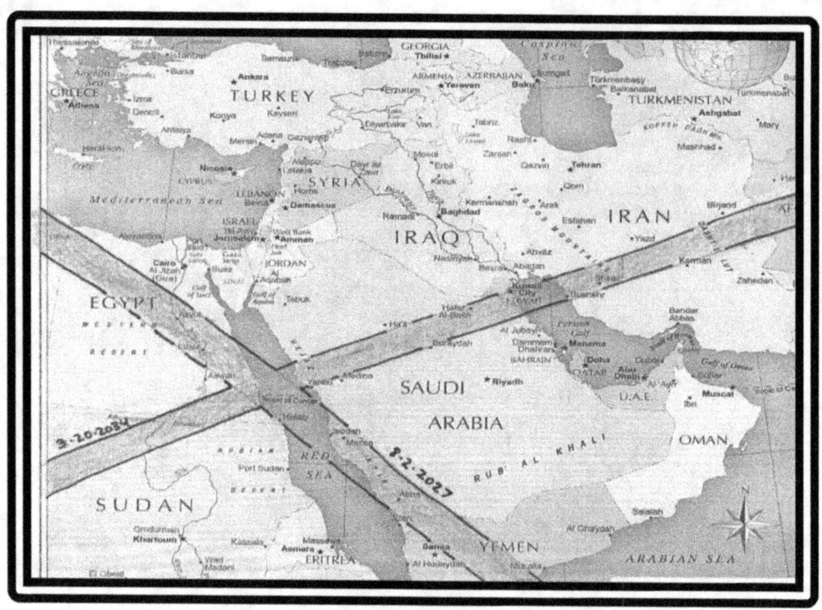

Close up of 2027/2034 Hepton Mideast Cross

Center of Apex of the two total eclipses is on the Tropic of Cancer near St. John's Island
2027 Aug 2 Valley of the Kings Total Eclipse (Saros 136) the first time the great Saros 136 appears in the 500 year survey of the Mideast, though it has been making complete eclipses around the world since 1504. A Hepton Cycle Monday Eclipse.

Exactly one lunar year (355 days) after Spanish Judgment of 2026 Aug 12.
1211 days after USA second line of Hepton Cross on 2024 Apr 8 .
3633 days from Great American Eclipse of Aug. 21 2017.
Exactly 7 eclipse years (2422 days) from Electoral College Eclipse of 2020 Dec 14.

Gibraltar, hugging the north Coast of Africa before diving down Egypt over the Valley of the Kings with Maximum Eclipse in Luxor, the necropolis of the sun gods, --into Arabia--including totality in city of Mecca.. It will be the longest eclipse on Earth, at 6 minutes and 22 seconds, until the year 2114.

2034 Mar 20 Spring Equinox Great Civilizations Eclipse Spring Equinox Total Eclipse. Judgment Eclipse. Monday Eclipse. Last possible year for 2,000th Anniversary of Crucifixion and Resurrection. 2422 days (7 eclipse years) from Mideast Total Solar of 8-2-2027. A tremendous conglomeration of converging calendars (it is new year's day in several calendars) and historical interest along the path. From the coast of Brazil (near Natal--"Nativity") back directly through the "slave coast"--Porto Novo gets its second eclipse in 7 years, across the heart of Northern Africa, just south of Luxor--into Saudi (totality scraping Medina), Kuwait crossing head of Persian gulf, across Iran, Afghanistan, Kashmir, India, Pakistan, (totality in Islamabad) , Tibet and ending in Western China.

Tropic of Cancer? Apex of Cross is near tiny St. John's Island off the coast of Egypt. This is directly on line of the Tropic of Cancer. Cancer is the Crab, or scarab beetle of Egypt which faces Leo in the Zodiac. Crab represents Egypt. Leo is the Lion of Christ. The eclipse of 2027 over Valley of the Kings features the sun and moon directly in constellation of Cancer facing Leo which holds Jupiter, the King Planet.

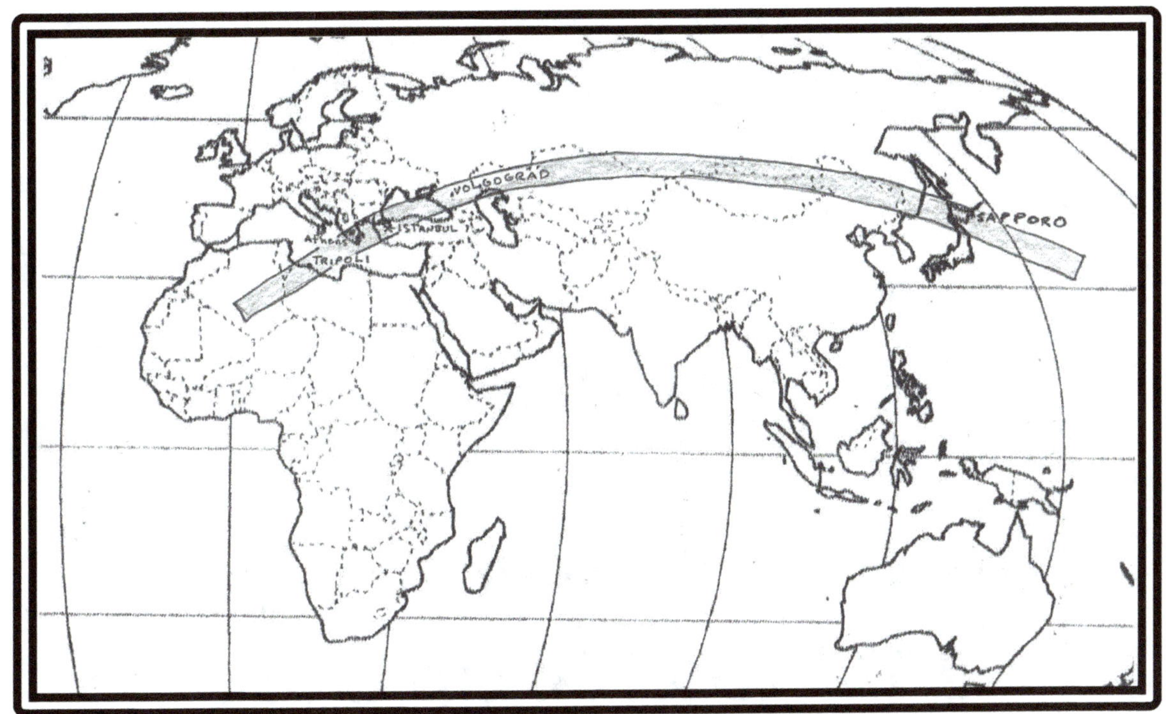

June 1 2030 Athens/Constantinople Ring of Fire.

2030 Jun 1 Athens/Constantinople Ring of Fire. Another extraordinary path beginning in Algeria and ending in Japan. Great cities along the path include Tripoli, Athens, Istanbul (Constantinople), Sevastopol, Volgograd and Sapporo. This Ring of Fire occurs 3 eclipse years after August 2 2027 Valley of the Kings.

Here's a color representation of the three eclipses of 2027, 2030 and 2034.

Chapter Six:

The Great Crossing Eclipses 2017 to 2034 Up Close

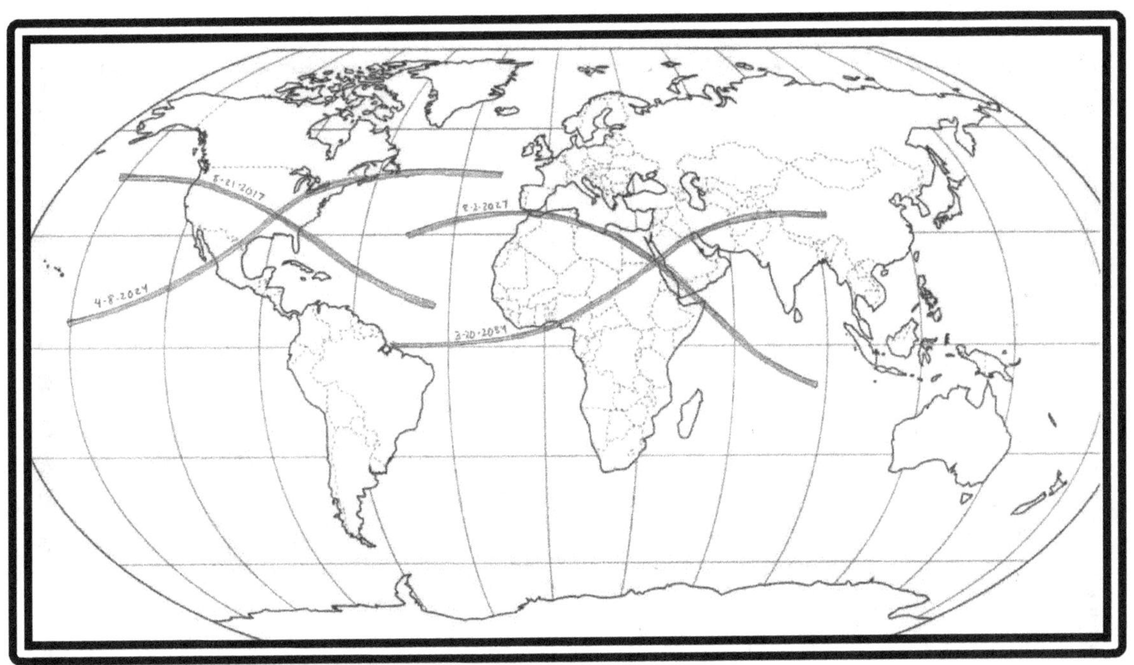

Above, The Hepton Crosses of the Spring Equinox Metonic (2017 to 2034)

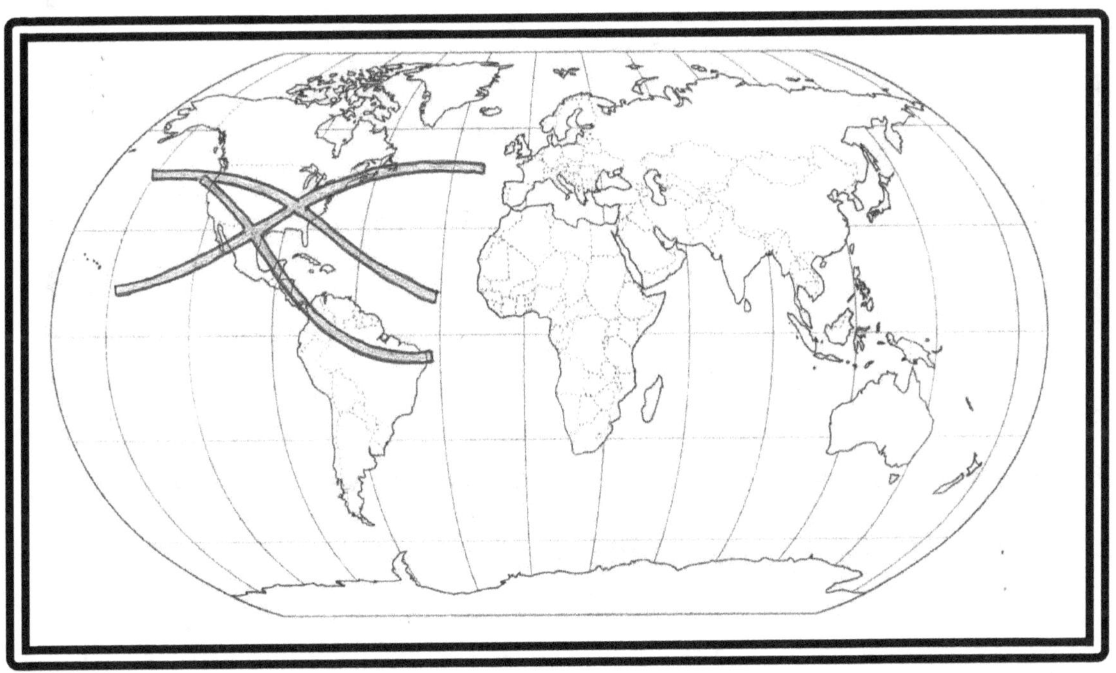

Above, the American Aleph of 2017, 2023 and 2024

Great American Eclipse 2017 Aug 21

Hailed as the "Great American Eclipse", the eclipse of 2017 Aug 21 was by far the most watched solar eclipse in American history. There was a great run-up in publicity to the eclipse, and many towns along its path held watch parties and celebrations. It has been said that the movement of people towards the narrow path of totality was the greatest single short-term migration in the nation's history, with millions positioning themselves for the sight of a lifetime, as daylight turned to darkness with the passing of the moon in front of the face of the sun.

We can use this eclipse as a test case for the idea of "reading" an eclipse like a song or a poem maybe. Three Main Aspects; The Time (History), The Space (Geography) and The Main Characters.

Astronomical facts about the Aug 21 2017 eclipse:

- Type of eclipse: Total
- Saros Number 145
- Duration of totality along center line of maximum eclipse: 2 minutes and 40 seconds
- length of path: perhaps 8000 miles long or so
- width of path, almost exactly 70 miles wide
- Locating: At eclipse, the Sun was in the paws of the lion of the constellation Leo. The star Regulus (the "King Star") was in conjunction with and clearly visible near the eclipsed Sun.
- Maximum Eclipse: Giant City State Park, near Carbondale, Illinois.

Geographical Peculiarities of the August 21 2017 Total Eclipse:

USA Divided: The USA was closely divided geographically by the eclipse. Approximately 2 million square miles of USA were to the north of centerline (including Alaska) and around 1.8 million square miles below. It was also closely divided well along the contiguous 48 states. 23 states had center line of totality either pass through them or were north of the line of totality. 25 states had the center line of totality either pass through them or were south of the line of totality.

Odd Spot of Maximum: Maximum Eclipse of the entire 9000-mile path occurred near Makanda, Illinois. Remarkably, Makanda (known as the "Star of Egypt") is also at the apex of the cross, the intersection of the two eclipses of 2017 and 2024.

Cross in the Cross: In the heart of both eclipses, 10 miles from Makanda, is the Bald Knob Cross of Peace, a 111 foot cross that was once the tallest cross in the western hemisphere.

Relation to 1918 Hepton: First cross-country USA eclipse in 99 years, since June 8 1918. That eclipse had a similar path but slightly convex to this. It featured totality at Mount Saint Helens. The 1918 eclipse was also a Hepton eclipse with a corresponding US eclipse exactly 2422 days later, on Jan 24 1925. Those paths did not touch.

Rare Cross-Country: It was one of only three Pacific to Atlantic USA total solar eclipses since the nation's founding (2017, 1918 and 1806)

Only in the USA: The eclipse touched no other national boundaries besides the USA. The last solar eclipse to have that geographic distinction occurred in 1257 AD (519 years before nation's founding) and it will not happen again until 2316 (199 years after the 2017 eclipse).

Touched most states ever: The eclipse touched 14 states, the most ever up to that time for an American Total eclipse. The April 8 2024 American Eclipse exceeds it, touching 15.

Near the center of the USA: The line of totality went within 50 miles of the geographic center of the contiguous USA (Lebanon, Kansas at 98.4% totality) and within a hundred miles of the population center of the US (Hartville, Missouri at 97.5% totality)

Nice Weather: The weather was exceptionally fair for most of the nation, making for good viewing.

Extended Peculiarities of the August 21 2017 Eclipse

First in 38 years: The 2017 Total Eclipse was the first in the contiguous USA in 38 years. The last one before 2017 was on 1979 Feb 26, with totality in the northwest, including Mt. Saint Helens one year before her eruption. The 2017 eclipse brought 98.4 % coverage to Mt. Saint Helens.

Seven Cities named Salem: Salem, Oregon (the capital of Oregon) was the first major city in the US to see totality. Salem is shorthand for Jerusalem, means "peace" in Arabic and is related to the Hebrew "shalom". The eclipse path passed over 7 towns named Salem in The US. They were nearly evenly spaced along the line —in Oregon, Wyoming, Idaho, Nebraska, Missouri, Kentucky and South Carolina

Echoes of Government and Lincoln: The eclipse was first visible on US soil at Government Point, Oregon. The first town in totality was Lincoln City, Oregon, It also went over Lincoln Point Wyoming and Lincoln Nebraska (the state capital) --99.6% at Lincoln Missouri and 98.4% totality at Lincoln's Birthplace in Kentucky.

Nashville: First total eclipse over area of Nashville, Tennessee since 1478. But, of course everyone knows real country music had already been eclipsed for some time. Country Music Fact: #1 country song at time of eclipse. *"Body like a Back Road"* (*"I know every curve like the back of my hand"*)

Leaves USA at Slave Port of Charleston: Eclipse exited the USA with center line at Charleston, South Carolina. Charleston was the largest slave port in North America. More than 200,000 Africans passed through the port of Charleston in chains, between the founding of the Carolina colony in 1670 and the prohibition of the Trans-Atlantic slave trade in 1808. Interesting juxtaposition entering US at Salem and exiting at Charleston.

Interesting Historical Yearly Anniversaries of 2017

- 500th anniversary of Protestant Reformation (Began 1517 Oct 31)
- 500th anniversary of first defeat of Spanish Conquistadors in the Americas at Battle of Champton in what is now Mexico (1517 Mar 25)
- 500th Anniversary of first official European delegation to China (1517 Aug 15)
- 100th anniversary of Balfour Declaration proclaiming Jewish homeland (1917 Nov 2)
- 100th anniversary of US entrance into World War One (1917 Apr 6)
- 100th Anniversary of Russian Bolshevik Revolution (Oct 1917)
- 100th anniversary of birth of John F. Kennedy (1917 May 29)
- 50th anniversary of Jewish control of Jerusalem (6-7-1967)
- 50th Anniversary of American "Summer of Love" (1967)

Sevens and 70's

- 7 months exactly after Donald Trump's first full day in office (1-21-2017), who won the presidency by 77 electoral votes, and assumed his first full day when he was 70 years, 7 months and 7 days old. (Trump was born the day of a lunar eclipse Jun 14 1946)
- 70th anniversary of beginning of Israeli war of modern nationhood (November 1947)
- 70th anniversary of the discovery of the Dead Sea Scrolls (early 1947)

- 70th anniversary of founding of CIA (1947 July)
- 70th anniversary of Roswell "Flying Saucer" incident (7-7-1947)
- The eclipse path was 70 miles wide.
- Saros 145, the character eclipse, is composed of 77 events. 2017 Aug 21 is its 22nd.
- 2017 was Hebrew Year 5777.

Historical Calendar Peculiarities—Calendar date August 21st

Two other great eclipses occurred on the date of August 21 in the 20th Century.

August 21 1914 World War One (Saros 124)--One of the most clearly ominous eclipses in modern history. Fighting begins in worst slaughter in human history to that point just as total solar eclipse divides Europe, Ottoman Empire, and Russia.

August 21 1933 Jerusalem Ring of Fire--As Nazis seize control in Germany, ring of fire eclipse in Jerusalem and Baghdad (Babylon).

Saros 145 Recent History and Character

Saros are repeating characters (every 18 years 11 days and 8 hours- or 19 eclipse years). Saros 145 creates the August 21st 2017 Great American Eclipse. Here are last 5 eclipses for Saros 145 before 2017- see them go backwards 18 years at a time:

August 11 1999- Europe/ Mideast/ South Asia---Saros 145 created the last complete solar eclipse of the second millennium. One of the greatest eclipses of all time. Totality from Nineveh, Iraq to Hitler's birthplace in Austria to France and England.

Jul 31 1981 USSR-The last decade of the Soviet Union's power begins with a cross-country eclipse that touches no other nation.

Jul 20 1963 Japan ***to USA-*** 18 years after atomic bombs ,total eclipse begins in Japan six years to the day before moon landing

Jul 9 1945 *USA to USSR Atom Bomb Eclipse-*Total solar eclipse travels from the USA to USSR one week before the detonation of the first atomic bomb.

6-29-1927 England to USA Beatles Premonition Eclipse – Liverpool England to USA exactly 14 eclipse years (4851 days) before the birth of the Beatles oldest member, John Lennon, on 10-9-1940. All the Beatles will be born in Liverpool.

What's next for Saros 145? It is scheduled to return on September 2 2035, when it brings totality for the first time in centuries to Beijing, the capital of China. Totality also in Pyongyang, North Korea and north metropolitan Tokyo.

Great American Ring of Fire 2023 Oct 14

Astronomical facts about the eclipse:

- Type of eclipse: Annular (Ring of Fire)
- Saros Number 134
- Duration of totality along center line of maximum eclipse: 5 minutes 17 seconds
- length of path: perhaps 8000 miles long or so
- width of path, roughly 125 miles wide
- Zodiac: in constellation Libra

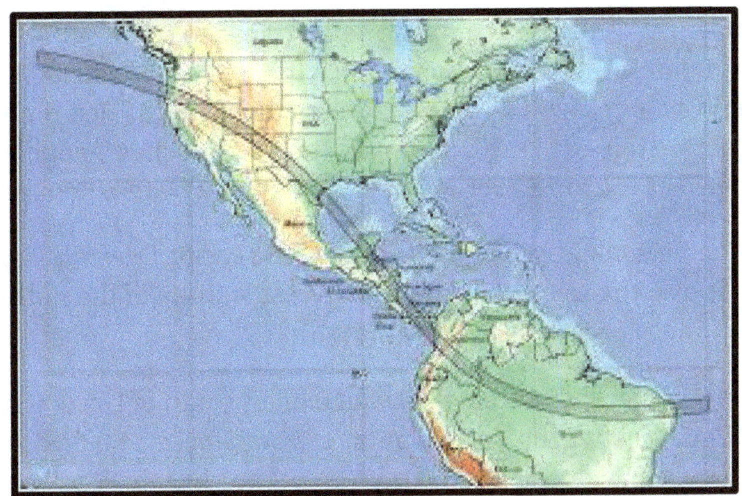

Great American Ring of Fire October 14, 2023
Map courtesy of eclipseophile.com

2023 is the Second of Three American eclipses in Seven Years

The above illustration shows how balanced the path itself appears in relation to North and South America. It is the most balanced eclipse path between the continents since Europeans arrived in 1492.

October 14 Facts and Peculiarities

Almost Columbus Day: October 14 2023 is 531 years and two days from the "discovery" of the Americas by Christopher Columbus. 48% sun coverage at his landing site in San Salvador Bahamas.

Anniversary of First Crime of Conquistadors: Oct 14 2023 is the exact 531st anniversary (Oct 14 1492) of Columbus' abduction of 7 native youth to serve as guides; the first colonial crime in the Americas.

The Metonic just left Columbus Day: The last complete eclipse on this eclipse Metonic occurred on Oct 12 1977, the 485th anniversary of Columbus Day. That total eclipse began over open ocean and traversed no land until it crossed Colombia, the only nation on earth named for the explorer.

Eisenhower vs. Hitler Rematch: 2023 Oct 14 is the 133rd Birthday of Dwight Eisenhower, born October 14 1890. Another complete eclipse occurred six months earlier on Adolf Hitler's 134th birthday on April 20. He was born April 20 1889. Eisenhower and Hitler were enemies in World War Two.

Peculiar distance from Mayan Apocalypse: 2023 Oct 14 is exactly 555.5 eclipse years from the "Mayan Apocalypse" total eclipse of August 8 1496 (Saros 127) which brought terror to the Aztec and Mayan tribes just as Spanish Conquistadors were arriving.

Extended Peculiarities of October 14 2023 Ring of Fire

North and South America: This is the first complete solar eclipse to connect the USA and South America (including full annularity at Panama Canal) since the USA's founding in 1776. The path connects North and South America with a complete Solar Eclipse for the first time since Oct 13 1632—391 years and one day before this Oct 14 2023 eclipse. The path touches 9 nations, a difficult geographic tightrope, perhaps the most ever for an eclipse in the Western hemisphere.

Latitude coincidence: The beginning spot of the eclipse in the Pacific Ocean is approximately 49.5 degrees latitude, which is also the same as the northernmost latitude of the northern boundary of the contiguous USA, 49.5 degrees at Angle Inlet, Minnesota.

Lincoln Anniversary: The 2023 eclipse enters the USA 10 miles south of Lincoln City Oregon exactly 167 eclipse years (57,879 days) after the assassination of Abraham Lincoln on April 14/15 1865.

Oregon: Newport, Oregon and a few other small settlements occupy the only tiny strip of land to receive complete eclipses in both the 2017 Aug 21 total and 2023 ring of fire solar eclipses. Salem just misses annularity. Eugene is in path.

Nevada: Annularity misses Las Vegas by a hundred miles (but still has 82 % coverage).

Utah: The eclipse crosses Utah, staying south. where it passes over several national parks and monuments including Bryce Canyon, Capitol Reef, Natural Bridges, Canyonlands and Bears Ears. Only small towns are in the path.

Colorado: Extreme SW Colorado in path, including Cortez. Anasazi ruins of Mesa Verde are in path.

Navajo Nation: Eclipse path crosses 3 of the four sacred Navajo Mountains ("corners of the Navajo Universe")- Navajo Mountain in Utah, Mount Taylor in New Mexico and Hesperus in Colorado. Passes over Window Rock, Arizona—capitol of the Navajo Nation.

Path covers the entirety of **Lake Powell** in Utah, site of the damming of the Colorado River canyon system, perhaps the greatest ecological catastrophe in the history of America.

It passes over the center of the cross of the **Four Corners**, the only place in the USA (and the world?) where state political boundaries form a cross.

New Mexico. Annularity includes Chaco Canyon; ceremonial center of the lost Anasazi civilization. Acoma Pueblo is in Annularity, the oldest continuously inhabited village in North America (since 1150 AD). Eclipse passes over Santa Fe—the oldest and highest capitol city in America. Annularity in Albuquerque. Path crosses Los Alamos NM Laboratory (birthplace of Atom Bomb) and Roswell, NM where "flying saucer" event occurred in 1947, and real space aliens have shops on Main Street.

Oct 14 2023 in Texas:

Midland Again: Path heads over West Texas, crosses Midland, home of George HW Bush's oil business in the 1950's and the boyhood home of George W. Bush. Second USA annular eclipse in a row to touch Midland (and third in a row to directly impact it).

Columbus and Midland: May 20 2012 ring of fire ended at sunset over Midland, making Columbus Death day (May 20) and Columbus Crime Day (Oct 14) anniversaries enshrined in consecutive Ring of Fire eclipses both crossing Midland. The two Midland rings of fires are exactly 12 eclipse years apart.

Intersection with 2024 Total: The path intersects with upcoming 2024 April 8 eclipse path in the hill country west of San Antonio. Apex of intersection is just north of the town of Utopia. Exact apex appears to be Sabinal River south of Lost Maples State Park. Sabinal means "Cypresses" in Spanish. Exact Apex is near Piece of Heaven RV park and Hacienda Del Sol (Plantation of the Sun) Cabins.

Uvalde: The town of Uvalde, where 19 children were massacred in 2022, is in path of both eclipses.

San Antonio: is in Annularity, the largest city in the nation to receive annularity in this eclipse.

The Body of Christ: Perhaps most striking of all, precise centerline of eclipse path exits the USA directly over Corpus Christi, or "Body of Christ", one of the only towns named that in the world.

Central America: Annularity in every nation in Central America except El Salvador (the Savior).

South America: Of course, being a Columbus Anniversary Eclipse, the path bisects the nation of Colombia 46 years and two days after their unique Columbus Day eclipse of Oct 12 1977

Brazil: The Ring of Fire crosses the northern entirety of the Brazilian state of Amazonas to exit the Americas at city of Natal (Nativity), very near the extreme northeastern tip of South America, the closest spot in the Americas to Europe. The eclipse also covers city of Joao Pessoa, the easternmost spot in all the Americas and the 3rd oldest city in Brazil.

Saros 134 Recent History and Character

Saros are repeating characters (every 18 years 11 days and 8 hours- or 19 eclipse years). Saros 134 creates the October 14 2023 Great American Ring of Fire.
Here are the last 5 eclipses for Saros 134 - see them go backwards 18 years at a time:

- **2005 October 3-Madrid Ring of Fire**: Madrid Spain directly in Totality, along with Algiers...path cuts through Africa, including Kenya.
- **1987 Sep 23- USSR Fall Equinox:** 7 weeks before Estonia became the first state of USSR to declare its independence. Eclipse also over China.
- **1969 September 11- South America:** Just missing the northern border of Chile, where a CIA coup ousts and kills President Salvador Allende four years later on September 11th 1973. 9-11 Metonic's first appearance in centuries.
- **1951 Sep 1 -Jamestown, Virginia to Africa Ring of Fire:** Eclipse path traces path of the first slaves brought by British colonists on 8-20- 1619 back to their homeland of Angola 331 years and 11 days later -see also **7-11-1619 Luanda Angola Eclipse**
- **1933 Aug 21 Jerusalem Ring of Fire** -mentioned earlier.

Great North American Eclipse: April 8 2024

Astronomical facts about the eclipse:

- Type of eclipse: Total
- Saros Number 139
- Duration of totality along center line of maximum eclipse: 4 minutes and 27 seconds
- length of path: perhaps 10,000 miles long
- width of path: 124 miles wide
- Maximum eclipse occurs near Nazas, Mexico
- Locater: Sun and Moon eclipse in Pisces between Mercury and Venus

Peculiarities Apr 8 2024 Total Eclipse

Easter: Occurs eight days after Easter, the closest any solar eclipse can astronomically occur to Easter. Easter is March 31, the first Sunday after full moon after Spring Equinox.

Buddha Birthday: April 8 is traditional birthday celebration of Buddha (Japan)

All three Nations of North America: April 8 2024 eclipse is the first total solar eclipse to cross all three nations of Mexico, USA and Canada overland since their founding. You have to go back to 664 AD for any path like the 2024 path. It will be the last total eclipse to touch all three nations until 2316. It is the only total solar eclipse to travel south to north overland border to border in US history.

The Most states ever for Total: The 2017 USA eclipse touched 14 states, the most states of any total eclipse in US history. The 2024 eclipse outdoes it by one, touching 15 states.

Last US Eclipse for a long While: It will be the last complete solar eclipse in USA for 20 years (until Aug 22 2044). This will be the longest stretch without an eclipse in USA since 1782 to 1806 24-year stretch.

New Madrid Fault: Totality over New Madrid, Mo, epicenter of largest earthquakes in USA history (1811/1812)

The Most People Ever in USA: The eclipse will cover far more metropolitan cities and go over more Americans in one path than any total eclipse in the nation's history.

Lots of Cities: Just some of the cities in the path of Totality-: Del Rio, Dallas-Ft. Worth, San Antonio, Austin, Waco, Texarkana, Little Rock, Terre Haute, Indianapolis, Dayton, Akron, Cleveland, Buffalo, Syracuse, Rochester, Burlington, Montpelier.

Near the Center again: The path passes within 50 miles of the geographic population center of the USA (Hartville, Missouri—with 99% totality)

Second line of Tav: It forms a balanced curved cross or x (Hebrew Tav) with the August 21 2017 eclipse

Mexico and Canada too: Crosses more metropolitan populations across nations of Mexico. USA and Canada than any other complete solar eclipse in the history of the western hemisphere.

Mexico: the eclipse first enters the North American continent with centerline directly over resort city of Mazatlán, it also traverses cities of Durango and Torreon ("tower"), site of the 3rd tallest statue of Christ in Latin America. Maximum Eclipse (4 minutes 27 seconds) of entire eclipse occurs near city of Nazas, an Arabic word for evil eye and shorthand for Nazareth. The eclipse exits Mexico at Ciudad Acuna to enter USA at Del Rio.

Canada: the eclipse scrapes the northern Toronto Metro area, and fully covers Montreal (founded in 1642 as Villa Marie or City of Mary).

Exits at St.John's: Eclipse exits North American north (99.24% eclipse) at St. John's, Newfoundland. St. John's is the oldest city in Canada, and one of the oldest European settlements in the New World, founded by the British in 1497. Saint John as in John the Apostle, writer of the Book of Revelation.

First Canadian Total in 45 years: This is the first Total Eclipse visible from Canada since February 26, 1979, more than 45 years.

Specific US Geographical Peculiarities of the April 8 2024 Total Solar Eclipse

Let's trace the path from Texas to Maine and look at peculiarities in each of the 15 states

Texas Historical and Geographical Peculiarities April 8 2024

Big Texas Cities: Path manages to pass over 3 of the four largest cities in Texas—San Antonio, Austin and Dallas-Fort Worth. Together they are home to more than 12 million people. Also in the path are several other large cities, including Plano, Del Rio, Waco, and Texarkana.

Texas Tav: the two eclipses of 2023 Oct 14 (Ring of Fire) and 2024 Apr 8 make a geographical X over south Texas. Texas is the only state in the USA with the letter x in its name. The center of the X of those two paths is near town of Utopia.

Uvalde Again: The eclipse passes over the town of Uvalde, site of the massacre of 19 schoolchildren in 2022 just as the 2023 Ring of Fire does.

Odd Waco Anniversary: Eclipse passes over the site of the Mount Carmel Branch Davidian Massacre in Waco, Texas, where 76 people, including 25 children, were burned to death exactly 30 solar years plus one lunar year before this eclipse (30 Years 11 months and 20 days) on 1993 April 19.

Dallas and JFK: The eclipse passes directly over Dallas, their first total solar eclipse since 1878 and only their third total eclipse in nearly 500 years. Dallas is the site of the most notorious murder in modern American history, the assassination of John F. Kennedy on 1963 Nov 22. Total eclipse occurs in Dealey Plaza, site of the shooting, 60 years and 137 days later at 1:40 pm Daylight Savings Time, very near the time of the shooting, which was at 12:30 pm Standard Time.

LBJ in the Path: The eclipse passes over both the birthplace/boyhood home (Stonewall, Texas) and death place (Johnson City, Texas) of Lyndon B. Johnson, Kennedy's successor to the Presidency.

Interesting anniversary of JFK Eclipse: The eclipse of April 8, 2024 is separated by exactly 110 eclipse years (which is 220 eclipse seasons) from the Kennedy Death Premonition Eclipse of 11-22-1919. That amazing 1919 eclipse occurred exactly 44 years before the murder of JFK and passed over both the Texas Capital of Austin and Stonewall, Texas, home to a then ten-year-old Lyndon B. Johnson, who would replace Kennedy on 11-22-1963. See USA Eclipses and Kennedy section.

More Bush in the Path: The Eclipse passes over ranch of George W. Bush in Crawford, Texas and his family's current residence in Dallas. The preceding Ring of Fire 177 days earlier on 2023 Oct 14 passes directly over his boyhood Home of Midland/Odessa Texas.

Space Shuttle Columbia: The eclipse passes over site of Columbia Space Shuttle disaster near Dallas on 2003 Feb 1, during George W. Bush's Presidency, seven weeks before the US invaded Iraq in March of 2003. Columbia is the female national personification of the United States, a historical name also applied to the Americas and to the New World.

Arkansas and Clinton: Totality over the birthplace/boyhood home of President Bill Clinton (Hope) as well as his adult and governor's home in Little Rock. Also pPasses over the town of Clinton, Arkansas.

Tennessee sees totality in only a few square miles, touching only a tiny corner of the state, the port of Cates Landing, one side of an oxbow of the Mississippi River across from New Madrid, Missouri...epicenter of the great New Madrid earthquakes (Dec 16 1811 to Feb 7 1812)

Little Egypt Apex: Sections of Missouri, Illinois, and Kentucky will see their second total eclipse in 7 years. That area is known as "Little Egypt". Center of the apex of the USA total eclipse cross is very near the township of Makanda, Illinois—once known as the "Star of Egypt". The exact apex appears to be on the east side of Cedar Lake at Salem Road.

Missouri is the state crossed most dramatically by the x created by the two eclipses of 2017 and 2024. The 2024 eclipse touches no major cities in Missouri, though many smaller towns.

Kentucky: Paducah is eclipsed for the second time in 7 years, the largest city in USA to receive totality in both eclipses. Farther to northeast, Abe Lincoln's birthplace receives 98% coverage.

Illinois and Bald Knob Cross: Carbondale and Chester Illinois receive totality for the second time in seven years. As in the 2017 eclipse, the path passes over Bald Knob Cross, near Alto Pass Illinois, once the largest cross in North America and still one of the biggest at 111 feet tall. It is situated on a hill at the highest point in southern Illinois, near Makanda.

Indiana: Most of the state sees Totality, including Columbus, birthplace of Trump's Vice-President Mike Pence. Also in path is Indianapolis, home of ex Vice-President Dan Quayle and politician Pete Buttigieg. Totality over site of Battle of Tippecanoe (1811 Nov 7) in Battle Ground, Indiana at Prophetstown State Park. That battle was one of the turning points in the wars for domination of the American Indians. Prophetstown was founded by Tecumseh, the great Shawnee warrior and Chief.

Ohio: the eclipse hits or scrapes nearly all major cities (totality in Cleveland), though barely the north Cincinnati Metro and bisecting Columbus Metro area. Eclipse occurs exactly 233 eclipse years after the founding of Ohio on 1803 Mar 1. Crosses site of Kent State killings of 1970 near Akron. Ohio was the home of 8 US Presidents.

Mormon Associations: In **Ohio**, path goes over Kirtland, the early home of the Mormon Church and Joseph Smith from 1831 to 1837. Path goes over **Hiram, Ohio** where Joseph Smith was famously tarred and feathered. Eclipse Totality goes over **Manchester, New York,** where Smith claimed to have found the golden tablets from which the Mormon religion is founded. The path very nearly goes over Joseph Smith birthplace of **Sharon, Vermont** (99.13%). 97% coverage at Brigham Young Birthplace of **Whitingham VT**.

Michigan sees even less totality than Tennessee, and only the barest few square miles experience total eclipse, in the area of Lost Peninsula.

Pennsylvania: crosses city of Erie. 94.7% at biden birthplace of Scranton.

New York State: In New York, the path follows entire northern border of the state. Totality at Niagara Falls. Centerline of totality in Rochester, Buffalo and Watertown, home of Allen Dulles who headed the CIA from 1953 to 1961. Dulles was born April 7 1893, 131 years and one day before eclipse.

Vermont: Total eclipse in Vermont's two largest cities, Burlington and Montpelier.

New Hampshire: Crosses the little populated extreme north.

Maine: Maine is one of the states most crossed by eclipses in the history of USA. Caribou, Maine is the last major town in the USA to experience totality in the 2024 eclipse.

US/ Canada Border: The eclipse path straddles the USA/ Canada Border for nearly 700 miles-from Toledo Ohio to just south of Quebec City Canada. The path exits Maine, about 100 miles from the easternmost point of the USA, then travels across extreme eastern Canada. The eclipse will have spent only about an hour and a half crossing North America.

Date Peculiarities of April 8 2024 and 2024 in general

Monday: Like 2017 Aug 21 USA Eclipse and both Mideast Cross eclipses, this eclipse also occurs on Monday. It is a Hepton Cycle Eclipse.

Anniversary of Titanic Eclipse: The Titanic Hybrid Eclipse of 1912 Apr 17 (60 hours after the sinking) and the 2024 Apr 8 total eclipse are exactly 118 eclipse years apart—40,902 days

Anniversary of New World Eclipse: The New World Eclipse (1492 Oct 21), which occurred nine days after Columbus landed in the Americas and the 2024 Apr 8 eclipse are exactly 560 eclipse years apart (194,104 days). Eclipse coverage in 2024 is 31% at Columbus landing site in Bahamas

500th Anniversary of sighting of Manhattan: Apr 8 2024 is nine days shy of exactly 500 years from the sighting of Manhattan Island by the explorer Verrazano on Apr 17 1524. New York City sees 90% coverage in the 2024 Eclipse. **400 Years** since the first Dutch colonists arrive in New York (1624)

100th Anniversary of end of Indian Wars: 2024 is 100 years after the final US Army skirmish with the Apache tribe, which ended more than 400 years of the North American conquest. (1924)

The character of Saros 139 History and Peculiarities

Saros 139 creates the April 8 2024 Great American Eclipse. **Here is the recent past of Saros 139,**

Turkish Cross March 29 2006: Began at sunrise over east coast city of Natal ("Nativity"), Brazil— It crossed the Atlantic, Africa, Turkey and Georgia, Kazakhstan, including totality at their unusual Capital of Nur-Sultan, home of the 203 ft Palace of Peace and Reconciliation, a pyramid devoted to reconciliation of the world religions. Paired with **Aug 11 1999 End of Millennium Eclipse.** The extraordinary design they create is nearly identical to the one they recreate over America between 2017 and 2024. See next page.

1988 Mar 18 Total eclipse in Indonesia, Malay and southern Philippines. Mount Pinatubo of the Philippines erupts 3 years, 3 months and 3 days later in the 2nd largest eruption of the 20th century.

1970 Mar 7 End of 60's USA eclipse... Total Eclipse across southeastern seaboard of USA. 93% totality at Cape Canaveral/ Cape Kennedy 234 days after first rocket to the moon. Totality at Charleston South Carolina, the greatest Slave Port in North America (totality again in 2017). The end of the sixties.

1952 Feb 25 Iranian Shahrud Eclipse ---one of the eclipses that make up the great **Shahrud Iran Cross of 1952 and 1954**, bordering the US led coup that established the reign of Shah Reza Pahlavi (see historical eclipses for more info).

Turkish Aleph formed by Saros 145, 134 and 139 as they appeared 1999 to 2006.

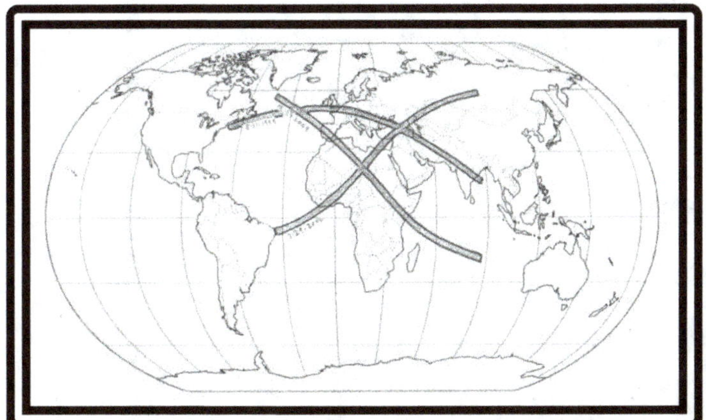

At left: Turkish Alef 1999 to 2006
Compare with American Aleph in color below formed
by same Saros members

American Aleph 2017, 2023 and 2024

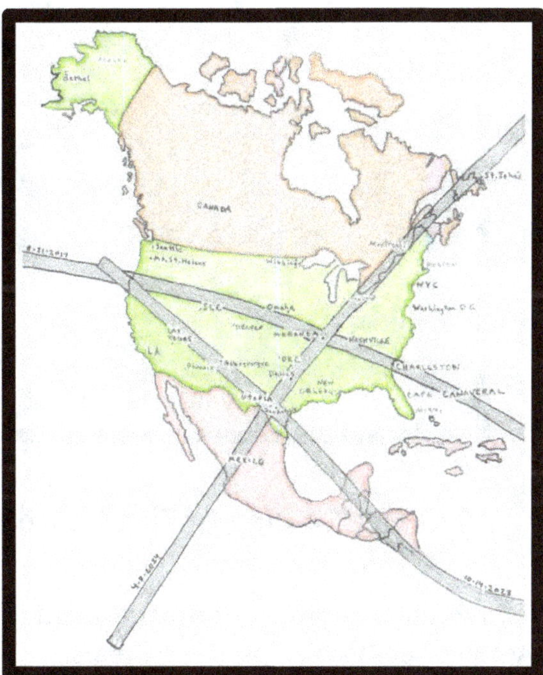

The Texas Tav: 2023 and 2024

2023 Oct 14 Ring of Fire (Saros 134) and 2024 Apr 8 Total (Saros 139)

Two extraordinary eclipses intersect in extraordinary fashion over South Texas.
San Antonio sees two eclipses in six years.

Exact apex apparently near Lost Maples State Park; near Texas town of Utopia, in the hill country west of San Antonio.

American Total Hepton Cross 2017 and 2024

Two unique total eclipses- Aug 21 2017 and Apr 8 2024- cross exactly 7 eclipse years apart over the area known as Little Egypt. Apex near the town of Makanda ("Star of Egypt") 8 days after Easter at the most critical time in the history of the United States. These two eclipses are separated precisely at midpoint by a total eclipse in the Southern Hemisphere on Dec 14 2020 (not pictured), which occurs simultaneously with the electoral college of the USA electing the Biden/Harris ticket on the 221st anniversary of George Washington's death.

The next Hepton Cross begins in Egypt at the Valley of the Kings 1211 days later on August 2 2027.

The Great Mideast Cross 2027 and 2034

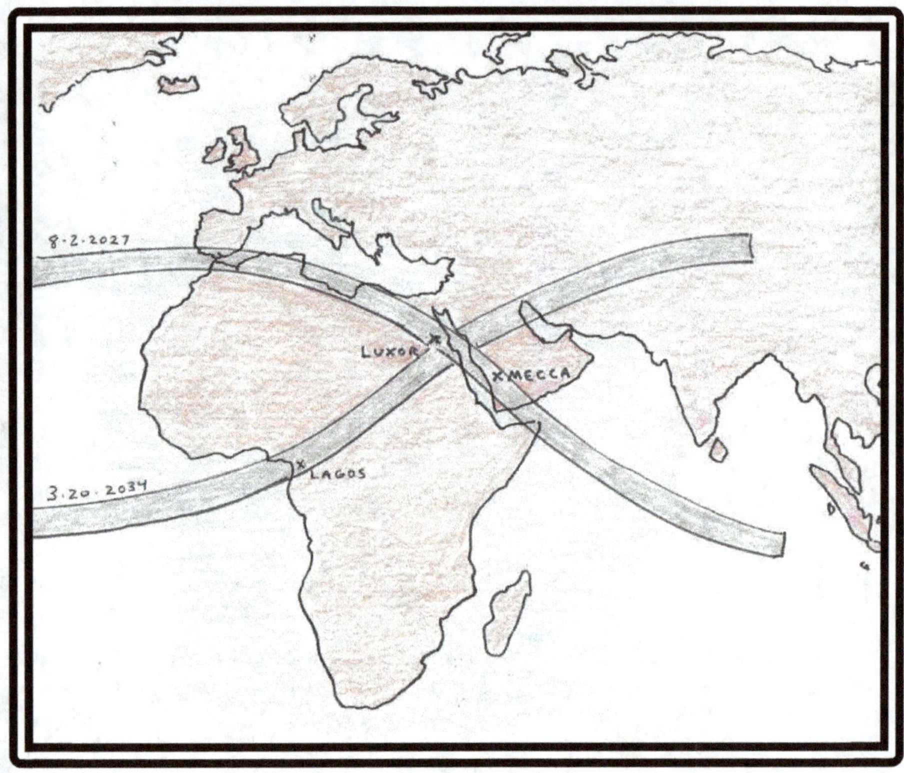

The Mideast Cross is formed by two Hepton Cycle eclipses of Aug 2027 and Mar 20 2034. They are part of the Hepton Cycle which creates the two Total American eclipses we just studied but are different Saros personalities. Like those two eclipses, these two happen on Mondays.

The precise apex of the two paths is southeast of the Valley of the Kings in the Red Sea near St. John's island almost precisely on the Tropic of Cancer. This cross mimics the nearly identically shaped Hepton Cross formed by the American eclipses of 2017 and 2024.

Like the American cross, these two eclipses are 2422 days apart and both appear on a Monday. Also, like the American cross, they are separated at the 1211-day midpoint by another significant Total Eclipse in the same hemisphere, in this case the Nov 25 2030 British Empire Eclipse from South Africa to Australia, which we will study later. They also have a significant Ring of Fire event in their midst, in this case a Jun 01 2030 Ring of Fire through all Eurasia, including Athens and Istanbul.

The 2027 Valley of the Kings Eclipse is 1211 days from the 2024 Easter North American Eclipse.

There are great peculiarities in the relations of all these eclipses, but first we will take a quick look at the individual eclipses themselves.

2027 August 2 —The Valley of the Kings and Mecca

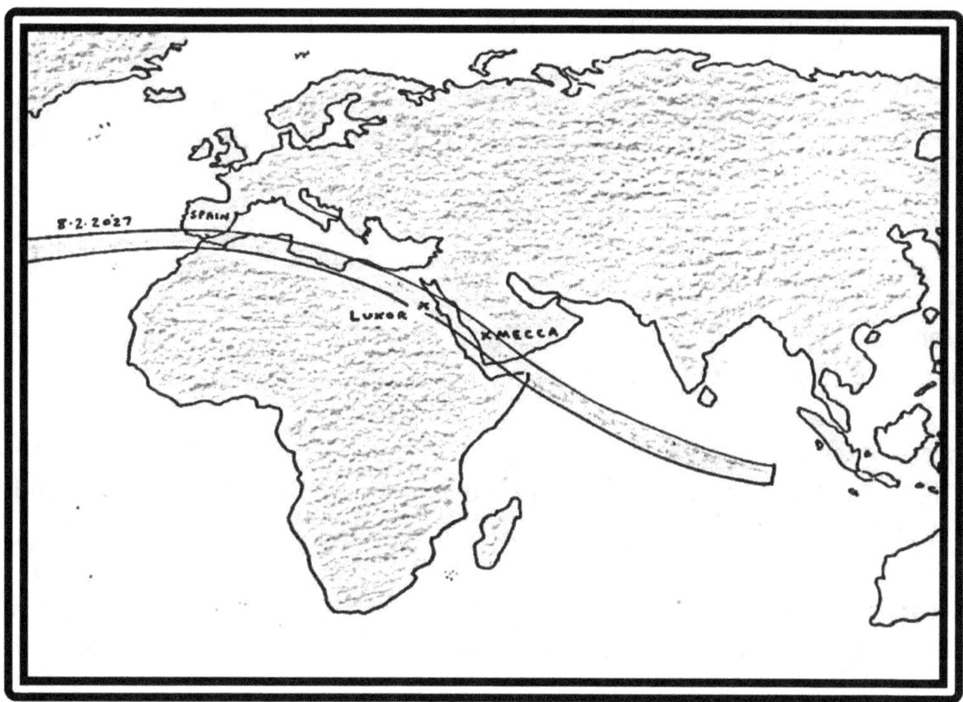

Introduction

The most powerful Saros Character of them all, Saros 136, which is creating the longest eclipses of the last thousand years, makes a powerful statement across the center of the world.
Maximum Eclipse occurs almost directly over Luxor, Egypt the Valley of the Kings, perhaps the greatest symbol of ancient Imperial Power, magic arts and mystery religion in the history of the world.

Prior to reaching there, the path straddles the Straits of Gibraltar and follows the coastline of North Africa. After exiting Egypt in the Red Sea, it passes directly over the Holiest City of Islam, Mecca.

Not long after, it crosses Yemen and the Gulf of Oman. It exits land at the easternmost tip of Africa at the Horn of Somalia and dies at sunset in the Indian Ocean.

Astronomical facts about the Aug 2 2027 eclipse:

- Type of eclipse: Total
- Saros Number 136
- Duration of totality along center line of maximum eclipse: 6 minutes and 23 seconds—this will be the longest eclipse on Earth until 2114
- length of path: perhaps 8500 miles long
- width of path, 160 miles wide
- Maximum eclipse occurs 37 miles southeast of Luxor, Egypt
- Zodiac Locater: Eclipse occurs in Cancer

Geographical and Historical peculiarities of August 2 2027 Total Eclipse

Two continents at once: Eclipse begins over open Atlantic Ocean and first hits land at the Straits of Gibraltar, simultaneously straddling Europe and Africa at their closest point.

North Africa: Totality at or very near most major cities on the north coast of Africa except Egypt. The eclipse touches every nation on the North coast of the African continent, Morocco, Tunisia, Algeria, Libya and Egypt.

Close to Total at Pyramids: Cairo Egypt has 95% totality. 94% at Pyramids of Giza.

Spanish Swarm: Extreme Southern Spain experiences a total eclipse exactly one lunar year after Spain's first Total eclipse in 128 years (which occurred 2026 Aug 12)

The Rock of Gibraltar: perhaps the most famous rock in the world. Rock of Gibraltar is one of the two traditional Pillars of Hercules. According to ancient myths of Greeks and the Romans, it marked the limit to the known world.

Algeria: path cuts just south of Algiers with totality over Constantine, Oran, Sidi Bel Abbes and others, covering virtually all the major cities of Algeria besides Algiers.

Tunisia: Eclipse bisects Tunisia, including ruins of Carthage.

Libya: Crosses Libya at Tripoli then Benghazi. Libya's 3rd eclipse in 22 years.

Egypt: crosses the famous Siwa Oasis, the most remote settlement in Egypt, populated for 3,000 years, also known as the Oasis of Amun-Ra, home to Oracle of the Egyptian god Ammon, visited by Alexander the Great in quest to conquer the Persian Empire.

- Crosses Asyut, Egypt ("guardian" in ancient Egypt) a 5,000 year old city founded by the banks of the Nile, now containing the largest concentration of Christians in Egypt. Mummies of wolves have been found there and apparently the god Osiris was worshiped there as a wolf.

Valley of the Kings: Maximum eclipse of the path occurs almost directly on Luxor, Egypt, (ancient name Thebes), the "Valley of the Kings". This necropolis has frequently been characterized as the "world's greatest open-air museum", the first center of human empire. The Egyptian temple complexes of Karnak and Luxor stand within the modern city. All of it is in total eclipse. More than 65 Royal burials took place in the Valley of the Kings.

Pharaoh Ramses of the Biblical story of Exodus and his extended family were buried in the Valley of the Kings, along with many structures built by the Hebrew slave labor depicted in the Bible. Ramses II is the Pharaoh confronted by Moses and the plagues of God. Nearly all burials have since been "excavated". Ramses tomb was apparently robbed beginning in late 1800's, and his body taken away in 1881 and now resides in a Cairo, Egypt museum.

The Valley of the Kings is also home of the tomb of 19 year old Pharaoh Tutankhamen. In the greatest authorized grave robbery of the 20th century, his tomb was opened on 1922 Nov 4, 7 months and 7 days after a Ring of Fire Eclipse bisected Egypt on 1922 Mar 28. That eclipse was the last great complete eclipse to cross Egypt before this one.

Mecca: Path crosses Red Sea into Saudi Arabia. Mecca, the holiest city of Islam, receives totality for the first time since Aug 30 1905.

Yemen: Path crosses ancient city of Sanaa, Yemen, then the Horn of Africa and exits into Indian Ocean. The last land touched at sunset is Nelson's Island in the Indian Ocean, the most northern and easternmost island of the Great Chagos Bank, the world's largest coral atoll structure.

Metonic and Historical Peculiarities for August 2nd 2027

Jesus: 2027 is the first reasonable possible 2,000th anniversary of the Crucifixion and resurrection of Jesus Christ, which necessarily occurred sometime between 27 and 34 AD.

Been a Long Time: This is the first appearance of an August 2 Solar Eclipse since Aug 2 1674.

Iraq War Anniversary :August 2 2027 is the 37th Anniversary of the invasion of Kuwait by Iraq on 1990 Aug 2, the beginning of the USA's attempted occupation of Babylon. Eclipse has 60% coverage at Kuwait City. 56% in Baghdad.

Declaration of independence Signing: August 2 2027 is the 251st anniversary of the final signing of the Declaration of Independence. Most signers signed it on Aug 2 1776.

Hitler becomes Fuhrer: August 2 2027 is the 93rd anniversary of the ascension of Adolf Hitler to the title of Fuhrer upon the death of Hindenburg. (Aug 2 1934)

Einstein promotes A-Bomb: August 2 2027 is the 88th anniversary of Albert Einstein's letter to Franklin Roosevelt encouraging the development of an atomic bomb (8-2-1939)

JFK and PT 109:August 2 2027 is the 84th anniversary of the sinking of PT-109 in the Solomon Islands on 8-2-1943, where the legendary heroics of future President John F. Kennedy saved the lives of his crew mates. That event is in itself shadowed in eclipses (see Kennedy Connections)

Gulf of Tonkin: August 2 2027 is the 63rd Anniversary of the Gulf of Tonkin "incident", which was used as the pretext for massive involvement by the USA in Vietnam.

Zodiac Peculiarity

- The Valley of the Kings eclipse of 2027 sees the sun and moon eclipse in Cancer, where Venus will be as well.
- Cancer faces Leo the Lion in the Zodiac. The eclipse of 2017 featured sun and moon in the paws of Leo.
- The constellation Cancer (the Crab) was considered a Scarab Beetle by the Egyptians. The Egyptian scarab (or dung beetle) was linked with the god Khepri, who presided over the sun, sunrise, and the renewal of life.
- Cancer is next to Leo in the Zodiac.
- The constellation Leo was home to the 2017 Aug 21 Eclipse. Leo is the Lion as symbolized in Judeo/Christianity as the Lion of Judah. To Jews, this Lion is yet to come. To Christians, The Lion is Jesus who will return again. These two great symbols, Lion and Crab, face each other with matching Total solar eclipses ten years minus 19 days apart.

Saros 136 Recent History and Character

Saros are repeating characters (every 18 years 11 days and 8 hours- or 19 eclipse years). Saros 136 creates the August 2nd, 2027 Valley of the Kings Eclipse. For visual depiction, see Saros Section. For detailed descriptions of these individual eclipses see World Section.

Solar **Saros 136** produced the six longest total solar eclipses of the 20th century, three of them over seven minutes long. It also produced the longest total eclipse of the 21st century at 6 min 38.86 sec on July 22 2009 (The **Wuhan Eclipse**). It will produce this century's three longest total eclipses. Saros 136 will ultimately produce a total of 44 total eclipses. It produced the most central total eclipses of any Saros between the years 1209 and 2718 and produced the greatest magnitude of any eclipse since the year 540 on July 11 1991. ***Clearly this is a peculiar Saros.***

Here are last three Saros 136 eclipses, going backwards in time

2009 Jul 22 The Wuhan Eclipse: Wuhan, China experiences the longest total eclipse of the 21st century along with Shanghai and many other great cities. Totality over many big cities in India. Ten years before pandemic.

1991 Jul 11 Hawaii and Central America: Big Island of Hawaii has totality. Totality in Mexico City over the Aztec Pyramids and most of the nations of Central America, many embroiled in civil war.

1955 June 20 Southeast Asia: Saros 136 produces the longest total eclipse (7 minutes, 7 seconds) since the 11th Century with center over Southeast Asia, totality bisecting Vietnam and Cambodia. Centerpiece of a six-eclipse swarm over that region of the world-- just as it plunges most deeply into the nightmare of modern war..

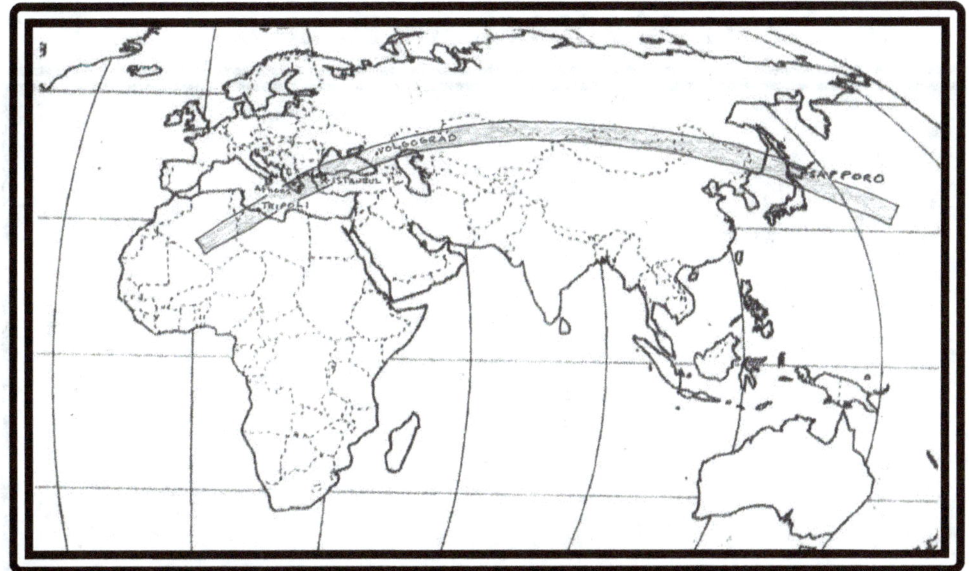

June 1 2030 Athens/Constantinople Ring of Fire.

2030 Jun 1 Athens/Constantinople Ring of Fire. Extraordinary path beginning in Algeria and ending in Japan. Great cities along the path include Tripoli, Athens, Istanbul (Constantinople), Sevastopol, Volgograd and Sapporo. This Ring of Fire occurs 3 eclipse years after August 2 2027 Valley of the Kings

Here's all three Mediterranean Eclipses Together

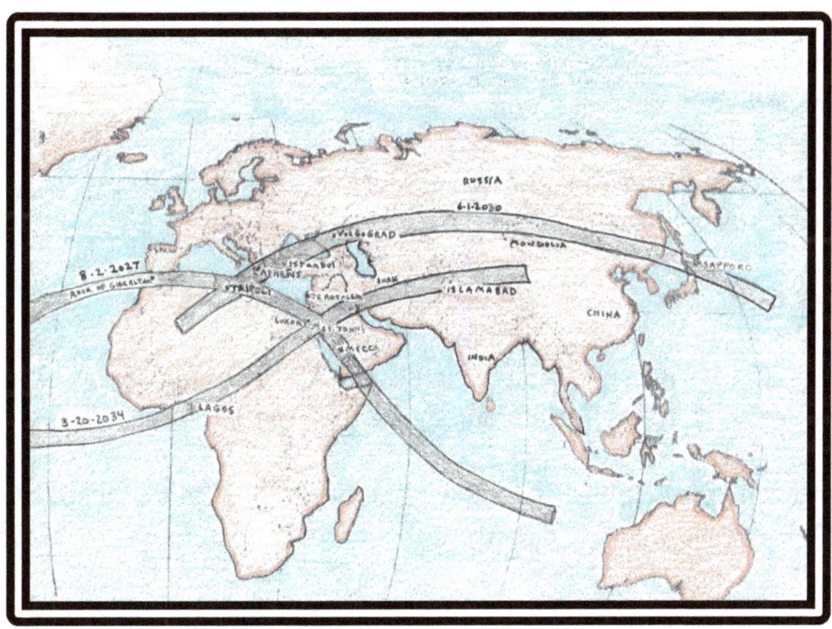

March 20 2034 Spring Equinox

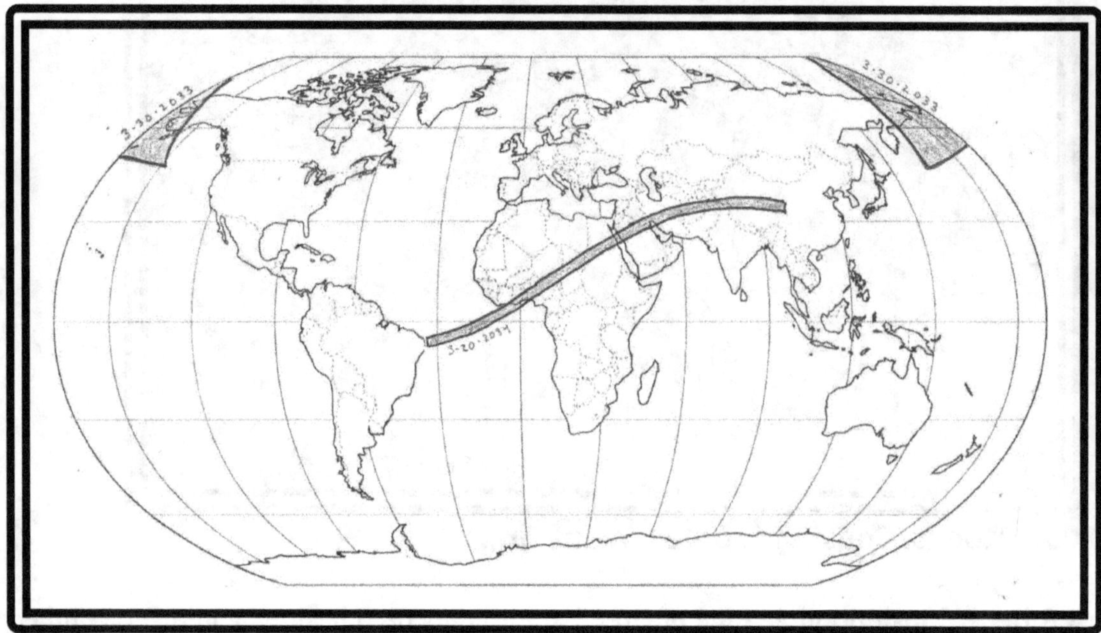

The Great Civilization Eclipse of March 20 2034 completes the Spring Equinox Metonic we will look at later which began in 2015. It is pictured in the center of map. Let's look at its peculiarities.

Astronomical facts about the eclipse:

- Type of eclipse: Total
- Saros Number 130
- Duration of totality along center line of maximum eclipse: 4 minutes and 9 seconds—
- length of path: perhaps 8500 miles long
- width of path, 99 miles wide
- Maximum eclipse occurs over Tougouni Mountain in nation of Chad
- Zodiac: eclipse occurs in Pisces

Date Peculiarities March 20, 2034

Spring Equinox: March 20^t 2034 is the Spring (Vernal) Equinox. The Spring Equinox is the traditional first day of the year for many cultures, ancient and modern.

Nineteen Years after North Pole Eclipse: There was a Spring Equinox Total Eclipse (Saros 120) 19 years prior on 2015 Mar 20. That eclipse ended directly at the North Pole, perhaps the first eclipse in human history to do so.

Happy New Years: March 20 2034 marks New Year or New Year's Eve for Muslim, Hindu, Zoroastrian and Bahai Calendars. 2034 Mar 20 marks the 307th anniversary of the death of Sir Isaac Newton, probably the most famous scientist of all time (1727 Mar 20 Julian Calendar). Newton calculated the Crucifixion of Jesus at Apr 23 34 AD, which would make the 2034 Mar 20 eclipse occur 34 days prior to his calculation of 2000th anniversary of Crucifixion and Resurrection.

March 20 Eclipses are Rare: There have been only 7 complete eclipses on the exact March 20 date between 403 BC and 2014 AD (2417 years). Of those: 3 were total, 2 hybrid and two annular. Only two touched major civilizations and empires of earth. There was a Mar 20 total eclipse in 1662, prior to that, there were two total eclipses on March 20 since the birth of Jesus. One was on 1140 Mar 20 over England and Germany, near the beginning of the war called "The Anarchy", which featured a total breakdown in law and order.

Jesus: The most obvious Solar Eclipse near possible Jesus Crucifixion was on March 19/20 33 AD had a total eclipse (only visible in Antarctica and Indian Ocean) one day prior to Spring Equinox and 14 days prior to the traditional Crucifixion date of Friday 33 April 3 AD, when a partial lunar eclipse was visible from Jerusalem. In 2034, exactly 2,001 years later, there will be both a March 20 solar eclipse and an April 3 Lunar eclipse

Temple Destruction: On Mar 20, 71 AD, the day of Spring Equinox, a hybrid total crossed Greece and Turkey, 143 days after the destruction of the Jewish Temple by Rome featuring totality in Greece (very near Athens—and directly over Artemida, named for the goddess of the moon. In other words, The Crucifixion of Jesus Eclipse (33 AD) and the destruction of the Temple Eclipse (70 AD) are framed by March 19/20 Spring Equinox eclipses occurring 38 years apart. 38 years is two Metonic Cycles.

One Lunar Year after Great Bering Strait Eclipse: Exactly one lunar year prior to the Mar 20 2034 eclipse is a Total Eclipse on Mar 30 2033 which straddles the Bering Strait between the USA and Russia, linking both hemispheres simultaneously. That eclipse is only four days from the traditional 2,000th anniversary of the Crucifixion and 7 days from the traditional Resurrection.

Path of the Mar 20 2034 Solar Eclipse.

Brazil: Begins just offshore of Brazil near Fortaleza-" Fortress" (90% coverage)

Africa: Crosses Atlantic and enters Africa, rapidly crossing 3 national capitals along the "slave coast". The first two capitals are Porto Novo, capital of Benin, the infamous Portuguese slave port, the second is Lome, capital of Togo. The third African capital is Lagos, Nigeria, the largest metropolitan area in Africa (nearly 25 million people), also once a slave port. Its native name is "Eko" which means "war camp". Path crosses Chad, with maximum eclipse in Chad near Tougouni Mountain—then continues through war-torn Darfur and Sudan, before entering southern Egypt.

Egypt: Path crosses southern Egypt

Saudi Arabia: The eclipse crosses the Red Sea (85% eclipse at both possible Mt. Sinai locations in Sinai Peninsula and Saudi Arabia) and enters Saudi at Yanbu, an important site in Islamic history just north of Medina. Totality scrapes northern metro area of Medina, Saudi Arabia, where the Prophet Mohammed died and the second most holy city in Islam. He died on June 8 632 AD- almost exactly 1477 eclipse years and 1445 lunar years prior to this eclipse.

Kuwait: Path crosses southern Kuwait City, 44 years after Iraq invasion of 1990.

Iran: Path bisects Iran, entering at Bandar Ganauch and crossing Shiraz, one of Iran's oldest cities. Path bisects southern Afghanistan, including city of Farah, founded by Alexander the Great.

Pakistan: Path crosses northern Pakistan, including Peshawar, one of the oldest cities in south Asia and over the planned city of Islamabad (means "City of Islam"), the capital of Pakistan.

Kashmir: Path continues over Kashmir, the disputed mountain region of Pakistan and India, including its largest city, Srinagar.

Tibet: Eclipse ends in Tibet at sunset, in an area now controlled by China.

Crosses: 2027 and 2034 Hepton Eclipses

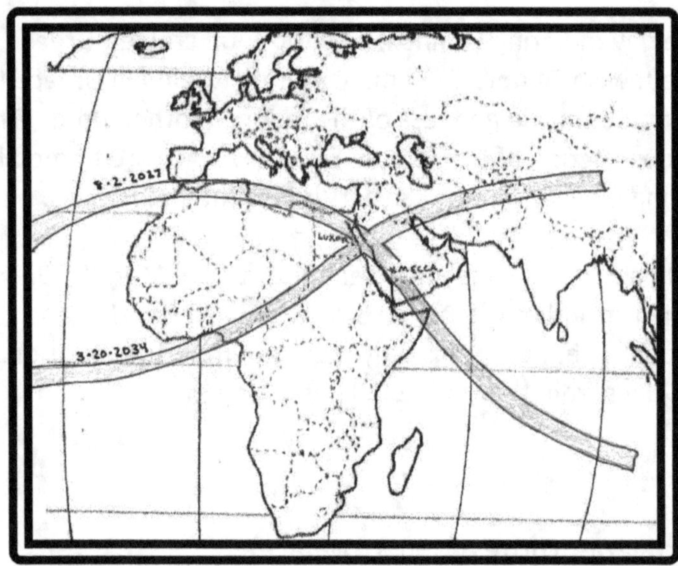

The 2034 path crosses the 2027 eclipse path just south of the Valley of the Kings and creates apex in the Red Sea over St. John's Island just east of Egypt. The 2030 path crosses the 2027 path over Tripoli.

Only two towns of any decent size are in both eclipse paths, Ras Banas and Al Shalateen, both in Egypt though Shalateen is a disputed territory with Sudan. Ras Banas is on a peninsula north of an inlet called "Foul Bay", by the ancient port of Berenice, founded by Ptolemy in the 3rd century BC and home of the famous troglodytes. Berenice was a critical trading port for several empires. Its most famous ruin is that of an Egyptian style Temple featuring the name of Tiberius. Tiberius was Caesar during the last days of Jesus of Nazareth. Both eclipses of 2027 and 2034 cover this same temple in totality. Little has been written about Al-Shalateen, but apparently, they have several markets there, and it is known as a place of camel trading.

The true apex of the two eclipses occurs in the Red Sea near St. John's Island (Zabargad), a small yet somewhat famous island known for its now submerged topaz mines. It is presumably named after Saint John, who wrote the Book of Revelation. St. John's, Canada was the last town touched by the 2024 USA eclipse.

Further Peculiarities during 2027 and 2034 Solar Eclipses

Apex is Right on the Tropic of Cancer: Apex of the two eclipses of 2027 and 2034 is directly on the Tropic of Cancer (23.5 degrees North). The Tropic of Cancer is the northernmost latitude on Earth where the Sun can be directly overhead, which occurs on Summer Solstice. It was named in the last centuries before Christ, when the sun was in constellation Cancer. It has since shifted to Gemini.

Eclipse of 2027 is in Sign of Cancer: The Eclipse of 2027 sees the sun and moon eclipse in Cancer, near the planet Venus. Cancer faces Leo the Lion in the Zodiac. The eclipse of 2017 Aug 21 featured sun and moon in the paws of Leo.

Cancer is the Egyptian gods Khepri and Atum-Re: The constellation Cancer (the Crab) was considered to be the Scarab Beetle by the Egyptians. The Egyptian scarab (or dung beetle) was linked with the god Khepri, who presided over the sun, sunrise, and the renewal of life. Ancient Egyptians believed that scarabs were the reincarnations of Khepri himself, and depictions of the god often show him with the head of a scarab. The scarab beetle was also associated with the gods Atum and Re, who represented primordial creation and the sun, respectively. Together, the gods formed Atum-Re, which illustrated the joint power of the sun and creation.

2034 Eclipse in Pisces: In the March 20 2034, the sun and moon eclipse just below Pisces near the planet Jupiter. Pisces is considered the sign of the fish. The age of Pisces apparently began around 68 BC, and thus close to the birth of Christ. It's one of many reasons Christians are associated with fish.

Saros 130 Recent History and Character

Saros 130 began its life with a partial eclipse on Aug 20 1096, five days after the First Crusade began on August 15, 1096. Saros 130 brought the first total eclipse on Earth after the discovery of the New World—Apr 16 1493. Exactly 570 eclipse years separate that eclipse and the March 20 2034 eclipse.

Saros 130 brought the longest total eclipse of the 17th Century, on Jul 11 1619, at 6 minutes and 41 seconds in duration. That path connected the Americas and Africa. It crossed the port of Luanda, Angola close to the exact day that African Ndongo slaves loaded there were brought to Jamestown. They were put on the Spanish slave ship San Juan Bautista probably bound for South America. Some of those slaves was taken aboard an English ship, White Lion, bound for Jamestown in North America. They arrived on August 20 1619. That arrival was 398 years and one day before the Aug 21 2017 USA eclipse, which featured totality in Charleston SC, North America's largest slave port. This was likely Saros 130's most extraordinary moment until 2034.
Recent history-last two events:

March 9 2016 Nyepi Eclipse was seen almost exclusively in Indonesia though it traveled for thousands of miles of open ocean. The eclipse occurred on Nyepi, a public holiday in Indonesia and the end of the Balinese Saka calendar. Nyepi is a day of silence, intended for self-reflection.

February 26 1998 Panama Isthmus Eclipse occurs over a tiny portion of land, at isthmus of the Americas 19 years to the day after the Mt. Saint Helens Eclipse of February 26 1979.

Chapter Seven

2015 to 2034
Spring Equinox Eclipse Metonic

2015 to 2034 Spring Equinox Eclipse Metonic

This is a summary of the Solar Eclipse Chapter we are living in.

March 20 is the current day of Spring Equinox.
There was a total solar eclipse on March 20 2015.
There will be a total solar eclipse on March 20 2034.
These dates are exactly 19 years apart, or 6,940 days.

There were 26 complete solar eclipses during that 19 year time span. 12 are total. 12 are Ring of fires.
2 of them are hybrids.

Two solar eclipses show up on Spring Equinoxes in the time we live. The time we live in is the 2,000th anniversary of the adulthood and ministry of Jesus. As it happens, our eclipses happen to mirror that time. There was a solar eclipse on March 19/20 two weeks before Jesus traditionally was crucified.

Welcome to Draco-Metonic reality, God's cosmic timekeeping. More about this in the next chapter.

In this chapter we briefly look at all complete solar eclipses (and a few famous lunar eclipses) between the Spring Equinoxes of 2015 and 2034. I have chosen the Spring Equinox Eclipses to define this Metonic. Any eclipse date would do as a starting point, because eclipses come in cycles, not in some straight line. But since there happen to be Equinox Eclipses in our current history, that makes a logical place to start for a number of reasons.

Why is the Spring Equinox Important? The Spring Equinox is an ancient signpost for cultures of the northern latitudes. It is the first day of the year for billions of people, or signals when that first day starts. It helps define the religious calendars of half the world. Symbolically, it is where day and night are equal length. A new moon on Spring Equinox is a somewhat rare occasion in and of itself. A solar eclipse added to that is tremendously rare.

There have been only 7 complete eclipses on March 20th between 403 BC and 2014 AD. Of those: 3 were total, 2 hybrid and two annular. Only two of these touched major civilizations.
Between 2015 and 2053 there are 3 March 20 eclipses.

Let's survey all complete solar eclipses on earth in the 2015 to 2034 Spring Equinox Metonic. We begin with an overview and then break down the individual eclipses into smaller units.

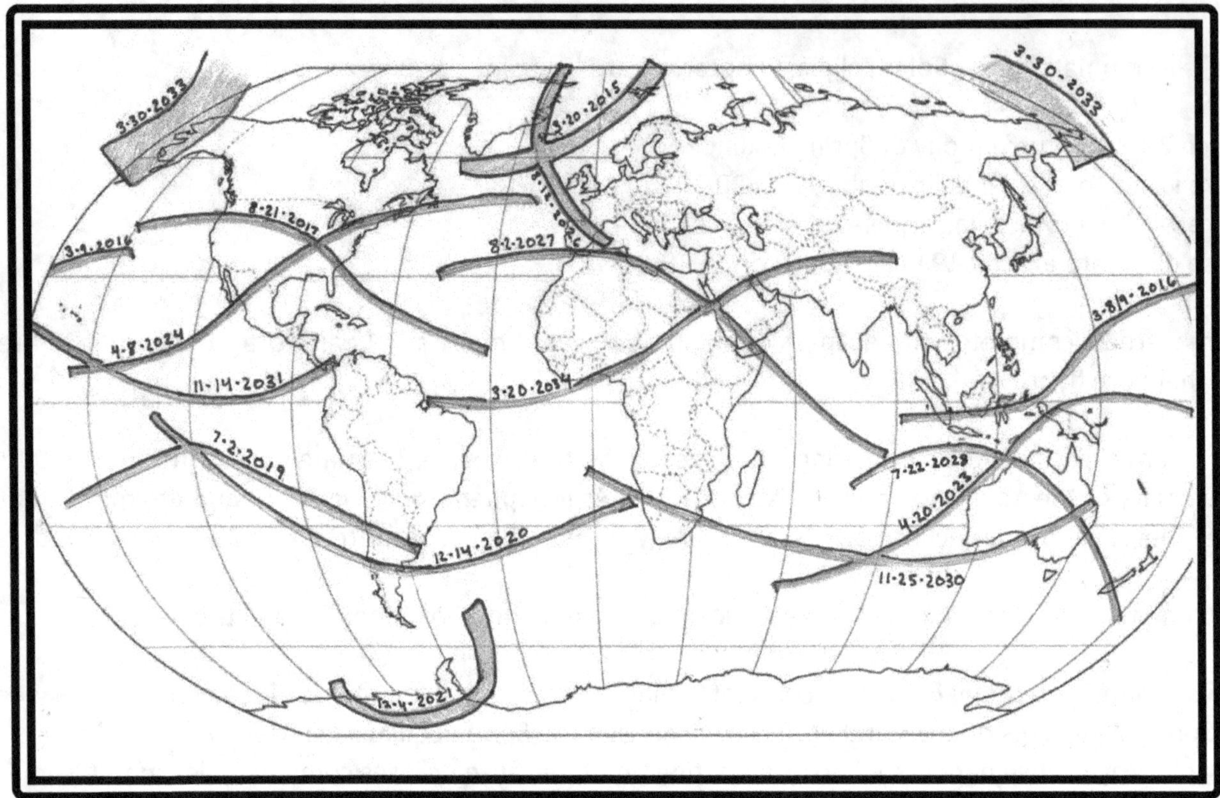

19 Year Spring Equinox Metonic Totals 2015 to 2034

2015 Mar 20 North Pole Spring Equinox (Saros 120)
2016 Mar 9 Indonesia/ Malaysia (Saros 130)
2017 Aug 21 Great American Eclipse (Saros 145)
2020 Dec 14 Electoral College Eclipse (Saros 142)
2021 Dec 4 Antarctica (Saros 152)
2023 Apr 20 Hitler Birthday Hybrid Australia/New Guinea (Saros 129)
2024 Apr 8 Great North American Eclipse (Saros 139)
2026 Aug 12 Spain/Reykjavik (Saros 126)
2027 Aug 2 Valley of the Kings (Saros 136)
2028 Jul 22 Botany Bay/ Australia Judgment (Saros 146)
2030 Nov 25 British Empire/ JFK Jr. 70th Birthday (Saros 133)
2031 Nov 14 Panama Isthmus Hybrid (Saros 143)
2033 Mar 30 Bering Strait 2,000 Anniversary of Christ Entry to Jerusalem (Saros 120)
2034 Mar 20 Great Civilization (Saros 130)

This is the first Mar 20 Metonic since the Mar 20 Metonic of 1643 to 1662. There are twelve totals and two hybrids mapped. Looking at the dates, you will notice many of them are exactly one lunar year apart. You will see around 11 days removed from one solar year eclipse date to the next solar year eclipse date. For instance, look at the first two on the list, and then the others. The first eclipse is Mar 20 2015. The second is Mar 9 2016. One lunar year is 354.3 days. Same phase of the moon, of course. Solar eclipses only happen at new moon.

All Ring of Fires 2015 to 2034

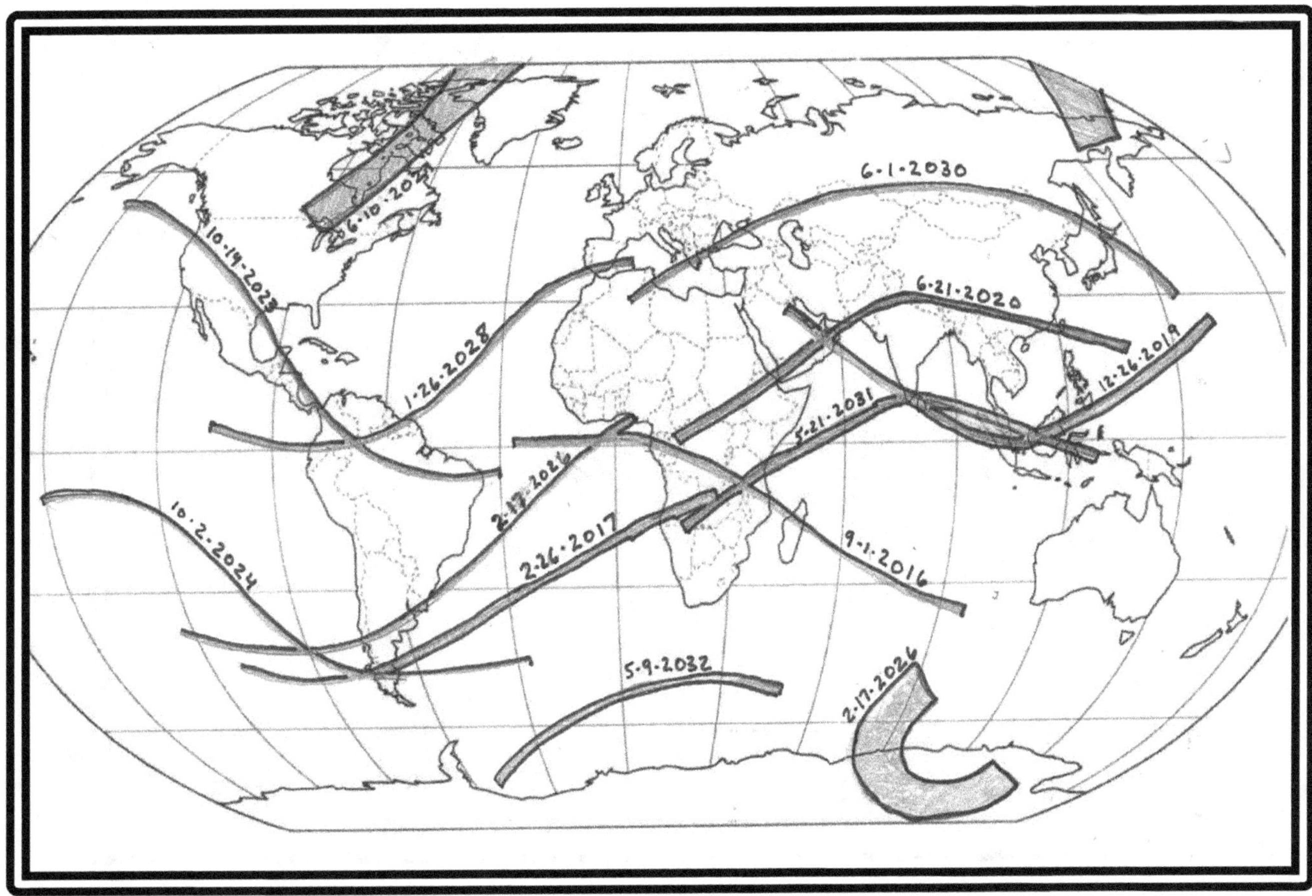

Ring of Fires 2015 to 2034

2016 Sep 1 Africa (Saros 135)

2017 Feb 26 South America to Africa (Saros 140)

2019 Dec 26 Tsunami Anniversary Indonesia/Sri Lanka (Saros 132)

2020 Jun 21 Pandemic Summer Solstice Father's Day (Saros 137)

2021 Jun 10 Canadian Judgment Thunder Bay (Saros 147)

2023 Oct 14 Great American Ring of Fire (Saros 134)

2024 Oct 2 Pacific to Patagonia (Saros 144)

2026 Feb 17 Antarctica (Saros 131)

2027 Feb 6 Slave Trade Route Eclipse Rio to Africa (Saros 131)

2028 Jan 26 Spain Ring of Fire (141)

2030 Jun 1 Athens/Istanbul Great Eurasian Ring of Fire (Saros 128)

2031 May 21 Dar Es Salaam/Thiruvananthapuram Temple of the Sun/ Indonesia (Saros 138)

Here are the 12 Ring of Fires in 19 years. There is a Ring of Fire across South America on Sep 12 2034, but that is outside the Spring Equinox Metonic, which ends on Mar 20 2034.

The 2015 Spring Equinox North Pole Eclipse

Spring Equinox Eclipse March 20 2015 ends directly at the north pole.
It's the only complete eclipse of 2015. It occurred two weeks before a Passover Lunar Eclipse of Apr 4 2015, exactly 47 years after murder of Martin Luther King and 1982 years and one day after traditional date of Jesus Crucifixion (33 AD Apr 3). This is the first solar eclipse after the famous Oct 8 2014 Feast of Tabernacles Blood Moon Lunar eclipse.

Visible from Svalbard Islands, one of the northernmost inhabited places on earth. The Eclipse ended immediately at the North Pole, but I have found no evidence of anyone traveling to the North Pole to see it. Of course, that can be a rough place to get to. Maybe Santa and Rudolph saw it.

It is unclear whether any documented total solar eclipse has ever ended directly at the North or South Pole or if any will happen again. It was an extraordinary geographic and cosmic occurrence.

This is the Solar beginning of the Metonic which ends 19 years later with the 2034 Equinox Eclipse. This is also the second to last complete eclipse of the Saros 120 personality (more on what that means later), which has been a star of eclipse history for centuries. Saros 120's last eclipse, on 2033 Mar 30, is its coup de grace.

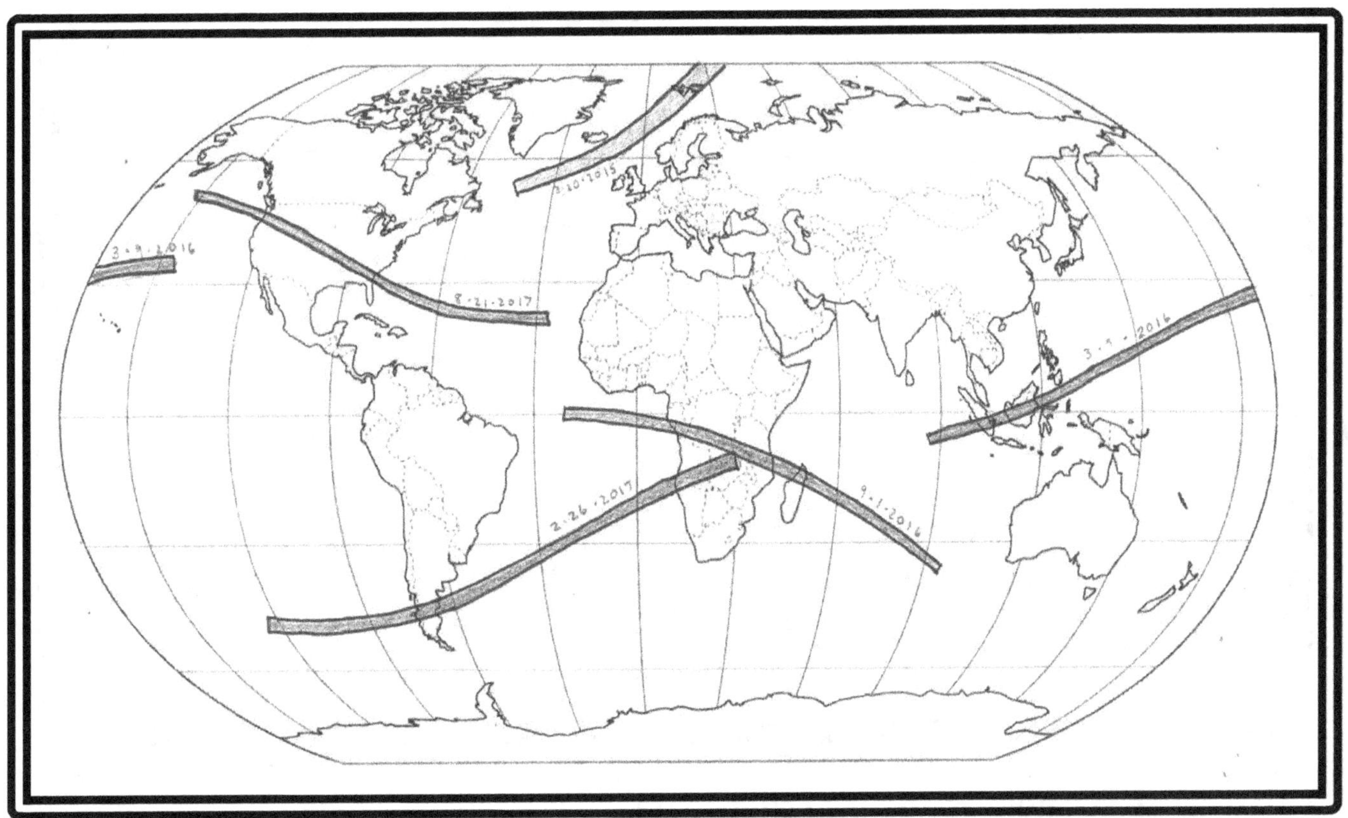

2015 to 2017 Lunar Year Triad and Duo

The Triad: Three eclipses each separated by one lunar year. See dates go 11 days back each eclipse.

2015 Mar 20 Spring Equinox North Pole Eclipse Total (Saros 120)
2016 Mar 9 Nyepi Day Indonesia and Malaysia Total (Saros 130)
2017 Feb 26 South America to Africa Ring of Fire (Saros 140)

Duo: Two eclipses separated by one lunar year
2016 Sep 1 Africa Ring of Fire (Saros 135)
2017 Aug 21 Great USA Eclipse Total (Saros 145)

2015 Mar 20 North Pole Eclipse Total (Saros 120)-discussed at beginning of chapter.

2016 Mar 9 Nyepi Day Indonesia and Malaysia Total (Saros 130) occurred on holiday of Nyepi, when most Indonesians take a day of silence.

2016 Sep 1 Africa Ring of Fire (Saros 135) From Gabon to Congo, Tanzania through Madagascar

2017 Feb 26 South America to Africa Ring of Fire (Saros 140)

2017 Aug 21 Great USA Eclipse Total (Saros 145) discussed at length elsewhere. Hepton Cycle Monday Eclipse. The most viewed American eclipse of all time. America Divided. First major city in totality was Salem, Oregon--the largest US city named after Jerusalem. The last city in the path is Charleston, South Carolina --the apex of the American slave trade. Occurs 38 years after the last total eclipse visible in the US, which occurred on Feb 26, 1979. Note date of prior eclipse in 2017.

Partials occur on 9-13-2015, 2-15-2018, 7-13-2018, 8-11-2018, 1-6-2019.

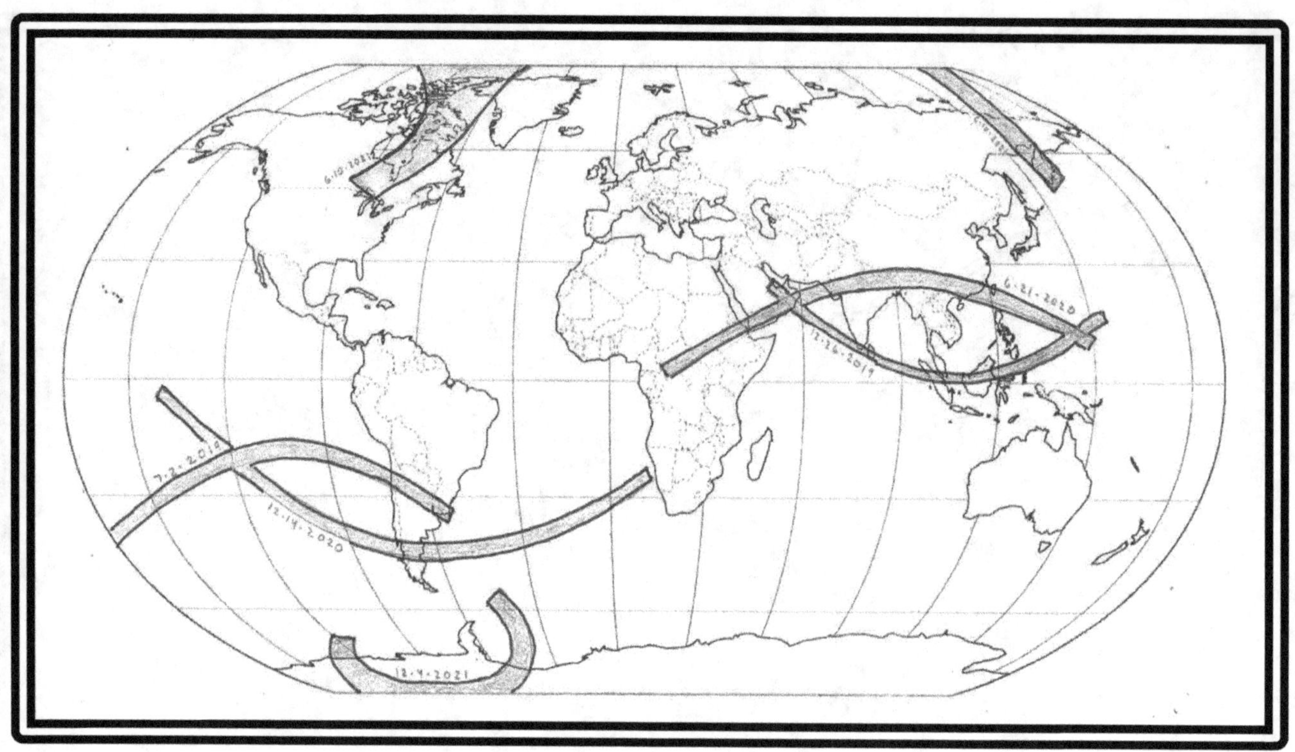

2019 to 2021 Double Pandemic Eclipse Triad

2019 Jul 2 Pope Francis Buenos Aires Hepton Total (Saros 127)
2019 Dec 26 Tsunami Anniversary on Tsunami Path Ring of Fire (Saros 132)
2020 Jun 21 Summer Solstice Father's Day Pandemic Ring of Fire (Saros 137)
2020 Dec 14 Electoral College Total (Saros 142)
2021 Jun 10 Canadian Judgment Ring of Fire (Saros 147)
2021 Dec 4 Antarctica Total (Saros 152)

In the two years of the big pandemic, interesting artistic balance and geographic/ Metonic peculiarities. South Asia is framed nicely in an eye or fish created by 6-month eclipses. In the western hemisphere, a similar duo is formed by eclipses over almost 18 months.

The 2019 Dec 26 and 2020 Jun 21 eclipses cross in two places. The far eastern cross has apex on southern tip of Guam. Its western cross has an apex in the desert just south of city of Nizwa, Oman.

2019 Jul 2 Pope Francis Middle of Year Buenos Aires Hepton Total (Saros 127) The second eclipse of the Monday Hepton Series. At the exact midpoint of Calendar year, the 183rd day of the Gregorian solar calendar year with 182 left to go. Crosses South America (Totality entering at La Serena ("calm"), Chile. Totality over San Juan (Saint John), Argentina, ending at sunset at Pope Francis birthplace of Buenos Aires, Argentina, 2302 days after his becoming Pope. Leaves nation at Dolores (sorrows),

2019 Dec 26 Tsunami Anniversary on Tsunami Path Ring of Fire (Saros 132) On the 15th anniversary of the greatest recorded Tsunami in human history, an annular solar eclipse follows the Tsunami path of Dec 26 2004 backwards between Arabia and Indonesia. Path is the one on the most lower right of map. Beginning in the Middle east at sunrise-44% coverage at Baghdad (Babylon), 60% coverage at Tehran, Iran. Annularity begins just north of Saudi capital of Riyadh, with 87% coverage. First true annularity is seen in ancient oasis city of Hofuf, Saudi Arabia, near the largest oil field in the world. Hofuf is said to the burial place of Laila and Majnoon, stars of the most popular love story in the Arab and Muslim world. The Queen of Sheba is said to have visited this city from her kingdom in Yemen. Eclipse crosses Oman, where tsunami waves went ashore on Dec 26 2004. Eclipse annularity crosses Indian Ocean, southern India and Sri Lanka, where more than 30,000 people died in the tsunami. Eclipse passes over Aceh Province of Indonesia, close to the epicenter of the earthquake. Crosses Malaysia and Singapore. Eclipse ends after Guam.

A truly strange and remarkably unnoticed synchronicity of calendars, eclipses, geologic events and geography. This eclipse occurs in the very days that the first reports of what became known as Covid-19 are being analyzed in Wuhan, China (Wuhan has 18% coverage around sunset)

2020 Jun 21 Summer Solstice Father's Day Pandemic Ring of Fire (Saros 137) Summer Solstice. Father's Day. Very close to total eclipse with 99% coverage at many places. Begins in Africa at sunrise in the Congo, crosses Sudan and Ethiopia. Through the Red Sea and Yemen. Pakistan (91% in Karachi—annularity in Sukkur), Northern India (94% in New Delhi). Bisects Tibet (94% in Lhasa). Crosses nearly 2,000 miles of China, exiting directly over city of Xiamen, then crossing Taiwan (92% at Taipei) as the Pandemic shuts the world down. 83% annularity in Wuhan, China.

2020 Dec 14 Electoral College Total (Saros 142) The precise midpoint of the 2017 and 2024 USA Hepton eclipses. Dec 14 2020. That day (which is also the 221st anniversary of George Washington's death on 1799 Dec 14), between 10am and 5pm Eastern time, the Electoral College certifies Joe Biden and Kamala Harris as President and Vice-President of the United States of America.

5,000 miles away, this total solar eclipse--same hemisphere, same time. Full eclipse occurred between 9:30 am and 1pm Eastern Standard Time. Very little of the eclipse path crossed land. Still, maximum eclipse did occur on land, in Argentina near the town of Neuquen (a palindrome). If followed north, longitude of Neuquen intersects with eastern Maine. Eclipse crosses Atlantic and stops just west of Namibia, dying in the ocean before it can reach Africa.

The Hepton Cycle in action. A significant Monday Total Solar eclipse perfectly spaced between two other Total Solar Monday eclipses. All of them western hemisphere. All of them with significant relations to the USA. 7 eclipse seasons--1211 days-- in either direction brings you to the USA Eclipse Judgments of 2017 Aug 21 and 2024 Apr 8 (which themselves are exactly 7 eclipse years apart).

2021 Jun 10 Canadian Judgment/ Canada to Siberia Ring of Fire (Saros 147) Eclipse begins at sunrise in southern Ontario, Canada, literally on the border of the US and Canada, near Thunder Bay. It traverses south to north through Canada and Greenland, then ends in extreme Northeast Russia. Occurs in the midst of the beginnings of the Canadian lockdown.

2021 Dec 4 Antarctica (Saros 152) A few hardy souls took the trip to Antarctica to witness the last true total eclipse for two and a half years. This is the fourth complete eclipse for Saros 152. Its first complete eclipse occurred on 1967 Nov 2, in the midst of the Jerusalem Blood Moons.

Partial occur on 4-30-2022 (the 70th anniversary of Hitler's death) and on 10-25-2022, 77th Anniversary of United nations + one day

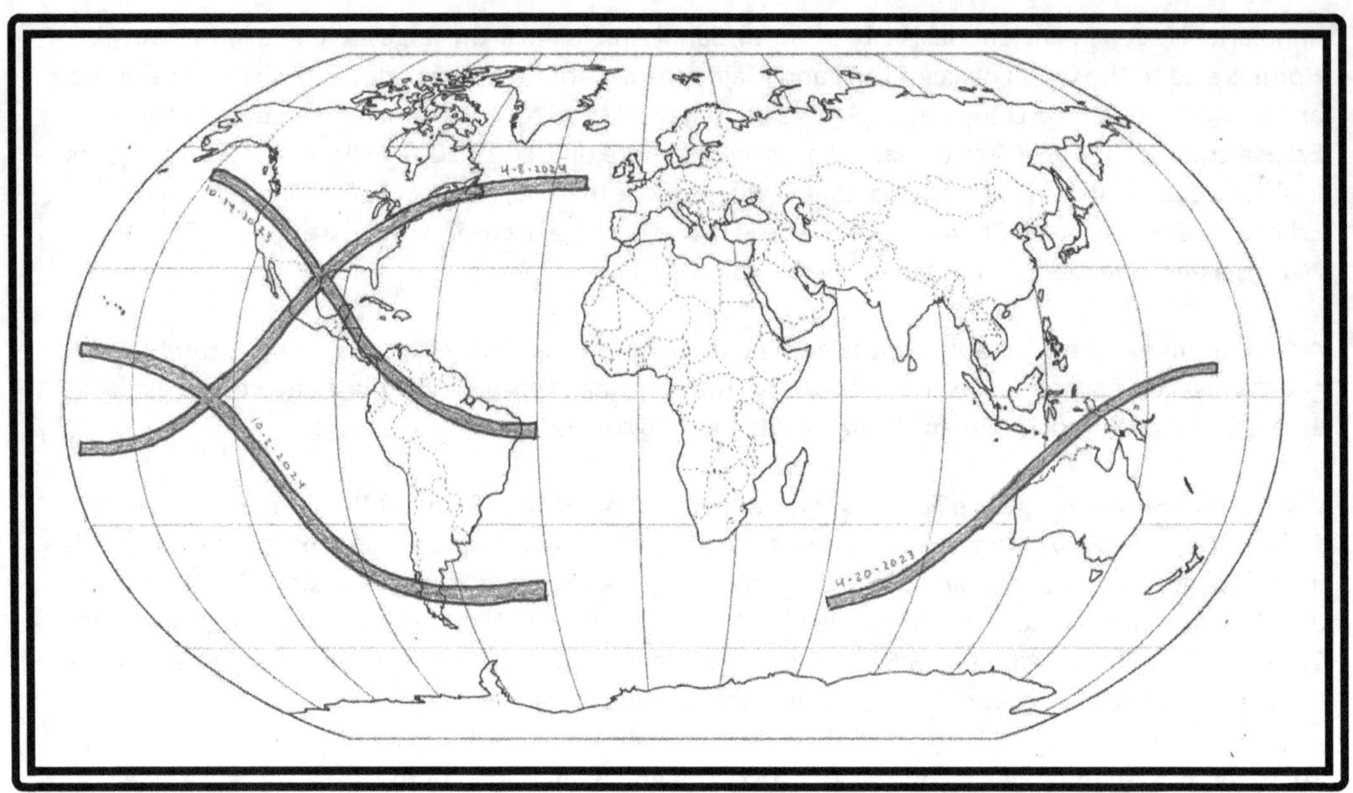

2023 and 2024

2023 Apr 20 Hitler Birthday Metonic Hybrid (Saros 129)
2023 Oct 14 Great American Ring of Fire (Saros 134)
2024 Oct 2 Pacific to Deseado Massif Ring of Fire (Saros 144)
2024 Apr 8 Great North American Eclipse Total (Saros 139)

Almost no land in the Eastern hemisphere is affected by eclipses.

Only one true total here, the April 8 2024 eclipse. That separates it out dramatically.
In fact, it's the only total eclipse on earth between the 2021 Antarctica and the 2026 Spanish
Judgment, nearly five years.

The April 8 2024 is also the first total eclipse to cross an inhabited area since 2019 July 2 Buenos Aires
eclipse, which would mean it stands out as the only total eclipse over established populations
between 2019 and 2026, more than seven years. The April 8 2024 American eclipse goes over more
population centers than any eclipse in the history of the western hemisphere.
The peculiarity appears accented.

The South Texas Tav is visible in upper left. It is the cross formed by the intersection of the 2023 and
2024 eclipses. Its apex is in the hill country near Utopia, Texas, which is southwest of San Antonio
(which also is in the path of both eclipses). Much more detail is found elsewhere in this work.

2023 Apr 20 Hitler Birthday Metonic Hybrid (Saros 129) Adolf Hitler's birthday (April 20) and death day (April 30) are on a perpetually linked Metonic, around 10 days apart, they show up in linked eclipses one lunar year apart throughout history at certain points around every 500 years. In this case, a partial solar eclipse occurred 2022 Apr 30, on the 77th anniversary of the death of Adolf Hitler. That partial brought with it a total lunar eclipse visible from the Americas on Nov 8, USA election day. In 2023, after a century of eclipses lingering on the April 19 (Darwin Death Day) eclipse occurs on the exact day of Hitler's birth 134 years before on April 20 1889. The next eclipse after this (Great American Ring of Fire) will be Saros 134 (note the 134's) but that's getting almost silly! Very little land experiences totality. Eclipse begins in Indian Ocean and avoids land, heads north outside Australia . Only a few hundred square miles of the extreme northwestern edge of the Australian land mass experiences true eclipse, and it is an astonishing 100.00% coverage, over the one true town of any size, the town of Exmouth, home to interesting military and scientific infrastructure as well as tourism and kangaroos. Eclipse travels over narrow strip of Timor and Papua before again avoiding nearly all land and dying south of Micronesia. Note 4-20 numeric, which has become inexplicably popularized in USA in recent years.

2023 Oct 14 Great American Ring of Fire (Saros 134) Looked at in much closer detail in the USA section. Bisecting the western US, totality entering Oregon, crossing the Four Corners and exiting the nation at Corpus Christi ("Body of Christ") Texas. Crossing the Yucatan and much of Central America. Exiting Brazil at the city of Natal, translated "Nativity" in Portuguese. Exactly 555.5 eclipse years after Mayan Apocalypse Eclipse of August 8, 1496. Occurs on exact 531st anniversary (10-14-1492) of first colonial crime in western hemisphere, the abduction and enslavement of 7 native youths by Columbus on Island he named San Salvador ("Holy Savior") with 48% coverage.

2024 Apr 8 Great North American Eclipse Total (Saros 139) Looked at much closer in USA Eclipse section. Certainly, the most significant eclipse, population-wise, in the history of the United States. Hepton Cycle. Monday Eclipse. Total Solar Eclipse. Judgment Eclipse. First true total eclipse to cross Mexico, USA and Canada overland since any of those nation's founding. Finishes both Tavs and the Aleph over the USA (2nd Tav pictured above). 8 days after Easter, 2422 days from USA eclipse of 2017, 1211 days after Election Eclipse of 2020, 177 days after Ring of Fire eclipse of 2023 Oct 14. Path of totality enters at Mazatlan, Mexico; it crosses Mexico, the USA (dozens of cities and towns) and Eastern Canada-- exits near St. Johns in province of Newfoundland and Labrador

1024 Oct 2 Pacific to Deseado Massif Ring of Fire (Saros 144) Like the 2023 Eclipse, almost no land except a small strip of extreme South America, including the Deseado Massif, supposedly some of the oldest rock on earth. Only town of size in path of Annularity appears to be Puerto San Julian on the eastern flank of Argentina.

Partials occur on 3-29-2025, 9-21-2025.

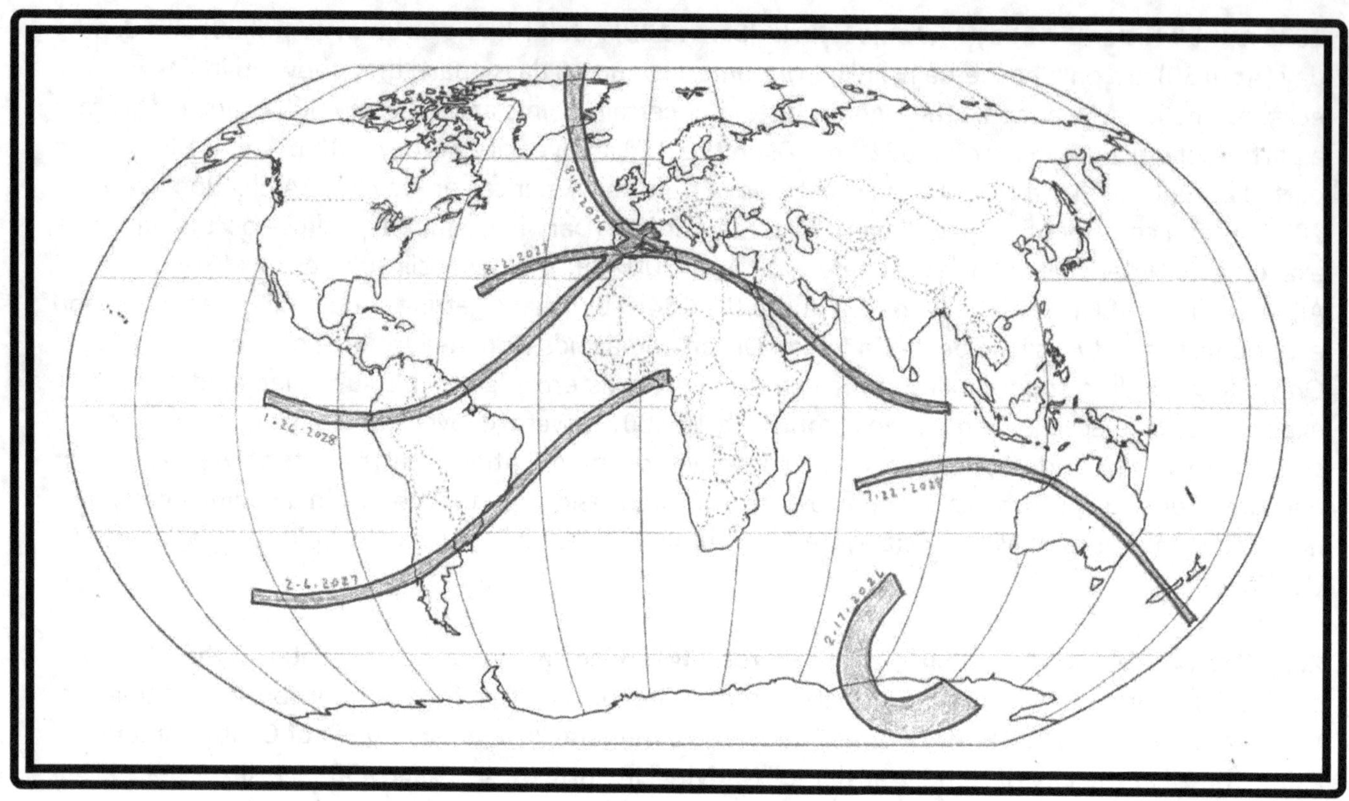

2026 to 2028 Spanish Judgment/ Botany Bay

2026 Feb 17 Antarctica Ring of Fire (Saros 121)
2026 Aug 12 Spain/Reykjavik Total (Saros 126)
2027 Feb 6 500th Anniversary Slave Trade Judgment/Rio de Janeiro to Africa (Saros 131)
2027 Aug 2 Great Valley of the Kings Total/ Spain (Saros 136)
2028 Jan 26 Spanish Third Eclipse in 18 months on "Australia Day" Ring of Fire (Saros 141)
2028 Jul 22 Australia- Botany Bay and Sydney/New Zealand (Saros 146)

Five of the six eclipses in this time frame are quite extraordinary and they occur in a row 177 days apart. Coming off two partials in 2025 and a nondescript Antarctica Ring of Fire in early 2026, the next five eclipses hit like a ton of bricks of meaning. They create a very unusual balanced triad with Spain in the crosshairs. Also, Australia begins its 15-year eclipse swarm in earnest, and an interesting Rio Slave Trade eclipse. Beautiful symmetric swirl is formed by the three colliding Spanish eclipses.

2026 Feb 17 Antarctica Ring of Fire (Saros 121) The second to last complete eclipse for Saros 121. Its last is scheduled for 2-28-2044. It began life in 944 and had its first complete eclipse in 1070.

2026 Aug 12 Spain/Reykjavik Total (Saros 126) First total eclipse in Spain for last 112 years. They last had a total eclipse in 1905. Spain's first of 3 complete eclipses in 18 months, two of them totals. 500 years after Spain began the African Slave Trade to the New World (1526). Eclipse occurs 128 years to the day of the peace treaty that ended the Spanish-American War on Aug 12 1898, basically ending the 600 year old Spanish Empire. The extraordinary path of this eclipse avoids almost all human populations, curling around the British Isles—91% coverage in London, 92% in Paris. The only other nation to experience totality besides Spain is Iceland. Amazingly that occurs only on the bare sliver of western coast that includes its capital of Reykjavik (whose metro area contains 60% of Iceland's population). Eclipse manages to enter Iberian Peninsula at just the right angle, without crossing the border of neighbors Britain, Portugal or France. For instance, the village of St. Jean de Luz in southern France has 99.7% coverage but no place in France has true totality. In Portugal, the village of Labiados in the extreme north experiences 99.99% coverage. A few square miles of apparently uninhabited Portuguese territory in the Pyrenees has totality but no established towns. Lisbon, Portugal has 95%.

Spain is another story. The eclipse bisects the northern half of the nation, bringing totality to a third of its land mass. There is totality in parts of metropolitan Madrid. Toledo has 99.4% Eclipse. Enters Spain at city of La Coruna ("crown") and ends in the Mediterranean just past Spain's Balearic Islands.

2027 Feb 6 500th Anniversary Slave Trade /Rio de Janeiro to Africa Ring of Fire (Saros 131) Just after Spain appears in the crosshairs, the 500 year old transatlantic slave trade is highlighted by this Ring of Fire. 500th Anniversary year of the importation of African slaves to South America by Portuguese slave traders, the eclipse path traces backwards the maritime routes followed by slave traders from South America to the northern "Slave Coast of Africa"--Benin, Ghana and Nigeria--ending totality over notorious port of Porto-Novo in Benin, one of the centers of the Slave Trade. Totality just offshore of Rio De Janeiro, Brazil - the largest slave port in the history of the world.

2027 Aug 2 Valley of the Kings Egypt/ Spain/Mecca Total (Saros 136) Looked at closer in Middle East section. The great Saros 136 comes to the most famous Necropolis on earth. Hepton Cycle Monday Eclipse. Judgment Eclipse. Exactly one lunar year after Spanish Judgment of Aug 12 2026. 1211 days after USA Little Egypt Tav. Passing directly over the straits of Gibraltar, hugging the north Coast of Africa (totality in Tripoli) before diving south through Egypt over the Nile and the Valley of the Kings with Maximum Eclipse in Luxor (ancient Thebes),the metropolis of the sun gods and necropolis of the Pharaohs. Next into Saudi Arabia-including totality in the holy city of Mecca. Exiting into Gulf of Oman through Yemen, passing over the most eastern part of Africa, the Horn, where scientists claim humanity originated, before ceasing in the southernmost Maldives in the Indian ocean.

2028 Jan 26 Spain's Third Eclipse in 18 Months on "Australia Day" Ring of Fire (Saros 141) They just keep coming! Spain's 3rd eclipse in 18 months. Across the breadth of the Amazon of South America, including Brazil. Crossing the Atlantic, entering Portugal at historic port of Sagres ("holy"). Eclipse just south of Lisbon, one of the oldest cities in the world. Like 2026, Ending just beyond border of Spain, stopping dead at border of France. Occurs on Australia Day, or 240 years after first convicts arrived in Botany Bay, Australia (Jan 26 1778). Australia hosts the next eclipse 177 days later.

2028 Jul 22 Australia/ Botany Bay and Sydney-New Zealand Total (Saros 146) 177 days after "Australia Day" Eclipse and one lunar year from Mideast Eclipse of 2027 Aug 2. 19 solar years to the day after 2009 July 22 Wuhan Total eclipse. Australia is bisected with centerline of totality directly over Sydney and Botany Bay, where the first prisoners were brought to colonize the continent 240 years earlier on the date of the last eclipse on earth 1-22, in 1788. Also crosses southern island of New Zealand, exiting over city of Dunedin, one of the oldest European settlements in New Zealand.

Partials occur on 1-14-2029, 5-12-2029, 7-11-2029, and 12-5-2029.

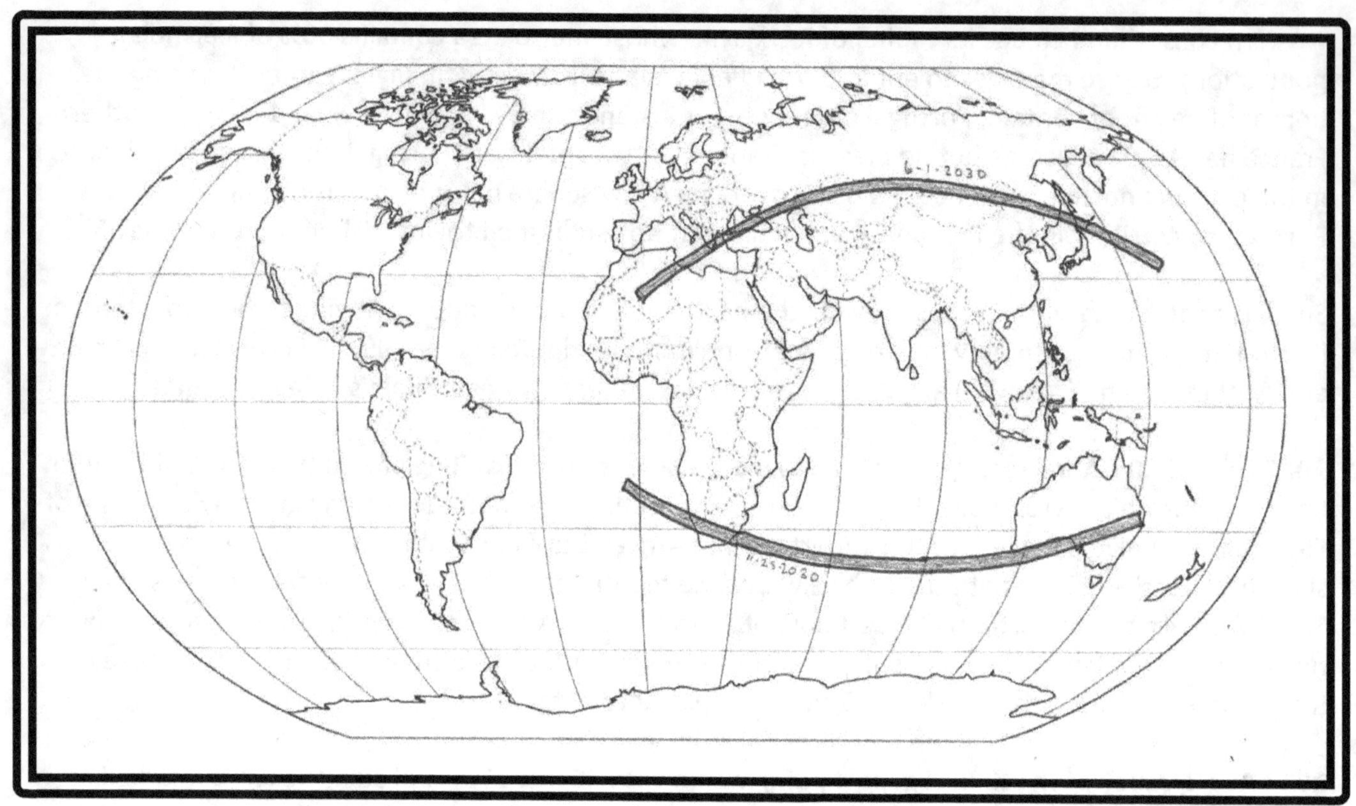

2030 Eclipses-Both Occur in Eastern Hemisphere

2030 June 1 Athens/ Istanbul Ring of Fire—Africa to Japan (Saros 128)
2030 Nov 25 British Empire-South Africa to Australia Total (Saros 133)

The 2030 eclipses have unusual symmetry. The two eclipses begin and end very near to each other on similar longitudinal lines.

2030 June 1 Athens/ Istanbul Ring of Fire—Africa to Japan (Saros 128) An extraordinary path beginning at sunrise near Gara Samani, Algeria (which has a dinosaur named after it). Eclipse crosses Libya (again!), the Mediterranean, Turkey, war-torn regions of Crimea and Southern Ukraine (annularity in Mariupol), follows the entire southern border of Russia and exits after crossing Japan. Great cities along the path include Tripoli (second complete eclipse in 3 years), Athens, Istanbul, (Constantinople). Sevastopol and Volgograd in Russia and Sapporo, Japan. 71% in Moscow, 75% in Kyiv, 63% in Jerusalem. 32% in Wuhan.

2030 Nov 25 British Empire-South Africa to Australia (Saros 133) Hepton Cycle Monday Eclipse. 1211 days from 8-2-2027 Mideast Eclipse. 2422 days from USA 4-8-2024 eclipse. Creating an extraordinary cup of Empire (see all land masses to north of eclipse) Almost all open ocean except crossing entirety of South Africa (totality in Durban) and southern Australia. Enters first at tiny town of Mount Mary. Reaching almost perfectly to the Gold Coast just north of Brisbane.

Occurs on what would have been JFK Jr's 70th Birthday.

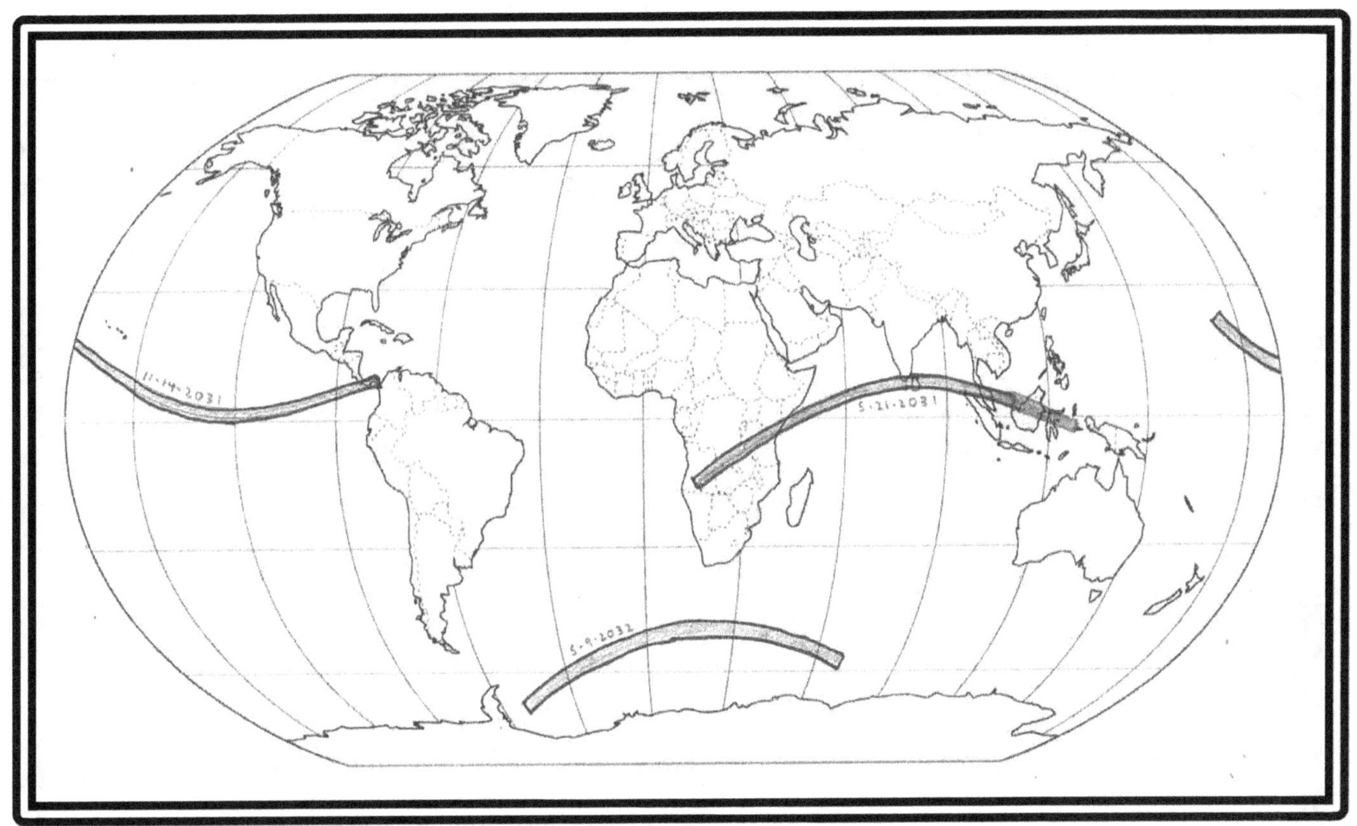

2031 and 2032 Last complete eclipses before 2033

2031 May 21 Dar Es Salaam Ring of Fire. Africa to Indonesia/Malaysia (Saros 138)
2031 Nov 14 Hybrid/ Pan-American Isthmus Totality (Saros 143)
2032 May 9 Antarctica (Saros 148)

2031 May 21 Dar Es Salaam Ring of Fire. Africa to Islands (Saros 138) Ring of Fire--Africa to Indonesia...across the middle of Africa, beginning in Angola, exiting with annularity at Dar Es Salaam (place of peace) in Tanzania (3rd time in 60 years). Crosses Indian Ocean and grazes southern peninsula of India, grazing the Temple of the Sun in Thiruvananthapuram for the third time in 21 years. Crosses Sri Lanka, once again to trace the tsunami path backwards towards Indonesia. Annularity in Ernakulum (which means "the abode of Lord Shiva") and Jaffna, Sri Lanka. Goes just north of Kuala Lampur over border of Thailand and Malaysia before ending in Indonesia.

2031 Nov 14 Isthmus Totality (Saros 143) Open ocean eclipse (without touching a single island) ending with totality at sunset just a few miles before the Atlantic Ocean directly over the absurdly small patch of land that constitutes the most extraordinary land bridge on planet earth, the Panama Isthmus, The northern and southern hemispheres are connected in the last total eclipse before the eastern and western hemispheres will be connected in the Great 3-30-2033 Total Eclipse.

2032 May 9 Antarctica Ocean Ring of Fire (Saros 148) Is there any significance to this little eclipse that touches no land in the middle of nowhere? Or is it a cosmic re-alignment to provide the necessary positioning set-ups for the ones that do bear meaning? The 530th anniversary of Columbus setting sail on his fourth and last voyage in search of a passage to Asia (May 9th 1502). This is the last eclipse before the Great Bering Strait eclipse of 2033.
partial occurs on 11-3-2032

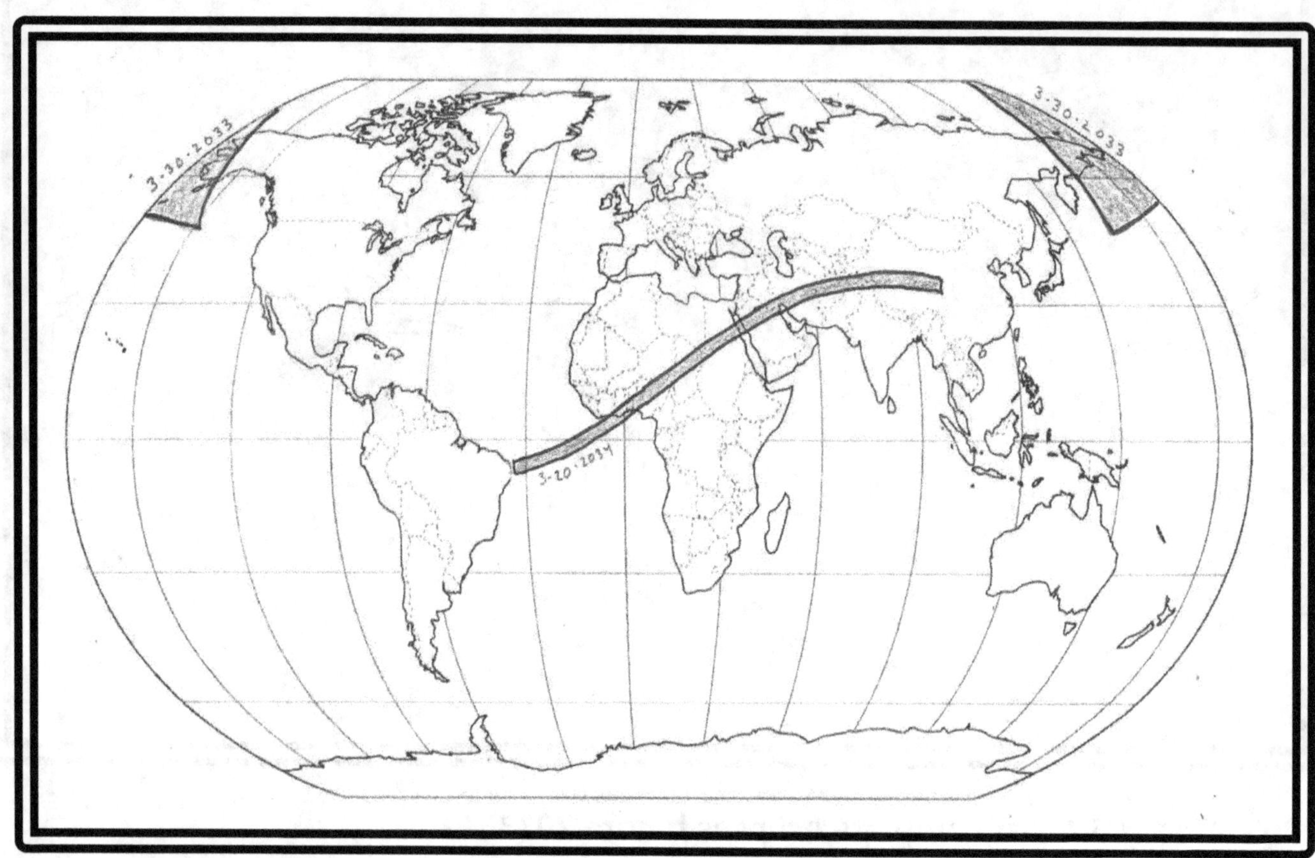

2033 and 2034- The Last Eclipses of Spring Equinox Metonic

2033 March 30 Bering Strait 2,000 Anniversary of Christ Entry to Jerusalem (Saros 120)
2034 March 20 Spring Equinox Great Civilizations Egypt Cross (Saros 130)

These two incredible eclipses are considered together for a number of reasons.

Firstly, they stand independent of the eclipses around them. The March 30th eclipse is the first complete eclipse in 325 days, since May of 2032. It is the only complete solar eclipse in 2033- just as the 2015 North Pole eclipse was the only complete solar eclipse in 2015. Both belonged to Saros 120, which performs its last complete eclipse right here after centuries of stunning work.
It is the first total eclipse since 2030 Nov 25.

The 2034 eclipse is the next solar eclipse of any kind on earth and occurs on the Equinox.

These two eclipses are exactly one lunar year apart and occur around the 2,000th anniversary of the traditional dating of the Crucifixion and Resurrection of Jesus Christ, Apr 3 33 AD.
As you will see, the details of the March 30th eclipse are extraordinary.

Just as interesting is the aesthetic balance of the two eclipses when placed on a map. The Bering Strait eclipse necessarily must occupy both corners of the map at once, while the Great Civilizations Equinox eclipse is plainly positioned almost perfectly in the center of the world.

Mar 30 2033 Bering Strait 2,000 Anniversary of Christ Total (Saros 120)

4 days before the traditional 2,000th Anniversary of the death of Jesus Christ on April 3 33 AD, a total solar eclipse perfectly crosses the Bering Strait, giving totality to both eastern and western hemispheres at the same time. This is the only place this could possibly happen, with both Russia and the USA having a total eclipse simultaneously. No other nation will experience totality.

Not only does eclipse cover two hemispheres simultaneously, it also straddles the international date line, so it simultaneously occurs on two days at once! The eclipse happens on 3-30 in the USA and 3-31 in Russia. Totality in Russian town of Provideniya ("Providence") and American town of Nome.

Definitely a Numeric oddity in the date of 3-30-2033. It is a completely balanced palindrome number, perhaps the only one in the entire survey. Abbreviation 3-30-33 is also balanced.

This eclipse is peculiar in its separation too. It is 18 months removed from the last total eclipse on earth (11-14-2031) which crossed perfectly the northern and southern hemispheres at the Panama Isthmus. It is one lunar year away from the next complete eclipse, the 2034 Spring Equinox Eclipse.

The only other solar eclipse in 2033 is a partial on 9-23

Mar 20 2034 Spring Equinox Great Civilizations Egypt Cross Total (Saros 130) Discussed in much greater detail elsewhere. The Spring Equinox Eclipse travels from the coast of South America across Africa to Egypt, scraping northen Medina (the last hometown of Mohammad) as it crosses Arabia, onward through Iran, Afghanistan, Pakistan (totality in Islamabad), India, Tibet and ending in China.

Jerusalem will see 74% coverage of the sun. Occurs of course on the Spring Equinox, only the ninth Spring Equinox eclipse in the last 2500 years. 19 years after North Pole Spring Equinox eclipse.

It is New Year's Day for Persian, Muslim, Zoroastrian and Bahai Calendars and New Year's Eve for India. It defines the coming of first month of Nisan for Jewish calendar, which begins the next day.

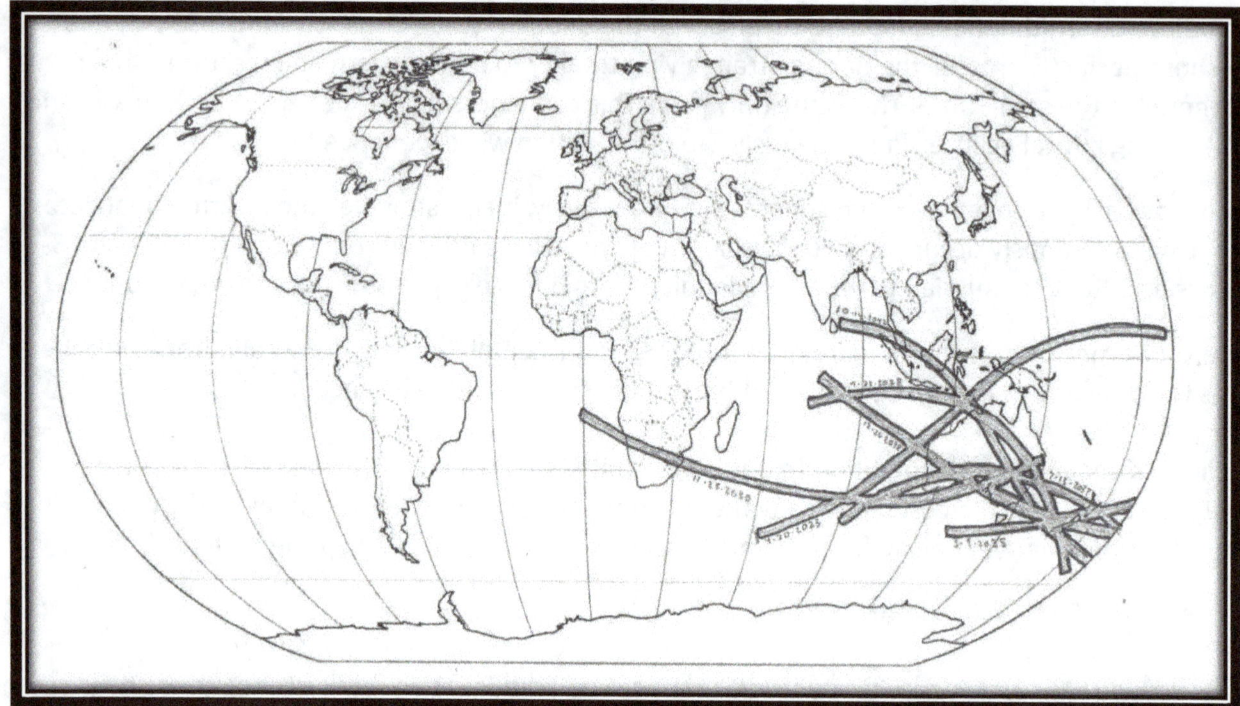

Lunar Eclipses

This work is not about lunar eclipses, but we should talk about them a bit. A lunar eclipse is a very different phenomenon than a solar eclipse, except for the fact that they both occur during eclipse seasons as the spheres begin to align on the perspective node. They often accompany each other on the eclipse nodes about 14 or 15 days apart.

Solar Eclipses must happen on the new moon. Lunar eclipses must happen on the full moon.

Since there is a certain plane, a node, where the earth, sun and moon align to lesser or greater degrees twice a year, when there is a solar eclipse, there will be a lunar eclipse. Lunar eclipses can be Total, Partial or Penumbral. A Penumbral Lunar Eclipse involves the lighter fringe shadow of earth. It is quite faint and can often happen without being noticeable.

In a lunar eclipse, the moon enters the shadow of the earth. Eclipses tend to last several hours, with maximum eclipse up to an hour and a half possible totality. Compare that to about 7 minutes maximum totality for a solar eclipse. That's because the solar eclipse is such a different phenomenon. In a solar eclipse, the moon must go in front of the sun, perfectly blocking it out. In the lunar eclipse, the moon enters the shadow of the earth.

In other words, complete solar eclipses involve a pinpoint shadow being cast upon the earth. Lunar eclipses involve viewing the moon entering the earth's rather large shadow which is being cast into space.

People have a much greater chance of witnessing a lunar eclipse during their lives than a solar eclipse. Lunar eclipses can easily cover half the world, and often more than that. Any place on earth will have a total lunar eclipse a few times per decade. Still, some folks don't see many of them due to the fact that they often happen when we're sleeping! Also, clouds can obscure a lunar eclipse. Many people don't know when lunar eclipses are happening or care about them when they do.

A single year with two total lunar eclipses happens about an average of every three years. Very occasionally, like 1982, you can have 3 total lunar eclipses in a single year.

Most lunar eclipses are not total. In fact, more than 60% are partial or penumbral. Penumbral lunar eclipses can often happen without being noticed. The impressiveness of a partial depends on the eclipse.

A few words about Blood Moon Tetrads (2014 and 2015/ 2033 and 2034):

Astronomers use the name Tetrad when there are four consecutive eclipse seasons which each contain a total lunar eclipse. That usually means two years in a row each with two Total lunar eclipses. That is four totals in a row, separated typically by around 177-day intervals.
Tetrads live in cycles of around 600 years. For about 300 years, there are no tetrads. This is followed by a period of roughly 300 years where a tetrad occurs roughly every 15 to 20 years.

Currently, we are about halfway through the second half of a 300-year cycle.
There have been 7 Total Lunar Tetrads since 1900, and an 8th on the way. Here they are:
1909 to 1910.
1927 to 1928.
1949 to 1950.
1967 to 1968.
1985 to 1986.
2003 to 2004.
2014 to 2015.
2032 to 2033.

Far more unusual is when Tetrads land on Jewish High Holy Days. A Tetrad of four Total Lunar Eclipses landed on the high holy days of Judaism in both 2014 and 2015. That is, there was a total lunar eclipse each of those years on both Passover in the Spring; and Sukkot, the first day of the Feast of Tabernacles, in the Fall. These bookend a total solar eclipse on the Spring Equinox March 20 2015.

Here are the dates of the famous Blood Moon Tetrads:
April 14/15 2014—Passover Total Lunar Eclipse-October 8 2015—Sukkot Total Lunar Eclipse
March 20 2015 –Spring Equinox North Pole Eclipse (Saros 120)
April 4 2015—Passover Total Lunar Eclipse-September 28 2015—Sukkot Total Lunar Eclipse

These eclipses created quite a bit of excitement and some doomsday prophecy, due to the unusual alignment with the Jewish calendar. Little attention was paid to the Spring Equinox Eclipse itself, and virtually none was paid to the incredible 19-year Solar Eclipse Metonic that it signaled (and which is the focus of this work).

The Blood Moon Tetrad of 2014

2014 Apr 14/15 Passover Total Lunar Eclipse. The first eclipse, on the Jewish Passover, occurred April 14th and 15th 2014. It was visible primarily in the western hemisphere. Abraham Lincoln was shot at 10 pm the evening of the 14th of April 1865 on Good Friday/Passover and pronounced dead the morning of the 15th exactly 149 years before this eclipse. See personalities section for more. April 14th and 15th are also the 102nd anniversary of the sinking of Titanic, which struck an iceberg on the night of April 14 1912 and sank a few hours later in the early morning of April 15th.

October 8 2014 Sukkot The second lunar eclipse occurred on the first day of the Feast of Tabernacles. Visible through Americas and Asia.

March 20 2015 –Spring Equinox North Pole Eclipse (Saros 120) This extraordinary solar eclipse was looked at earlier in this section. It is an isolated eclipse, with no other complete solar eclipses in 2015.

April 4 2015 Passover Lunar Eclipse Visible throughout Americas and Asia. 47 years to the day after Martin Luther King (Apr 4 1968) was murdered in Memphis, Tennessee. King was killed 8 days before a Passover Total Lunar Eclipse *only* visible from the Americas April 12th and 13th, 1968. April 4th 33 AD is the traditional day for Jesus resting in the tomb on the Sabbath 1982 years before, on the day after he was crucified (by traditional dating) on April 3rd, 33 AD.

September 27/28, 2015 Lunar Eclipse Sukkot Visible in Europe, Mideast and Americas (including Jerusalem)

Interestingly, the 1949/1950 and 1967/1968 Tetrads also landed on Jewish high holy days. It is hard to overstate how rare this phenomenon is. Apparently, it happens on average only once every 500 years, and it has now happened 3 times in 75 years. Another High Holy Day Tetrad occurs in 2033 to 2034.

Notably, the High Holy Day Passover eclipse of 1949 Apr 13 occurred within the year that Israel declared independence, on 1948 May 14 and 98 days before the war was concluded on July 20 1949.

The 1967 April 24 Passover Eclipse occurred 40 days before the Six-Day War when Israel gained control of Jerusalem on June 7 1967.

A High Holy Day Tetrad also occurred in 1493/1494, beginning the next year after the expulsion of the Jews from Spain in 1492 and the "discovery" of America.

Two Total lunar eclipses occur in 1968, also on Passover (April 13, eight days after King murder) and Sukkot, October 6th.

The Jewish Temple in Jerusalem was fully destroyed by the Romans on August 10th, 70 AD after a several month siege. *Exactly* 2000 eclipse years from that date was August 21st 1968. 1968 was an exceptionally important year in the history of the world, for many reasons, not the least of which were the assassinations of both Martin Luther King and Robert Kennedy.

In the current 19-year period which contains the anniversary of the Crucifixion/Resurrection, a Tetrad of Total Lunar eclipses reappears in 2032 and 2033 and then in 2033 and 2034, there are also four lunar eclipses (though not all total) on the High Holy Days. 19 years after they did the same on the same days in 2014 and 2015. These upcoming lunar eclipses on holy days are:

April 14/15 2033 Passover Total Lunar
October 8, 2033 Sukkot Total Lunar
April 3 2034 Passover Penumbral Lunar
September 28 2034 Sukkot Partial Lunar

This is essentially a Blood Moon Tetrad except two of the eclipses are not "Total".

The Passover eclipse of April 14 2033 would be 2,000 years and 11 days after the Crucifixion if the Lord died on April 3, 33 AD (the traditional date). It would also be exactly 168 years after the murder of Abraham Lincoln and 121 years since Titanic sank. The Passover lunar eclipse of April 3/4 2034 is on both the exact 2001st traditional anniversary of the Crucifixion and simultaneously 66 years after the murder of Martin Luther King.

The Tetrads of 2033 and 2034 sandwich a Spring Equinox Total Solar Eclipse, just as Tetrads sandwiched the Spring Equinox Eclipse of March 20 2015.

Chapter Eight

The Metonic Calendar/ The Cosmic Clock

The Metonic Calendar. There is a Cosmic Clock.

The Metonic Calendar is the proof of the Cosmic Clock. To appreciate the Metonic Calendar, we need to literally take our time. It is a simple phenomenon, but it's still easy to get mixed up in simple phenomenon. If we take a few minutes to process each aspect, however, you may learn something that few in the history of the world have known, a secret held close by only the most learned monks, astronomers, magi and mystics.

The Metonic Cosmic Clock repeats at 19 solar year intervals, 19 eclipse year intervals and at exactly 391 year intervals and then approximately 490 year intervals.

This Metonic moment's position of the moon and sun in relation to earth are almost identical to what they were 19 solar years ago and 19 years from now. This position repeats even more perfectly at the 391 year mark.

This Saros moment's position is also almost identical gravitationally as it was 19 eclipse years ago (18 years 11 days and 8 hours—which is the same as 19 solar years minus one lunar year.) It will again attain such perfection 2,290 years from now.

None of this is conjecture. None of this is theory. It is real. But first we need to confront Time.

First: Eclipses Happen in Time and Space, like Stories do

Time is space. Time is a space where stories take place, where maximum expression can exhibit its function of the creation of growth and identity. Eclipses occur in time and space, and are one of the only repeatable phenomenon that could be used as a benchmark of great time spans, larger than individual months or years, or somewhat arbitrary groupings of time such as decades or centuries. Eclipses occur independently of observers but obtain their meaning *from* observers. It's a perspective, beauty in the eyes of beholders, a relational reality.

A solar eclipse is a brief cosmic story, containing bold and important characters. It has action, suspense, hints of the danger of an eternal and universal death, a mystic climax scene of pure union, a revelation of hidden secrets, final victory and a moral. Of course, it also has love. It's a love story.

The plot: Light is unexpectedly attacked by a consuming darkness. Darkness seems to conquer for a while. Not that long ago, people were struck with terror when the life-giving Sun appeared to be eaten alive. Animals respond perplexed and anxious. Finally, the disc of the sun is covered. Night falls in the day. In the midst of eclipse, a surreal heavenly glow (the Corona-The Crown) is revealed around the sun. The sun's blinding glare is subdued in a transcendent rarity. Humans can actually stare at the edges of his glorious light for the first and maybe last time. We see the crown, the angelic light that drifts from his edges-not his face, which remains obscured by that strange little dead moon of ours. Finally, bit by bit, the tension eases as Light returns victorious. Life returns to normal.
Was a lesson learned?

In this work I present the idea that it's not only the physical eclipses that bear significance, that fleeting cosmic moment, but also that meaning is found in the eclipse paths they took, the residue of story they left behind. The last chapter on The Saros made the case that these paths are predictable for a reason. They are too easy to understand to be chance, and the proof is the in the co-ordination of three separate cosmic calendars at the 19-eclipse year mark. In this last part of human history, a few interested souls took the time to draw out these paths. Now we have these maps. There are all these shadows that were written on the earth. Do they say something?

I say eclipses tell stories, or perhaps they sing. I think it's obvious eclipses they do, but stories are not a matter of scientific logic or argument. An eclipse story is like music. The meaning can't be explained without damaging the story it seeks to express. Instead, a song is presented. It is played. I present the circumstances of complete eclipses and let the story come to the listener.
I don't claim the gift of eclipse interpretation. I do hope to present them faithfully.

I can't explain a Bob Dylan song. I can listen to it. I can memorize it and play it back for you on the guitar. There it is. It's not perfect, but I learned all the words. You can decide if you like it or not. That's all we can do with eclipses. The stories of eclipses, just like all stories, are not cut and dry science lessons. They don't belong to anybody. Rather, they belong to everybody, just like music. They can be odd, confusing, often intense. Sometimes they provoke emotion. The paths of history remind us that real people were there under those eclipses. When we begin to learn about those people and what they went through, and we start thinking about Light and Darkness and His story...well...

Fortuitously, as it so happens, complete solar eclipses last about as long as a short concert or play or movie. A couple of hours. It all comes down to that climax in the middle, when either totality or annularity comes to call. Totality can last anywhere from a few seconds to more than 7 minutes. Annularity (Ring of Fire completeness) can last up to 12 minutes or so, the length of a long song (*bye bye miss American pie*? Or maybe *In a Gadda da Vida*? -the album cut).

Second: Let's Review Conventional Calendar Time

The Metonic Cycle is Cosmic Timekeeping, time that intersects with time. It is dazzlingly four dimensional. To conceive of the Cosmic Clock, let's consider in the time we already know best.
I start with basics: conventional calendar time.

Calendar Time? Calendar time is the agreement that time can be a quantifiable space measured by an accumulation of days. Calendar Time can be dissected into chunks like hours, days or weeks; it can also be spread out like peanut butter over terms or decades or centuries. But at the foundation of time is the day, the month, and the year. Natural, repeating phenomenon create our basic notions of day, month and year. While there are a few different calendars, most typically revolve around combinations of the lunar month and solar year. There are exceptions, such as with the Bahai, who have 19 months of 19 days, still adding up to close to a solar. For most of the rest of the world, people usually try to make a year out of the motions and appearance of moon and sun.

First, the Good Old Fashioned Day. The rotation of the earth allows for day and night. This is generally the first notion of time recognized by small children and animals. From morning to evening and back is one day. That, I believe, even in these challenging times, is something we can all agree on. Or can we? Then why did George Van Den Bergh say *"a day is 48 hours"*? Think globally.

Second, the Lunar Month. The revolution of the moon around the earth, and its accompanying visible phases, allows for the idea of month ("moon"-th). It takes 29.53 days from New Moon to New Moon, the generally accepted beginning and end of the lunar month for nearly all human societies. It is calendar time based on a visible and agreed-upon cosmic reality.

Third, the Solar Year. The revolution of the earth around the sun creates the solar year. Egypt created the first known Solar calendar with 12 months of 30 days each and 5 days added at the end as early as 4,000 B.C. Did you know the Egyptian New Year is September 11th? Weird, huh?

Most early civilizations recognized the reality of the solar year early on. The observation of seasons and the rising of the sun in the exact same spot (i.e. from your village you can see the sun come up again over that mountain or rock or tree) after 365 days are clear, repeatable phenomenon that can be agreed upon. Solar Years encourage such collective observances as Solstices and Equinoxes naturally. "New Years'" typically begin for most ancient societies on Spring Equinox or Winter Solstice.

I know this is all primary school stuff, but we must build towards the Cosmic Clock.

Conventional Calendar Time Subdivisions

Each basic unit of Calendar Time has important subdivisions and amalgamations. Let's briefly touch on those, and how they relate to our study of eclipses and the Metonic Cycle.

The Hour: the hour is an interesting human creation established by sun and constellation movements. It was made more precise with the advent of the clock in the 14th Century. Greenwich mean time (Greenwich England) was an agreed upon Universal Human Clock established 11-1-1884. Universal Atomic Time was established 1-1-1960. Since the world now had atoms at their disposal, the new world coordinating the world's clocks to the same radioactive beat.

The use of 60 minutes and 60 seconds in an hour is a vestige from the Babylonian, Greek and other culture's 360 day year. The 360-day year, with a five day leap added on at the end was a typical solar year. There are 3600 seconds in an hour. 3600 is the number the Saros was mistakenly named for.

The Month is divided into weeks and amalgamated into a year. The lunar month of 29.53 days is a precise representation. The months of Rome (which we live under) are grouped around the 30 day mark.

The Seven Day Week originated in Mesopotamia and is established in Genesis, the first Book of the Bible. Roman Emperor Constantine officially adopted it in 321 AD. The week is the most logical division of the cycles of the moon with its four defined geometric appearances. New Moon/ Quarter Moon/ Half Moon/Full moon/ Quarter Moon. 7 days per each arrives at 28 days, within a day and a half of a full lunar month. Not sure what to do with the next day and a half? Me either.

52 weeks x 7 (364 days) is very close to the true solar year. There have been other lengths for weeks in history for other societies but the 7 day week won out across the world in the end.

The year is divided into seasons or months and amalgamated into generations or decades or lifetimes.

Twelve "Moons" brings the Lunar Year of 354.43 days, which is 11 days off of a solar year. To follow 12 lunar months as a lunar year requires the insertion of an 11-day leap week to accommodate solar calendar reality. The Luni-solar calendar has been a feature of human societies for Millennia. In the luni-solar calendar, the month is still created by the moon, retaining its individual obvious time marking characteristics. The visible changing of the moon's phases add a reality to the month not necessarily available in Roman solar calendars. Current lunar-solar calendars include the Chinese, Muslim, Jewish, Hindu and Vietnamese cultures.

The current division of the solar calendar familiar to western culture takes the "true" solar year, 365.25 days, and shapes months of various lengths with no regard to lunar phases. It had its origins in the Julian Calendar first decreed by Julius Caesar in 45 B.C. which honored various Roman gods and such. Julius Caesar even got the month of July named after him- and a knife in the back. The Julian calendar had a leap year, adding a single day every 4 years.

This became cemented more accurately by the Gregorian Calendar in 1582 which gave us the proper leap year arrangement we currently enjoy, skipping fourth year leap years at the turn of each century.

To recap calendar time. A lunar year is 354.3 days. That is 12 full cycles of the moon.
A solar year is 365.25 days. That is one full revolution of the earth around the sun.

The Lunar Year appears in Eclipse repetitions.

Eclipses often occur exactly one lunar year apart from each other in triads.
Example: There was a Ring of Fire Eclipse from Indonesia to Sri Lanka on Dec 26, 2019.
One lunar year later, there was a Total Solar Eclipse in South America on Dec 14, 2020.
One lunar year later, there was a Total Solar Eclipse in Antarctica on Dec 4 2021.

These eclipses are separated by 12 Moons (354.3 days). Of course, each lands on the new moon.

There was a partial solar eclipse on April 30 2022—77 years after Hitler's death day. One lunar year later there is a solar eclipse on April 20 2023—134 years after Hitler's birth day.
These are just two examples. The phenomenon is persistent, but the pattern skips around.

The lunar year repetition happens because a Saros Cycle is exactly 19 eclipse years long. 19 eclipse years is 19 solar years minus one lunar year, believe it or not.

The Saros is its own cosmic time impulse interacting with the larger Metonic Cosmic Clock.

The Metonic Cycle itself is the phenomenon of **repeating dates in the eclipse calendar** due to God's precise alignment of lunar, solar and eclipse cycles calendar every 19 solar years.

Metonic dates (repeating eclipse dates) can show up at 19 year, 38 year, or 57 year intervals, then it lurches 8 years, and sometimes comes back. Rarely a repeating date can show back up at 84 years, and even 103 years! But they don't repeat forever. Eventually the date moves on until the next time they re-align, around 500 years later.

Example: There was an August 21 1914 eclipse that announced World War I.
19 years later on August 21 1933 there was the Jerusalem Ring of Fire.
84 years after that there was an August 21 2017 eclipse that announced the division of America.
Between these eclipses were a couple eclipses on August 20 and one on August 22.
Were they all related? Absolutely. They are repeating Metonic dates which represent real repetitive cosmic alignments of sun, earth and moon. dignified by the irrefutable appearance of solar and lunar eclipses, clicks of a cosmic clock which also acknowledges solar and lunar years, as well as Saros Cycles. The dates must change. They move on. Time marches on. By 2044, The Aug 21 Metonic has moved forward to Aug 23.

What's a Metonic Day? It's the time around the day in question. It signals the heartbeat of the cosmic clock finding common ground among three calendars in real space and time.

A Day is 48 Hours: One would think that a day would be a concrete phenomenon. Sun comes up. Sun goes down. It turns out, however, as Copernicus and Kepler discovered, that this thing we call a day could also be described as a local illusion.
Or, as George Van Den Bergh points out in his book, *The Universe in Space and Time*
"A *day is 48 hours.*"
Or as country singer Alan Jackson croons, " *It's five o'clock somewhere."*

The earth is spinning, chasing its terrestrial tail. When it's supper here, it's breakfast somewhere. Even in a single nation it can be two days at once. We've even drawn an International Date Line on the Pacific Ocean, where you can step on one side of the ship and its today and then step on the other side of the ship and its tomorrow, or maybe yesterday. It can be difficult to fathom the reality of a day on a rotating host, which is subject to our old friend, perspective.
Van Den Bergh says flatly and confidently, *"A Day is 48 hours".*

Myself? I found a globe. I spun it around. That's the only way it made sense, and it did... I guess.

Metonic Days and Metonic anniversaries

I bring up Metonic anniversaries, solar year anniversaries and eclipse year anniversaries at many places throughout the work. The Metonic day is flexible one day in either direction.

For example, take President John F. Kennedy, whose birth and death days are reflected distinctly in the 20th Century Metonic. His actual Massachusetts birthday is May 29 1917. Still, I have never found mentioned in any records the hour of his birth. It could have been 2 in the morning. It could have been 11 at night. It may very well have been that everyone in his vicinity agreed it was truly May 29 when he poked his head out. But whatever the clock said when he arrived, it is still true that other people on earth were living in either May 28 or May 30. What does that mean?

Time is a construct, but it's not entirely illusion. Time is a space for story created by cosmic reality. It's flexible but not irrational. This flexibility has its limits. May 29 maybe the same as May 28 because they exist simultaneously on earth. But I can't logically say that June 2nd is the same as May 29. Why? Because it's not. Even two days before a date (May 27) and two days after a date (May 31) is not a reasonable, logical anniversary for May 29. No way can you confuse a date two days apart with an anniversary of the date itself. It just feels false even considering it.

But the day before? The day after? That's a different story. Why? Because time has a fluid nature as it tries to mark a single day on a spinning planet. I believe we actually feel this property of time, though we may never have described it. I have personally found it particularly affects the senses as you *approach* an anniversary. There is almost a doppler effect, where an approaching date has more significance than a departing one. The same sensation is common in eclipses, as people and animals are far more excited as the eclipse approaches climax than as it leaves.

In this work, I often say that something is "on" a Metonic if it's in that in that fluid one-day place. If I am discussing the May 29 Metonic, a May 28 or May 30 date is clearly within that parameter. I won't ever say it's on the Metonic if its two days away in either direction.
When the Metonic truly lines up, you can feel it.

OK, so here's a perfect Metonic line-up, with a catch. Let's continue with the JFK example. May 29 1919 was JFK's second birthday. On that day there is a very famous total solar eclipse in South America ("Einstein's Eclipse"). Now, in this case, we find it to be exactly May 29 in the exact same time zones. Also, it's only two years from JFK's birth, with no leap year in between. That's a nearly perfect anniversary! That's an easy one, it seems. Well, it's great and all, but it's not the Metonic reality we are considering. All we learned is that two years went by and hey, now there's an eclipse! That's not cosmic timekeeping.

Cosmic timekeeping is when you travel 19 years from May 29 1919 and find another eclipse, again on May 29 1938. Which we do. That's a Metonic date. Then it's 1984 (65 years after 1919) and we have a May 30 eclipse over New Orleans, Washington DC and Atlanta. It would have been JFK's 77th birthday just hours before this eclipse if he hadn't been murdered years before. The date is officially May 30, not May 29. But 1984 was a leap year. If we hadn't added a day 12 weeks before, it would still be May 29. See? There are too many artificial variables to impose a human clock-driven 24-hour prison on the concept of a single day. Real Time is too fluid a property to be so constrained.

For the purposes of this work, if it's within a day of the Metonic in question, it might be worthy of mention. Maybe it is, maybe it isn't. I'm just piling up peculiarities. What interests me most is the apparent resonance of anniversaries of various kinds in the Metonic Calendar. We consistently find the same dates appearing in eclipses every 19 years. Then 38 (2x 19) years intervene and we might find the same dates appearing or maybe it's started to move.,.

The Metonic Cycle-How Peculiar is it? The numbers.

Hold on to our hats. If we get this right, we know more about the cosmic clock than nearly every soul in history. This is the greatest conglomeration of coincidences in the reality we call Time.

The Metonic Cycle is named after Meton of Athens (5th century B.C.) who first wrote about it, though it was also known by ancient astronomers of both hemispheres. The Metonic Cycle is the intersection of the lunar and solar calendars which occurs almost precisely every 19 year Solar Years.

19 solar years is the same as 235 lunar months.

Say it again, 19 solar years equals exactly 235 lunar months. Whatever phase of the moon it is today, it will be that same phase 19 years from now.

The Metonic cycle synchronizes the solar year and the lunar month. Check it out:

A Solar year = 365.24 days. 365.24 x 19 = 6,939.60 days

It is usually rounded to 6,940 days. The difference between 19 solar years and 235 Lunar months is only about *two hours*. In other words, by extraordinary Cosmic arrangement, these two primary cosmic patterns align with each other in nearly perfect arrangement every 19 solar years. What does that mean? Simply put: if there is a new moon tonight, there will be a new moon 19 years from now on this same date, give or take two hours. Likewise, 19 years ago on this date, there was the same phase of the moon as there is tonight. After a couple of centuries, the date shifts slightly ahead due to slight offsets of calendar time and orbital reality.

It is not some kind of mathematical necessity that the two primary astronomical calendars of humanity should align so precisely in such an accessible time frame as every 19 years. They could just as easily have aligned only once every 214 years or every 672 years given different orbital realities. Indeed, it's unclear if *anyone* has calculated *any* number where sun and moon synchronize so closely at a solar year anniversary!

The Metonic cycle is a convenient synchronicity still utilized for calculating movable feasts such as Easter. The Metonic cycle was known to the Celts, who used it as their "Great Year". It was used by the Babylonians, Persians, Chinese, Hebrew, Mayan and other ancient societies.

But the synchronicity is not nearly over yet. As cool as it is to have such an alignment, the synchronicity becomes downright astonishing as we approach the deeper secrets of Draconic-Metonic Reality. Ladies and gentlemen, here comes the Eclipse year.

The Draconic Month, Draconic Year and the Eclipse Season

There is a kind of year that hardly anyone knows about except astronomers. But it is a year, nevertheless.. It is the eclipse year or Draconic Year. The dragon year.

Here's the full scientific explanation from Encyclopedia Brittanica:
The draconic, or nodical, month of 27.212220 days (27 days 5 hours 5 minutes 35.8 seconds) is the time between the Moon's passages through the same node, or intersection of its orbit with the ecliptic, the apparent pathway of the Sun. The moon's draconic period of revolution (the so-called draconic month of 27.2 days) has important significance in the theory of solar and lunar eclipses. The period of time it takes the Sun to travel from the Moon's North (or South) Node around the zodiac and back is called the "Draconic Year". Because the Moon's Nodes move backwards 19-20 degrees a year, the Draconic Year is shorter than the usual calendar year by several weeks. Its average length is 346.62005 days.

The difference between a solar year and a draconic year is very close to 19 days. In the draconic month of 27.2 days, sun and moon are crossing the ecliptic plane. When they return around to about the same spot, an eclipse year has elapsed.

The term "draconic" (or dragon-like) is connected with an idea of the ancients, according to which the sun and moon are devoured by a dragon during an eclipse.

The facts:

Is there a Draconic Day? I don't think there is a draconic day.

Draconic week? There appears to be a Draconic/Metonic week, about 11 Days 8 hours long. 33 of these weeks fit in a year. You can see it appear in these calendars, but it's difficult to pin down.

The draconic month: The draconic, or nodical, month of 27.212220 days (27 days 5 hours 5 minutes 35.8 seconds) is the time between the Moon's passages through the same node, or intersection of its orbit with the ecliptic, the apparent pathway of the Sun. It takes less time than the phases of the moon take to complete (29.5 days).

The Draconic Year/ The Eclipse Year is the time taken for the Sun (as seen from the Earth) to complete one revolution with respect to the same lunar node (a point where the Moon's orbit intersects this ecliptic plane). Eclipses occur only when both the Sun and the Moon are near these nodes.

In layman's terms, an eclipse year is the time it takes for the moon and sun to sort of line up in the same place in relation to each other and the earth again. It is a year based in relations of three cosmic entities from a cosmic perspective.

There are two eclipse seasons every eclipse year, approximately 173 days apart. The duration of the eclipse year is 346.62 days. Eclipses within an eclipse year often occur 177 days apart, which is exactly half of one lunar year. An eclipse season is one of only two periods during each year when eclipses can occur, due to variations in the angle of the moon in relation to the sun. Each season lasts about 35 days and repeats just short of six months later; thus two full eclipse seasons always occur each year. That does not mean there are always complete eclipses during that time, but often there are.

The Eclipse Year is a very real alignment of these three great objects-Sun, Moon and Earth.
Again, I know I am repeating. There are three types of years:
The one we most recognize: **Solar Year** of 365.24 days: The time it takes for the earth to revolve around the sun.
The source of our idea of month: **The Lunar Year**-354.3 days: 12 Lunar Months of 29.5 days each
And the secret year: The **Draconic (Eclipse) Year:** The time required for the Earth, Moon and Sun to precisely realign on an elliptic plane. 346.6 days

Here we go for the greatest coincidence in existence, the Draco Metonic Calendar:

The Metonic Calendar aligns both solar and lunar calendars every 19 years. And we just found out there is also a secret eclipse year, the "draconic" dragon year.

Now it gets absolutely amazing. Remember that the eclipse year involves a completely different cosmic perspective than a lunar or solar year. The Eclipse Year is an alignment of three objects on an elliptic plane irrelevant to both solar and lunar orbital necessities. The orbit of moon around sun is also kinked about 5 degrees, which means eclipses don't happen in the exact same spot.

It lines up, doesn't it? We know that already. But how close?
As we saw, the eclipse, or draconic month is 27.2 days. So it turns out that 255 Draconic months comes out to 6,939.11 days. That's very, very close to the 6,939.60 days of 19 solar years and the 6,939.68 days of the 235 lunar months. Only hours separate these huge arrangements of time which all align at exactly 19 solar year spans. These dates, these three calendars, are closely aligned enough that ECLIPSES CAN REPEAT on the same days of the year every 19 years for up to a century!

Recap of the Draco-Metonic Calendar:

Months:
Lunar month: 29.5 days
Eclipse (Draconic) month: 27.2 days
Years:
Solar Year: 365.25 days
Lunar Year: 354.3 days
Eclipse Year: 346.6 days
19 Year Draco-Metonic Alignment
19 solar years: 6939.6 days
235 lunar months: 6939.68 days
255 Draconic months 6939.11 days

The figure is typically rounded to 6,940. An extraordinary precision of alignment considering nearly 7,000 days. The reality is so impossible that it's unclear if any similar alignment at any point has ever been calculated. Is there any other group of solar years other than 19 that matches phases of the eclipse and lunar calendar so precisely within hours? Does anyone know?
Again: The Saros Cycle Alignment: 18 years, 11 days and 8 hours (exactly 19 eclipse years)
19 eclipse years: 6585 days
223 lunar months: 6585 days
239 anomalistic months (perigee to perigee) months 6585 days

The Saros has a separate almost perfect precision alignment that allows for precise eclipse prediction and history. There is no scientific "reason" for this alignment of different cosmic timekeeping, just as there is no scientific "reason" for the sun and moon being the same size in the sky. Science tends to acknowledge these realities only on its farthest back pages, merely offering them the trifling distinction of "coincidence". 19 solar years provides one alignment. 19 eclipse years provides the other alignment. Within these amazing cycles are other cycles.

So what's left to do? Well let's see if we can find our birthday in the Metonic! Probably about one in eight people can do so. Here are the current Metonic repeating eclipse dates graphed so that you can see their progression over the course of the last 120 years or so.

Metonic Date Shifts last 120 or so years

Current solar eclipse dates to help us visualize movements of the Metonic clock over the course of decades. There are about three impulses, or days, a month which reside in 11 day "weeks". These go left to right. As you look down the row, you can see how the date itself slowly shifts over time.

1st Metonic Week	2nd Metonic Week	3rd Metonic Week
January		
1-1-1889	1-11-1899	1-22-1898
1-3-1908	1-14-1907	1-23-1917
1-5-2038	1-15-2010	1-26-2028
February		
2-1-1897	2-13-1896	2-23-1906
2-3-1916	2-14-1915	2-25-1952
2-5-2000	2-15-2018	2-26-1998
March		
3-6-1905	3-17-1904	3-26-1895
3-7-1932	3-17-1923	3-27-1960
3-9-2035	3-20-2015	3-29-2006
April		
4-6-1894	4-16-1893	4-26-1892
4-6-1903	4-17-1912	4-30-1957
4-8-2024	4-20-2023	4-29-2014
May		
5-7-1902	5-18-1901	5-28-1900
5-9-1967	5-20-1966	5-29-1917
5-10-2013	5-21-2031	6-1-2030
June		
6-8-1899	6-17-1909	6-28-1908
6-8-1918	6-19-1907	6-29-1927
6-10-2002	6-21-2020	7-01-2001
July		
7-10-1907	7-18-1898	7-29-1897
7-9-1945	7-20-1963	7-30-1916
7-11-2010	7-22-2009	8-2-2027

August

8-9-1896	8-20-1895	8-30-1905
8-10-1915	8-21-1914	8-31-1913
8-12-2026	8-21-2017	9-1-2016

September

9-9-1904	9-18-1895	9-29-1894
9-11-1969	9-21-1903	9-30-1913
9-11-2007	9-22-2006	10-3-2005

October

10-9-1893	10-20-1893	10-31-1902
10-10-1912	10-22-1911	11-2-1910
10-14-2023	10-25-2022	11-3-2032

November

10-31-1902	11-11-1901	11-22-1900
11-2-1910		11-21-1938
11-3-2032	11-14-2031	11-25-2030

December

12-1-1890	12-13-1898	12-23-1908
12-3-1918	12-12-1909	12-25-1954
12-4-2021	12-14-2020	12-26-2019

Example of pattern---Merry Metonic Month of May

We now have a chance to look at macro patterns of the Metonic progression.

The spaces between years of similar-dated eclipses will be either 19 years or 8 years.

Metonic Solar Eclipse Days in May last 130 years or so.

First Impulse/Early Day in May Metonic (eclipses on each of these days)

5-7-1902— (Saros 146) after this, the 8 year lurch -, jumps 2 days with 8 year lurch

5-9-1910 – (Saros 117)-it shifts back to 19 year Metonic for three eclipses

5-9-1929— (Saros 127) then 19 years

5-9-1948 - (Saros 137) then 19 years

5-9-1967 -- (Saros 147)-here comes 8 year lurch again

5-11-1975-- (Saros 118) -and then back to 19 year Metonic

5-10-1994-- (Saros 128) then 19 years

5-10-2013-- (Saros 138) then 19 years

5-9-2032 --- (Saros 148)(then 8 year lurch)

5-11-2040---(Saros 119) Back to 19 years

5-11-2059—(Saros 129)

See the pattern? The Metonic (19 year) eclipses are linked for 3 eclipses, then an eight year lurch to the next number…Now look at the next week in May over course of 170 years.

Second Impulse/Middle Day in May Metonic

5-17-1882—(Saros 126) afterwards 19 years

5-18-1901---(saros 135) then 19 years

5-18-1920---(Saros 146) then 8 year lurch

5-19-1928 --(Saros 117) 19 years 3 times in a row

5-20-1947—(Saros 127)

5-20-1966—(Saros 137)

5-19-1985—(Saros 147) then 8 year lurch

5-21-1993—(Saros 118) then 19 years 3 times in a row

5-20-2012—(Saros 128)

5-21-2031---(Saros 138)

5-20-2050—(Saros 148) then 8 year lurch

5-22-2058--- (Saros 119)

The Great Metonic- Repeating eclipse dates can appear for up to a century, though there is no exact figure, and some dates don't repeat, or sometimes even appear. These dates do their thing and then disappear, only to line up very closely again after about 500 years. Once the Gregorian calendar arrives, it is replaced by the more exact 391-year interval. But our current metonic is the same as Jesus' metonic either way! It is a mind-blowing exploration of perceived time when you face the reality of orbital and ecliptic truth.

The following table is an example of four very different times in history with similar Metonics. Follow the dates left to right and see how closely these times match up.

List of repetitive solar eclipse Metonic dates in four historical time frames:

2015/2034	14 to 35 AD	554 to 575 AD	488 to 468 BC
(current)	(Jesus)	(end of Rome)	(Buddha/ Pythagoras)
2015 Mar 20	0014 Mar 19	0553 Mar 19	-468 Mar 19
2016 Mar 9	0015 Mar 9	0555 Mar 09	-487 Mar 09
2016 Sep 01	0015 Sep 2	0555 Sep 01	-487 Sep 01
2017 Feb 26	0016 Feb 26	0556 Feb 26	-486 Feb 26
2017 Aug 21	0016 Aug 21	0556 Aug 21	-486 Aug 22/21
2019 Jul 02	0018 Jul 01	0558 Jul 01	-484 Jul 01
2019 Dec 26	0018 Dec 26	0558 Dec 25	-484 Dec 25
2020 June 21	0019 Jun 21	0559 June 21	-483 Jun 21
2020 Dec 14	0019 Dec 15/14	0559 Dec 14	-483 Dec 15/14
2021 Jun 10	0020 Jun 10	0568 Jun 11	-482 Jun 10
2021 Dec 4	0020 Dec 03	0568 Dec 04	-482 Dec 04
2023 Apr 20	0022 Apr 19	0562 Apr 19	-480 Apr 19
2023 Oct 14	0022 Oct 14	0562 Oct 14	-480 Oct 14
2024 April 08	0023 Apr 09	0563 Apr 08	-479 Apr 9/8
2024 Oct 02	0023 Oct 03	0563 Oct 3	-479 Oct 02
2026 Feb 17	0025 Feb 16	0565 Feb 16	-477 Feb 17
2026 Aug 12	0025 Aug 12	0565 Aug 11	-477 Aug 12
2027 Feb 06	0026 Feb 05	0566 Feb 06	-476 Feb 06
2027 Aug 02	0026 Aug 01	0566 Aug 01	-476 Aug 0
2028 Jan 26	0027 Jan 27	0567 Jan 26	-475 Jan 25
2028 Jul 22	0027 Jul 22	0567 Jul 22	-475 Jul 22
2030 Jun 01	0029 Jun 01	0569 May 31	-473 Jun 01
2030 Nov 25	0029 Nov 24	0569 Nov 24	-473 Nov 25
2031 May 21	0030 May 21	0570 May 20	-472 May 20
2031 Nov 14	0020 Nov 14	0570 Nov 13	-472 Nov 14
2032 May 9	0031 May 10	0571 May 09	-471 May 09
2033 Mar 30	0032 Mar 29	0572 Mar 29	-478 mar 29
2034 Mar 20	0033 Mar 19	0573 Mar 19	-469 Mar 20

These patterns are virtually identical, considering leap years and orbits and time zones.

The Dates we live in:

Here's a few famous event anniversaries on Metonic eclipse dates we live in and around. Some are interesting and some trivial.

March 20-Spring Equinox

Mar 09—Parthenon consecrated in Athens 432 BC, US napalms Tokyo in 1945

September 11- Egyptian New Year

Feb 26-First color movie, Johnny Cash birthday, Anniversary of first recorded date of history, the Feb 26 747 coronation of King Nabonassar of Babylon.

Aug 21- Great Eclipse Metonic of 20th Century (three eclipses 1914,1933, 2017)

July 02-Jim Morrison died, President Garfield shot, first Wal-Mart opens.

Dec 25/26- Christmas/ Isaac Newton's Birthday

June 21-Summer Solstice

Dec 14-George Washington's Death Day in 1799

June 10-first witch hanged in Salem (1692) Italy enters WW2 (1940)

Dec 4-not much happens...Oh wait,...Led Zeppelin breaks up in 1980!

April 19-Darwin Death Day (1882)

April 20- Hitler Birthday (1889)

April 30 - Hitler Death Day

October 14- First colonial crime in Americas, the kidnapping of natives by Columbus...1492

April 08- Traditional date of Buddha's Birth (Japan). 8th day of fourth month of lunar calendar elsewhere).

Oct 02-Gandhi's Birthday

Feb 16- King Tuts' Tomb opened in 1922

Aug 12- First IBM personal computer in stores 1981.

Feb 06-Microchip patented 1959, Bob Marley and Ronald Reagan birthday

Aug 02- Iraq invades Kuwait in 1990. Pt-109 with JFK sinks in 1943

Jan 26-first convict ship in Australia (1788) India becomes Republic (1950)

Jul 22- Wuhan Eclipse of 7-22-2009...77 killed in Norway Terror attack 2011

June 01- Marilyn Monroe born in 1926, CNN debuts in 1980.

Sep 22 -Fall equinox

Nov 25-JFK Jr. born in 1960

May 20- Columbus Death Day in 1506

Nov 14-unusually quiet day. Oh wait, Charles (now King) born in 1948. Apollo 12 takes off in 1969.

May 9- Fourth and final voyage of Columbus launched 1502

March 30- Norah Jones is born in 1979. Queen Elizabeth dies in 2022.

Of note, four of the most important days of human civilization are represented in our current Metonic of the last hundred years: Spring Equinox, Summer Solstice, Fall Equinox and Christmas

Hopefully we have a little better idea of the Metonic Cycle now. We'll watch it unfold in the eclipses.

Chapter 9: History and Empire

People Groups

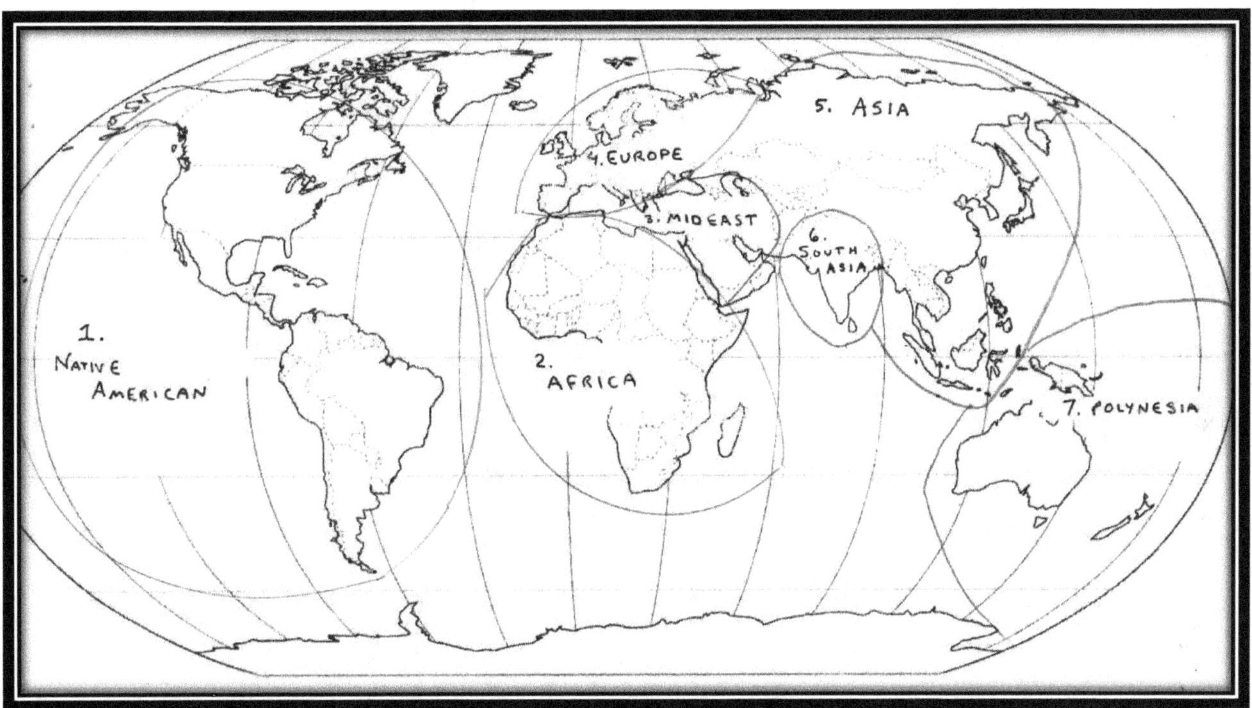

Seven Main People Groups and their geographies

1. **Native America.** Includes North and South America, which remained isolated by oceans for most of human history.
2. **Africa** Meaning Sub-Saharan Africa, the area isolated by the Sahara desert and Ocean. North Africa is essentially part of Mideast group.
3. **Mideast** Abrahamic, Egyptian and Persian people groups. Birthplace of empire.
4. **Europe** Home of the Caucasoid group, Barbarians, Vikings, Visigoths and such.
5. **Asia** China, Central Asia, Japan, Southeast Asia, Indonesia etc..
6. **South Asia** Indian Sub-continent, isolated by the Himalaya.
7. **Polynesia** The Australian aborigines, New Zealand Maoris and island native groups were the last on earth to be confronted with the phenomenon of modern empire. Isolated by oceans.

The law of sevens at work again. Seven essential people groups of the earth.

Seven Mediterranean and Mideast Empires

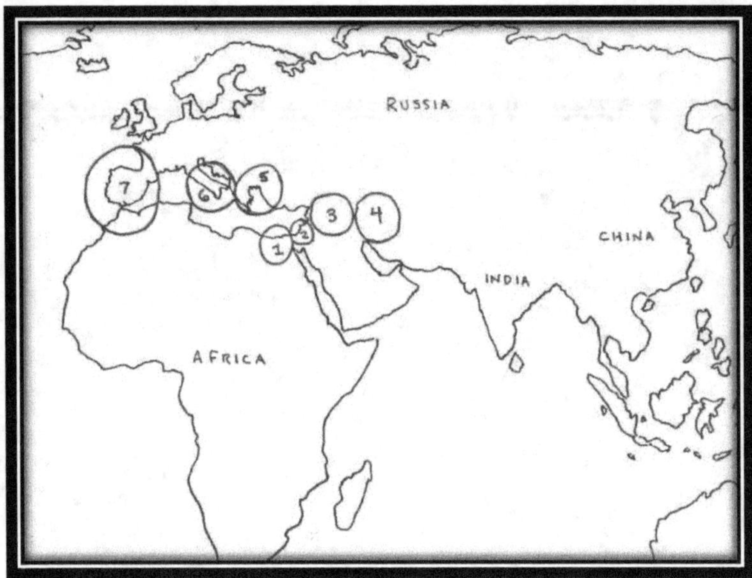

Seven primary Mediterranean and Mideast Empires arising prior to 600 AD

1. Egypt (National age 4,000 BC to present-Imperial age perhaps 3400 BC to 700 BC) Egypt is birthplace of empire, science, bureaucracy, priest craft and the mega-machine: the manipulation of large masses of slaves for the purposes of the elite. The oldest history on earth and the oldest calendar. It is currently Egyptian year 6264. Interesting, their New Year's is Sep 11. Egyptian Imperial energies and methods were apparently transferred to all later empires.

2. Israel (National age 1400 BC to 600 BC, vassal state to 70 AD-reformed 1948/1967) While only briefly a large geographic empire, Israel had the greatest influence over spiritual history.

3. Babylonian/Assyrian/ (origins 2000 BC-Imperial age 747 BC to 500 BC) Neighboring empires. Deeply pagan, scientific and aesthetic. Militaristic and organized. Babylon is the ultimate resting place of the Egyptian imperial energy and the Biblical symbol of decadence and enslavement.

4. Persian (National age 700 BC to present. Imperial Age 539 BC to 331 BC) modern day Iran.

5. Greek (National age 750 BC to present. Imperial age perhaps 490 BC to 150 BC) Culturally, in the secular meaning of the word, the Greek Empire was likely the most influential.

6. Roman (Imperial Age 300 BC to 476 AD) Rome was very much entwined with Greek ideals and organizations but added their own peculiar character and efficiency.

7. Holy Roman Empire (400 AD to Present?) The expression of state power as influenced by the Roman Church exists to this day. Maximum expression Spain is circled in the map so as to differentiate it from pagan Rome but of course, the Empire began in Rome and spread to France, Spain, Germany, Britain and elsewhere. The papal state included most of Europe and, after colonization, the whole of Central and South America.

Progression of Bureaucratic Model of Empire

This is a somewhat crude rendering of the advance of the Imperial Energy

1. **Egypt (b. 4000 BC)**
2. **Israel (b. 1400 BC-kings begin roughly 1040 BC)**
3. **Babylonian/Persian (2000 BC: modern era began 747 BC)**
4. **Greco-Roman (b. roughly 700 BC)**
5. **Holy Roman Empire -Europe-maximum expression in Spain (b. 600 AD)**
6. **Islamic -centered in modern Saudi Arabia- (b. 600 AD)**
7. **Ottoman -centered in Modern Turkey- (b. 1300 AD)**
8. **British (b. 1500 AD)**
9. **American (b. 1800 AD)**
10. **Chinese (Ancient 2000 BC -modern b. 1949-western imperial energy transferred 1972)**

Russian Imperial energy is real but its character is difficult to pinpoint. The conception of Marxism as a Global phenomenon is imperial in nature, but not mapped here. Many other empires have arisen i.e. Aztec, Mongolian, Japanese, Norse, Nubian and others and have transferred their imperial energies along the way. I believe these ten are unique archetype empires of mind and culture, as well as primary models of state power.

Imperial Movements in Certain Locations/Transfers of Energy

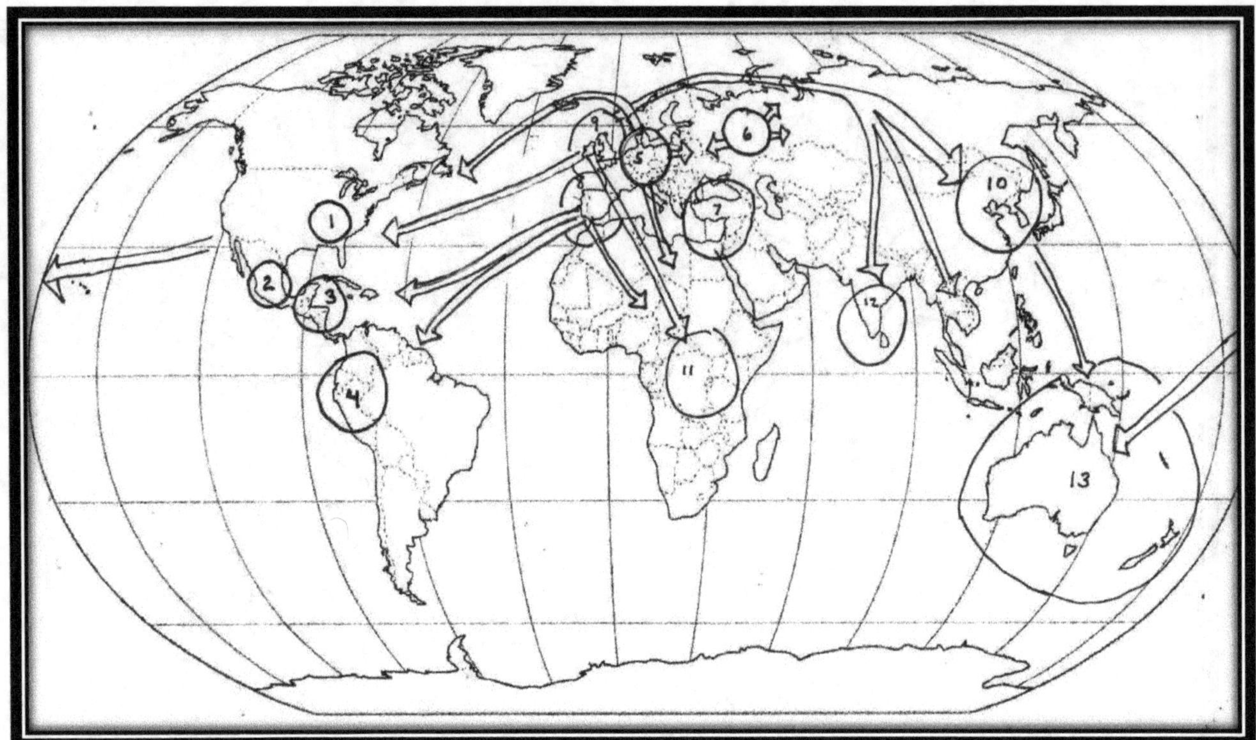

1. **Free Tribes of North America.** Separated by the Great American desert from the Aztec and Mayan slave raiding imperial energy, the tribes of North America were a maximum expression of free willed human energy for thousands of years.
2. **Aztec Empire.** Perhaps the maximum expression of organized religious human sacrifice. When the Great Pyramid of Tenochtitlan near modern Mexico City was consecrated in 1487, the Aztecs recorded that 84,000 people were sacrificed in four days.
3. **Mayan Empire.** The Mayans, located in Central America, were perhaps the greatest astronomers of the ancient world, and also practitioners of ritualistic human sacrifice.
4. **Inca Empire.** the Inca had a great empire in South America and created astonishing architecture. They also practiced large scale human sacrifice.
5. **Germanic/Visigoth Empire.** Movement of the Visigoth barbarian energy poured into the Roman Empire starting in the late 4th Century The Germans later became home of the Reformation and a great legacy of music, art and philosophy. Later on, they became a testing ground of Neo-Darwinist inspired extermination campaigns. A secondary energy transfer occurred at the end of the Second World War as when the US hired nazi scientists.
6. **Russia.** Huge and mysterious Russia pushes easily into the wilderness of the East but tends to stall out when it runs into the remains of the Holy Roman Empire as it pushes west.

7. **Ottoman and Islamic Empires** Muslim armies reached into Southern Europe at one point, and Islam still dominates Central Asia, Indonesia, the Mideast and North Africa.
8. **Spain.** Spain and Portugal were perhaps the maximum expression of worldwide colonial brutality. They founded the African slave trade (1526) and were the home of the Inquisition. Still I hear they are very pretty…plus they invented the guitar!
9. **Britain.** The grand paradox of Britain. Home of the Magna Charta, the King James Bible, free-willed protestants and abolitionists, Dickens and The Beatles. Creator of the penal colony, and perfector of Victorian hypocrisy and industrial subjugation. At its height in the early 1920's it was the largest empire in the history of the world, covering a quarter of the earth's land surface.
10. **China.** China is one of the oldest civilizations on earth, with a written history of 3500 years, similar in age to the first writings of Egypt, Babylon and Israel. Ancient China also excelled at astronomy. It typically did not engage in large scale empire building outside of its general locale. Its modern manifestation has excelled at the industrial techniques of modern empires. Home of Confucianism and perfector of state-run collectivism.
11. **Africa** Home to many kingdoms, innumerable tribes and the most beautiful land animals on earth. Since 1500, Africa has been used as a resource by European empires. A case study of the reality of imperial energy in every direction.
12. **India/South Asia** India is unique in all the world. It is unique religiously, ethnically, historically and geographically. It lies separated from invading armies by the Himalayan Mountains. Home to many warrior kingdoms, it has not apparently engaged in empire building outside of South Asia. Like virtually everywhere on earth, it received transfer of imperial energy from Europe. Home of Hinduism and Buddhism.
13. **Polynesia** The last outposts of native life were in Polynesia, including Australia and New Zealand. Conquered by the Dutch and British.

I did not mention every place in the world. It's just an overview of some main characters.

The main force of another kind of imperial energy, an energy of spirit, has come from tiny Israel. The Bible, though sometimes brought by unholy empires, provided in its Word a testament of the true nature and dignity of the eternal story, and a living method of interface with our Creator God. The Bible has now been translated into 2,800 languages in 157 nations.

Examples of these interesting transfers of energy and story will be seen in the eclipses of the World History Sections.

Chapter Ten

World Total Eclipses from 1900 to 2034 In 19 Year Metonic Maps

Now we'll begin to survey some eclipse history to see if the current Total Crosses are peculiar. Most people don't have any idea as to the frequency of solar eclipses. I sure didn't. First, let's quickly review the crosses that are coming up. Two balanced cross patterns are created by the current Hepton Monday Total Eclipse Series.

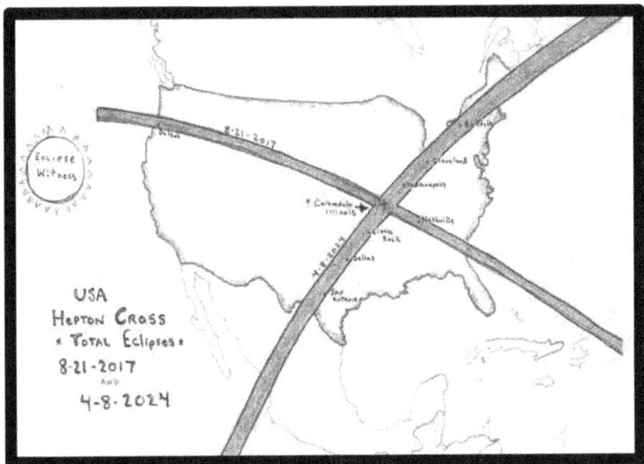

Above: USA 2017 and 2024

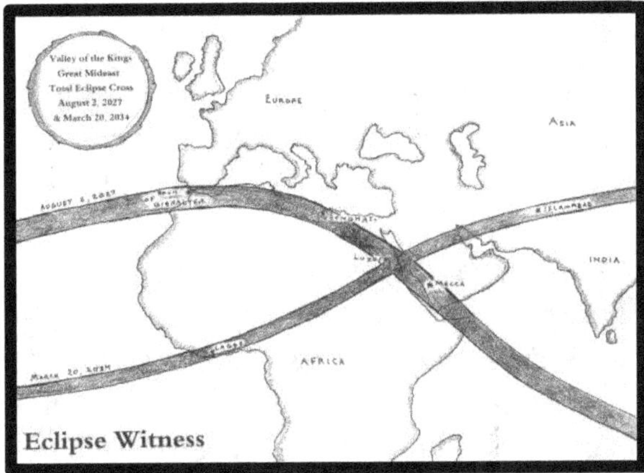

Above: Mideast 2027 and 2034

They appear together on a map in the image below, four total eclipses, all on Mondays: 2422 days separate the eclipses within each cross. Between last USA eclipse and first Mideast eclipse is 1211 days, or half of 2422.

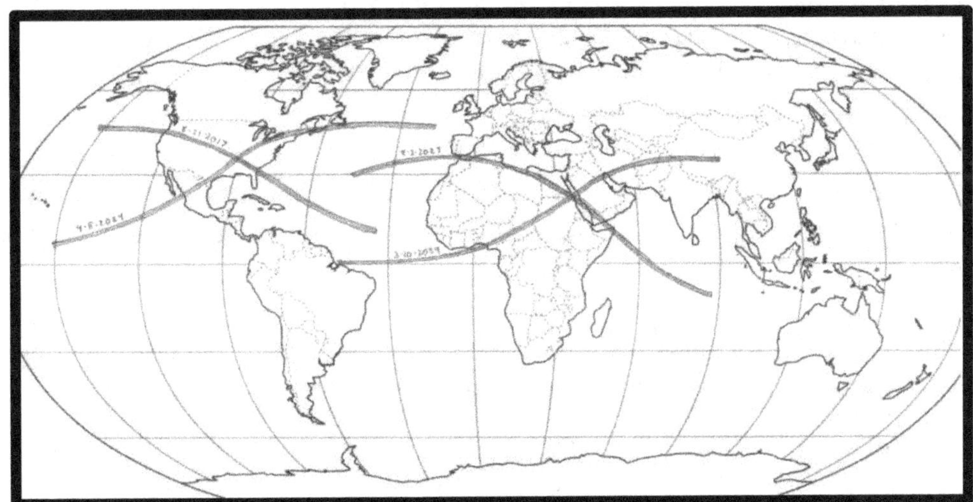

Above: Hepton Crosses of 2017 to 2034

Is that Peculiar?

To roughly determine the peculiarity exhibited by the two crosses of 2017/2024 and 2027/2034, we will first survey maps of all total eclipses on planet earth since the year 1900.

Are such True Hepton Total Crosses with an apex over land common or peculiar?

A Hepton Total Cross is two total eclipses exactly seven eclipse years apart (2422 days) which occur on the same day of the week. While the focus of this chapter's overview is on True Total Hepton Cycle crosses, we will also see lots of crosses which are not Heptons. There's also a couple that are Hepton crosses but in which one of the eclipses is a Hybrid, not a total.

I am not doing this as some trivial exercise. I maintain that the Hepton Cycle is the Maximum Expression of the eclipse phenomenon, which is in turn the maximum expression of the proof of the sovereignty of God. I believe they signal a message to earth. The full mapping of the modern Hepton Cycle occurs later in the book.

These maps represent time frames separated into natural 19-year Metonic Cycles. These have been established by simply going back in time from our current Spring Equinox Metonic of 2015 to 2034 to cover each 19-year period prior. The first period only covers 14 years, so as to start directly at 1900.

The organization of the eclipses into time frames is important for visual clarity. This presents a problem. If we are looking for crosses over 7-year periods, it's possible that crosses occur between periods, thus making some crosses not represented. Yes. I try and deal with this issue in the Hepton Chapter by mapping only Hepton Eclipses over the same period of time. I assure you that I did not find any overland true Total Hepton Crosses that cross over from one mapped period to the next. I am not leaving anything unmapped to make my point. God is making His point quite clearly.

Of all 26 complete eclipses in our current cycle, only seven are total Hepton eclipses.

I emphasize in this work crosses which occur in *eight years or less* apart. But other crosses occurring between eclipses much further apart in time will also appear on the maps. Given enough time, crosses appear almost everywhere. For clarity and rationality, a cross that is 17 years apart is not as interesting as one that appears three or seven years apart.

This chapter is intended to look at all total eclipses between 1900 and 2034 to determine one thing: do interesting crosses happen all the time? Are Total Hepton Crosses just another common phenomenon?

Let's look at all total eclipses of the last 120 years. It's a good opportunity to examine the crossing phenomenon in general, and the Hepton Cross in particular.

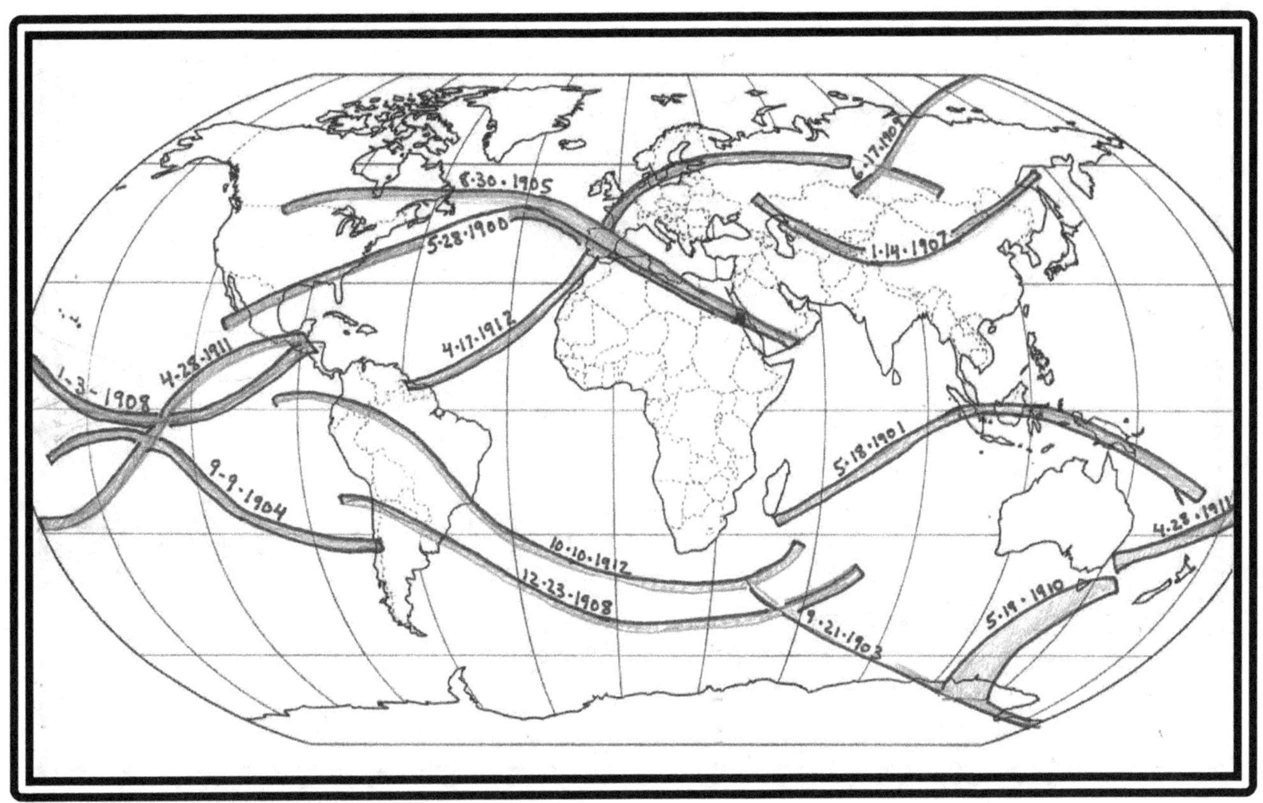

All Total Eclipses from 1900 to 1913 (includes Hybrids)

There were 12 total eclipses in this time frame and two hybrids.

Right off the bat, there are two Hepton Crosses...

- One is on the left side of map, part of a Hepton Swarm between 1904 Sep 9 and 1911 Apr 28, This is a True Total Hepton Cross with apex over open Pacific Ocean. The three eclipses barely touch any land, all of them ending on coasts of Central and South America.

- The other Hepton Cross is one of the most stunning examples in the whole survey outside of the ones we are currently experiencing. It's in the direct upper middle of map, part of three eclipses crossing Spain in 12 years. It's composed of the 1905 Egypt/Mecca and 1912 Titanic eclipse. The 1912 Titanic Eclipse is a Hybrid (with much of the path not creating totality), so it's **not** a True Hepton Total Cross like our current ones. Still, obviously a very interesting Hepton Cross occurs just before the first world war with apex in Spain.

- The three eclipse swarm over Spain abounds with interesting relationships discussed elsewhere.

- Non-Hepton intersections include 1909/1912 in Russia and 1903/1908 in Antarctic Ocean

Conclusion: No True Hepton Total Crosses over land, though great cross over Europe

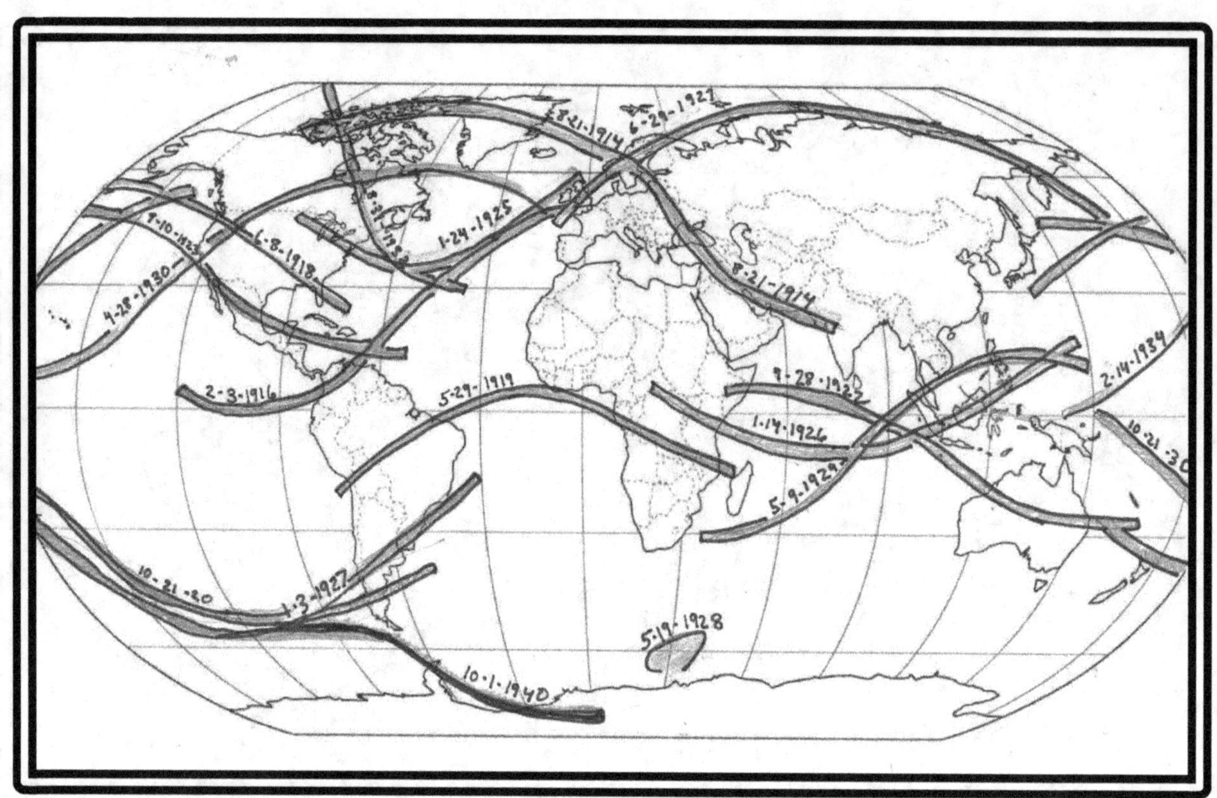

All Total Eclipses 1914 to 1935

Eclipse Swarm over North America. 19 Total Eclipses on earth during time frame.

Catalog of crosses which occur in less than eight years:

- True Total Hepton Cross lower right of map 1922 Sep 21 (mistakenly marked Sep 28) and 1929 May 9 with apex in Indian Ocean. Makes cool weaved symbol with 1926 Jan 14 Hepton, which occurs at precise mid-point between the other Heptons. Apexes of all occur in Ocean.

- Intersection in lower right off coast of South America comes from author accidentally including the 1927 Jan 3, which was a ring of fire. The European cross is a full 13 years apart.

- Hepton Cross upper left between 1923 Sep 10 Eclipse and 1930 Apr 28 San Francisco Hybrid. Has an intersection in the Pacific a few hundred miles off the coast of California. Not a true Total Hepton Cross. The San Francisco was a hybrid with only a kilometer of totality. The other apparent US cross (1918 and 1930) is 12 years apart.

- The 1925 Jan 24 NYC and 1932 Aug 31 New England intersect in the Atlantic. They are an Octon (7 years, 7 months and 7 days). 1916 and 1923 eclipses also make an Octon Cross in the Caribbean. 1930 and 1932 North American eclipses make an 856-day cross in Canada.

- Other intersections are more than 8 years apart. Most of Eurasia does not see a total eclipse.

Conclusion: No true Total Hepton Crosses over land during this time frame

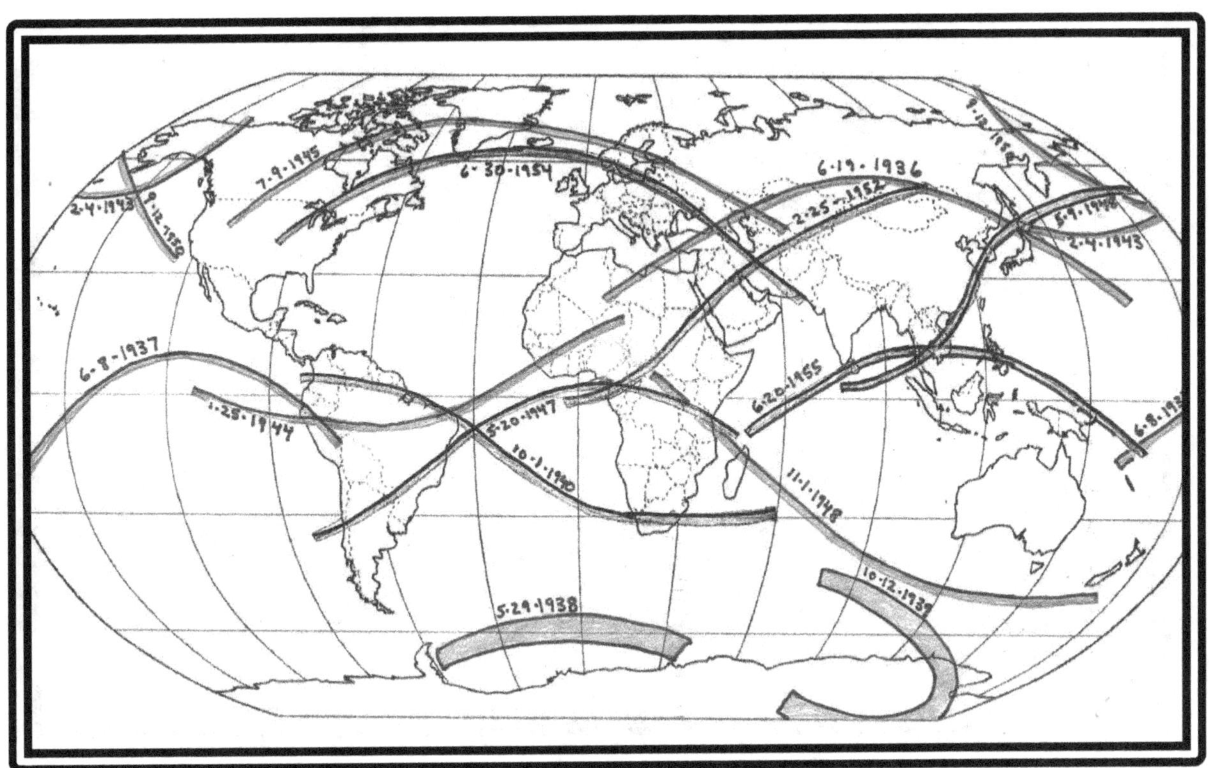

All Total Eclipses 1936 to 1955

- Crosses include: The Shahrud Iran Cross of 1952 and 1954. Two totals of Southeast Asia Eclipse Swarm, 1948 and 1955 with Bangkok in the apex. An Octon Cross in upper left of map between 1943 and 1950 off coast of Alaska. Three eclipses make an unusual pattern in Africa between 1947 and 1952. This pattern inversely reappears over Spain 2026 to 2028.

- In all, two Hepton Pairs touch each other. One (1937 and 1944) meets in an awkward intersection over Northern Peru, near headwaters of the Amazon River. The other is a True Total Hepton Cross, with an apex in the Atlantic several hundred miles east of Brazil.

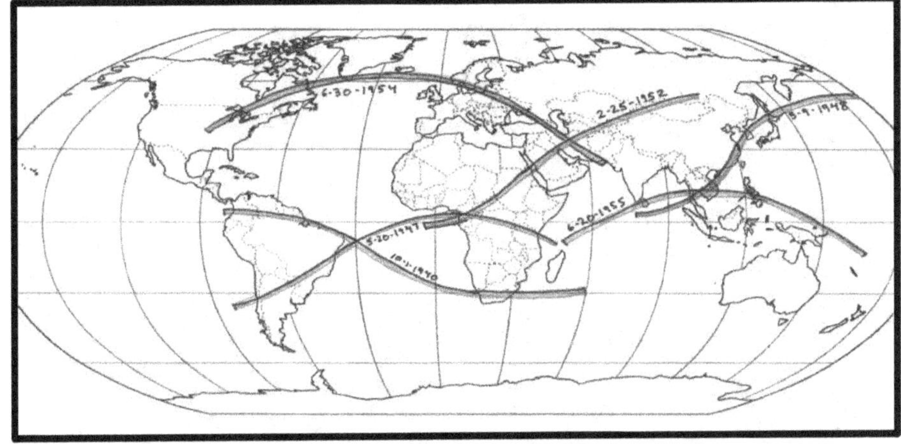

Three of the Crosses Highlighted

Conclusion: Only one true Total Hepton Cross, with apex over open ocean-this time off Brazil.

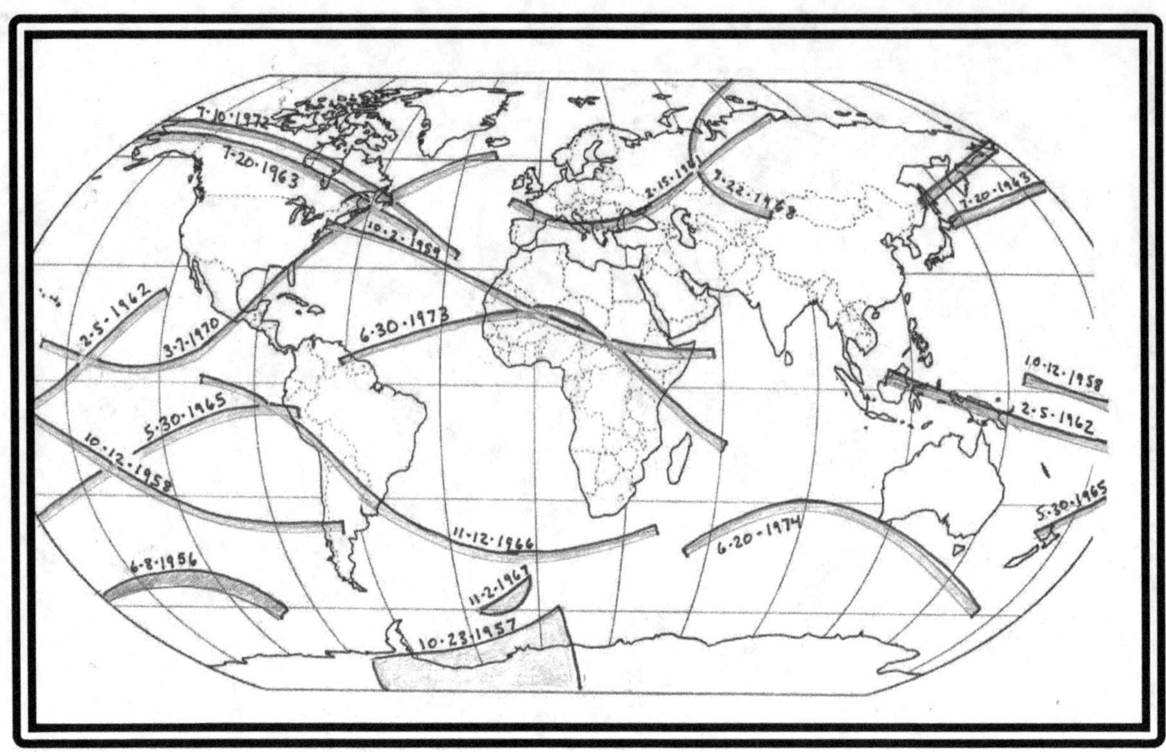

All Total Eclipses 1956 to 1975

- Most obviously significant cross is unusual Octon Trident Cross over USSR between 1961 and 1968. Apex near city of Ufa in Southern Russia, on the west side of Ural Mountains.

- There's one intersection off western South America (1965 and 1966) and in Pacific between 1962 and 1970. In general, a peculiar lack of land sees eclipses of this time frame.

- Hepton Total intersection between 1963 Jul 20 and 1970 Mar 7 "sixties eclipses", both of which touch the USA, but with apex several hundred miles off coast of Maine. These are Saros 145 and 139 which bring the USA Hepton Cross of 2017 and 2024. Also, True Total Hepton Cross over open ocean: 1958 and 1965 in lower left map over southern Pacific

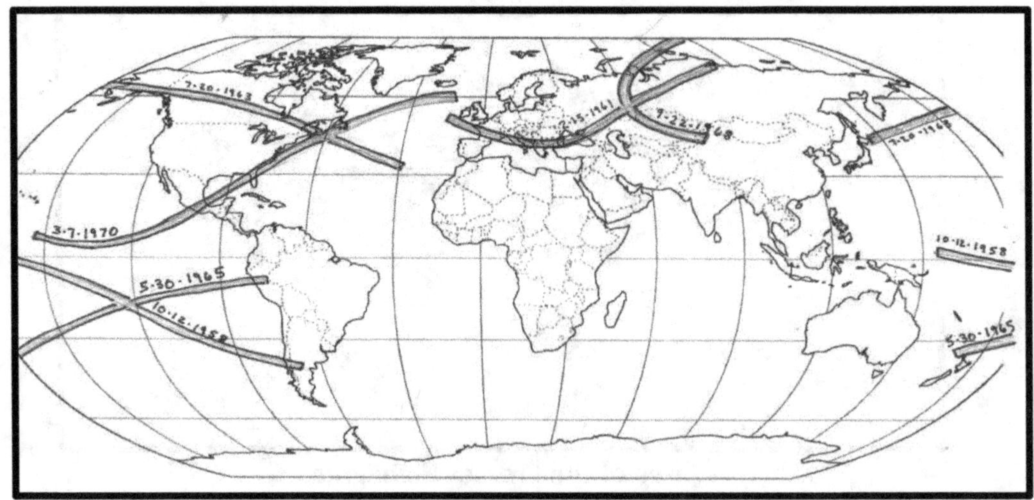

Crosses of 1956 to 1975 Highlighted

Conclusion: Two Total Hepton Crosses.Both intersect over ocean again. Octon Cross over USSR.

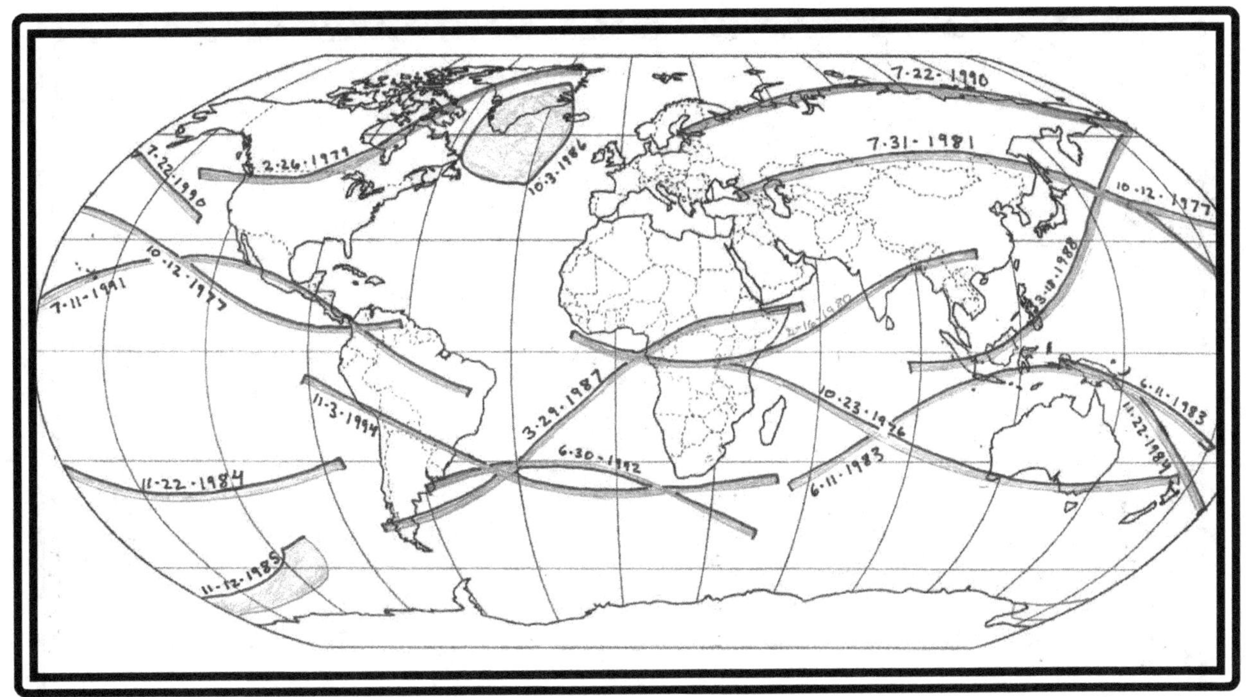

All Total Eclipses 1976 to 1995

- There are two True Total Hepton crosses in this time frame, each with intersections over open ocean. 1976 and 1983 in Indian Ocean and 1981 and 1988 in the Pacific off Japan.

- The USSR has two eclipses nine years apart which roughly define that nation's north and south boundaries during the decade leading up to its dissolution. The US has the Mount Saint Helens eclipse of Feb 26 1979 . There's an Octon Cross in South Atlantic between 1987 and 1994.

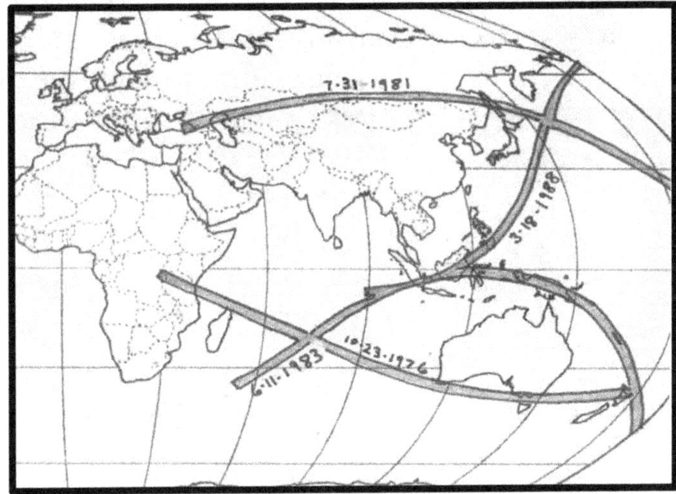

Closer view of the two Hepton crosses highlighted above

Conclusion: Once again, True Hepton Total Crosses appear over the ocean. The southern Hepton pair gives the distinct impression of a fish with Australia in its belly.

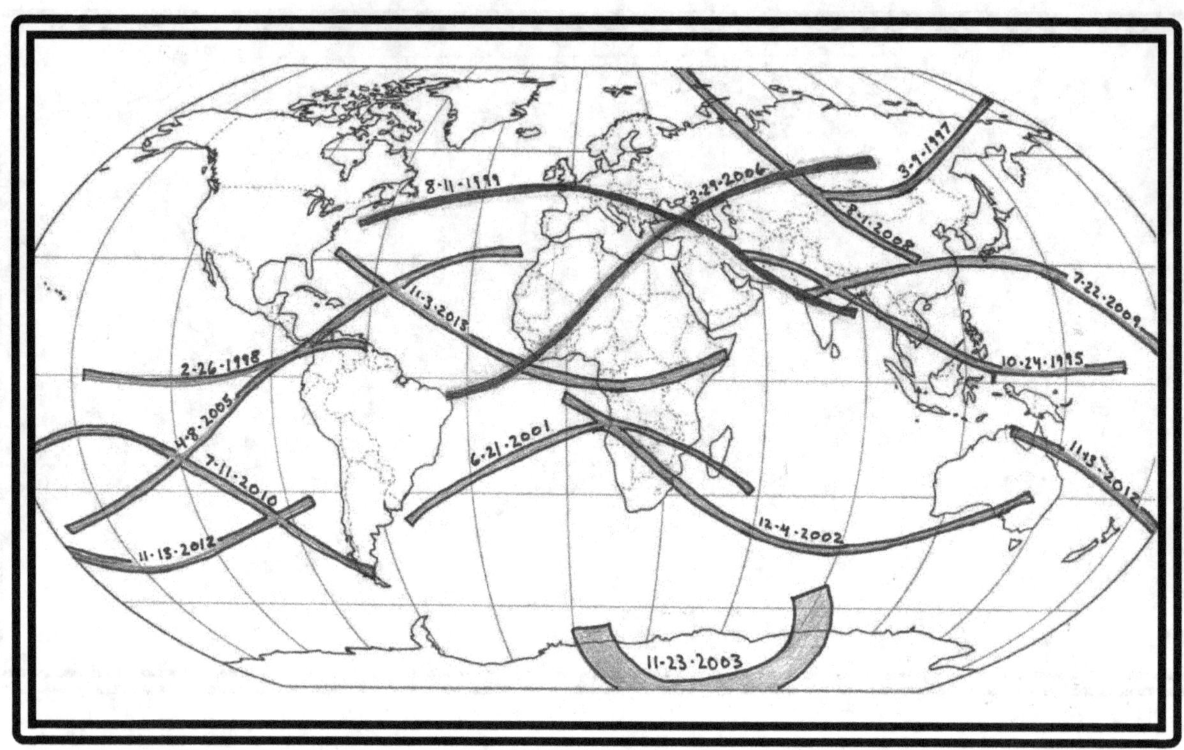

All Total Eclipses 1995 to 2014

- The primary highlight is the Turkish Tav of 1999 and 2006 in the center of the map. Created by two great eclipses, including the 1999 Aug 11 last eclipse of the millennium, this is a True Hepton Total Cross, the first overland in the survey since 1900. Apex is near Tokat, Turkey.

- Other Intersections include an 18-month cross in Angola, Africa (2001 and 2002), a 3 eclipse swarm between 1995 and 2009 in India (including the 2009 eclipse which brings Wuhan China totality). A one lunar year dead end in China from 2008 and 2009 and a sort of connection in Russia from 1997 to 2008. Also, two eclipses seven years plus 40 days apart overlap at the isthmus of Panama. Absence of eclipses over North America and virtually all South America.

Above, the Turkish Hepton Cross of 1999 to 2006 is Highlighted

Saros 145 and 139 Hepton is now organizing into a symmetrical cross, apex over Turkey. These two same Saros reappear over the USA in 2017 and 2024, with apex over "Little Egypt".

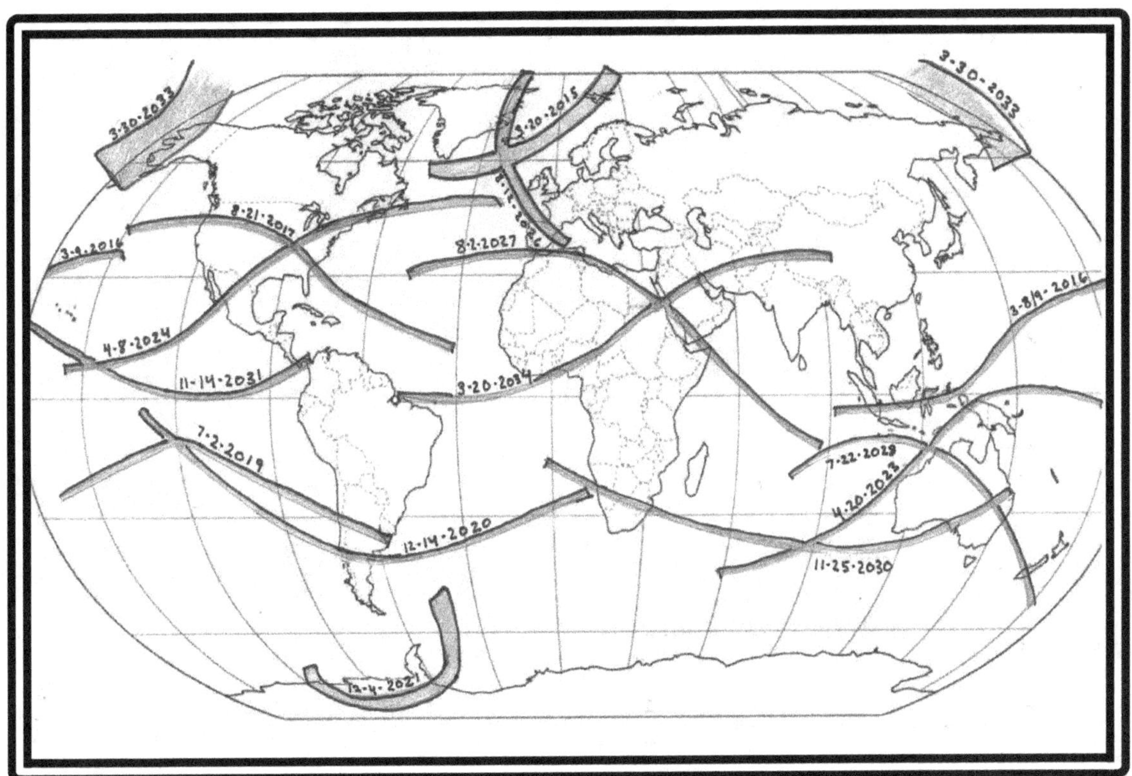

All Total Eclipses 2015 to 2034

- This map is to be compared with the former Metonic maps in this section. It shows all Total eclipses from the 2015 Mar 20 Spring Equinox North Pole Eclipse to the 2034 Mar 20 Spring Equinox Great Civilizations Eclipse. What do we see?

- In prior 120 years of survey, we found only one true overland Total Hepton Series cross, the Turkish Tav of 1999/2006. Here are two True Hepton Total Crosses overland, one over Egypt and one over the area known as "Little Egypt" in the USA. The crosses are symmetrical and isolated. Indeed, there seems a peculiar independence to all the "sets" of eclipses depicted.

- Also of note is a peculiar shape towards the North Pole formed by the eclipses of 2015 and 2026. There is also a Cross over Australia (apex in western New South Wales) formed by the 2028/ 2030 eclipses, and an Octon Cross just offshore of Northern Australia-2023 and 2028. These are part of an eclipse swarm over Australia and New Zealand between 2023 and 2038. Also, the Isthmus of Panama is perfectly covered in 2031 by an eclipse that has crossed the entire Pacific without touching any land before stopping there.

- The Great Bering Strait Eclipse of 2033 Mar 30 is illustrated on both corners because it connects both hemispheres (and International date Line) simultaneously at the USA and Russia. This happens four days before the traditional date of the 2,000[th] anniversary of the Crucifixion of Jesus Christ (April 3, 33AD).

Conclusion: The most unique pattern of eclipses ever mapped reside within a 19 year Metonic bracketed by Spring Equinox total eclipses occurring within the 2,000[th] anniversary of the adulthood, ministry, death and resurrection of the Lord Jesus Christ.

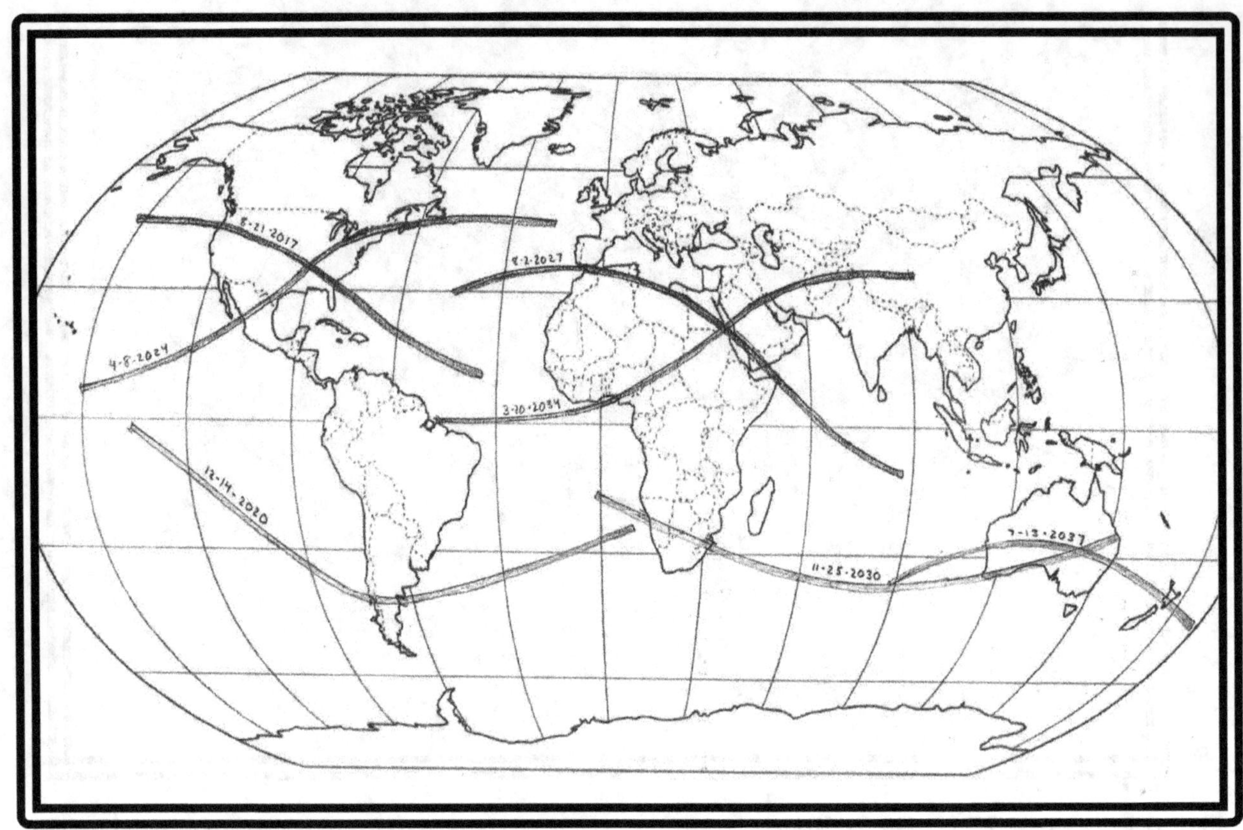

Highlighted: The Monday Hepton Series 2017 to 2037

These eclipses all occur on Mondays, in a series of 7 eclipses that occupy a space of 7266 days. 20 years minus 40 days or exactly 21 eclipse years.

- These eclipses all occur 7 eclipse seasons or 1211 days apart. Every other eclipse occurs 2422 days apart, or 7 eclipse years.

2017 Aug 21 Great American Eclipse/USA divided.
2020 Dec 14 Electoral College Eclipse—occurs while electoral college votes in the President and Vice-President of the United States.
2024 Apr 8 Great North American Eclipse—Final line of Hepton Cross—south to north
2027 Aug 2 Valley of the Kings Eclipse---maximum eclipse over Luxor. Also, Mecca is eclipsed.
2030 Nov 25 British Empire South Eclipse—South Africa to Australia
2034 Mar 20 Spring Equinox Eclipse---finishes Mideast Hepton Cross
2037 Jul 13 Australia and New Zealand—final Monday Hepton in the series.

All six inhabited continents are directly touched by this series, though Europe is touched only at Straits of Gibraltar. Two extraordinary balanced Hepton crosses are created overland for the first time in the survey, and a third unbalanced Hepton Cross occurs over Australia. These three crosses are roughly the same distance from each other across the sphere of the globe, occur in succession and declining latitude/longitude from left to right.

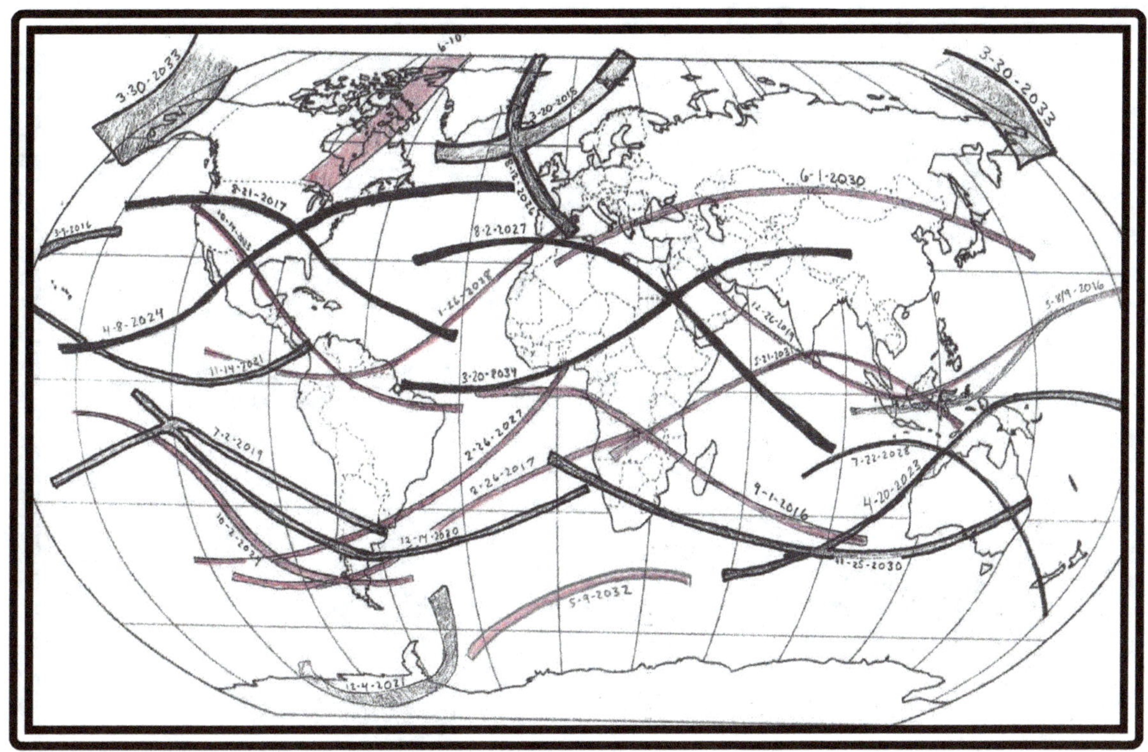

Messy! All Complete (Total and Ring of Fire) Eclipses 2015 to 2034

Let's briefly look at a messy image of all complete eclipses in the Spring Equinox Metonic of 2015 to 2034. Total eclipses are in black/gray and ring of fire (annular) eclipses are shaded red. The Jun 21 2020 pandemic eclipse is missing due to author error. You can see why it's difficult to map eclipses over long periods of time. It gets really messy. It's not necessarily because there are so many eclipses, but because the paths are so long that they end running all over each other.

- Nevertheless, the profound distinction of the USA eclipses remains, essentially isolated from the larger mess. You can see the extraordinary Aleph formed over North America.

- The Mideast Hepton Cross also maintains its general integrity, though it is crossed by ring of fires in three places and touched in another. One apex is the Straits of Gibraltar between Europe and Africa-2027 and 2028, less than six months apart. One apex is Northern Algeria (2027 and 2030) and another is in the Indian Ocean (2027 and 2031)

- Argentina and Chile see five eclipses in 13 years, with three intersections in a small geographic area. The Indonesia area sees five.

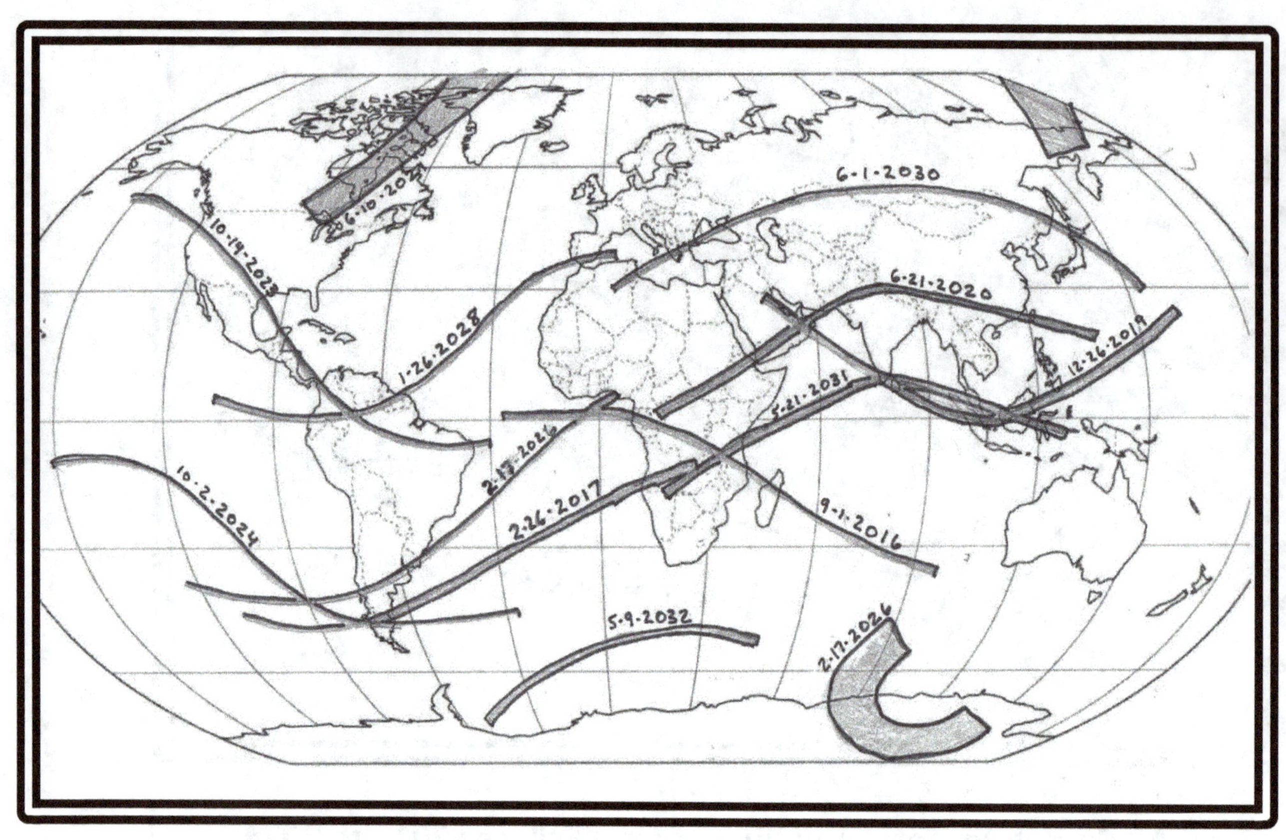

Above Image singles out all 2015 to 2034 Ring of Fires

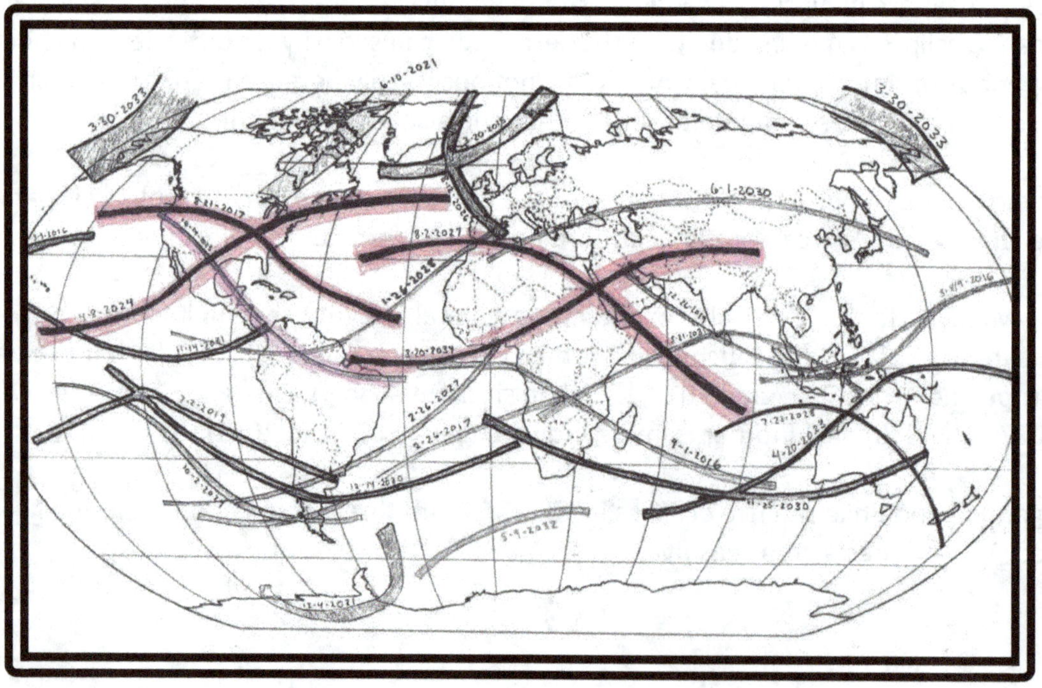

This above image shows all complete eclipses during the Spring Equinox Metonic with the USA/ Mideast **Alpha/Omega Highlighted**

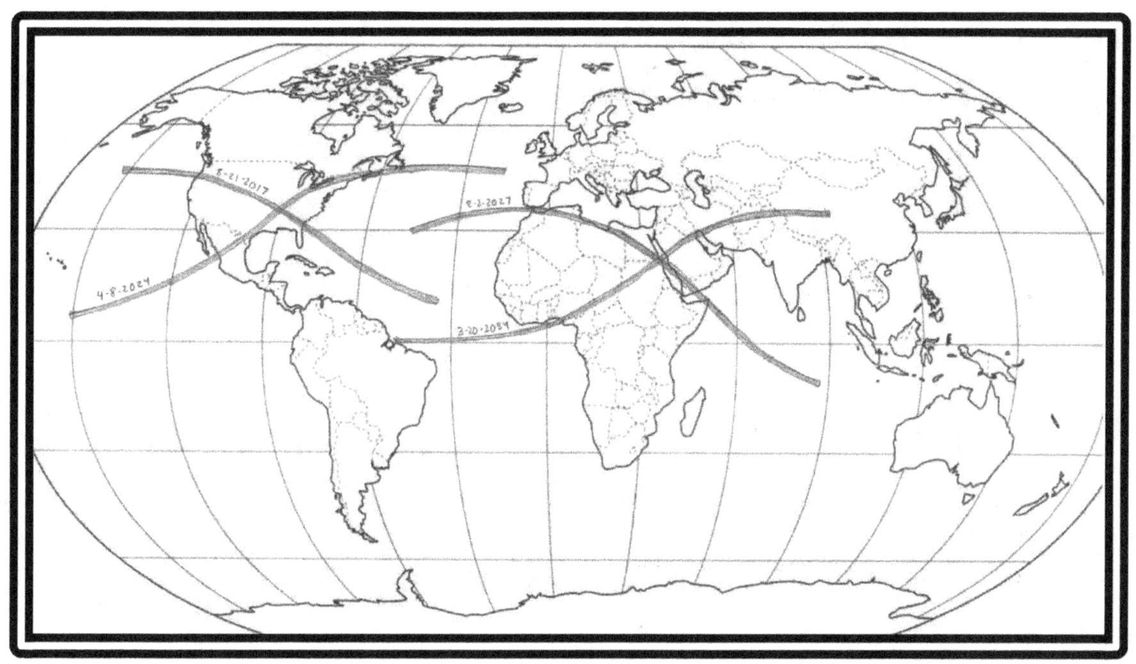

Above, The Hepton Tavs of the Spring Equinox Metonic (2015 to 2034)

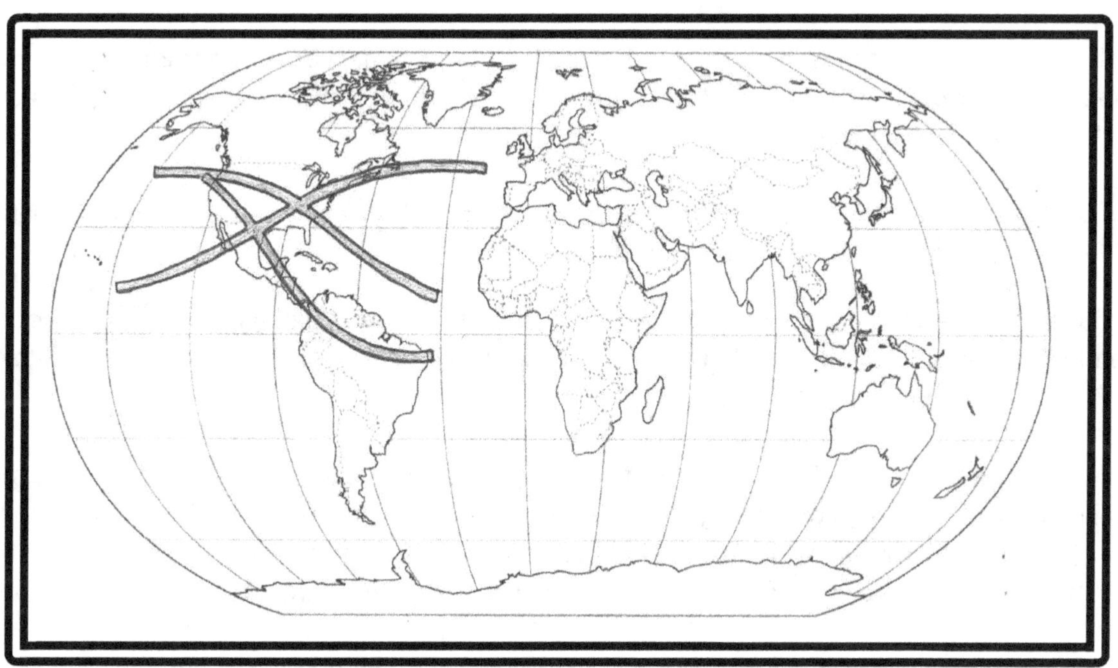

Above, the American Aleph of 2017, 2023 and 2024

We have reviewed 130 years of world total eclipse history and there is nothing like the current Hepton Series which creates balanced Crosses overland. We have established peculiarity.

The Aleph and the Tav. God's Signature on Planet Earth.

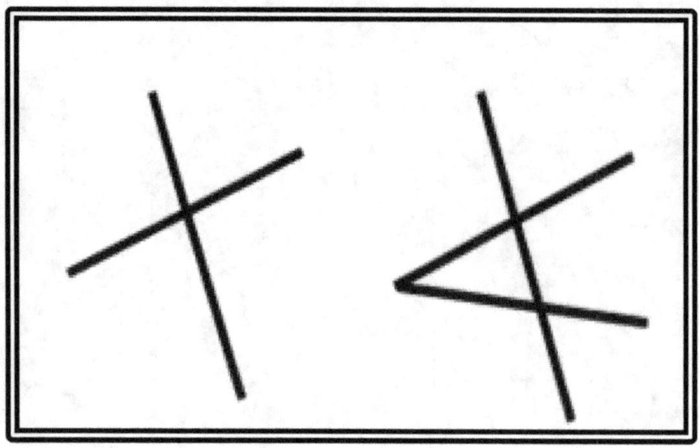

Egypt and and "Little Egypt" of the USA are in the cross hairs of an extraordinary alignment.
Please review the above drawing and then prior Alpha/ Omega USA and Mideast maps.

Paleo-Hebrew Writing (not to be confused with modern Hebrew, which arose after Babylonian Exile) is the original written word of the earliest Bible (Torah, the five Books of Moses) and is closely related to other early Middle eastern script.

Tav is the last letter (the 22nd) of the Hebrew alphabet. It is an X. This character x is essentially the same in Egyptian Hieroglyph, Phoenician, and Paleo-Hebrew.

Aleph is the first letter of Paleo-Hebrew, or our A. It again is similar through all the early alphabets, and resembled a bull's head in some. In modern times, it has interestingly become also the sign of "anarchy".

The Tav X is, of course, also the Cross. It is also the x of a signature (as in signing "x" on a document). It is also a nearly universal sign of "wrong answer", condemnation or refusal. The X letter is also the 22nd letter ("chi") of the Greek (which has 24 characters), and has historically represented Christ (i.e. xmas for Christmas). It was a sign of early Christians, creating the tail of the Christian "fish" symbol.

Tav (X) in Hebrew is understood to mean completion, mark, sign, omen, or seal. Tav is the last letter of the Hebrew word emet, which means 'truth'. The Jewish Midrash says that emet is made up of the first, middle, and last letters of the Hebrew alphabet (aleph, mem, and tav).

God has written on our planet with big simple letters that He is the Alpha and the Omega

Chapter Eleven:

World Eclipses 1869 to 1955

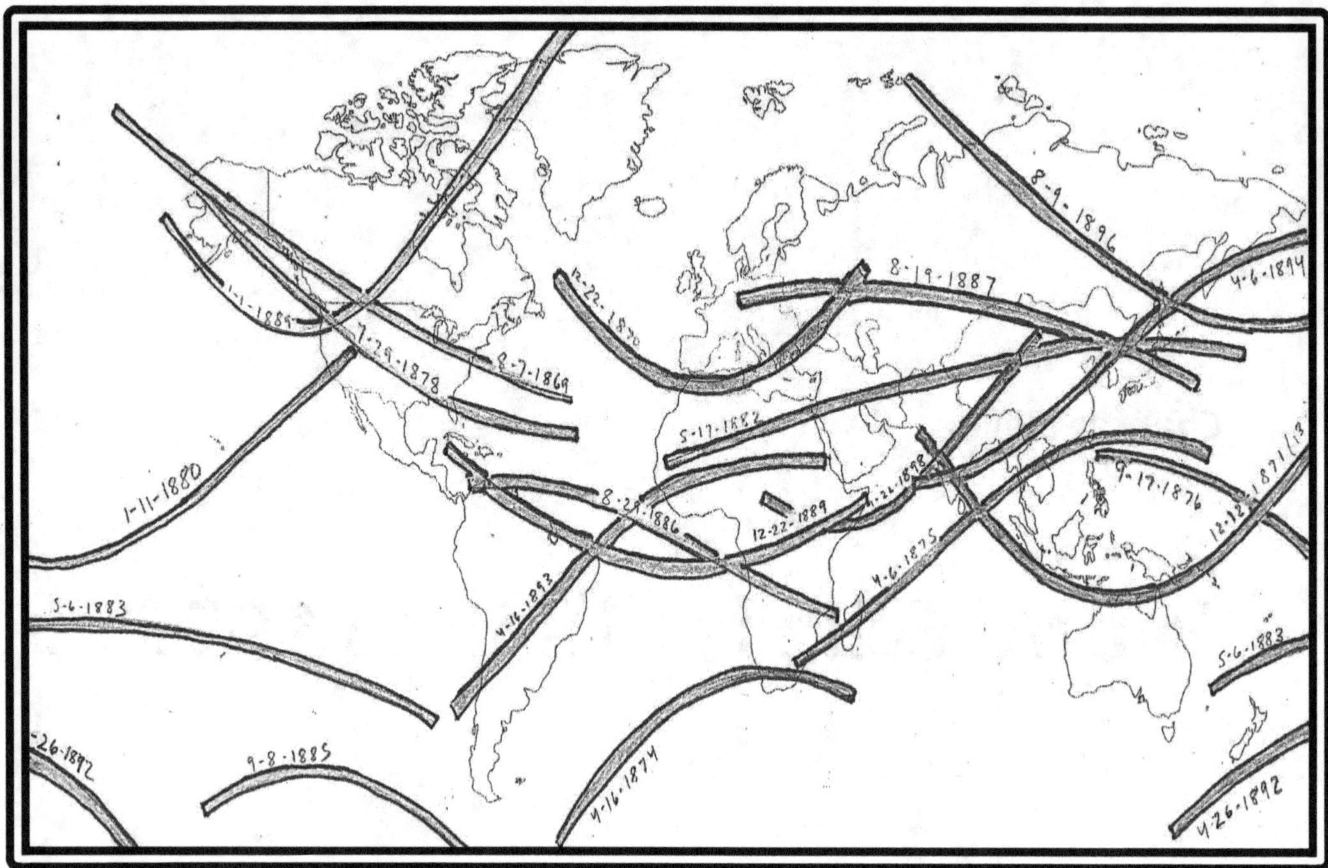

Total Eclipses 1869 to 1898

This 29-year span is presented as a comparison of the 1869-to-1898-time frame to our current time frame. We are looking for overland balanced Hepton Crosses similar to the ones we are currently experiencing. Careful study of the image reveals that there are no balanced overland Hepton Crosses such as appear in the Monday Hepton Judgment that is the main subject of this work.

In the map above, the most interesting clusters are found over the northwest USA (more closely mapped in the USA eclipse section) between 1869 and 1889, and in the northeast coast of China above the Korean Peninsula between 1882 and 1896. That cluster includes the Berlin Sunrise Eclipse of 1887 Aug 19, which we look at before we start the main survey. How are we to examine such a confusing-looking phenomenon? There has to be an editing mechanism, a way of eliminating clutter that maintains chronology. How to see the patterns complete eclipses make? To comprehend eclipses over large time spans requires us to divide them into slices of time. Fortunately, eclipses themselves provide the mechanism. They divide neatly into 19-year Metonic time frames.

19-year Metonics begin and end with eclipses that occur on the same day of the year, 19 years apart.

Using The Metonic to Create Chapters of Eclipse History

As we studied earlier, the 19-year Metonic is where the three separate calendars of solar year, lunar month and eclipse year perfectly connect at almost exactly 6,940 days. This nearly impossible arrangement allows for eclipses to occur on the same dates of the year for extended periods of time.

The main focus of our study is the eclipses occurring in the exactly 19-year solar period Spring Equinox Metonic of March 20 2015 to March 20 2034. As we've seen, there are profound total solar eclipses on both these March 20 dates, and lots of interesting eclipses in between. The Spring Equinox Metonic becomes our starting place. We then go backwards in time, and the eclipses themselves create chapters of eclipse Metonics, with bookend dates on eclipses, leaving no eclipse behind! Partial Eclipses often provide natural boundaries between the chapters- almost resets, if you will.

I name each Metonic to help distinguish them.

Feb 14 Valentine's Day Metonic: Feb 13/14 1896 to Feb 14 1915
Aug 10 Temple Destruction Metonic: Aug 10 1915 to Aug 10 1934
Dec 25 Christmas Metonic: Dec 25 1935 to Dec 25 1954
Jun 20 Summer Solstice Eve Metonic: June 20 1955 to Jun 20 1974
Apr 29 Hitler Death Eve Metonic: Apr 29 1976 to Apr 29 1995
Oct 24 United Nations Day Metonic: Oct 24 1995 to Oct 23/24 2014 (partial)
Mar 20 Spring Equinox Metonic: Mar 20 2015 to Mar 20 2034

These seven chapters in a 138-year story provide an overview of cosmic timekeeping. They contain every complete solar eclipse on earth during that time. In the middle of the chapters and between the chapters are often sets of partials. For instance, in the 22 month stretch between Jun 20 1974 and April 29 1976 are three partials. All partials are listed in the survey in small print, but not mapped.

Careful observation reveals a pattern of repetitive dates over the course of the entire survey. That's because we are only dealing with about 36 repeated Metonic clicks of the eclipse clock. These typically return a few times and then lurch forward.

The pattern for the Metonic date progression is usually this:
19 years + 19+19 then an 8-year lurch (where the date of eclipses jumps forward one or two days) then 19+19+19 then an 8-year lurch etc...
Dates can be one day forward or back due to the reality of time zones, date lines and orbital realities.

August 21 Metonic Example

For example, our current late-August Eclipse Metonic is August 21. That means all late August eclipses basically land on that date. Eclipses must land on August 20, 21st or August 22nd. If it is creating August 19th eclipses, it must belong to that prior Metonic date (either August 18, 19 or 20). We then go back in the eclipse record and note that, yes indeed, there was a solar eclipse on August 19 1887 from Berlin to China. If it is creating August 23rd eclipses, then it must belong to the Metonic between the 22nd and 24th. This occurs on August 23 2044, when the Metonic has officially moved on. It is a slippery yet oddly consistent phenomenon, made possible only by the incredible alignment of the Draco-Metonic Calendar. The cosmic timekeeping of the Metonic provides a history of the August 21st Metonic's 120-year reign during our current approximately 500-year cycle. The last Metonic prior to Aug 21 Metonic was on August 19. It showed up in **1887 Aug 19 Berlin Omen Total Eclipse.**

Then we had: an August 20, 1906 Partial eclipse then
August 21 1914 World War One Eclipse ...then 19 years...
August 21 1933 Jerusalem Ring of Fire...then 19 years...
August 20 1952 Nazca Lines Ring of Fire...then 19 years
August 21 1971 Partial solar eclipse ...then an 8 year lurch
August 22 1979 Ring of Fire near Antarctica... then 19 years
August 22 1998 Ring of Fire in Indonesia..then 19 years
August 21 2017 Great American Eclipse...then 19 years
August 21 2036 Partial Solar Eclipse
...then 8-year lurch to August 23 2044. By 2100 that same impulse is August 25.

See? It makes sense.

There are nine August 21-ish Metonic eclipses (including partials) in a 120-year span...Between 1906 and 2036 there were a grand total of 326 solar eclipses (including partials). This means that the August 21st Metonic occupies 1/36th of all eclipses that occurred—very close to the average of what the math should bring. There should be about 36 repeating dates sliding in and out of prominence.

You will find very little hard and fast about these dates. They slip and slide a bit. But even a child can see the repetitive date factor at work throughout the survey of eclipses. Importantly, there will be partial eclipses which interrupt the progression of complete eclipses. They seem to work as a sort of reset, or breather. These are listed in the progression, but not studied.
Even if you don't fully understand, try looking at the maps and reality should make itself manifest.

Some Things to look For:

- Repeating dates. Every 19 years look for the same or similar date to reappear.
- Lots of Crosses and clusters. Only a few of the longer seven and eight year crosses will be apparent in this chapter, as the maps generally represent shorter time frames.
- Lunar year triads and duos. Eclipses often occur exactly one lunar year apart from each other. A lunar year, or 12 complete cycles of the moon, is about 354 days. A lunar year separation will appear from one year to the next as minus 10, 11 or 12 days. Modern example One lunar year separated the total eclipses of Dec 14 2020 to Dec 4 2021.
- Hepton eclipses. There are two sets of 7 eclipses each in the Hepton cycle occurring throughout this chapter. These are 1211 days apart from each other (7 eclipse seasons) and almost always occur on the same day of the week.
- Octon eclipses. 8 eclipse years apart from each other (7 years, 7 months and 7 days)
- You'll see lots of patterns in Saros numbers. The Saros are related. They leap by fives and tens. For instance you will see Saros 119 (1896 Feb 13 ring of fire) followed one lunar year later by Saros 129 (1897 Feb 1), exactly 10 Saros numbers apart.
- Is there a plausible relation between the paths and the world events and empires which they seem to correlate with? Judge for yourself.

Berlin Sunrise 1887 Aug 19 Total Eclipse

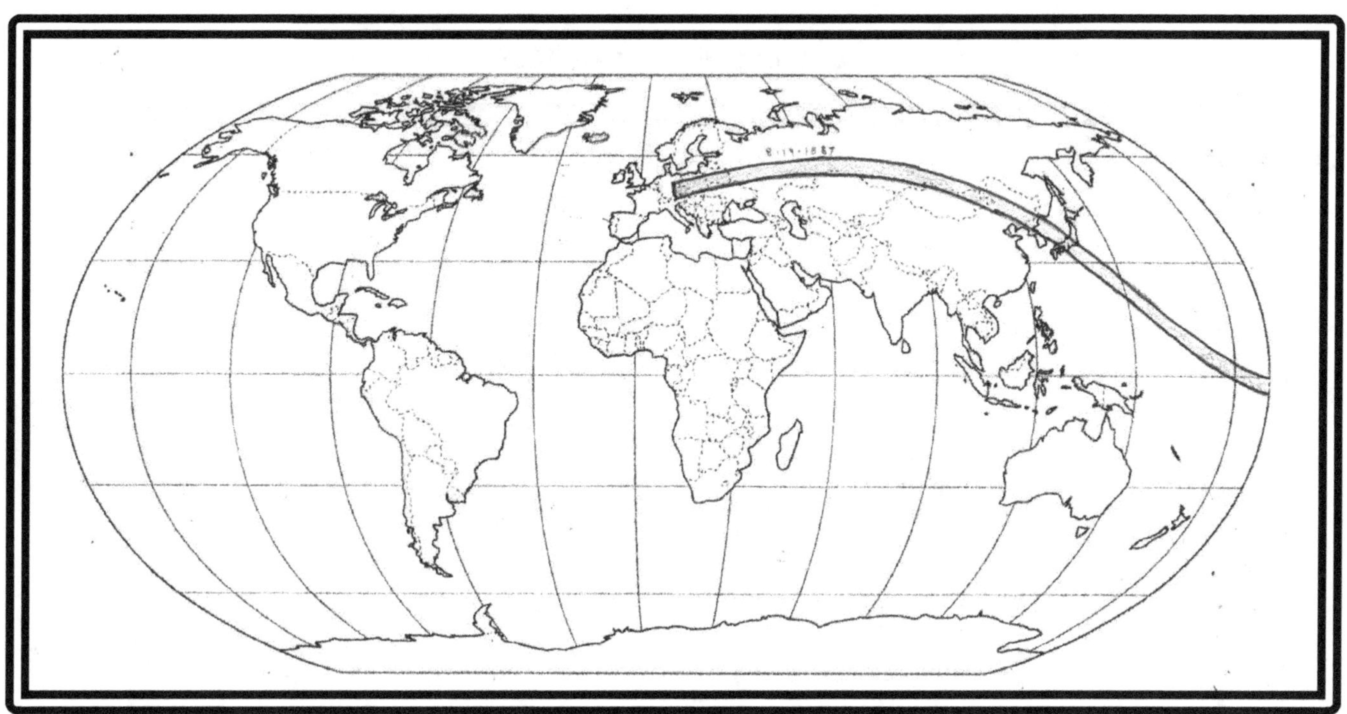

Prior to our full accounting of all world eclipses from 1896 to 2034, one peculiar eclipse stands out.

The Total eclipse of 1887 Aug 19 begins at sunrise just west of Berlin, Germany and brings Totality to that city for the first time since 1424 (463 years) and the last time until 2135. It was the first complete solar eclipse of any kind over central Berlin since 1547, 340 years prior. Berlin has only had 8 total eclipses since the birth of Jesus, including one in 29 AD, which also skirted the Holy land.

This 1887 eclipse occurs 610 days before birth of Adolf Hitler in Branau Aum Inn, Austria on Apr 20 1889, Probably about 95%+ coverage at Hitler Birthplace.
The eclipse is 27 years and two days before 1914 Aug 21 World War One Eclipse. A 19 year Metonic plus 8 year lurch. Essentially the same Metonic.

The path: Many would suffer terribly in the next 50 years. In path: Germany, Poland, Baltics, Russia, China, Korea and Japan

Germany: Germany had the highest military death toll In WWI with some two million dead (Hitler was wounded). More than four million more German soldiers and civilians died in WWII.
Poland: Eclipse includes totality at future Auschwitz. Six million Poles died in WWII alone.
Lithuania, Latvia and Belarus: All were battlegrounds in WWI and II
Russia: totality in Moscow. At least 3 million Russians died in WWI, 20 million died in civil wars and famines after the revolution of 1917 and 25 million more died in WWII.
China: Chinese deaths in WWII are around 20 million. Perhaps 30 to 50 million more died in the civil wars, purges and famines of the 20th Century.
Korea: Korea's brutal war of 1950 to 1953 may have killed up to 8 million Koreans.
Japan: Japan lost at least 3 million in WWII.

Beginning with 1896 to 1899

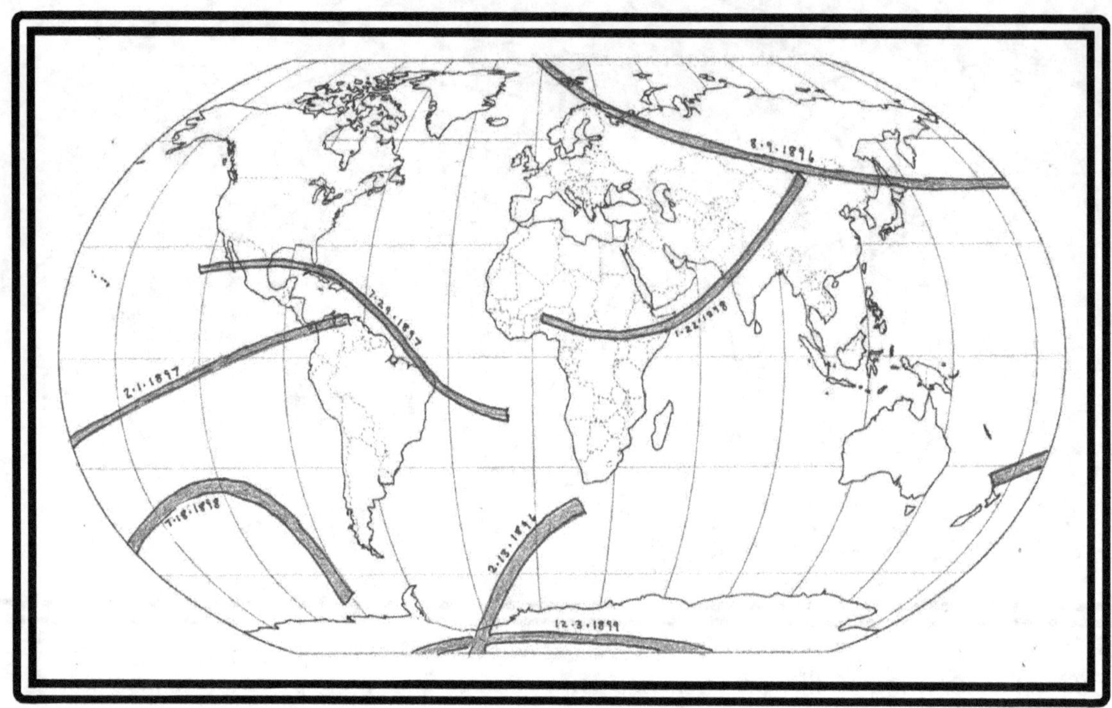

1896 to 1899-Last eclipses of 19th Century

1896 Feb 13/14 Ring of Fire: Antarctica - Saros 119
1896 Aug 9 Total: Russia/Japan -Saros 124
1897 Feb 1 Ring of Fire Pacific to Colombia -Saros 129
1897 Jul 29 Ring of Fire Mexico, Caribbean to Natal, Brazil -Saros 134
1898 Jan 22 Total Africa, India, China - Saros 139
1898 Jul 18 Ring of Fire South Pacific - Saros 144
1899 Dec 3 Ring of Fire Antarctica near south pole -Saros 121

The last eclipses of the 19th Century.

Russia and Japan are linked by the 1896 Eclipse, just as they had been by the 1887 Berlin Sunrise eclipse. They will be at war within 8 years.

1897 Feb 1 has an eclipse over Pan-American isthmus seven years before USA takes over construction of the Panama Canal. July 1897 sees eclipse over Cuba about six months before the explosion of Battleship Maine in Havana harbor in Feb 1898 ignites Spanish-American war.

China and India are linked by an eclipse in 1898, as both were under British Rule. Eclipse occurs as tensions between the Chinese and colonial powers were reaching a boiling point, about 18 months before The Boxer Rebellion.

First Four Eclipses of the 20th Century 1900 and 1901

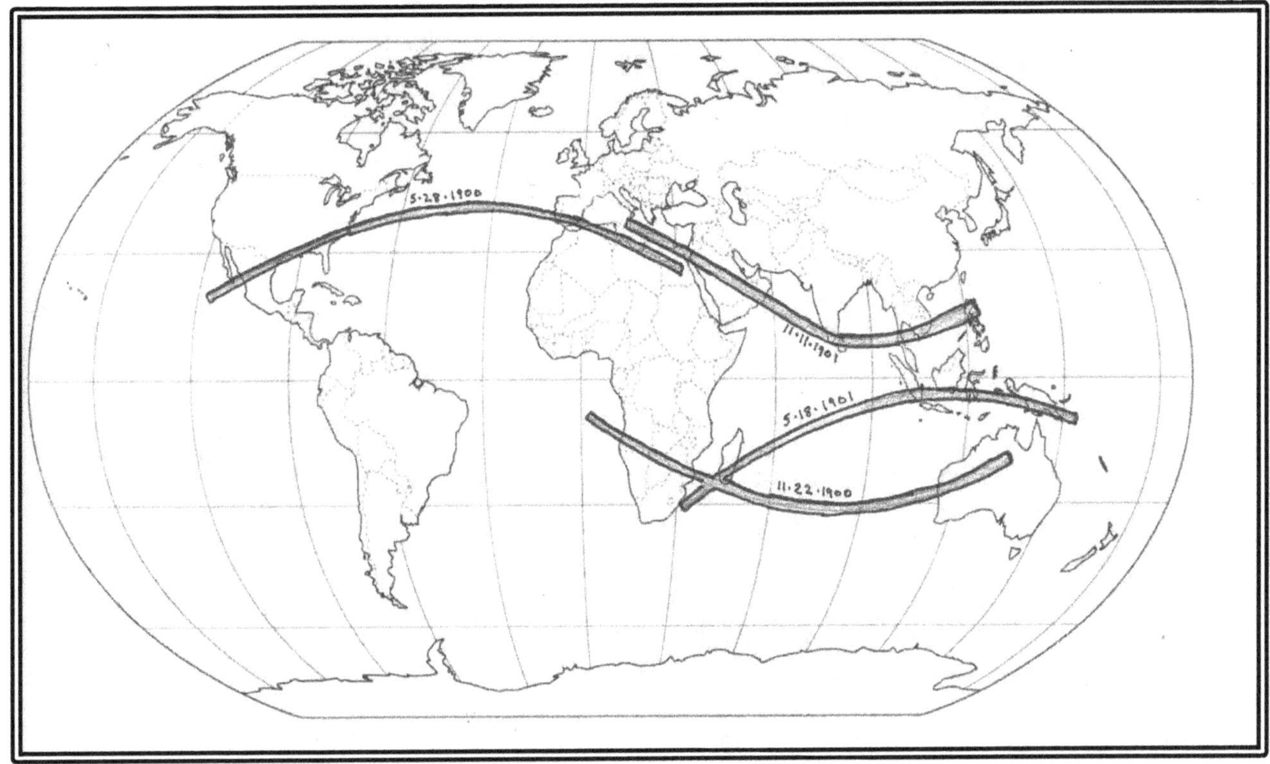

1900 and 1901

1900 May 28 Total America to Egypt -Saros 126: The first solar eclipse of the 20th Century connects empires past, present and future on the first JFK birthday Metonic in 500 years (he was born 17 years later on May 29 1917). Beginning in the Pacific, path crosses Mexico, hugs the US gulf Coast. Totality in New Orleans, Montgomery, and Raleigh--exiting at Virginia Beach. 98% totality in DC. 92 % in New York City, 90% in JFK birthplace of Brookline, Mass.

Eclipse bisects Portugal and Spain, the first colonizers of New World, almost exactly 402 years after Columbus left on his third voyage to the Americas in 1498 May 30, and 400 years after Columbus was sent in chains back to Spain for trial in 1500. The eclipse directly crosses Algiers, Algeria, and Tripoli Libya before ending a few miles south of the Valley of the Kings in Luxor Egypt, revisited by two eclipses in 2027 and 2034 - examined in depth at US and Mideast section.

1900 Nov 22 Ring of Fire Africa to Australia- Saros 131 –JFK Death Day Metonic (exactly 63 years before his death). The side-by-side JFK Metonic is examined in "Personalities" section.

1901 May 18 Total South Africa to New Guinea- Saros 136 Pope John Paul Birthday 19 years later. Mt. St. Helens erupts 79 years later. Only time May 18 appears as a complete eclipse date for centuries.

1901 Nov 11 "11-11" Ring of Fire Mideast to Southeast Asia-Saros 141 17 years to the day before WWI armistice. Bizarre numeric created by ones and elevens. According to NASA, maximum eclipse was 11 minutes and 1 second. Thereby creating an 11:01 minute eclipse on 11-11-01

Partial Solar Eclipses Occur 4-8-1902/ 5-7-1902/ 10-31-1902

1903 to 1905 Lunar Year Triads

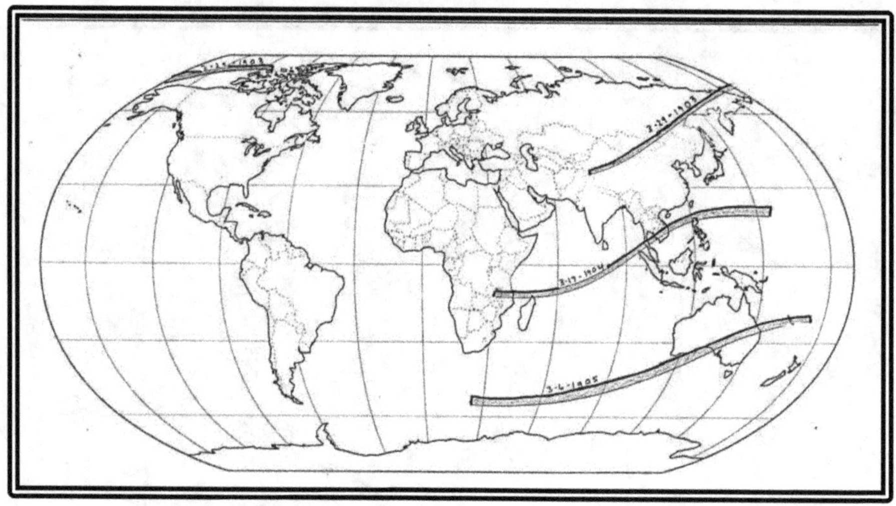

1903 to 1905 Ring of Fire Triad

1903 Mar 29 Ring of Fire China/ Russia -Saros 118
1904 Mar 17 Ring of Fire Africa to Southeast Asia - Saros 128
1905 Mar 6 Ring of Fire Australia Gold Coast -Saros 138

A good example of a balanced triad of annular eclipses. Each follows exactly one lunar year (354.3 days) after the other. In this case, the eclipses progress south in latitude in a very orderly fashion.

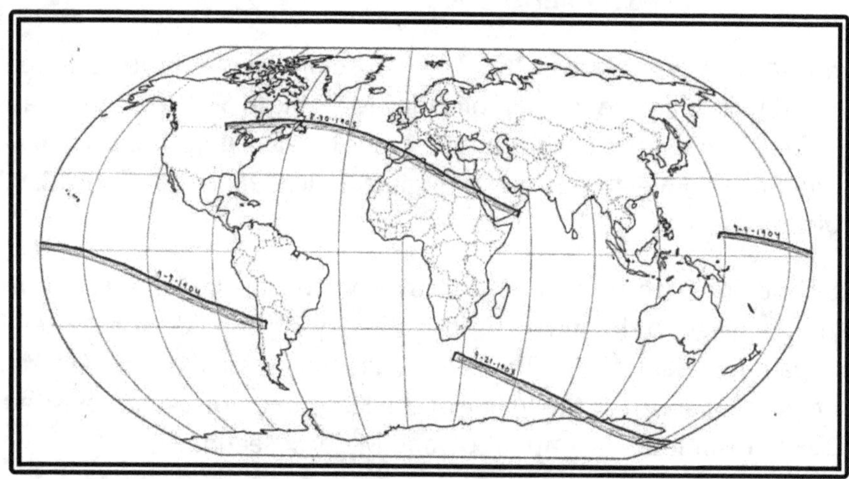

1903 to 1905 Total Eclipse Triad

1903 Sep 21 Total Equinox Indian Ocean to Antarctica- Saros 123
1904 Sep 9 Total Ocean to Chile (Atacama) - Saros 133
1905 Aug 30 Total North America to Spain/ Egypt (Luxor) /Mecca- Saros 143

Concurrently with the Ring of Fire Triad shown in the previous grouping, there is this total eclipse triad. Triad travels south to north over the course of the three years, in the opposite direction as the ring of fires. Notice the variation in the geographic location as compared to the last. Aug 30 1905 Total Eclipse (part of 1900 to 1905 Egypt Eclipse swarm) is looked at in detail in the Mideast Eclipse section.
Partial Solar Eclipses occur on 2-23-1906/ 7-21-1906/ 8-20-1906

1907 to 1910

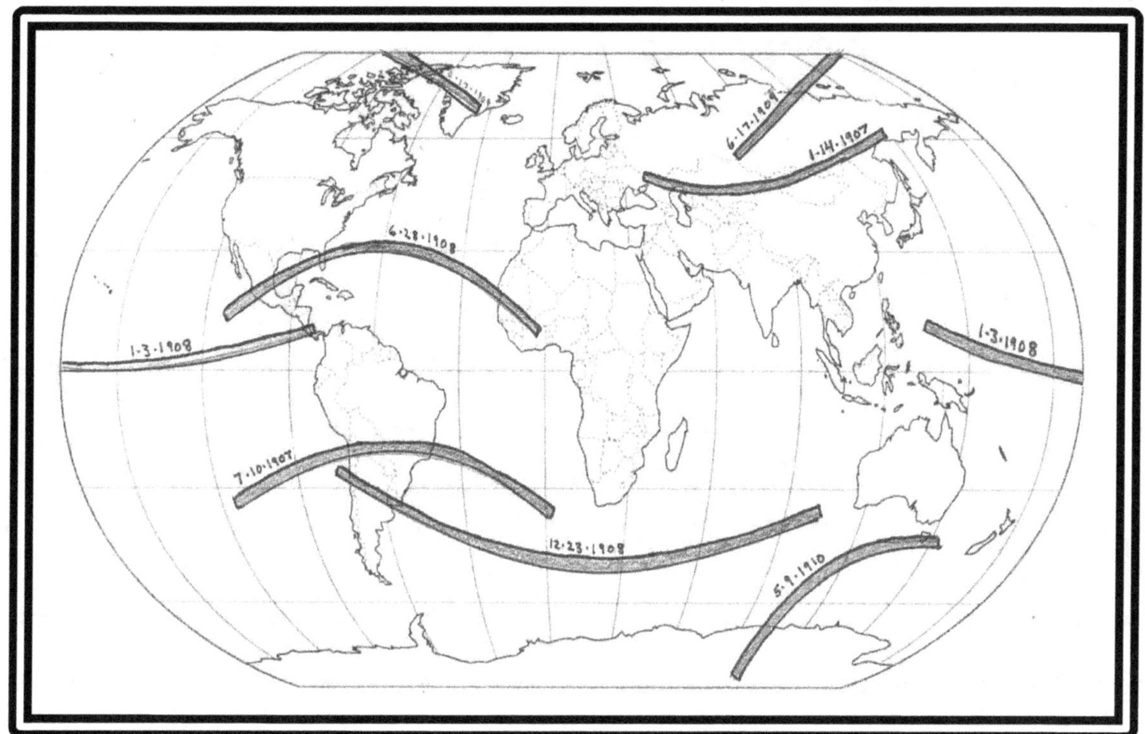

1907 to 1910

1907 Jan 14 Total Future USSR Southern Boundary- Saros 120
1907 Jul 10 Ring of Fire South America- Saros 125
1908 Jun 28 Ring of Fire Temple of Sun and Moon/Cape Canaveral -Saros 135
1908 Jan 3 Total Pacific Ocean to American Isthmus- Saros 130
1908 Dec 23 Hybrid South America to Indian Ocean- Saros 140
1909 Jun 17 Total Russia- Saros 145
1910 May 9 Total Antarctica to Tasmania- Saros 117

The 1907 eclipse starts on border of Ukraine, bisects Kazakhstan, Mongolia
The 1908 Ring of Fire, which connects the Aztec Temples of the Sun and Moon, Cape Canaveral and Africa is looked at closer in the USA section.
There are two Eclipses over Russia (there will be four eclipses within ten years prior of the 1917 Revolution).
1908 eclipse travels open ocean to die over Pan-American Isthmus in Panama, something that occurs with striking frequency in this survey.

Last Eclipses of Valentine's Metonic: 1911 to 1915

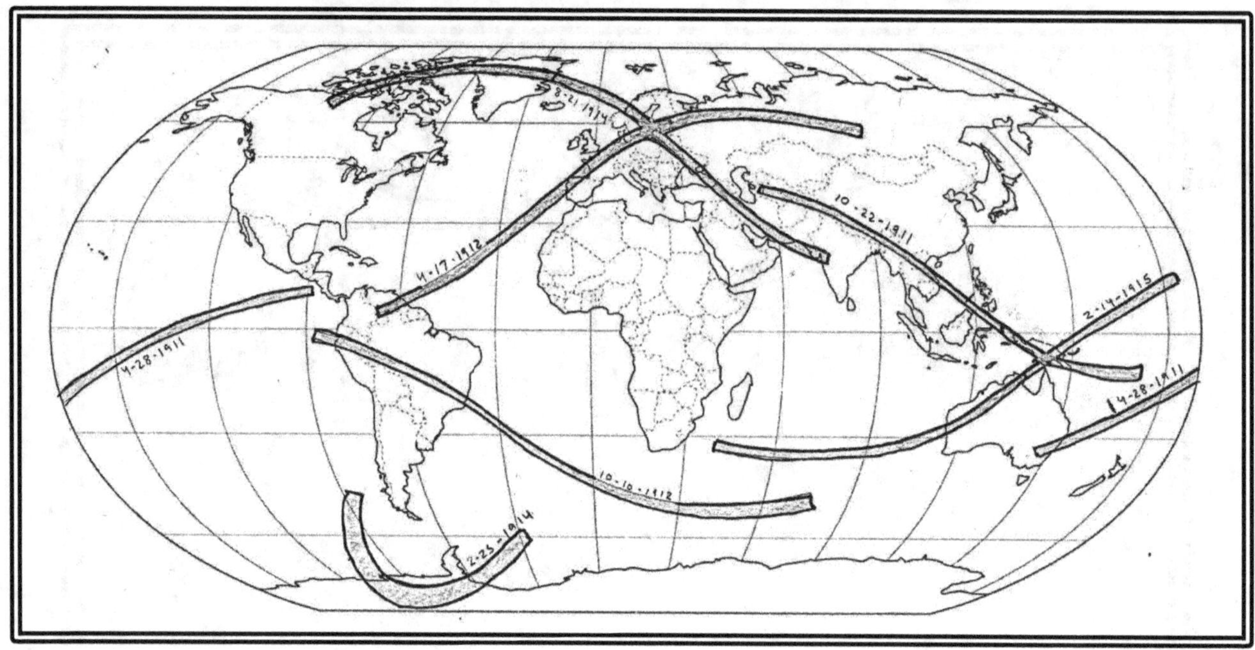

1911 to 1915 (end of Valentine's Day Metonic)

1911 Apr 28 Total Eclipse Australia almost to Central America- Saros 127
1911 Oct 22 Ring of Fire South Asia to New Guinea- Saros 132
1912 Apr 17 Hybrid Titanic Eclipse Colombia/ Europe/ Russia -Saros 137
1912 Oct 10 Total Peru/ South America/ Ocean -Saros 142
1914 Feb 25 Ring of Fire "Endurance" Eclipse Antarctica- Saros 119
1914 Aug 21 Total World War One Eclipse Mideast/ Europe/ North America- Saros 124
1915 Feb 14 Valentine's Day Ring of Fire Darwin Australia/New Guinea-Saros 129

Two obvious crosses in seven eclipses. One over New Guinea (far lower right of map) between 1911 and 1915. One over Europe with apex in Baltic Sea not far from St. Petersburg Russia. We will look in depth at a couple of the eclipses in this section.

The Titanic Eclipse April 17 1912 - Saros 137 60 hours after the sinking of the Titanic on the night of April 14/15 1912, a Hybrid solar eclipse occurs in the Atlantic, through Europe and deep into Russia. A great solar eclipse marking the most famous man-made disaster of all times, the legendary sinking of the Titanic, the largest ship ever launched, the symbol of modern technology, the "unsinkable". She was sent to a watery grave on her maiden voyage along with more than 1,500 people. Path goes just south of Southampton, England (90% coverage) where Titanic launched a week before. 35% coverage at the site of the disaster in Atlantic southeast of Newfoundland. 18% coverage at intended harbor of New York. Path cuts a foreboding shadow across Europe two years before World War One. Stripe of Totality divides metropolitan Paris. Larger path of annularity goes through France, Belgium, Germany, Baltics, and Russia.

The Titanic Eclipse creates a distinctive cross with August 21 1914 World War One Eclipse, looked at in more detail in next section. This is the end of The Valentine's Day Metonic of 1896 to 1915.

partials occur on 4-6-1913/ 8-31-1913/ 9-30-1913

World War One 1914 to 1918

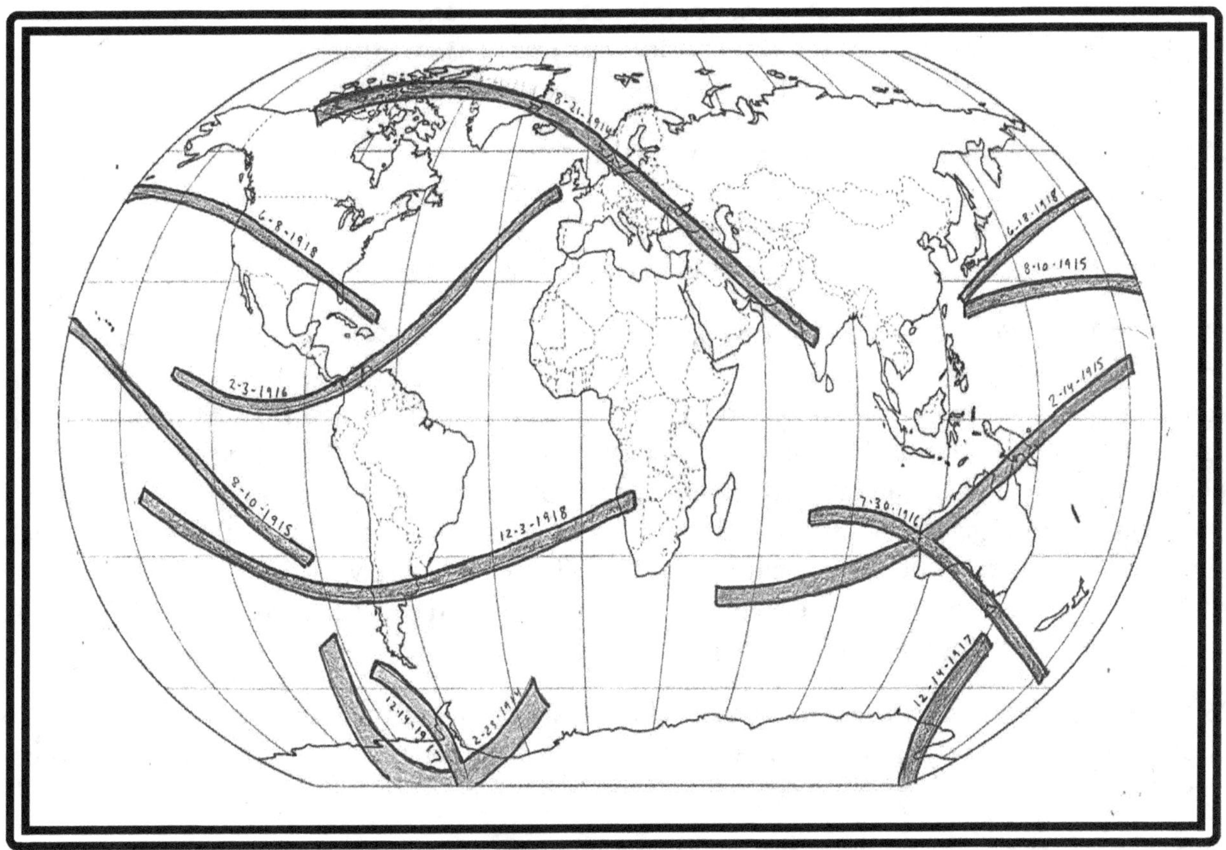

World War One Eclipses 1914 to 1918/ Temple Destruction Metonic

Aug 10 1915 to Aug 10 1934 is the Temple Destruction Metonic, August 10 was the Julian Calendar date Jerusalem was destroyed and the Temple burned by Rome in 70 AD. The Aug 10 Metonic begins one lunar year after the World War One eclipse of Aug 21 1914. In this section we will look at all eclipses that occurred during the years 1914 to 1918, the years of the first World War.

1914 Feb 25 Endurance Ring of Fire Eclipse Antarctica- Saros 119
1914 Aug 21 World War One Total Eclipse Europe/Mideast- Saros 124
1915 Feb 14 Ring of Fire Australia/New Guinea- Saros 129
1915 Aug 10 Ring of Fire Pacific Ocean- Saros 134
1916 Feb 3   Panama Isthmus almost to England Total Eclipse -Saros 139
1916 Jul 30 Ring of Fire Australia/Tasmania-Saros 144
1917 Dec 14 Ring of Fire Antarctica -Saros 121
1918 Jun 8 USA Cross Country Total Eclipse -Saros 126
1918 Dec 3 Atacama earthquake eclipse Ring of Fire- Saros 131

The WWI eclipses overlap Valentine's Day (Feb 14) and Temple Destruction (Aug 10) Metonic Cycles.

Nine complete solar eclipses occur between 1914 through 1918. All are pictured in the map on the previous page. Seven of the eclipses occurred during hostilities associated with the First World War, the single deadliest war in history up to that point. Due to peculiar orbital realities and calendars, complete eclipses occur in each of the six years in a row from 1914 to 1919, a fairly rare occurrence. The eclipse paths do not touch except in extreme southern hemisphere.

1914 Feb 25 Ring of Fire Endurance Eclipse Antarctica- Saros 119 The eclipse occurs on Feb 24 1914 in Antarctica, near the coming shipwreck of the ill-fated Endurance South Pole Expedition which left England on August 8 1914. One of the maximum expression survival stories of all time, the expedition of the crew of Earnest Shackleton lived and struggled for two years in the Antarctic wilderness.

1914 Aug 21 Total World War One Eclipse Europe/Mideast- Saros 124 One of the most clearly ominous eclipses of all time. In the first weeks of World War I, a total solar eclipse descends from the Northern latitudes above North America and divides the Eurasian land mass. The first shot fired by a British soldier in World War One came the next day on August 22 1914 in the village of Casteau in Belgium (60.7% coverage). Empires affected include the Ottoman, Persian, British, French, German, Russian and Austro-Hungarian empires. Scandinavia split in half. Great cities in totality include Kiev, Ukraine-Karachi, Pakistan and the ancient city of Nineveh (modern-day Mosul) in Iraq. Saros Cycle 124. The full path connects both hemispheres in both eclipses, the path intersects the Titanic Eclipse of Apr 17 1912 near St. Petersburg, Russia.

1916 Feb 3 Total Panama Isthmus almost to England -Saros 139 Fourteen months before entrance of the USA into World War One, a total eclipse crosses the Pan-American Isthmus to almost reach the shores of England.

1917 Dec 14 Ring of Fire Antarctica - Saros 121 An interesting example of the 103 year repeating Metonic. Exactly 103 years after this eclipse comes the Electoral College Eclipse of Dec 14 2020, which crosses Southern South America while the President and Vice-President of the USA are being elected.103 years is a long-lasting Metonic.

1918 Jun 8 Total USA Cross-Country- Saros 126 The first American Cross County eclipse in 112 years and the last for another 99. Happens in the final months of World War One, as American forces were fighting to a stalemate in the Battle of Belleau Wood, and the Spanish Flu had begun to rage across the country. This total solar eclipse marks the beginning of the American century, which would come to an end 99 years later with a very similar (but convex mirrored) eclipse on August 21 2017. Saros cycle 126, which began the Century on 1900 May 28 with USA to Egypt Eclipse and reappears in Spanish judgment on August 12 2026. Take a closer look in the USA section.

1918 Dec 3 Ring of Fire Atacama Earthquake eclipse- Saros 131 This eclipse is a startling example of an earthquake/eclipse conjunction. The ring of fire crosses the Atacama Desert in Chile on Dec 3. 24 hours later, an 8.2 magnitude earthquake leveled the city of Atacama. Of course, Chile is an especially earthquake-prone region. But still...

partials occur on 12-24-1916/ 1-23-1917/ 6-19-1917/ 7-19-1917

You will note that complete eclipses occurred in every calendar year from 1914 to 1919, essentially the outside limit for consecutive years to contain complete eclipses.

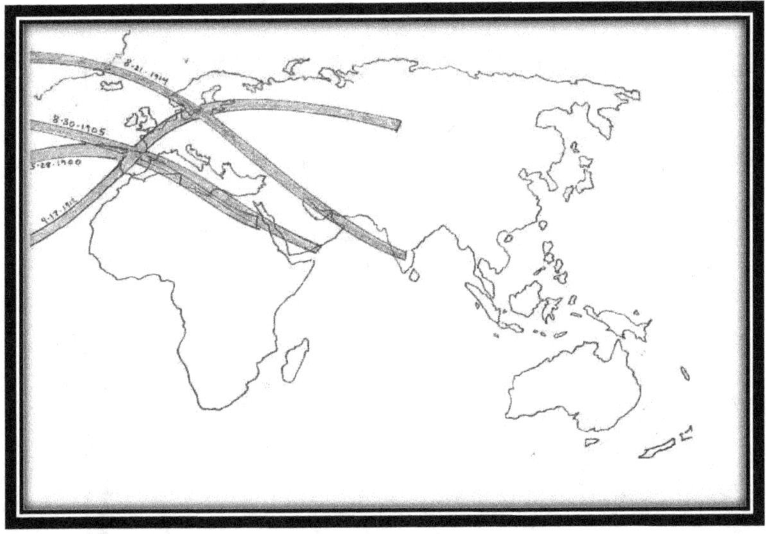

Above: Europe/Spain/Egypt 1900 to 1914

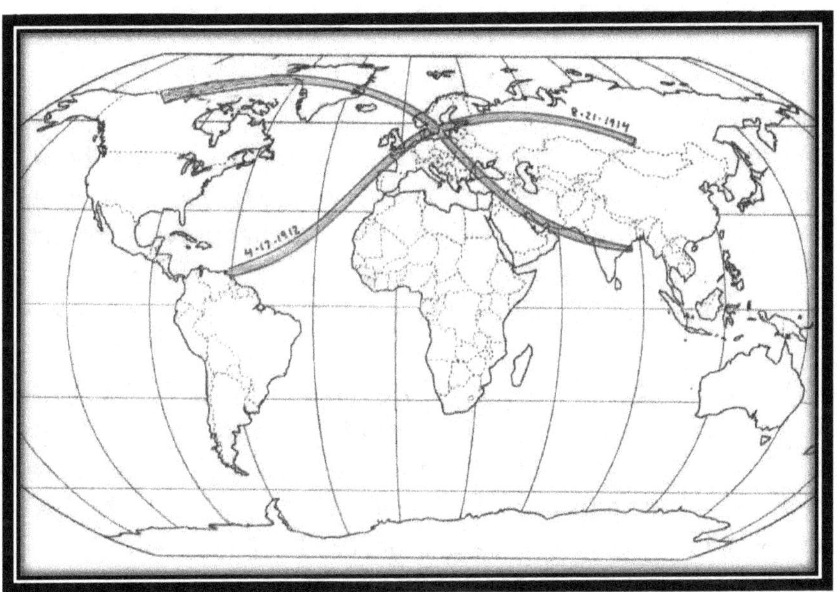

Above Titanic 1912 and WW1 1914

1919 JFK Birthday/Death day Eclipses

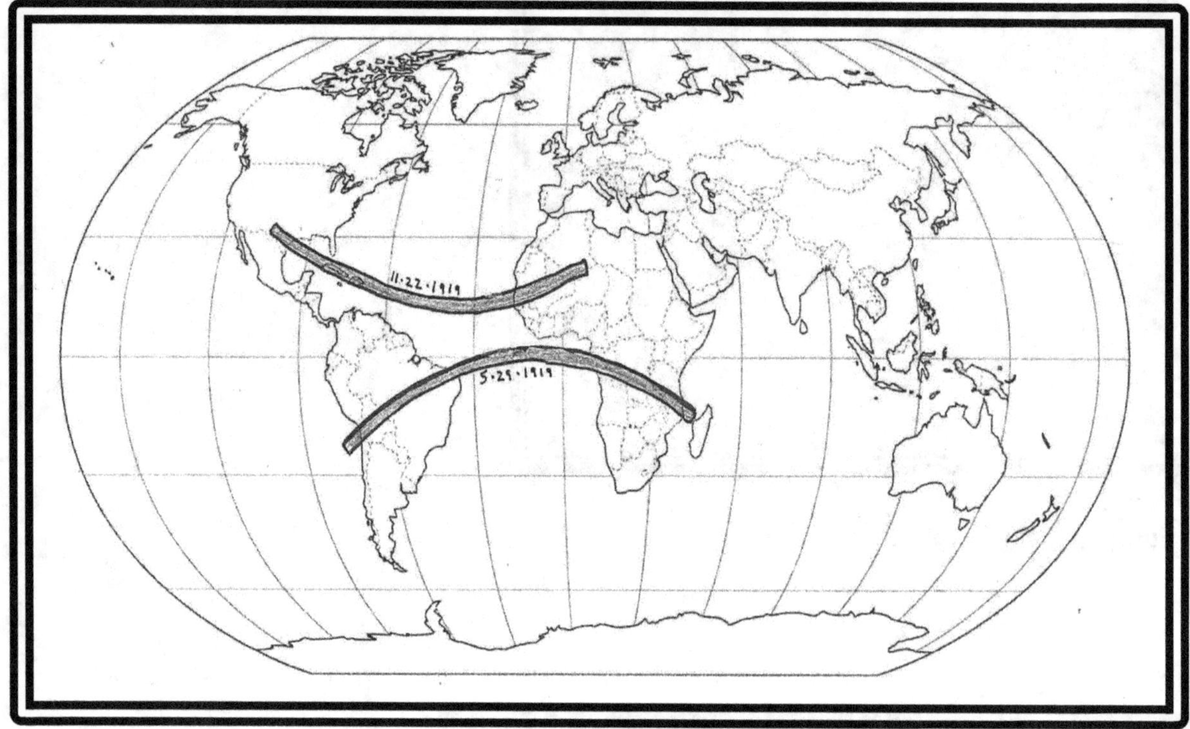

1919 JFK Birthday of May 29 and JFK Death Day of Nov 22 Eclipses

1919 May 29 Total Einstein's Eclipse (Saros 136) On John F. Kennedy's second birthday (he was born May 29 1917), Einstein's theory of Relativity is finally proved by observation of the total solar eclipse. Two teams, one in Brazil and one in Africa take the simultaneous measurements that prove it correct. The longest eclipse duration (at nearly 7 minutes) since 1416 AD. Saros Cycle 136, one of the stars of the Judgment series--reappearing notably in Mideast Valley of the Kings eclipse on August 2nd 2027

1919 Nov 22 Ring of Fire Texas Sunrise/ JFK Death day (Saros 141) 177 days later, on 11-22-1919, exactly 44 years before JFK assassination, thereby making the exact dates of JFK's birth and death commemorated in successive eclipses while he is alive in the same hemisphere. A Ring of Fire Eclipse in Texas, the state where JFK would be murdered 44 years later is the place where the eclipse begins at sunrise. Total annularity in Stonewall Texas, home to a then 10-year-old Lyndon Baines Johnson, who would replace JFK as president exactly 44 years later on 11-22-1963. Total annularity in the capital of Austin and in Houston. --path over almost entire island of Cuba , with which JFK's destiny would become intertwined, then across Haiti, St. Georges' , Kingston, Port of Spain--then across the Atlantic to Africa, where the path reached the countries of Senegal and Mali, both of whose presidents visited Kennedy's White House. Eclipse ends after totality in ancient city of Timbuktu. The amazing Saros Cycle 141, which starred in Farewell Abraham Eclipse of 1865 among many others. Saros 141 reappears in Australia Day Spanish Judgment of Jan 26 2028.

partial eclipses occur May 18, 1920, the day John Paul II is born (also 60 years to the day before Mount Saint Helens eruption) and on November 10, 1920

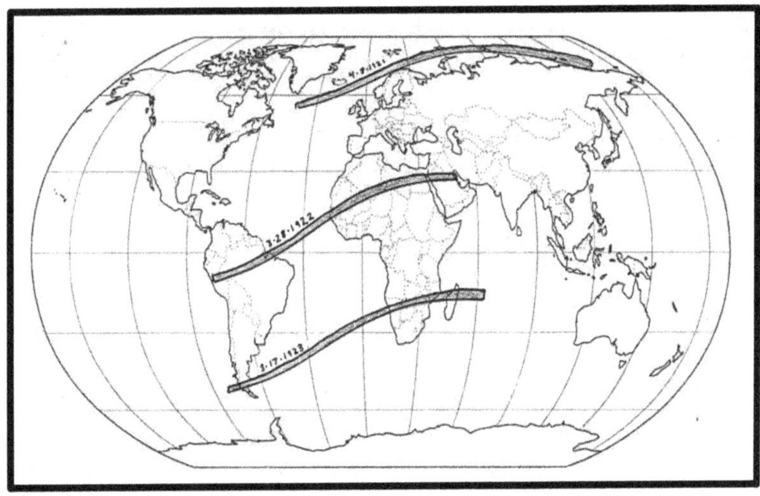

Eclipses 1920 to 1923—Ring of Fire Triad

1921 Apr 8 Ring of Fire Northern latitudes/USSR Buddha Birthday -Saros 118
1922 Mar 28 Ring of Fire Egypt Independence/ King Tut Eclipse -Saros 128
1923 Mar 17 Ring of Fire Patagonia to Madagascar - Saros 138

 A great example of how eclipses can progress by lunar year. Three rings of fires exactly one lunar year apart from each other (354.3 days). 1921 Buddha Birthday Eclipse is exactly 103 years from Apr 8 2024 USA eclipse, an example of the 103 year Metonic. The 1922 Egypt Independence / King Tut Eclipse is looked at closer in the Mideast Eclipse section. This Triad is extremely balanced as it progresses from north to south. Notice how the Saros number increases by tens from 118 to 128 to 138.

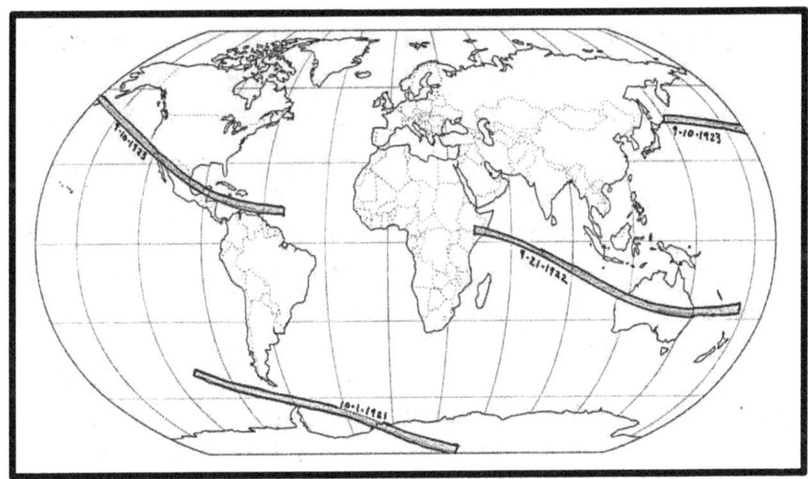

Eclipses 1920 to 1923—Total Triad

1921 Oct 1 Total Antarctica -Saros 123
1922 Sep 21 Total Equinox Africa to Australia -Saros 133
1923 Sep 10 Total Japan Earthquake/ Southern California -Saros 143

Here are the total eclipses of 1921 to 1923, also in a lunar year triad. The 1923 eclipse is examined in the USA section. It emanated from the northern islands of Japan 9 days after the worst natural disaster in Japanese history, the 1923 Tokyo earthquake and fire.
Partial eclipses occur on Mar 5 1924, July 3 1924 and August 30 1924

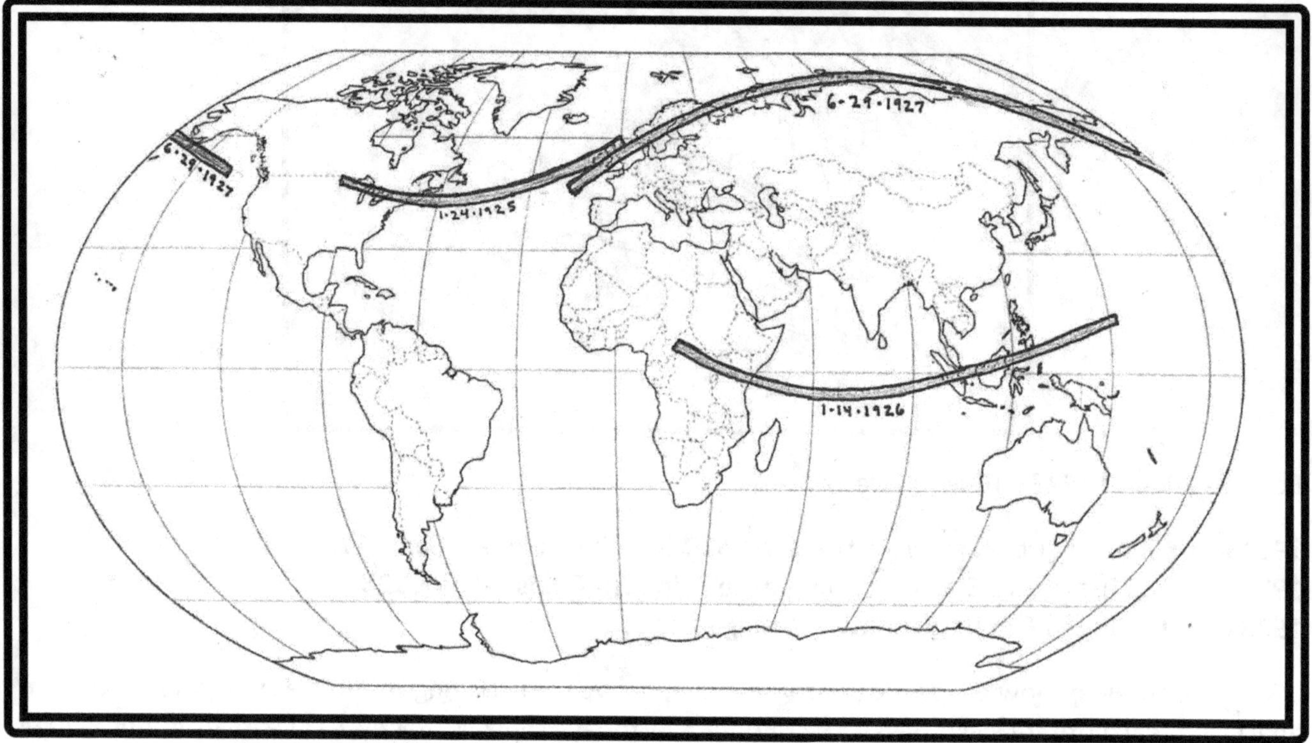

1925 to 1927 Roaring Twenties: Bob Dylan, Beatles and Martin Luther King Premonitions
Three personality associations in three total eclipses in three years.

1925 Jan 24 Total USA to England Roaring 20's -Saros 120 At the mid-point of the roaring twenties, a total solar eclipse across the northern USA brings a total solar eclipse to New York City for the first time since 1478. NYC is becoming the new epicenter of western civilization and empire ("empire state"). New York was home to the financial and cultural centers of the nation--Wall Street, Broadway, Vaudeville, Tin Pan Alley. The city was treated to rare clear skies and the eclipse was widely observed-- eclipse path crosses the ocean to extreme northern Britain. Path goes from Hibbing and Duluth Minnesota (birthplace and hometown of Bob Dylan about 16 years later) and crosses his adult homes of Woodstock and New York City. 36 years to the day before he arrived by bus to NYC from Minnesota on Jan 24 1961. Nearly a dozen important US figures are born in the months around this eclipse.

1926 Jan 14 Total Central Africa to Indonesia -Saros 130 Three years essentially to the day before Martin Luther King is born on Jan 15 1929. 36 years before the pinnacle of his fame.

1927 Jun 29 Total Liverpool to USA -Saros 145 At the height of the British Empire, The United Kingdom is bisected by an early morning total solar eclipse, including totality at sunrise in Liverpool 16 years before the last Beatle George is born there. All four Beatles were born in Liverpool. Eclipse occurs 36 years before their fame exploded in 1963/1964. Path travels across Scandinavia, extreme north latitudes of Russia before ending in the Aleutian islands of the USA...see Focus spotlight in personality sections...Saros 145, which is one of the greatest of the Saros, returning for USA atom bomb 1945, Great 1999 eclipse and USA 2017.

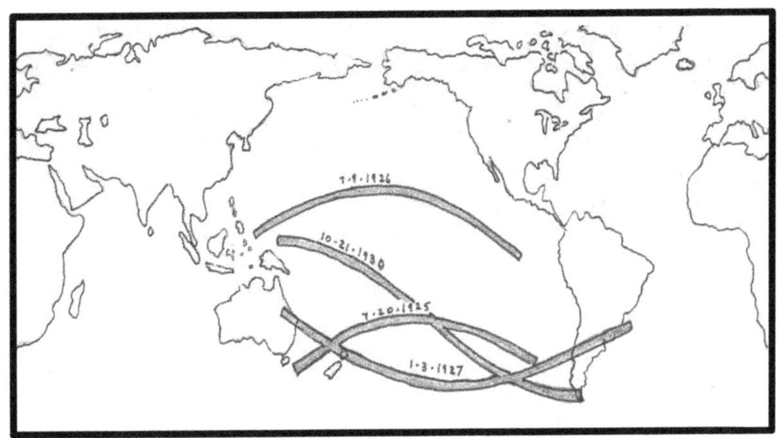

1925 to 1930 Pacific Cluster (above)

1925 Jul 20 Ring of Fire (Saros 125)
1926 Jul 9 Ring of Fire (Saros 135)
1927 Jan 3 Ring of Fire (Saros 140)
1930 Oct 21 Total (Saros 142)
A clump of eclipses cross the Pacific beginning at generally similar longitudes between 1925 and 1930.

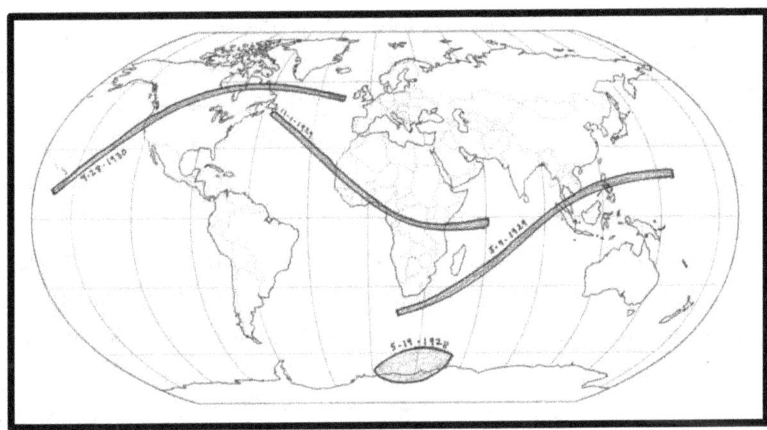

1928 to 1930 Total Triad and Stock Market Crash Ring of Fire (above)

1928 May 19 Total Antarctica (Saros 117)
1929 May 9 Total Indian Ocean to Indonesia (Saros 127)
1930 Apr 28 Hybrid San Francisco (Saros 137)
1929 Nov 2 Ring of Fire Stock Market (Saros 132)

A Triad of two totals and a hybrid precede the **1929 Stock Market Eclipse**, which occurred three days after Crash of Oct 29 1929, an annular eclipse beginning east of North America to traverse Africa, with maximum eclipse on the Slave Coast. Annularity in African cities of Lomé, Kinshasa and Dar es Salaam. The **1930 Apr 28 San Francisco Hybrid** featured totality of less than a kilometer wide and was the only totality in San Francisco in centuries, 36 years and two days before Anton Lavey founded Church of Satan in San Francisco on Apr 30 1966.
Eclipses of 1925, 1926, 1927, 1930 all have interesting associations 36 years later.
partials occur on 12-24-1927, 6-17-1928, 11-12-1928

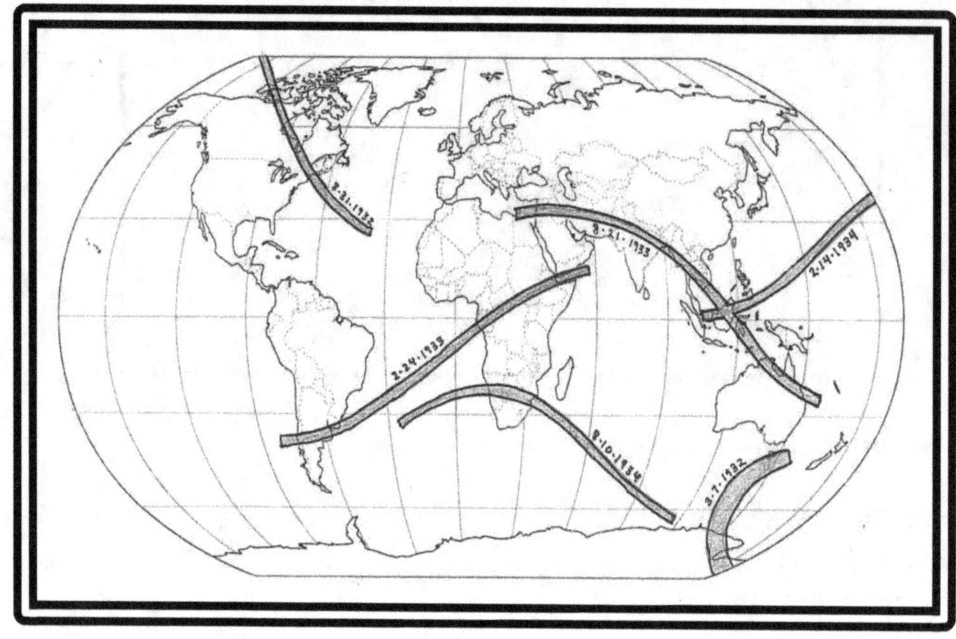

1932 Mar 7 Ring of Fire Antarctica to Tasmania -Saros 119
1932 Aug 31 Total New England Total -Saros 124 Looked at in USA section. The end of the USA 1918 to 1932 Eclipse Swarm. Two months before the Presidential election that brought Franklin Delano Roosevelt to power, 96 % at his home of Hyde Park, New York. Six weeks after lowest point of stock market in the Great Depression, Eclipse crosses Canada (totality in Montreal) and New England. Totality at Bush center of Kennebunkport and Cape Cod. 99.9% at Kennedy home.
1933 Feb 24 Ring of Fire South America to Arabia -Saros 129
1933 Aug 21 Ring of Fire Jerusalem -Saros 134 Looked at closer in Mideast section, As Nazis seize control in Germany, at the peak of a worldwide depression, a Ring of Fire across the Mideast- Palestine, Jordan, Iraq, Iran and Afghanistan-- then South Asia, India and Southeast Asia, total annularity in Jerusalem, Baghdad (Babylon), Amman, Gaza, Alexandria (Egypt) ---path also cuts through Bangladesh, Thailand, Indonesia before finally ending at the east coast of Australia.
1934 Feb 14 Total Valentine's Day in Borneo -Saros 139
1934 Aug 10 Ring of Fire End of Temple Destruction Metonic- Saros 144)

In 1933 two Ring of Fires frame Arabia. The year western oil interests first arrived in Saudi Arabia. It is somewhat rare to have the only complete eclipses in a year both be rings of fires. 1933 was the year Germany fully went from democracy to dictatorship--the Reichstag fire, the first anti-Jewish laws, the purge of the opposition, the construction of Dachau. August 21 1933, the day of the eclipse, was the start of the 18th Zionist Congress in Prague. In the Hebrew calendar it was 1st day of month of Elul.

In August 1933 in Iraq, thousands of Assyrian Christians were killed in the Simele massacre Also, the "great white spot " of Saturn was seen by astronomers for the first time in 30 years in early August, and stayed for 4 weeks, allowing for the measurement of Saturn's rotation. Mahatma Gandhi was arrested by the British in India on eclipse days of August 21 and began a hunger strike. Palestinian leader Yasser Arafat was born 3 days after Aug 21 eclipse in Cairo on August 24 1933

partials occur on 4-18-1931, 9-12-1931, and 10-11-1931

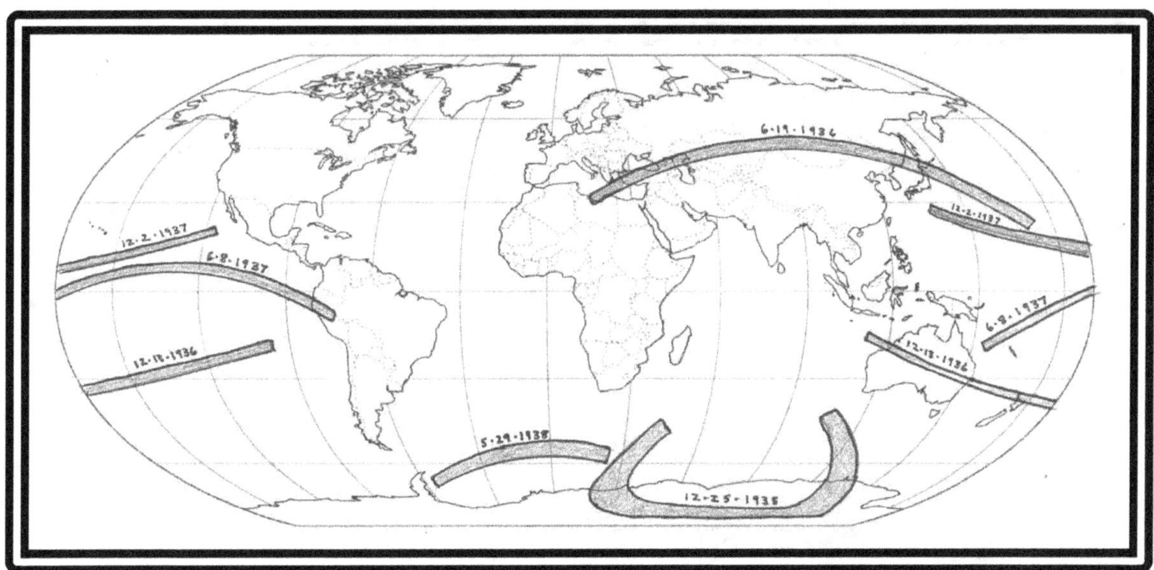

Christmas Metonic Begins with two Triads 1935 to 1938

The Christmas Metonic extends from Dec 25 1935 to Dec 25 1954. These are the first Christmas Eclipses since 1582 (the "Christmas Island" and 2 years of Christmas eclipses in a row.

Total Triads (each one lunar year apart):
1936 Jun 19 Total World War Premonition eclipse- Saros 126
1937 Jun 8 Total Pacific to Peru -Saros 136
1938 May 29 Total JFK 21st Birthday/ Endurance /Antarctic Ocean -Saros 146

Ring of Fire Triads (each one lunar year apart):
1935 Dec 25 Ring of Fire Christmas in Antarctica -Saros 121
1936 Dec 13 Ring of Fire Australia/ New Zealand -Saros 131
1937 Dec 2 Ring of Fire Open Pacific Ocean -Saros 141

1936 Jun 19 World War Premonition: Begins off Libya, crosses Asia to pass over Japan.   Path across great civilizations in days leading up to WWII. The path crosses Greece with totality in Athens (a few weeks before the Metaxas coup), eastern Turkey, across USSR (at beginning of Stalin's "Great Purge") and into Japan, a year before beginning of the second China/Japanese War (July 1937)
1935 Dec 25 Christmas in Antarctica: One of only two true Christmas solar eclipses in centuries.
1936 Dec 13 Australia/ New Zealand eclipse occurs four days before birth of Pope Francis and three days before birth of this author's mother. Eclipse occurs over northern New Zealand where author would be conceived 30 years later.
1938 May 29 Endurance /John F. Kennedy's 21st birthday, a total solar eclipse covers the site of Endurance Antarctic Expedition on JFK's birthday, 22 years almost exactly after Ernest Shackleton's daring May 1916 500 mile rowboat ocean crossing of the Ocean and desperate mountaineering traverse of South Georgia Island, one of the greatest survival stories in the history of humanity. Totality on South Georgia and 93% on Elephant Island, where his crew huddled for months.

Partials occur on 1-5-1935, 2-3-1935, 6-30-1935, 7-30-1935, 11-21-1938

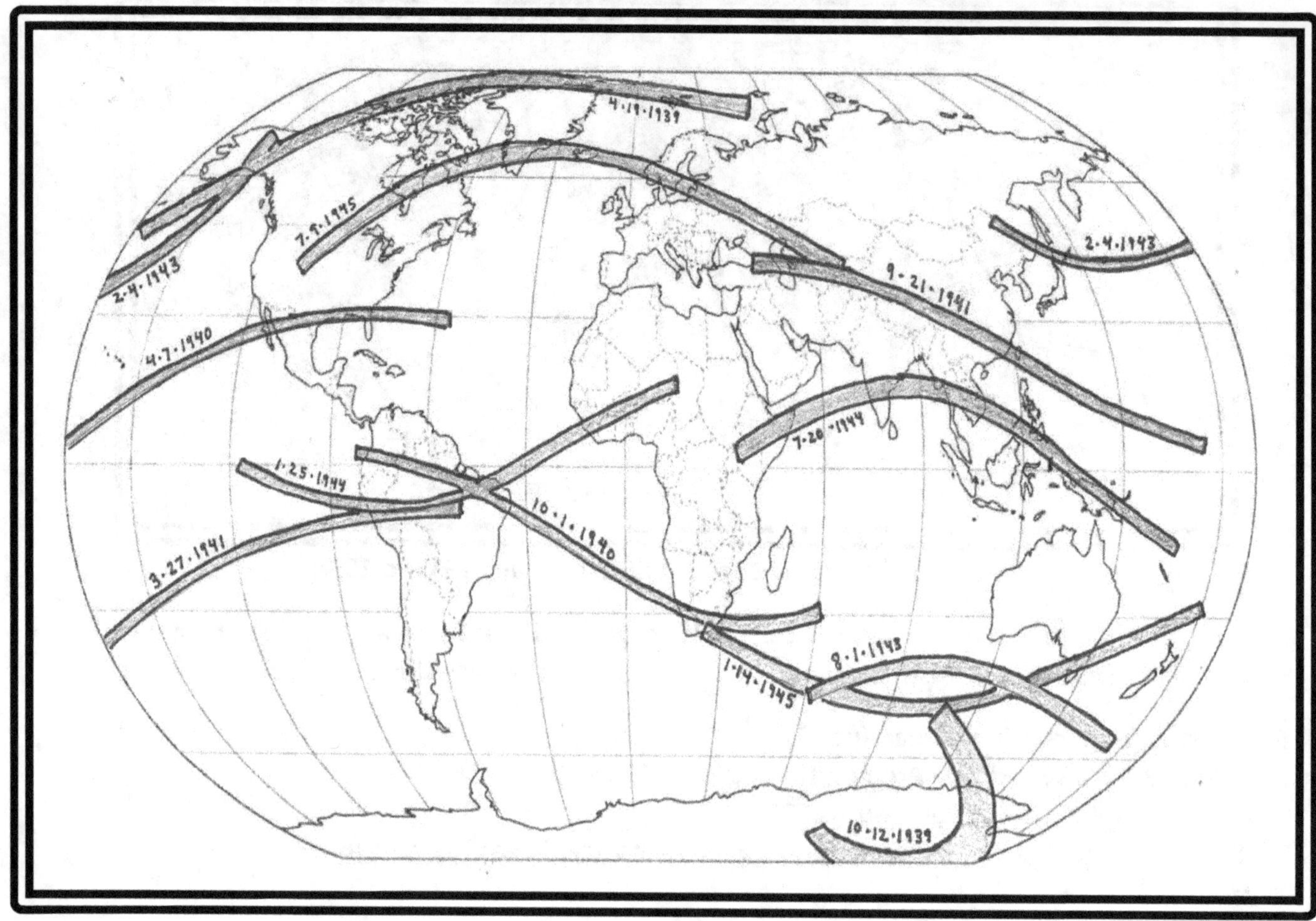

World War Two Eclipses 1939 to 1945

Twelve complete solar eclipses occurred in the seven years of World War Two.
 It is estimated that nearly 60 million people died in the war.

1939 Apr 19 Ring of Fire Hitler Birthday Metonic Northern Latitudes -Saros 118
1939 Oct 12 Total Columbus Day first weeks of WWII Antarctica -Saros 123
1940 Apr 7 Ring of Fire USA Deep South -Saros 128
1940 Oct 1 Total South America to Africa -Saros 133
1941 Mar 27 Ring of Fire Pacific to South America -Saros 138
1941 Sep 21 Total Black Sea to China Fall Equinox -Saros 143
1943 Feb 4 Total Japan to Alaska -Saros 120
1943 Aug 1 Ring of Fire South Indian Ocean -Saros 125
1944 Jan 25 Total South America to Africa -Saros 130
1944 Jul 20 Ring of Fire Africa, India, Southeast Asia -Saros 135
1945 Jan 14 Ring of Fire South Africa to Australia - Saros 140
1945 Jul 9 Total Atom Bomb Eclipse USA to USSR -Saros 145

1939 Apr 19 Ring of Fire Hitler Birthday/Darwin Death Day Metonic (Saros 118) in the hours before Adolf Hitler's 50th birthday on Apr 20 ,on 57th anniversary of Darwin's death (57 years is a complete Metonic triad 19x3); 135 days before invasion of Poland. Annularity in Anchorage, Alaska. Berlin 30% coverage. Last complete eclipse for Saros Cycle 118 after nearly 1,000 years of complete eclipses.

1940 Apr 7 Ring of Fire USA Deep South (Saros 128) As war rages in Europe, a Ring of Fire crosses northern Mexico, Texas (total annularity in Austin and Houston) and hugs the Gulf Coast (Baton Rouge, New Orleans, Mobile and Jacksonville) on the first new moon after Easter. Metonic similarity to upcoming Apr 8 2024 North American Eclipse.

1941 Sep 21 Total Black Sea to China Fall Equinox (Saros 143) 3 months after Germany turned on the Soviet Union and invaded them. 3 weeks after 23,000 Hungarian Jews are slaughtered by the Gestapo and 8 days before 30,000 more are killed at Babi Yar near Kiev. Total solar eclipse cuts across the USSR (starting just east of Ukraine) and China, during great battle of Changsha China (95% totality), Totality in Wuhan and Taipei.

1943 Feb 4 Total Japan to Alaska (Saros 120) Warring nations of Japan and USA are linked by a total solar eclipse. Totality in Sapporo Japan and Anchorage Alaska (which has second eclipse in 4 years). Path begins in eastern China and USSR. Eclipse 5 days before Japanese resistance ends at Guadalcanal. Saros 120, which returns with the Microsoft/Mount Saint Helens USA eclipse in 1979 Feb 26.

1944 Jan 25 Total South America to Africa (Saros 130) Peru and Brazil then across ocean to Modern day Sierra Leone and Guinea-including capital of Conakry.

1944 Jul 20 Ring of Fire Africa, India, Southeast Asia (Saros 135) Path crosses Horn of Africa and bisects India. Annularity in Brahmapur ("The abode of Brahma, the Creator"), also in Burma as war raged there. Path goes over Cambodia, Laos and Vietnam, 81% in Bangkok, Thailand. First eclipse in Southeast Asia Eclipse Swarm. Path over fighting in Philippines, ends over the Solomon Islands of New Guinea, site of sinking of JFK's PT-109 boat exactly one lunar year before, the night of August 2 1943. The next Jul 20 Metonic eclipse occurs 19 years later, leaves Japan at sunrise and reaches the USA four months before JFK's murder on Jul 20 1963.

1945 Jul 9 Total Atom Bomb Eclipse USA to USSR (Saros 145) This total eclipse occurs the very day White Sand s Missile Range in New Mexico is established, determining where one week later, on July 16th, the world's first atomic bomb would be detonated. The bomb would be first used on humanity beginning 27 days later in Hiroshima and then Nagasaki Japan.
Total solar eclipse travels from the USA to USSR. Path of Totality in Idaho and Montana (Butte), Canada, Greenland, Scandinavia and then bisecting Russia ---98% totality in St. Petersburg and 95 % in Moscow. Path ends at western border of China. 60% coverage at nuclear test site of White Sands. 70% at Los Alamos Laboratory, New Mexico.

Partials occur on 3-16-1942, 8-12-1942, 9-10-1942

World War Two from 1939 to 1945 is isolated in time, eclipse wise. There are no complete eclipses from May 1938 until Hitler Birthday in Apr 1939. After the 1945 A-Bomb Eclipse, nearly two years elapse until another complete eclipse, May 20 1947, the 441st anniversary of the death of Columbus.

1945 to 1959 USA Emanations—Special Focus map

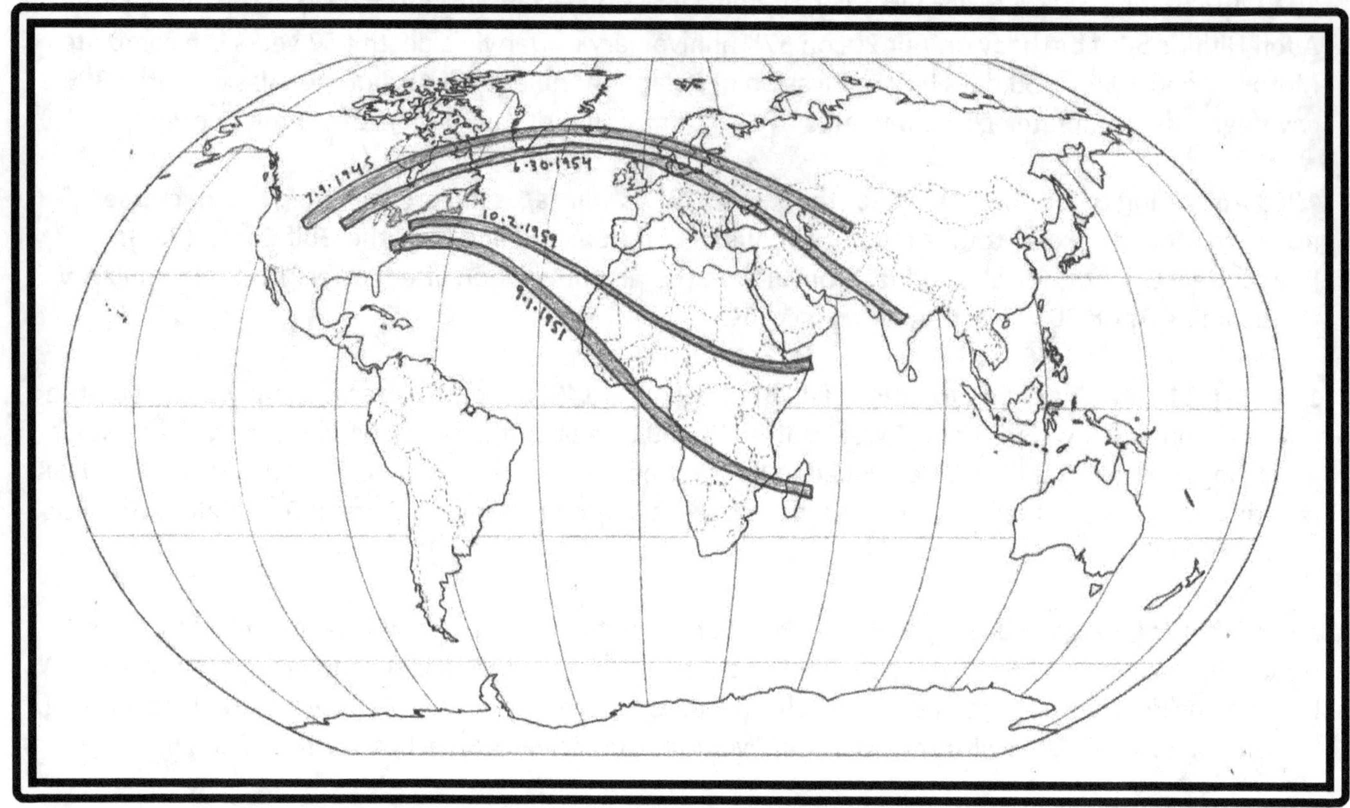

1945 to 1959 USA Emanations

A unique moment of eclipse History. In the space of 15 years, four complete solar eclipses originate in the USA and head out into the world.

1945 Jul 9 Total Atom Bomb Eclipse USA to USSR -Saros 145
1951 Sep 1 Ring of Fire Jamestown to Angola -Saros 134
1954 Jun 30 Total USA to Iran Shahrud Tav -Saros 126
1959 Oct 2 Total JFK Boston Single City across Africa -Saros 143

A focus on a unique moment of eclipse History.
In the space of 15 years, four complete solar eclipses originate in the USA and head out into the world.

1947 to 1951 All Complete Eclipses

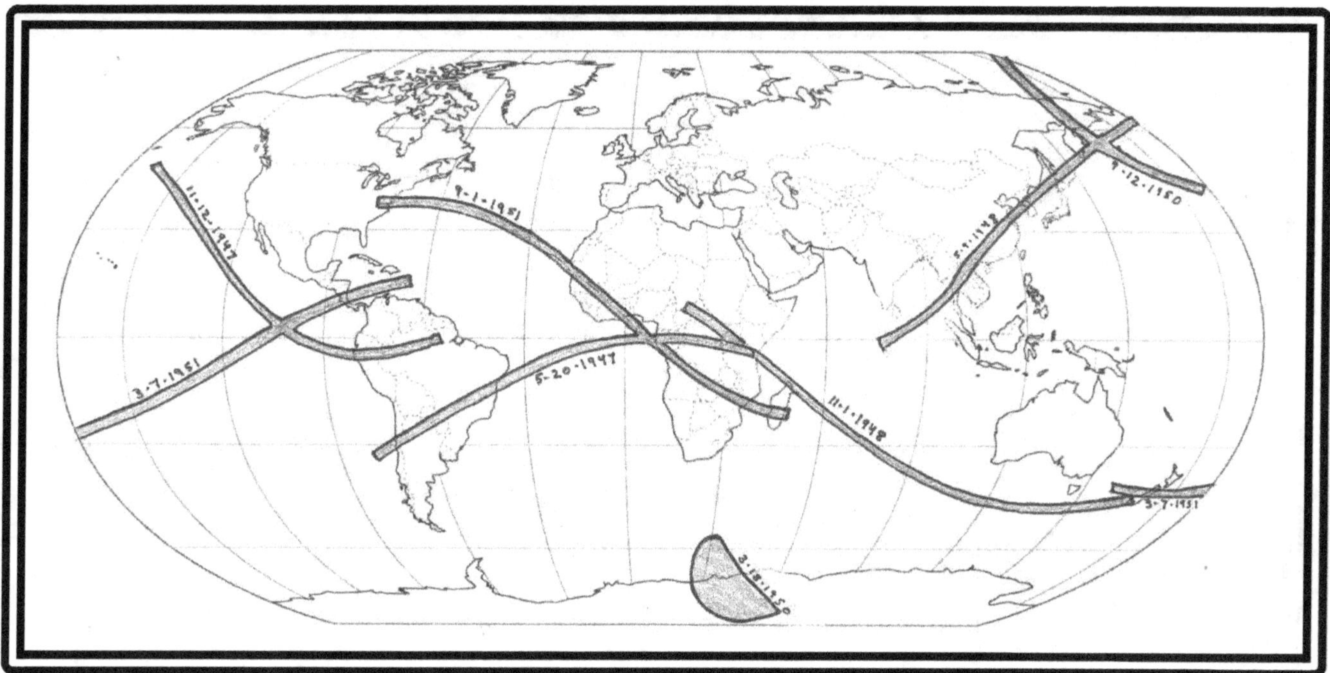

1947 to 1951

1947 May 20 Total Columbus Death Day South America to Africa -Saros 127
1947 Nov 12 Ring of Fire Pacific to South America -Saros 132
1948 May 9 Ring of Fire-almost Total Korea Divided/ Southeast Asia -Saros 137
1948 Nov 1 Total Central Africa to New Zealand -Saros 142
1950 Mar 18 Ring of Fire Antarctica -Saros 119
1950 Sep 12 Total Northeast USSR - Saros 124
1951 Mar 7 Ring of Fire New Zealand to American Isthmus -Saros 129
1951 Sep 1 Ring of Fire Jamestown USA single city to Africa -Saros 134

1947 May 20 Columbus Death Day Eclipse (Saros 127)--South America to Africa on 441st anniversary of Columbus death in 1506. Begins just west of Chile, totality in Santiago and Salvador, Brazil. Crosses Atlantic into Africa, dividing metropolitan Kampala, Uganda. Stops just after Nairobi, Kenya.

1948 May 9 Korea Divided-Southeast Asia (Saros 137) Sections of this eclipse reached as much as 99.999% total, one of the closest ring of fires to totality ever. Annularity in Bangkok, Thailand (first of four complete eclipses in ten years). Cuts Vietnam and Laos in half and crosses China. Straddles North and South Korea close to the 38th Parallel. The North invades the South 2 years later in 1950. Vietnam, also embroiled in war with its colonizer France, would be officially divided into two nations 6 years later. Eclipse eventually ends in Pacific after crossing Aleutian Islands of USA.

1951 Sep 1 Jamestown USA to Africa (Saros 134) Ring of Fire eclipse 332 years after the first slaves were brought to colonial Jamestown on August 20 1619. Path begins at sunrise in Virginia at Jamestown, crosses Atlantic to the Slave Coast, including Angola, which was the source of the 20 Africans brought on the first ship. The eclipse crosses Africa and ends in Madagascar at sunset almost precisely on capital of Antananarivo.
partials on 1-3-1946, 5-30-1946, 6-29-1946, 11-23-1946, 4-28-1949, 10-21-1949

1952 to 1954 Shahrud Cross/ End of Christmas Metonic

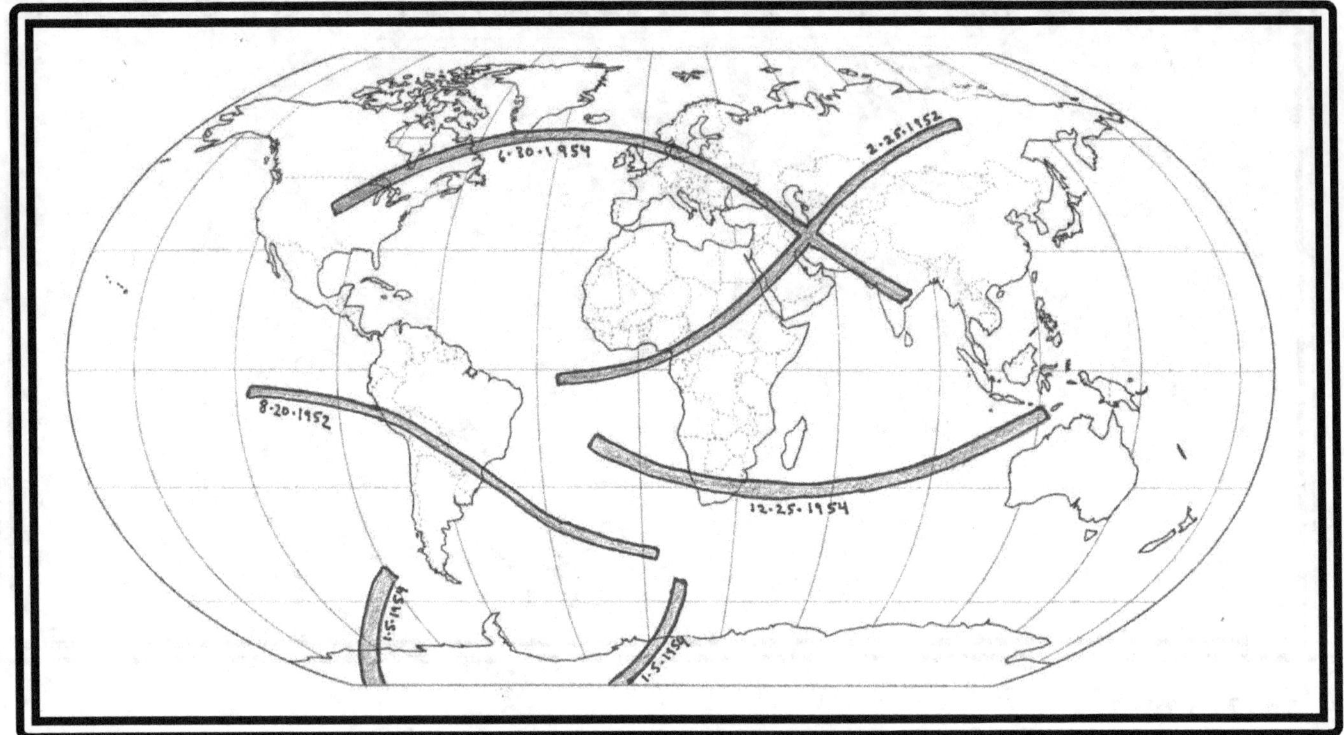

1952 to 1954 Shahrud Cross / End of Christmas Metonic

1952 Feb 25 Total Africa/Mideast/ Iran/ USSR -Saros 139
1952 Aug 20 Ring of Fire South America -Saros 144
1954 Jan 5 Ring of Fire Antarctica- Saros 121
1954 Jun 30 Total USA to Iran and India -Saros 126
1954 Dec 25 Ring of Fire Christmas in South Africa -Saros 131

1952 Feb 25 Africa/Mideast/ Iran/ USSR Total (Saros 139) Total eclipse traverses Africa (Totality in Khartoum, Sudan), then crosses Arabia (totality in Jeddah), crosses Iraq near Basra, and then Iran (totality in Shahrud) before it ends in Russia. 1952 saw upheaval in Iran, leading to the US backed coup of August 1953 that installed the Shah. Saros Cycle 139 returns for USA 2024 eclipse.

1954 Jun 30 USA to Iran and India (Saros 126) This next total eclipse after 1952 starts in USA and ends in Iran, intersecting the 1952 path over city of Shahrud in the first year after the Shah was re-installed by a US backed coup. The eclipse path starts in Nebraska, brings totality in Saint Paul, bisects Canada, crosses the Atlantic, dives through Scandinavia into Poland, Eastern USSR (totality in Kiev), Iran. Path then over Afghanistan, Pakistan (totality in Quetta) before ending in India. The western backed coup of August 1953 installed Shah Reza Pahlavi over the elected government of Iran. The coup lies between successive total eclipses with intersection over city of Shahrud (literally "City of the Shah").

12-25-1954 Christmas in South Africa (Saros 131) One of two Christmas eclipses in 20th century, first since 1600's and the last true Christmas eclipse over q populated region for next 500 years.

partials on 2-14-1953, 7-11-1953, 8-9-1953

1955 Southeast Asia/ Bangkok/Vietnam War Cross

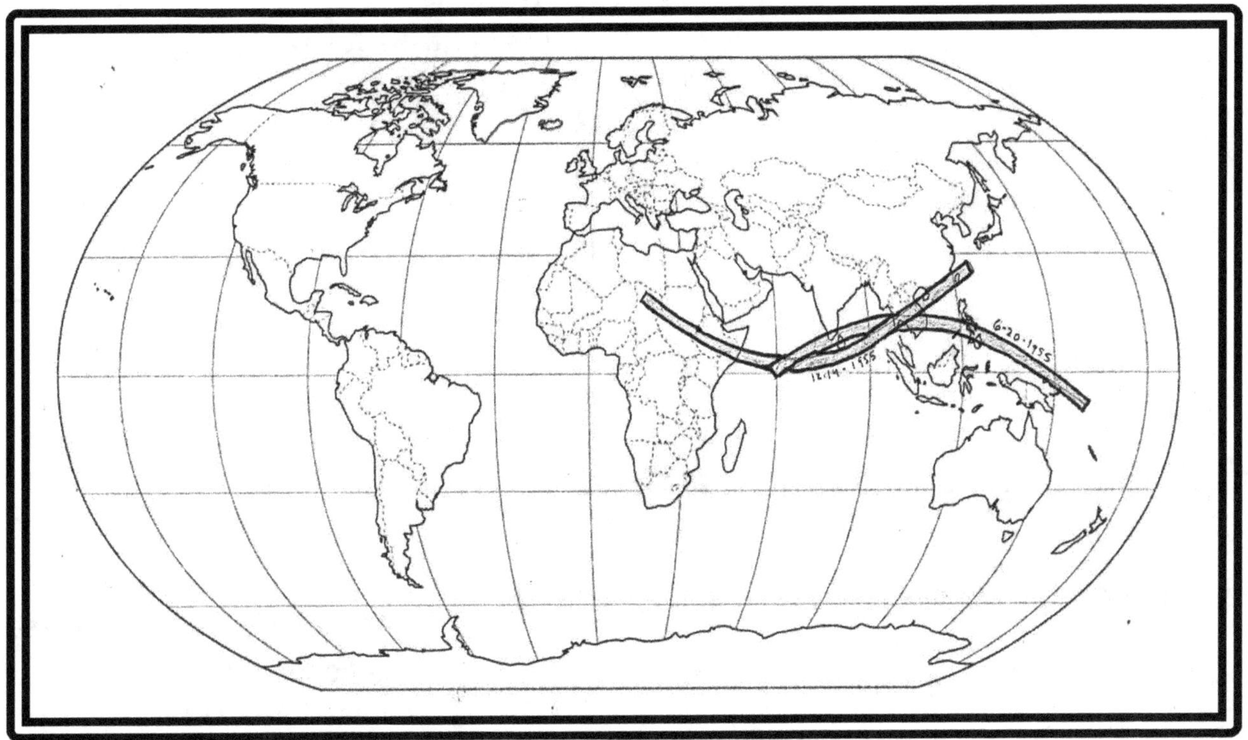

1955 Southeast Asia/Bangkok/ Vietnam War Cross (Vietnam war "officially begins" Nov 1 1955)

The eclipse cross which occurs over Vietnam in 1955 is composed of the longest total eclipse and longest ring of fire eclipse of the last thousand years.

1955 Jun 20 Total Summer Solstice Eve Southeast Asia (Saros 136) The longest duration total solar eclipse (at 7 minutes 7.74 seconds) since the 11th Century and longest until the 22nd Century. It's also the longest eclipse of Saros 136. Occurs day before Summer Solstice. Total solar eclipse begins over Indian Ocean and covers Sri Lanka before giving Bangkok their second eclipse in 7 years. It shadows ancient mystic metropolis of Angkor Wat in Cambodia and divides Vietnam 3 weeks after pullout of last French forces. Eclipse goes on to give totality to Manila (their last total eclipse until 2100. Totality over Mount Pinatubo almost exactly 36 years before its historic eruption on June 15, 1991.Saros 136 brings the upcoming Aug 2nd 2027 Valley of the Kings Eclipse.

1955 Dec 14 Ring of Fire Africa to Taiwan/ Southeast Asia (Saros 141) With a maximum annularity duration of 12 minutes and 9.17 seconds, it's the greatest duration ring of fire of the second millennium, and the longest duration in the life of Saros 141. It is quite literally one of the longest eclipses known to have happened on earth. Taking an opposite convex path 177 days after the summer solstice eclipse of 1955 Jun 20. The path begins in North Africa, dives just south of Sri Lanka and comes ashore incredibly at Bangkok, giving the city its second solar eclipse in less than 6 months-- an event with a less than one in 80,000 chance of occurring. It is also Bangkok's 3rd eclipse in 7 years (less than one in 2 million chance). Path continues on over Vietnam, Cambodia and Laos, bisecting those nations again for the third time in seven years . Also crosses Hong Kong and Taiwan. The 156th year anniversary of death of George Washington in 1799.

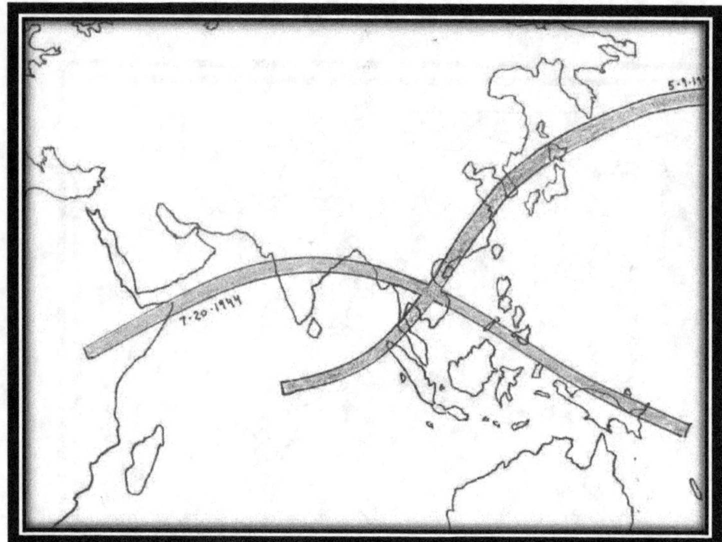

**At Left is
First Southeast Asia Cross
1944 and 1948**

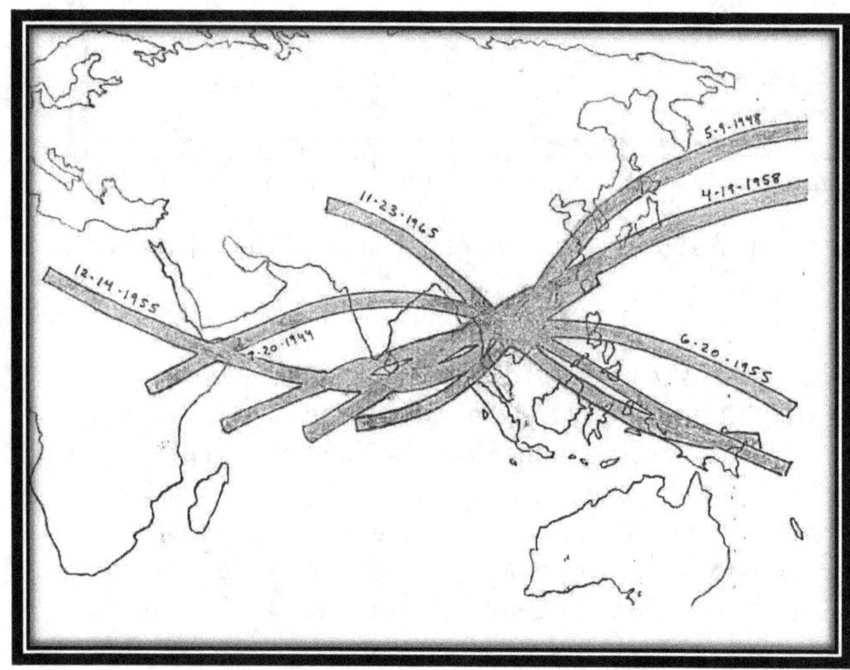

**1944 to 1965 Full Southeast
Asia Eclipse Swarm**
**Southeast Asia 21 Year Eclipse
Swarm 1944 to 1965**

Six solar eclipses intersect over Southeast Asia in 21 years, in probably the greatest known intersecting eclipse swarm in history. Meanwhile, in that region, a brutal cycle of violence is fueled by imperial struggles between Communist and Capitalist powers. War in Korea, Vietnam, Laos, Burma, Cambodia. Between 1945 and 1980, perhaps 15 to 20 million are murdered across the region.

1944 Jul 20 Africa to Pacific Ring of Fire (Saros 135)
1948 May 9 Korea Divided Ring of Fire (Saros 137)
1955 Jun 20 Vietnam/ Bangkok Total (Saros 136)
1955 Dec 14 Vietnam/ Bangkok Ring of Fire (Saros 141)
1958 Apr 19 Vietnam/ Bangkok Ring of Fire Two (Saros 128)
1965 Nov 22/23 Vietnam- JFK Death Metonic and PT 109 (Saros 132)

Chapter 12

All World Eclipses 1956 to 2014

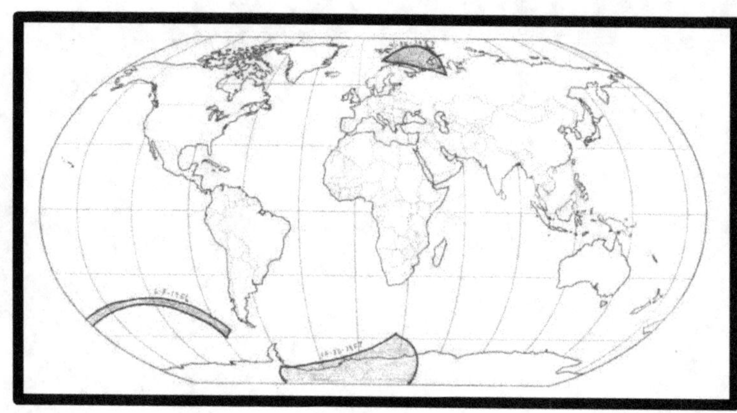

1956 to 1957 Not Much Going On

1956 Jun 8 Total Deep South Pacific - Saros 146
1957 Apr 30 Ring of Fire Hitler Death Day -Saros 118
1957 Oct 23 Total Antarctica -Saros 123

Proof that not every year something exciting is going on in the eclipse world. The Apr 30 1957 eclipse is the 12th anniversary of Hitler's suicide in Berlin.

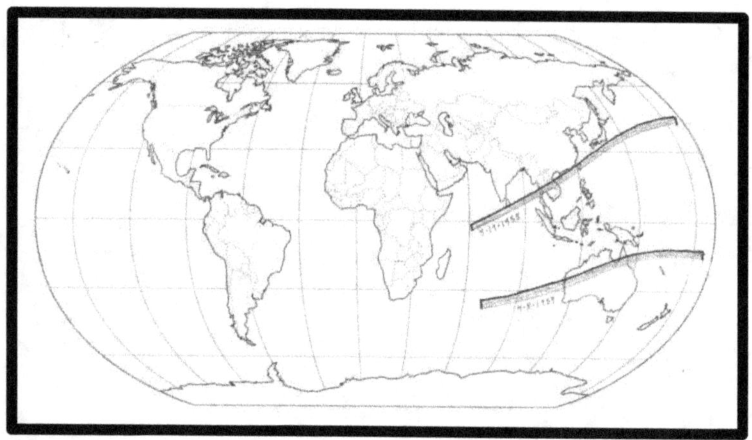

partial occurs 12-2-1956

1958 to 1959 Ring of Fires

1958 Apr 19 Ring of Fire Darwin Death Day Vietnam/ Bangkok Two -Saros 128 On the 4-19 Hitler birthday/ Darwin Death Day Metonic, yet another Ring of Fire eclipse heads straight for Southeast Asia. You will likely not find anything quite so remarkable as this eclipse swarm. Yes, Bangkok is again in the path of Annularity for the fourth time in 10 years (odds for this happening randomly I calculate at about one in 500 million, but its unclear whether anyone has any clue how to calculate such a thing!). The path then proceeds once again over Angkor Wat in Cambodia and divides Vietnam for the fourth time in 10 years. Total annularity again in Taipei, Taiwan.

1959 Apr 8 Ring of Fire Australia -Saros 138 Buddha's Birthday in Japan. The April 8 Metonic appears for the first time. It will appear in the USA 65 years later 4-8-2024.

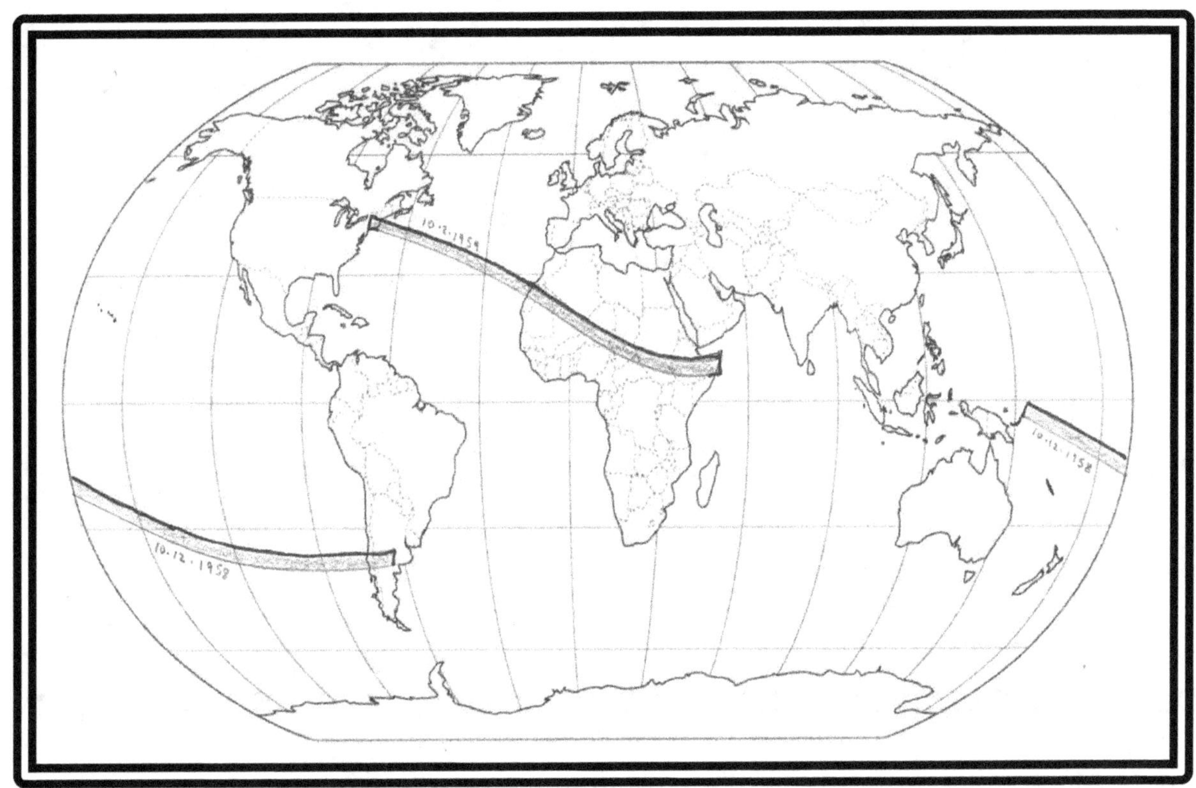

1958 and 1959 Total Eclipses (One lunar year apart)

1958 Oct 12 Total Santiago Chile Single City -Saros 133 Every now and then an eclipse only impacts a single city. On the 466th anniversary to the day after Columbus discovers America a total eclipse travels the entire Pacific Ocean without touching major land mass to end at sunset over the city of Santiago, Chile. Eclipse occurred 5 weeks after Presidential elections (the first that featured candidacy of Salvador Allende, who would be overthrown by a CIA backed coup on 9-11-1973)

1959 Oct 2 Boston Single City to Africa JFK Eclipse -Saros 143 Total Eclipse begins just west of Boston and gives that city a total solar eclipse for the first time since 1806 (it won't have another until 2200) just as its most famous modern son is campaigning for the Presidency of the United States. Kennedy gives a speech that night on the campaign trail in Rochester, New York. It is the only major city in America to witness the eclipse. Kennedy would be elected President the following year on Nov. 8th, the day after a transit of Mercury was visible across the Sun's face from the western hemisphere. Something that occurs about 13 times a century. There are repeated cosmic associations in the 20th century that seem to bring associations of John F. Kennedy. Total Eclipse heads across the Atlantic before reaching Africa, bringing totality to Senegal and Mali (both of whose Presidents visited Kennedy in the White House, a first for either of those countries) The shadow crosses the continent. Saros 143 (Kennedy would turn 43 seven months after eclipse). Last total eclipse in Saros 143 occurred Oct 24 1995 on the 50th anniversary of the United Nations, over Vietnam and Southeast Asia.

partials on 3-27-1960 and 9-20-1960

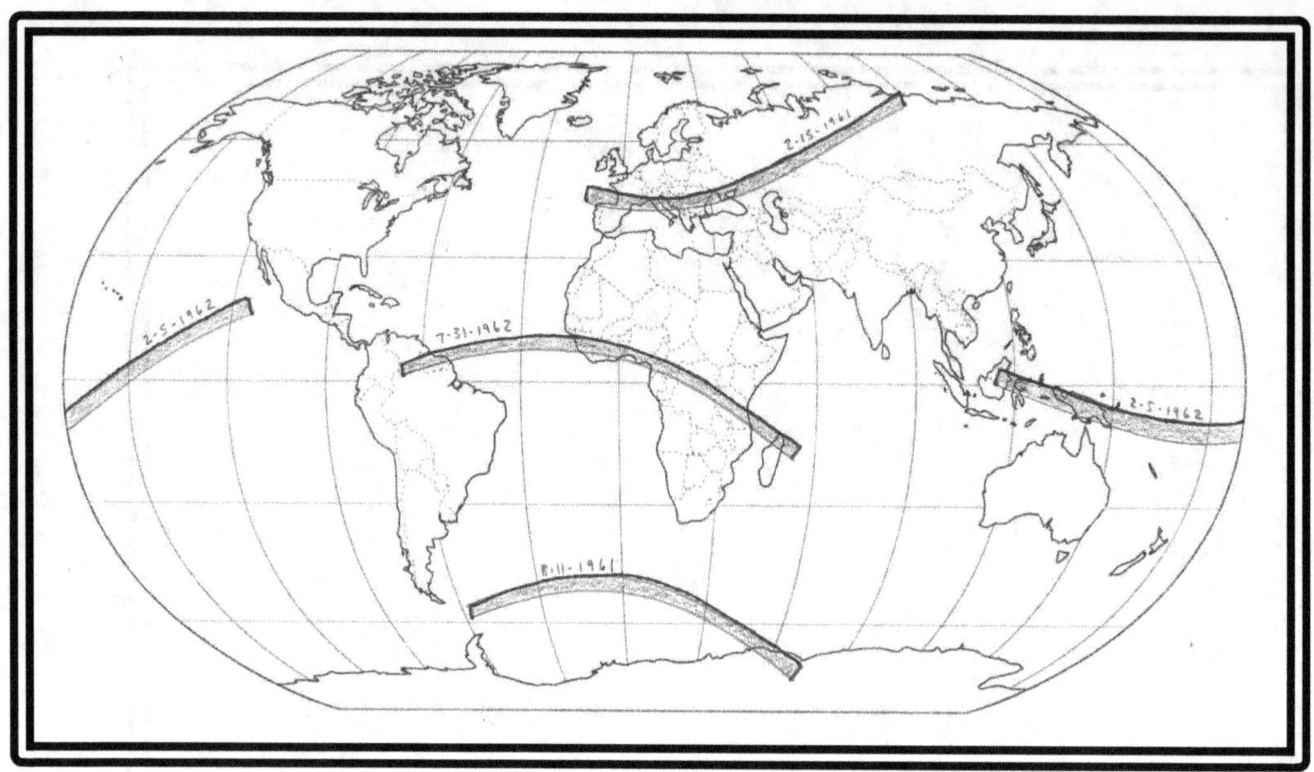

1961 and 1962

1961 Feb 15 Total Europe and USSR -Saros 120
1961 Aug 11 Ring of Fire Extreme South Atlantic -Saros 125
1962 Feb 5 Total Borneo/ New Guinea -Saros 130
1962 Jul 31 Ring of Fire South America to Africa -Saros 130

1961 Feb 15 Europe and USSR Total (Saros 120) Total solar eclipse begins at sunrise just off France, divides France, Italy, the Balkans and then bisects the USSR-- the amazing Saros 120. Great cities in totality include Bordeaux, Turin, Monaco, San Marino, Sarajevo, Sofia, Bucharest, Florence, Sevastopol, Mariupol, Volgograd, Ufa. Six months before beginning of construction of Berlin Wall (88% coverage in Berlin)

1962 Feb 5 Borneo/ New Guinea Total (Saros 130) It sure seems like New Guinea had a lot of eclipses. There are five in the island group between 1944 and 1965. This 1962 Feb 4 Total eclipse goes directly over the site of Kolombangara Island, where JFK's boat PT 109 sank almost 19 years before on the night of August 1 and 2nd, 1943.

Saros 120 and Saros 130 have prominent positions in Spring Equinox Metonic (2033 and 2034)

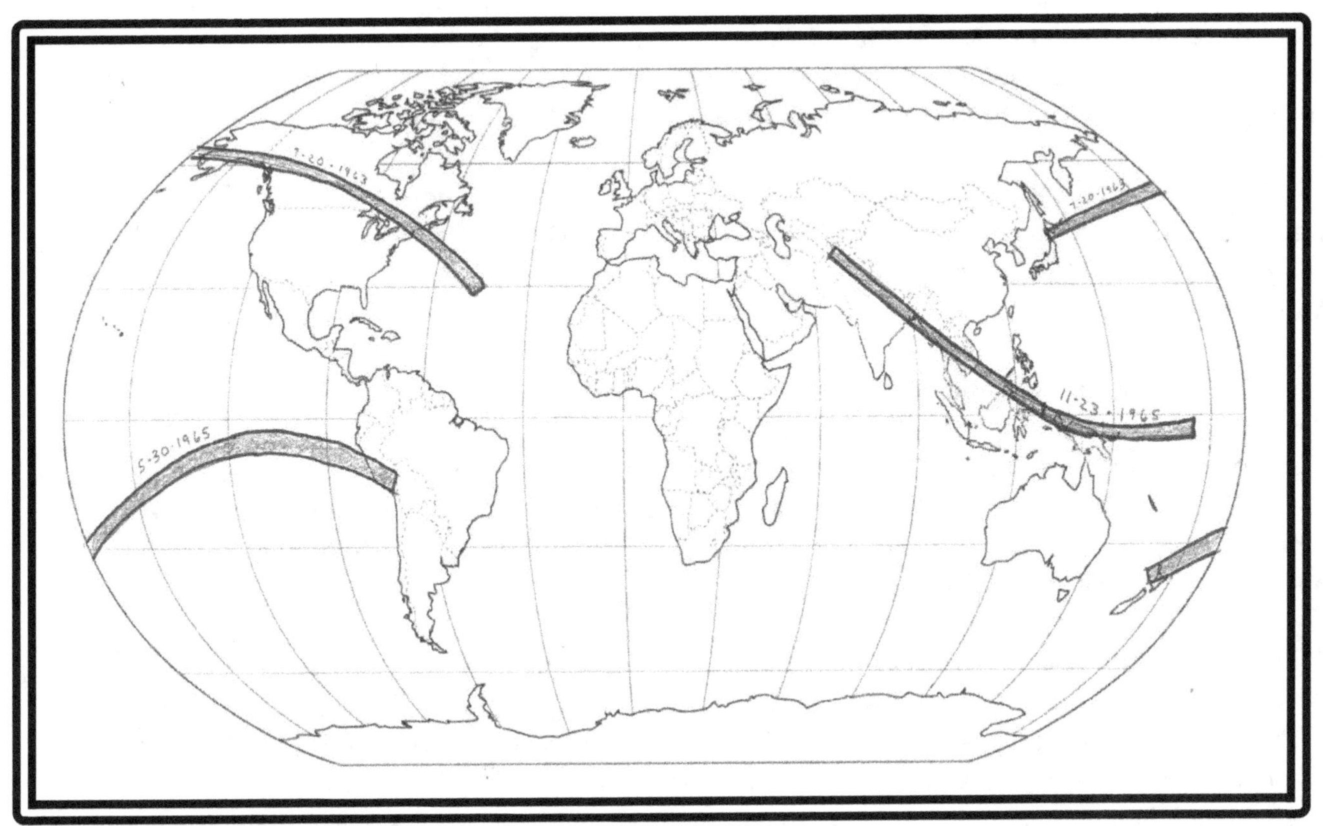

1963 to 1965 Death of JFK/ War in Vietnam (above)

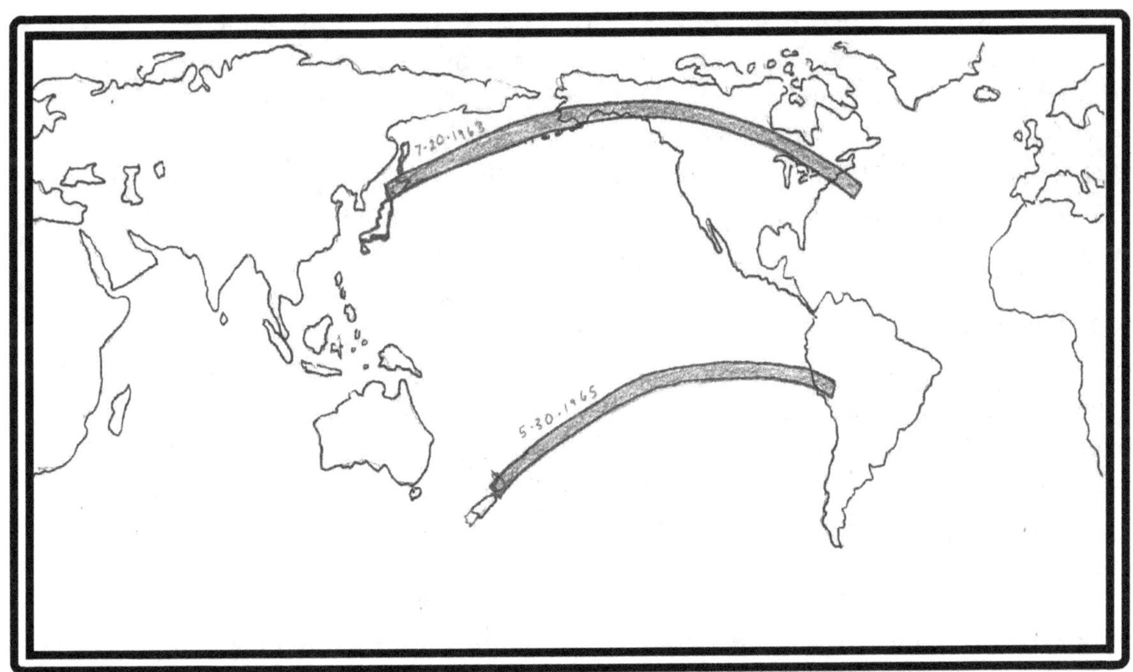

1963 and 1965 Pacific View (above)

1963 Jul 20 Japan to USA (Maine) Total JFK Premonition -Saros 145
1965 May 29/30 JFK Birth Metonic Total - Saros 127
1965 Nov 22/23 JFK Death Day Metonic Vietnam Ring of Fire -Saros 132

Jul 20 1963 Total Eclipse Only Maine (Saros 145) 6 years to the day before Apollo 11 astronauts stepped on the moon. 39 days before the March on Washington that featured both Martin Luther King and Bob Dylan. Dylan's "Blowin' in the Wind" was climbing the pop charts at the time (Peter, Paul and Mary version) and was at number 14. It would eventually reach number 2 on August 14.

The Eclipse occurs 125 days before the assassination of John F. Kennedy. It occurs 18 years (minus 17 days) after the twin atomic bombs were dropped by the US on Japan—this eclipse begins in Japan (the island of Hokkaido,) bisects the six-year-old state of Alaska before cutting across Canada. It touches the USA only in Maine, with totality in Bangor. The last point of land touched in North America is Bethel's Point, Nova Scotia. Bethel means "house of God" in Hebrew, where both Abraham and Jacob built altars in the Bible's Book of Genesis.

Four months later, John F. Kennedy would be shot dead in Dallas Texas (35 % coverage). His hometown Boston saw 95% coverage. 97+% totality at Bush family haven of Kennebunkport, Maine. Washington DC had 77% coverage. Eclipse marks the beginnings of the 60's, as far as cultural and solar eclipse associations go. The first Beatles record to be released in America happened two days later on Jul 22 1963. It was "Please Please Me". The decade culturally will end 7 eclipse years later with the March 7, 1970 USA eclipse. Saros 145 appeared at the Atom Bomb eclipse of July 9 1945, Liverpool 1927 and again for USA Divided Aug 21 2017 ,then with coda at Beijing/Pyongyang/Tokyo on Sep 2 2035

1965 May 29/30 JFK Birth Metonic New Zealand to Peru Total (Saros 127) Only the tiniest portion of extreme northern New Zealand sees totality at sunrise 48 years and one day after the birth of John F. Kennedy. JFK was born on May 29 1917 in Brookline, Massachusetts. The eclipse travels only over a couple of small islands as it crosses the Pacific to end just after reaching Peru at sunset. It shadows the Nazca Lines (75% coverage at sunset). This is the first complete eclipse on earth after the murder of JFK on 11-22-1963. Eclipse occurs approximately the same time the author of this work's family emigrated to New Zealand where he was conceived approximately Sep 1966 in Wellington.

1965 Nov 22/23 JFK Death Metonic Vietnam Ring of Fire (Saros 132) Essentially the two year anniversary of the execution of JFK. Due to time zones and Metonic reality it is typically marked as Nov 23 (UCT) , but it was Nov 22 in USA when eclipse began. A ring of Fire Eclipse again heads to Southeast Asia from Turkmenistan across Afghanistan and Pakistan, Kashmir into Thailand, Cambodia and Vietnam (their 6th solar eclipse in 17 years) then out through Indonesia and Asia. By the end of 1965 more than a half million US troops were in Vietnam. In the week before the eclipse 240 US servicemen had been killed, the most ever up until that time.

A partial solar eclipse occurs 1-14-1964 MLK Birthday Metonic, the day before his 35th birthday, the first solar eclipse after JFK assassination and two weeks after Dec 30 1963 total lunar eclipse over USA which occurred 38 days after the assassination.

Also partials on 6-10-1964, 12-4-1964

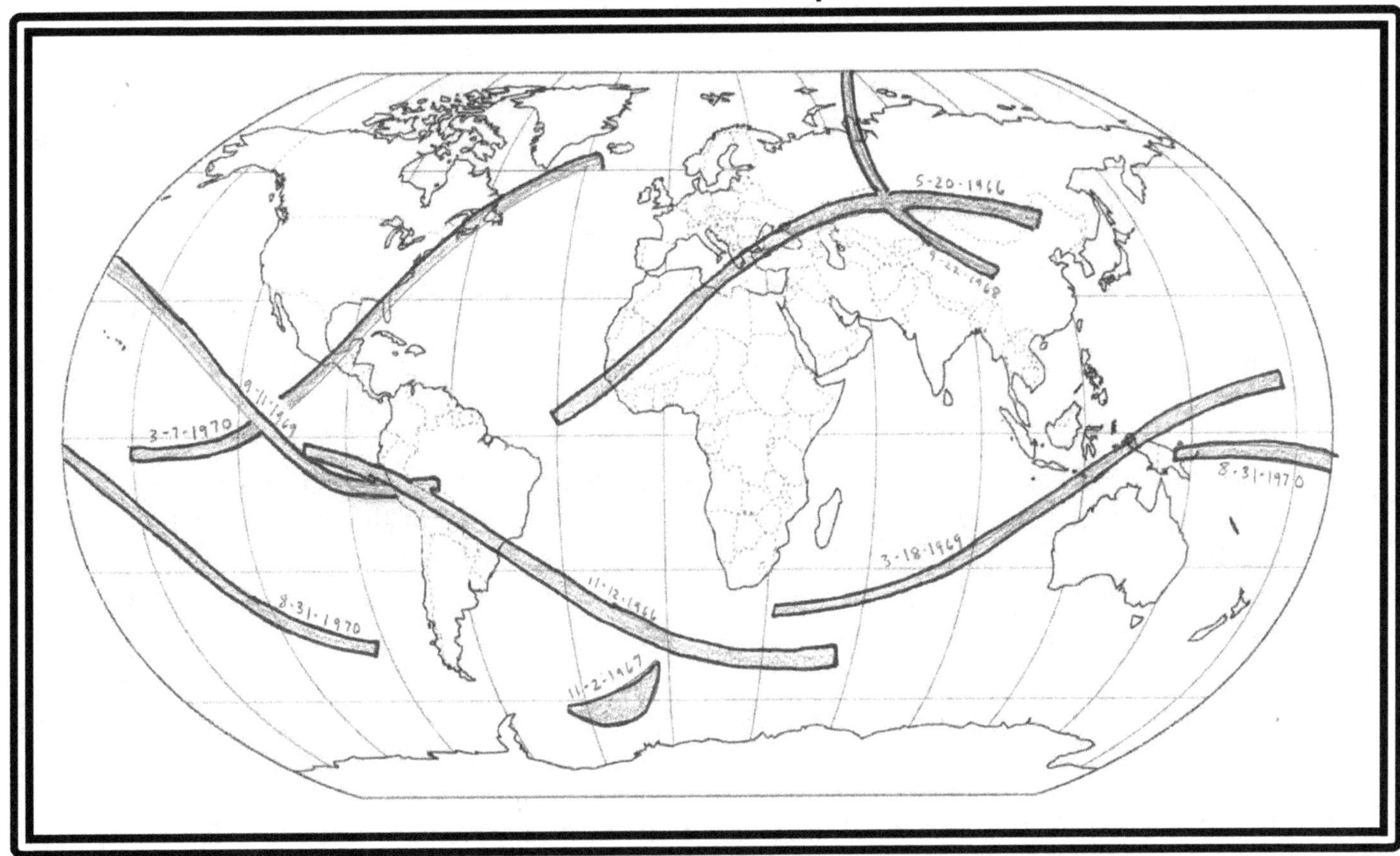

1966 May 20 Ring of Fire Columbus Death Day Africa/ Greece/ USSR -Saros 137

1966 Nov 12 Total Peru and South America -Saros 142

1967 Nov 2 Total Antarctic Ocean -Saros 152

1968 Sep 22 Total Fall Equinox USSR to China -Saros 124

1969 Mar 18 Indian Ocean/ New Guinea -Saros 129

1969 Sep 11 Ring of Fire Pacific to Peru -Saros 134

1970 Mar 7 Total Mexico/USA/Canada -Saros 139

1970 Aug 31 Ring of Fire New Guinea to South Pacific (Saros 144)

A six-year span from 1965 to 1970 saw complete solar eclipses in each of the six years, the first time that had happened since World War One era 1914 to 1919.

1966 May 20 Ring of Fire Columbus Death Day Africa/ Greece/ USSR (Saros 137) Very nearly a total, with greater than 99% coverage on the centerline. It occurred on the 460th anniversary of the death of Christopher Columbus. Annularity over Tripoli Libya, Athens Greece and Istanbul Turkey. Sevastopol and Astrakhan in USSR. Ends over China.

1966 Nov 12 Peru and South America Total (Saros 142) One of three eclipses in Peru 1965 to 1969.

1967 Nov 2 Antarctic Ocean Total (Saros 152) One lunar year later, the only complete eclipse of 1967. 1967 saw the Blood Moon Lunar Tetrads surrounding the six-day war in Jerusalem.

1968 Sep 22 Fall Equinox USSR to China Total (Saros 124) The last complete eclipse of Saros Cycle 124, which brought the World War One eclipse of 8-21-1914. A month after The Soviet Union invades Czechoslovakia, an Autumn equinox Eclipse bisects the USSR from north to south and ends in China. Eclipse occurred one week after USSR performed the first orbit of the moon with unmanned space craft. The craft was recovered in the Indian Ocean the day before eclipse.

1969 Mar 18 Indian Ocean/ New Guinea Ring of Fire (Saros 129) The seemingly endless New Guinea eclipses keep on coming!

1969 Sep 11 Pacific to Peru Ring of Fire (Saros 134) The 9-11 Metonic arrives on the last of the Peruvian three eclipse swarm of 1965 to 1969. It will appear once more in 1988 Mogadishu sunrise, Saros 134, note resemblance of path to 10-14-2023 American Ring of Fire

1970 Mar 7 Mexico/USA/Canada Total (Saros 139) Total Eclipse. The southeastern seaboard had mostly cloudy weather though Florida remained fair. 93% totality at Cape Canaveral 230 days after men landed on the moon. Totality at Charleston South Carolina, the greatest Slave Port in North America.95 % totality in Washington D.C., 96% in New York and 97% in Boston. The end of the sixties, eclipse wise. Nine days after beginning of US bombing of Cambodia, which killed more than 100,000. The Beatles announced their break-up a month later.Eclipse is in a Hepton pair with July 20 1963 USA eclipse. Closer look in USA section.

1970 Aug 31 New Guinea Sunrise Ring of Fire (Saros 144) As the title suggests, yet again in New Guinea! The island, which is slightly larger than Texas, receives 6 complete eclipses between 1955 and 1970, and 9 complete eclipses between 1944 and 1984!

partial eclipses occur on 5-9-1967, 3-28-1968, 2-25-1971, 7-22-1971, 8-20-1971

Focus on USSR Eclipses

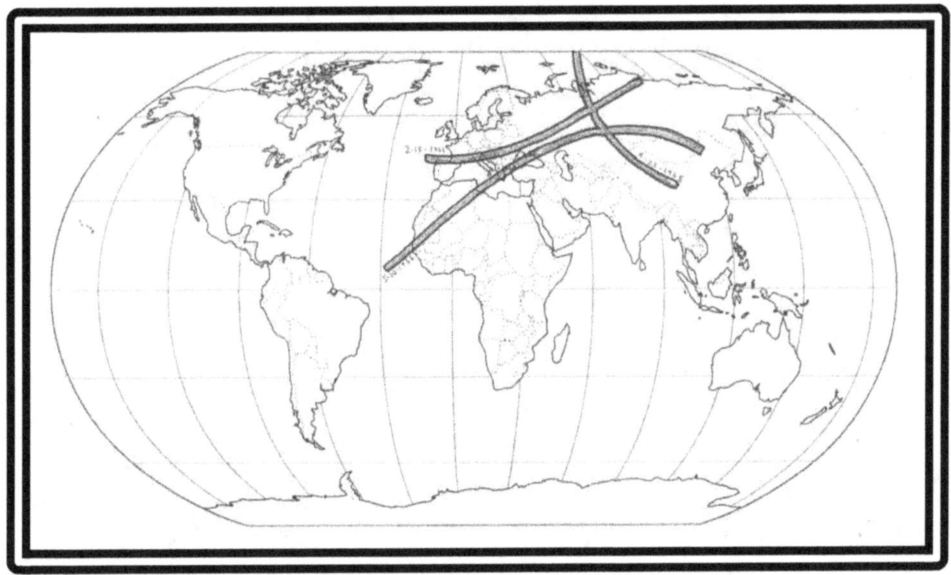

1961 to 1968 USSR Octon Cross (above)

1961 Feb 15 Total France through USSR -Saros 120
1966 May 20 Ring of Fire Columbus Death Metonic- Saros 137
1968 Sep 22 Total Fall Equinox USSR to China (Saros 124)
A double Octon is 8 eclipse years or 2776 days. That is 7 years, 7 months and 7 days in solar years.

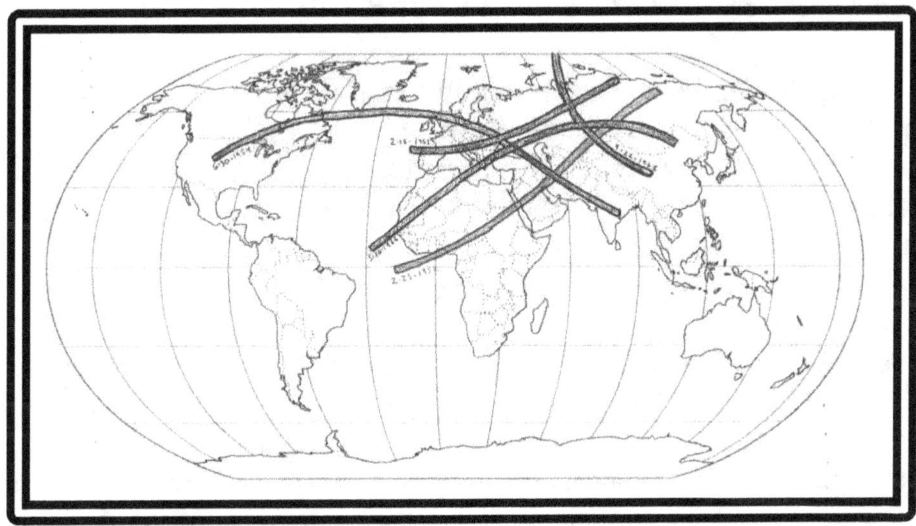

Full USSR Swarm 1952 to 1968 (above)

1952 Feb 25 Africa/ Iran/ USSR Total (Saros 139)
1954 Jun 30 USA to USSR Total (Saros 126)
1961 Feb 15 France through USSR Total (Saros 120)
May 20 1966 Columbus Death Metonic Ring of Fire (Saros 137)
Sep 22 1968 Fall Equinox USSR to China Total (Saros 124)

Four of the five complete eclipses in the 16-year period are totals. Includes Iran Shahrud Tav. By comparison, in 52 years between 2008 and 2060, the former USSR receives no total eclipses.

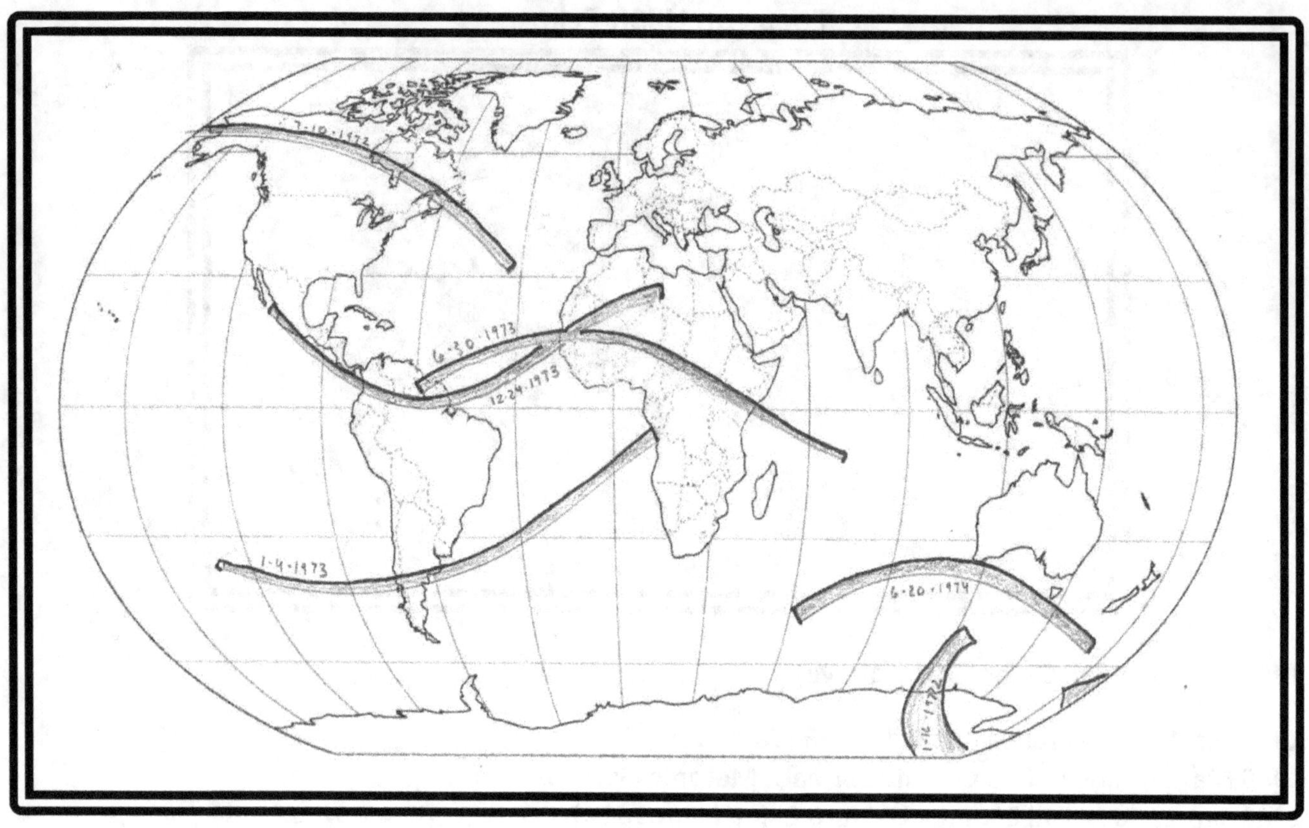

1972 to 1974 Last eclipses of Summer Solstice Eve Metonic

1972 Jan 15/16 Ring of Fire MLK Birthday Metonic Antarctica -Saros 121
1972 Jul 10 Total "You're So Vain" Nova Scotia -Saros 126
1973 Jan 4 Ring of Fire South America to Africa -Saros 131
1973 Jun 30 Total South America through Heart of Africa -Saros 136
1973 Dec 24 Ring of Fire Christmas Eve Central America to Africa -Saros 141
1974 Jun 20 Total Eclipse Island Summer Solstice Eve -Saros 146

The sixties are over and a new world struggles to re-shape itself. Africa will be at the epicenter of a great eclipse swarm in the mid-1970's.
On a lighter note, two very famous songs address eclipses in this time frame. "You're so Vain" by Carly Simon goes to number one in 1972 while name-checking the 1972 Nova Scotia Total Eclipse.
Also, Pink Floyd's song "Eclipse" finishes up the album "Dark Side of the Moon", released Mar 1 1973. It goes on to become one of the biggest selling pieces of music in human history.

And all that is now
And all that is gone
And all that's to come
And everything under the sun is in tune;
But the sun is eclipsed by the moon.

Eclipse. Roger Waters

1971 Jan 15/16 MLK Birthday Metonic Antarctica Ring of Fire (Saros 121) Notably on the Martin Luther King Birthday Metonic. Eclipse begins on January 15th local time just before midnight in the Antarctic regions where daylight lasts nearly all day and extends into full eclipse on January 16 local time. The eclipse occurs on what would have been Martin Luther King's 42nd birthday back in the USA. He had been murdered less than four years prior. There had also been a partial solar eclipse on his birthday Metonic four years prior to his death on Jan. 14/15 1964, the first solar eclipse on earth after JFK assassination. The 1972 eclipse is the first complete eclipse to land precisely on MLK birthday in centuries. Very few likely witnessed this eclipse.

1972 July 10 You're So Vain Nova Scotia Total (Saros 126) July 10 1972 sees only one total solar eclipse. It travels across the extreme northern latitudes of USSR and Canada. The eclipse is notable for inspiring the lyric Carly Simon's number one hit "You're So Vain" (also one of the biggest hits of the entire decade of the 1970's). The lyric says famously, " *Then you flew your Learjet up to Nova Scotia to see a total eclipse of the sun"*

1973 Jan 4 South America to Africa Ring of Fire (Saros 131) Annularity ends just a few miles from African Coast of Gabon.

1973 Jun 30 South America through Heart of Africa (Saros 136) Eclipse begins in Guyana just south of Georgetown. 99% coverage at famous massacre/mass suicide site of Jonestown, which occurred 5 years later on 11-18-1978 and killed 909 people. Crosses Atlantic and Africa. At 7 minutes and 4 seconds, this was the last total eclipse to last more than 7 minutes until 2150. It was a carefully studied eclipse. A specialized commercial cruise by the S.S. Canberra traveled from New York City to the Canary Islands and Dakar, Senegal, observing totality in the Atlantic. That cruise's passengers included notables in the scientific community such as Neil Armstrong, Scott Carpenter, Isaac Asimov, Walter Sullivan, and a 15-year old Neil deGrasse Tyson. The great Saros 136.

1973 Dec 24 Christmas Eve America to Africa Ring of Fire (Saros 141) Eclipse occurs on Anthony Fauci's 43rd birthday. It is the third eclipse to connect South America and Africa in 1973. Annularity in Central America and Colombia including Bogotá, southern Venezuela, Brazil, Guyana again. Path crosses ocean and touches Cape Verde islands. It intersects the Jun 30 1973 path outside of Mauritania. including the capital city Nouakchott. Then to Mali and Algeria. The duration of annularity at maximum eclipse was 12 minutes, 2.37 seconds. It would be the last eclipse with duration over 12 minutes until 3080 AD.

1974 Jun 20 Eclipse Island Solstice Eve Total (Saros 146) The last eclipse of the Summer Solstice Eve Metonic of 1955 to 1974. This is the only complete eclipse on planet Earth between Christmas 1973 and April 29, 1976. During a time frame of 679 days, no land on Earth experiences a complete eclipse except these few thousand square miles of the extreme corner of Southwest Australia several hundred miles south of Perth. The last land touched is Eclipse Island, named in 1791 by George Vancouver who observed a partial solar eclipse there. 1974 to 1975 is perhaps the quietest time for eclipses in this study. The next total occurs on Oct 23 1976. It passes just north of this one, bringing 97% coverage to Eclipse Island.

Partials occur on 12-13-1974, 5-11-1975, and 11-3-1975

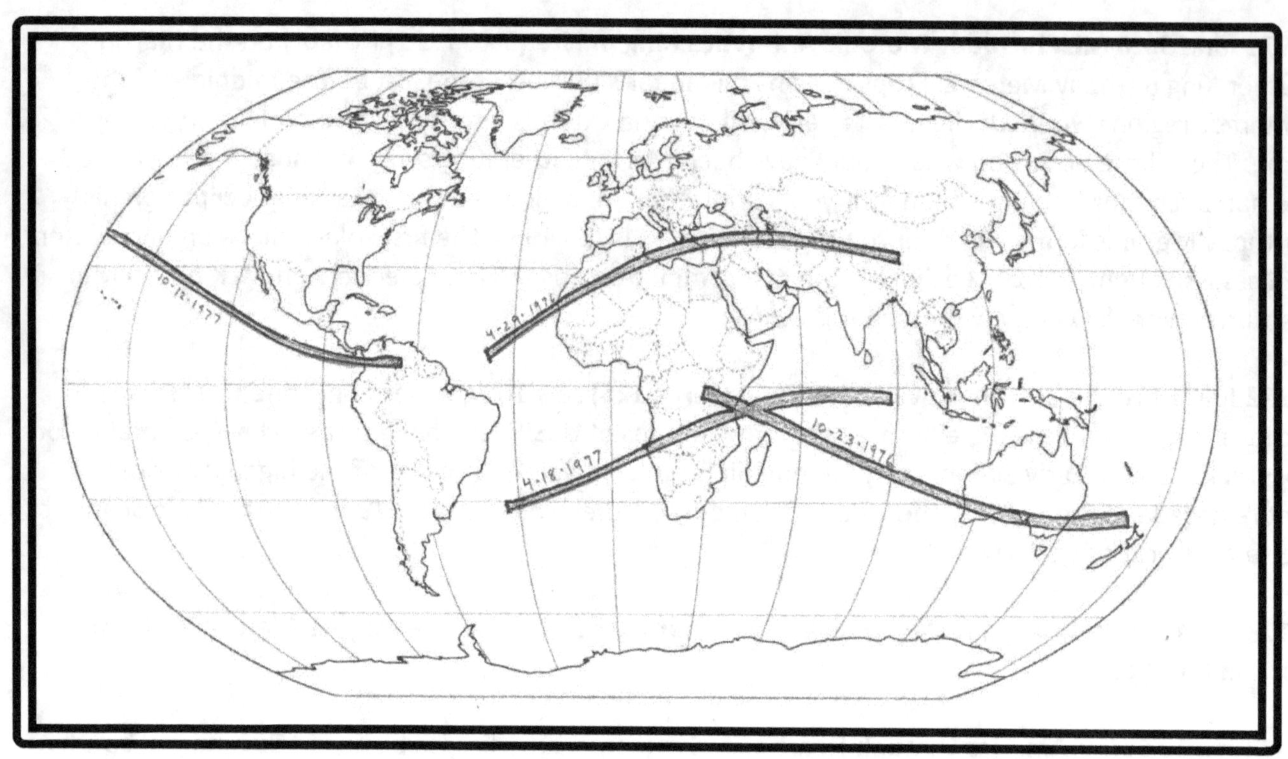

1976 and 1977 Dar Es Salaam Cross/ Beginning Hitler Death Eve Metonic

1976 Apr 29 Third Mauritania in 3 years/ Mideast to China Ring of Fire (Saros 128) The Hitler Death Metonic (4-30) slips backward a day 19 years after an Apr 30 1957 eclipse. West coast of Mauritania Africa gets its third complete eclipse in 3 years (it had two in 1973). Tripoli, Libya sees total annularity 3 weeks after student protests erupted on April 7 1976 which led to public executions a year later. The path crosses Mediterranean, bisects Turkey and Southern Soviet Republics, ends in China.

1976 Oct 23 Dar Es Salaam Africa to Australia Total (Saros 133) Interesting path connects city of Dar Es Salaam, Tanzania ("place of peace") with Melbourne Australia on October 23 1976.

1977 Apr 18 Dar Es Salaam Ring of Fire (Saros 138) The 4-19 Darwin Death/ Hitler Birth Day Metonic slips backwards to 4-18 for the first time since 1931 after two eclipses on 4-19 date. The only major city to experience annularity in the eclipse is Dar Es Salaam, Tanzania-- which was just eclipsed 6 months before--Tanzania adopts its modern constitution a week later on Apr 25 1977.

1977 Oct 12 Columbus Day in Colombia Total (Saros 143) Almost no country in totality except Colombia, the only nation on earth named after Columbus, on the 485th anniversary of the discovery of the New World on 10-12-1492. Totality in Bogota. The eclipse accompanied national civil strife. This is last complete eclipse on earth before 1979 Mt. Saint Helens Eclipse, and had 17% totality at Mount Saint Helens. Maximum eclipse of the 1977 eclipse occurred at 123 degrees west longitude, within a single degree of the crater of Mount Saint Helens and Seattle's longitude (122 degrees). Eclipse ended just east of Puerto Ayacucho ("soul or spirit" in native tongue) in Venezuela.

Partials occur on 4-7-1978 and 10-2-1978

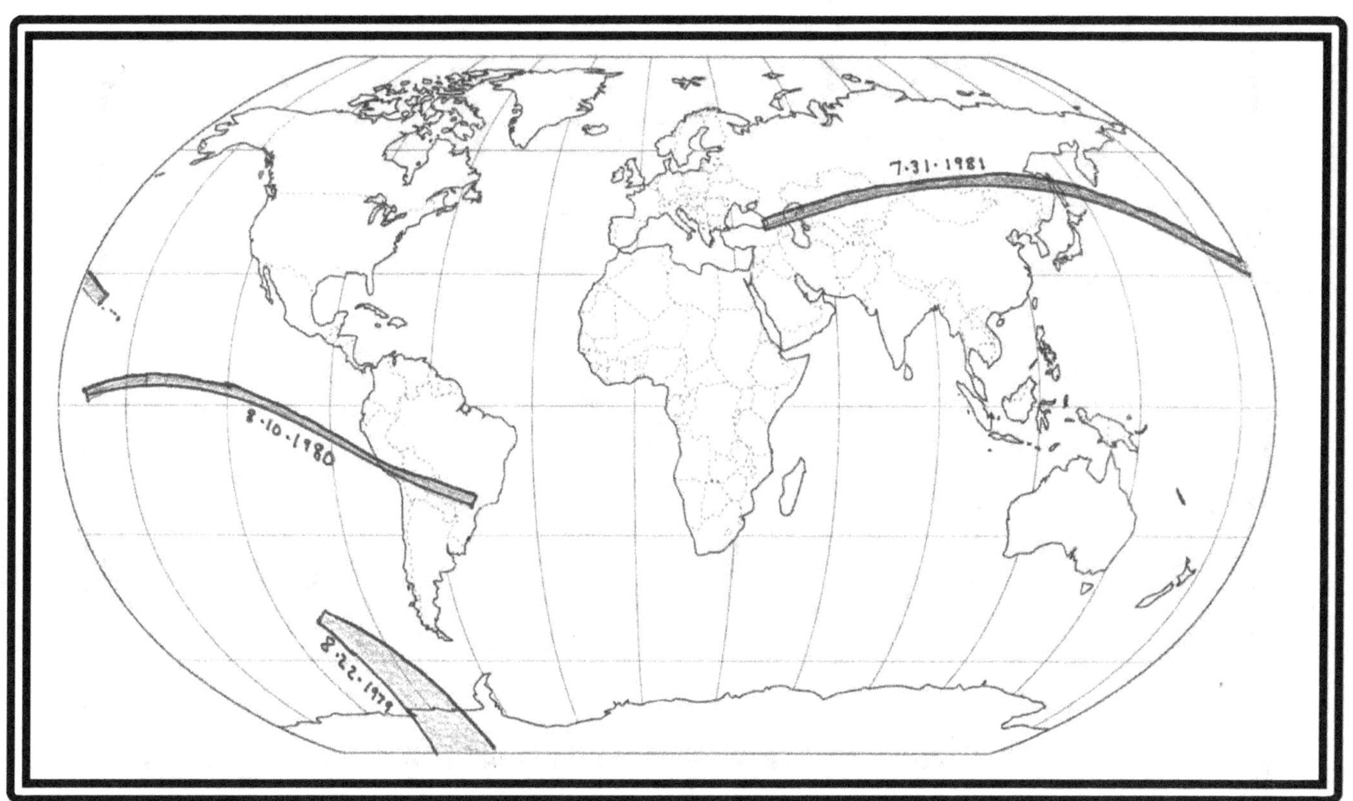

1979 to 1981 Triad One—August Lunar Years

For comparison's sake, we separate the two Lunar year Triads of 1979 to 1981. The "August" one is composed of two annulars and a total, each separated by exactly one lunar year.

1979 Aug 22 Antarctica Ring of Fire (Saros 125) The 8-21 Metonic leaps a day forward. It returns to its 8-21 position for the USA Divided Aug 21 2017 eclipse. Two days after Bob Dylan releases "Slow Train Coming", his first Gospel Album (and 19th Studio record) on August 20 1979 (a date also on the Aug 21 Metonic)

1980 Aug 10 Temple Destruction Metonic South America Ring of Fire (Saros 135) The 1910th anniversary of the final destruction of the Hebrew Temple in Jerusalem. From Pacific across Peru, Bolivia and into Brazil, total annularity in La Paz (The Peace), Ends just before Sao Paulo, Brazil.

1981 Jul 31 USSR Judgment Total (Saros 145) The last decade of the Soviet Union's power begins with cross-country eclipse roughly defining its southern border that touches no other nations. Seven years later, Lithuania will secede. Within a decade both Georgia and Kazakhstan, where eclipse bisects, will declare their independence. Another cross-country eclipse defines the northern border in 1990. USSR will officially dissolve on December 26, 1991 (12-26 is also a current Metonic date). The last complete eclipse on earth for 680 days, until Jun 11 1983.

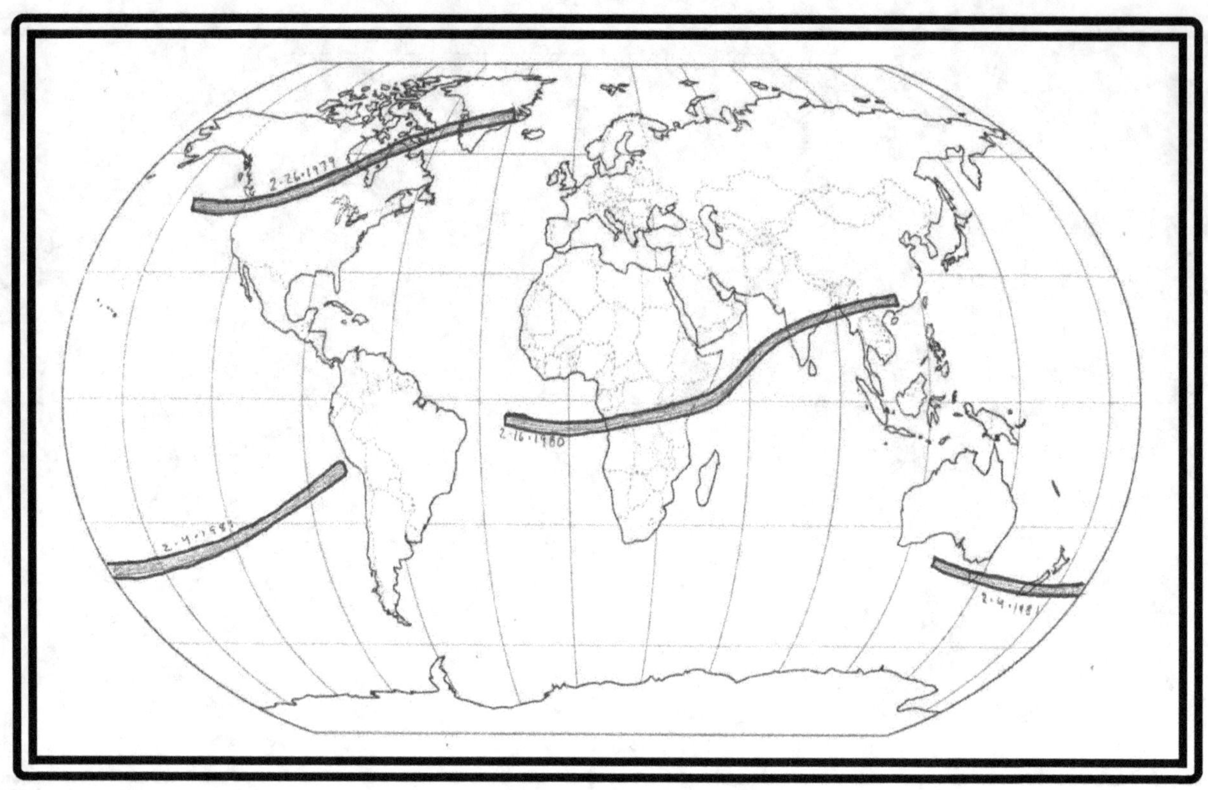

1979 to 1981 Triad Two-Brave New World-February

Notice that exact lunar years separate the three eclipses.

1979 Feb 26 Total Brave New World Microsoft/ Mount Saint Helens Eclipse Total (Saros 120) Totality over Mount Saint Helens. 388 days before she awoke with an earthquake on the Spring Equinox of March 20 1980, and 447 days before she exploded to the due north in the direction of Seattle and Microsoft on May 18 1980 in the largest volcanic eruption in modern North American History. Eclipse occurs seven weeks after Bill Gates and Paul Allen move Microsoft Corporation to Gate's hometown of Bellevue, Washington (99.64% Totality) The beginning of the technological era. The final era of the American Empire and the creation of the last world empire of the machine. The number one song in America was "Do you think I'm Sexy" by Rod Stewart. The number one show was "Three's Company". Salem and Portland Oregon in Totality. Crosses 5 northwestern states before bisecting Canada (Totality in Winnipeg). Saros Cycle 120. See USA eclipses.

1980 Feb 16 Africa/ Tanzania's third in 3 years Total (Saros 130) One lunar year after USA eclipse, Totality crosses Africa from the slave coast across Africa, bisects India (centerline of totality in Brahmapur-"*abode of Brahma*") and Myanmar before ending in China . Part of Dar Es Salaam swarm.

1981 Feb 4 Tasmania/ New Zealand Ring of Fire (Saros 140) Annularity in Hobart, Tasmania.

Partials occur on 1-25-1982, 6-21-1982, 7-20-1982, and 12-15-1982

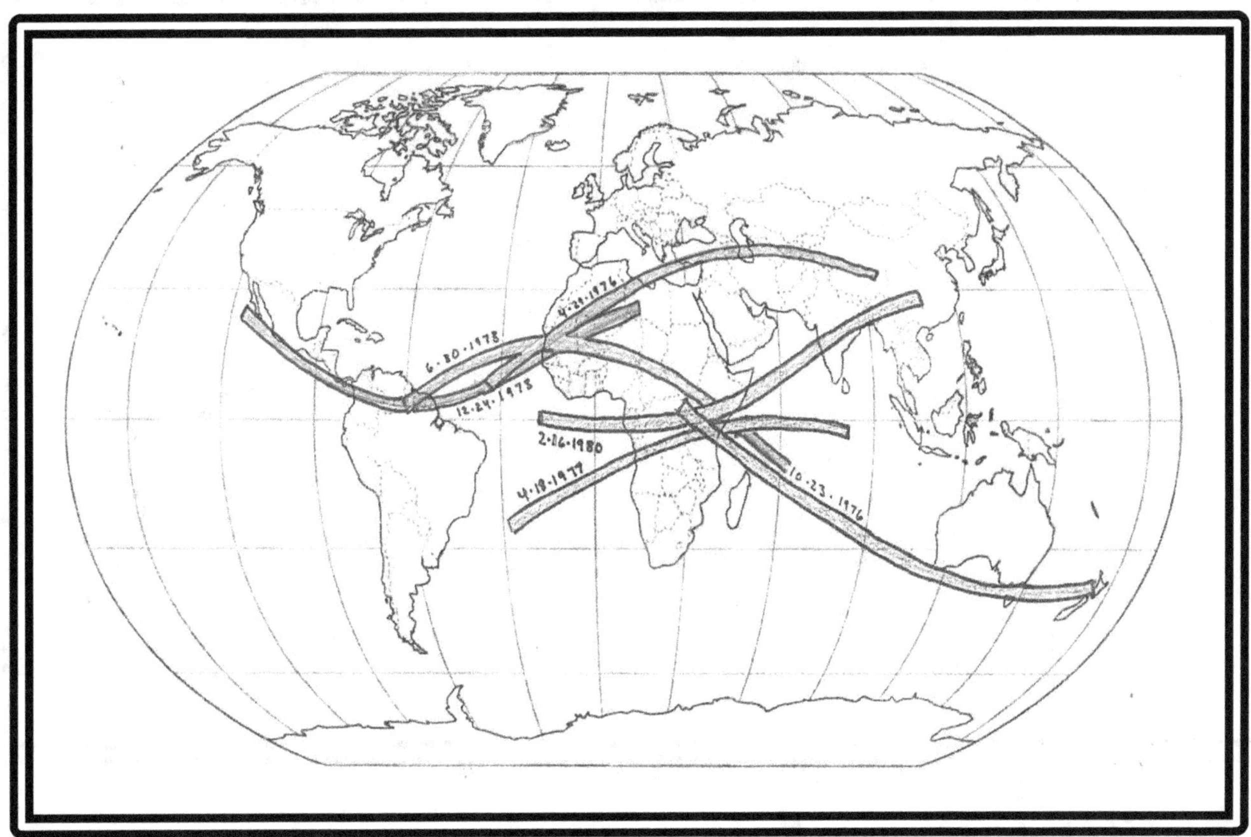

1973 to 1980 African Hepton Cluster

1973 Jun 30 South America through Heart of Africa (Saros 136)
1973 Dec 24 Christmas Eve America to Africa (Saros 141)
1976 Apr 29 Third Mauritania in 3 years/ Mideast to China Ring of Fire (Saros 128)
1976 Oct 23 Dar Es Salaam Africa to Australia Total (Saros 133)
1977 Apr 18 Dar Es Salaam Ring of Fire (Saros 138)
1980 Feb 16 Africa/ Tanzania's third in 3 years Total (Saros 130)

In the space of 7 eclipse years (2422 days) two clusters occur over Africa. One (between 1973 and 1976 over the course of 1033 days) has an epicenter on the western edge of the nation of Mauritania, near capital of Nouakchott. The other (between 1976 and 1980 over 846 days) has an epicenter near the capital of Tanzania, Dar Es Salaam.

The 6-30-1973 and 2-16-1980 solar eclipses (Saros 136 and 130) are a matching Hepton Pair exactly seven eclipse years apart, both occurring on Saturday. Together they bookend the larger cluster. These two eclipses also cross each other, in the Indian Ocean several hundred miles east of Kenya.

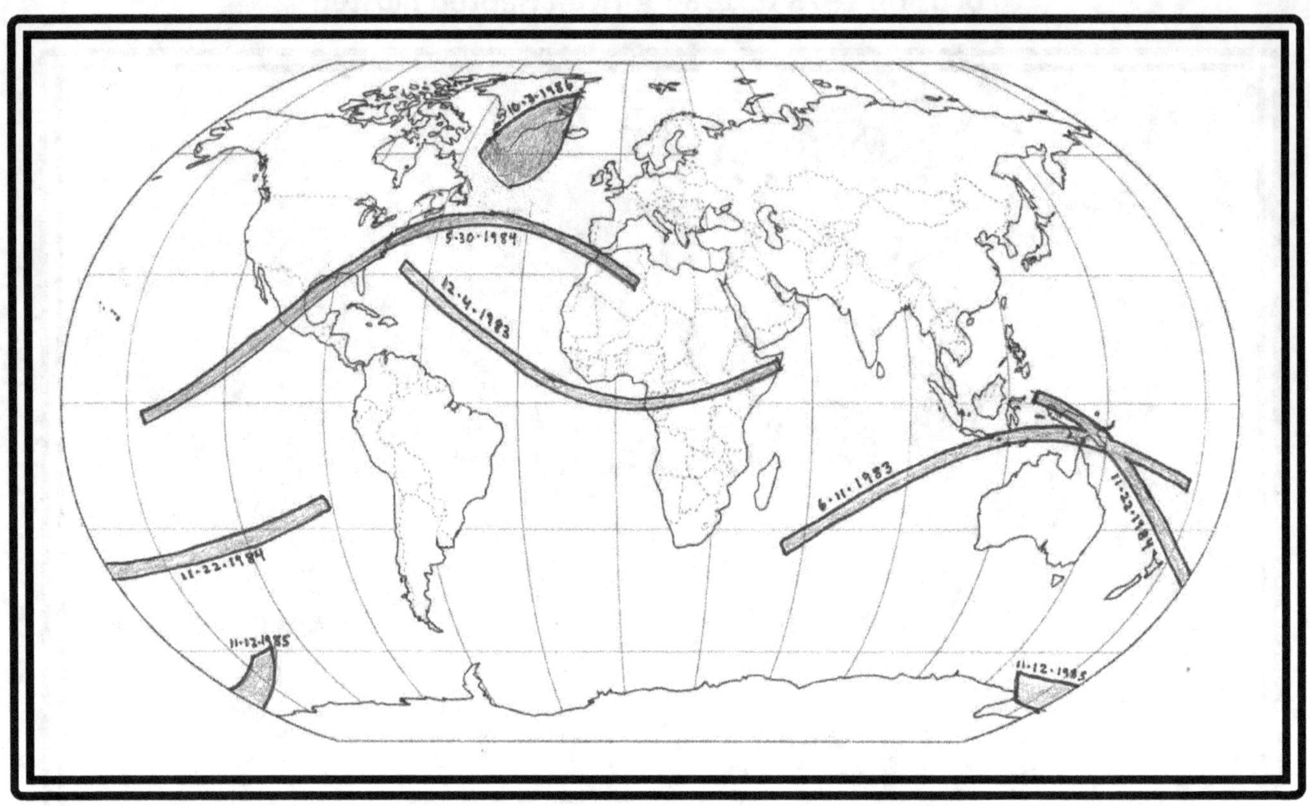

1983 to 1986 Last JFK Metonic/ New Guinea Cross

1983 Jun 11 Total New Guinea- Saros 127
1983 Dec 4 Ring of Fire Africa -Saros 132
1984 May 30 Ring of Fire JFK 67th Birthday USA -Saros 137
1984 Nov 22 Total JFK 21st Anniversary of Death Day New Guinea -Saros 142
1985 Nov 12 Total Antarctica -Saros 152
1986 Oct 3 North Atlantic -Saros 124

1983 Jun 11 New Guinea Total (Saros 127) Eclipse over capital of Port Moresby, New Guinea, part of New Guinea Cross. Totality also in Java, Indonesia.

1983 Dec 4 Africa Ring of Fire (Saros 132) Ring of Fire crosses continent to end precisely at sunset over the easternmost point of the continent, Ras Hafun, Gobolki Bari, Somalia.

1984 May 30 JFK 67th Birthday USA Ring of Fire/ Almost Total (Saros 137) Almost as close to a total eclipse as you can get without being a total, featuring a peak of 99.77% in Virginia. JFK Birthday Metonic. His actual birthday is May 29 but that's how Metonic works, sliding on the hours around a date. He would have been celebrating his 67th Birthday 12 hours before the eclipse began if he'd not been murdered 21 years before. Besides 1984 was a leap year, and if we hadn't leapt 12 weeks before, it would be May 29.

Reagan and George HW Bush in the White House. 95% coverage in Washington DC.

Ring of Fire begins in Mexico and Crosses eastern USA before heading to Africa. Perfect Annularity in Atlanta, Martin Luther King's home, with 99.5 % coverage exactly 5900 days after he was murdered on 4-4-1968 in Memphis (85% coverage). He would have been 55 in 1984.
90% in JFK's birthplace of Brookline Massachusetts, 80% in Dallas, Texas, where he was murdered. 93% in New York City.

Perfect annularity along thin path in the Deep South goes directly over New Orleans, birthplace of Lee Harvey Oswald and site of famous JFK conspiracy trial at 99%. Exits USA at Assateague Island, Maryland. The eclipse crosses the Atlantic Ocean and enters Africa over Casablanca ("white house") Province of Morocco, just north of Marrakesh ("Land of God") --ends in desert of the Sahara

Nov 22 1984 JFK 21st Anniversary of Death Day/ New Guinea Cross Total (Saros 142) Just as the solar eclipse prior brings the JFK birthday Metonic for the last time, so the 21st anniversary of the Kennedy Assassination is cosmically acknowledged on 11-22-1984. These linked 5-30 and 11-22 eclipses are the only two solar eclipses of 1984, and the last over inhabited terrain until 1987. The only nation in the 11-22 path is Papua New Guinea, with totality in the capital of Port Moresby for the second time in a year and a half.

New Guinea? One of the most famous war stories in American history involves the heroics of young Lieutenant John F. Kennedy, who miraculously helped save his crew after their small boat was rammed by a Japanese destroyer on August 3 1943 in the Solomon islands of New Guinea, which is inside the eclipse path of the Nov 22 1984 eclipse. The story of JFK's heroism as he helped save his shipwrecked crew was enshrined in both book and movie, and Kennedy's indisputable bravery helped propel him to the Presidency 20 years later.

An extremely powerful Nor'easter began impacting the United States' November 22, 1984, beginning unusually south between Florida and Cuba. The Presidential elections happened 16 days prior, on Nov. 6, one of the largest landslides in US history, with Reagan/Bush defeating the Walter Mondale ticket 525 electoral votes to 13, even winning JFK's home state of Massachusetts with 51.22 percent of the vote. The number one song in America was "Wake me up before you Go Go" by Wham. 1984 would be the last year that the JFK Birthday and Death day metonics would appear in the same year. They began in 1919, when he was 2 years old. Both brother Robert Kennedy (11-20-1925) and son John F. Kennedy Jr. (Nov 25 1960) were born around that same November Metonic where he died.

1985 Nov 12 Antarctica Total (Saros 152)
1986 Oct 3 North Atlantic (Saros 124)

1985 and 1986 "eclipses over nowhere" are the only complete eclipses of their respective years, and it's likely very few witnessed them. Isolation peculiarity: After the extraordinary paired JFK eclipses of 1984, there are no complete eclipses over any populated region of the planet until March 29, 1987...a total of 857 days or more than two years without a population center witnessing a complete eclipse. In fact, it will be nearly four years between true Total eclipses, Nov 12 1985 to Mar 19 1988.

Partials occur on 5-19-1985 and 4-9-1986

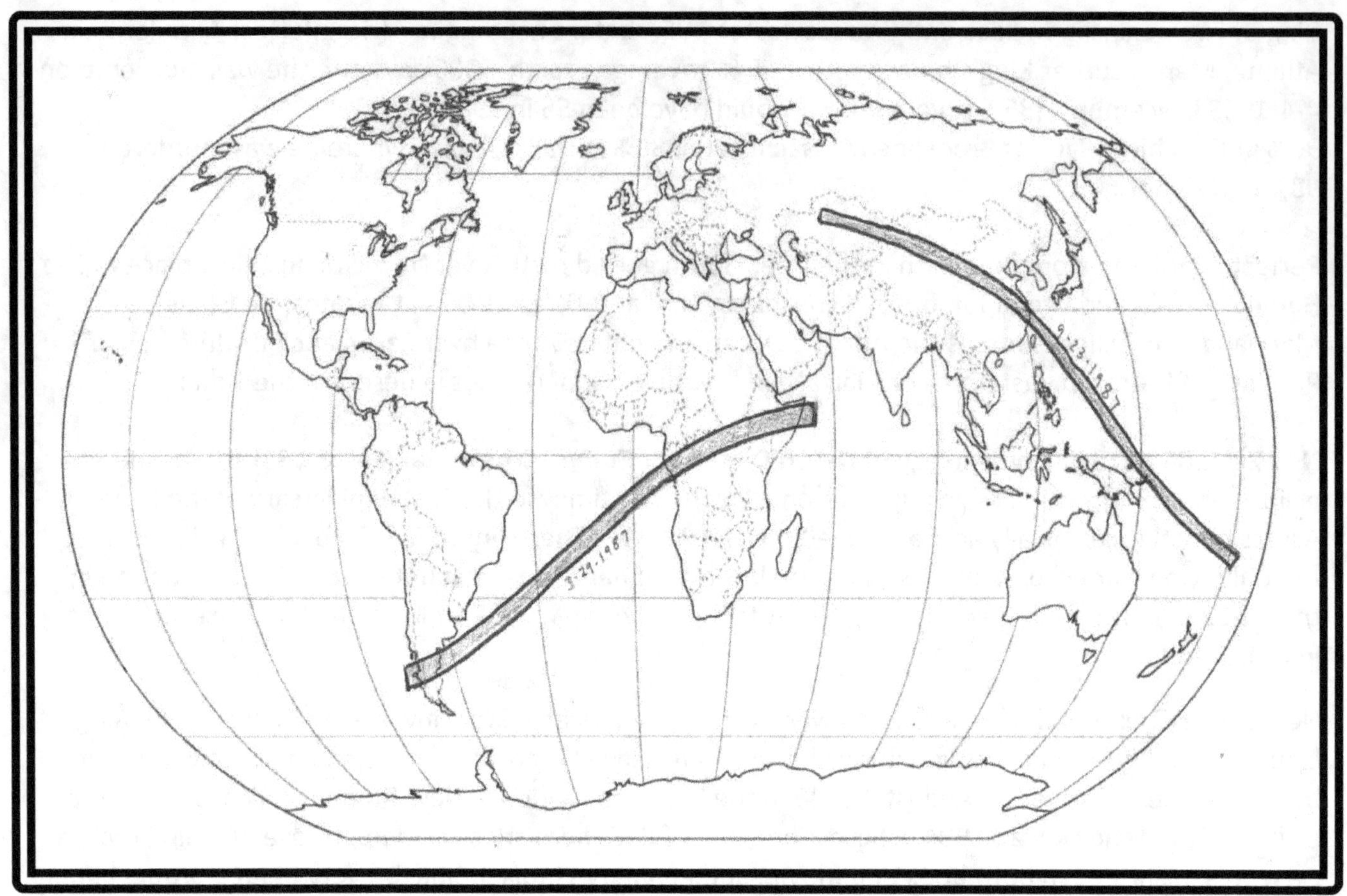

1987: The Palindrome Eclipses

3-29-1987 Patagonia to Horn of Africa Hybrid (Saros 129)
9-23- 1987 Kazakhstan/ China/ South Pacific Ring of Fire (Saros 134)

The first eclipse is on 3-29 and the second is on 9-23, which is a palindrome between the only complete eclipses of 1987. Speaking of numeric of 3-29-87, it so happens 3 x 29 = 87. 3-29-87 is also 87 days from New Years Day 1987 and 9-23-1987 is 100 days from New Year's Day 1988. Of course 3+ 29 =32 and 9+ 23 also equals 32 (32 is found in the first two numbers of the first eclipse and backwards in the second and third numbers of the second eclipse). But of course, it's just number fun and impossible to conclude anything from, except that numbers sure are funny, sometimes. These eclipses bare a distinct geometric similarity to 1933 Ring of Fires.

3-29-1987 Patagonia to Horn of Africa Hybrid (Saros 129) A Total solar eclipse exits South America at Caleta Olivia (Olive Branch) Argentina and crosses Africa for the second time in four years.

9-23-1987 Kazakhstan/ China/ South Pacific Fall Equinox Ring of Fire From Kazakhstan (Independent 4 years later) through China with total Annularity in Shanghai. Ends past northern Solomon Islands, place of JFK death day eclipse less than 3 years before. (10th eclipse over nation of New Guinea in 43 years). Saros cycle 134, Jerusalem ring of fire of Aug 21 1933 and USA ring of fire Oct 14 2023

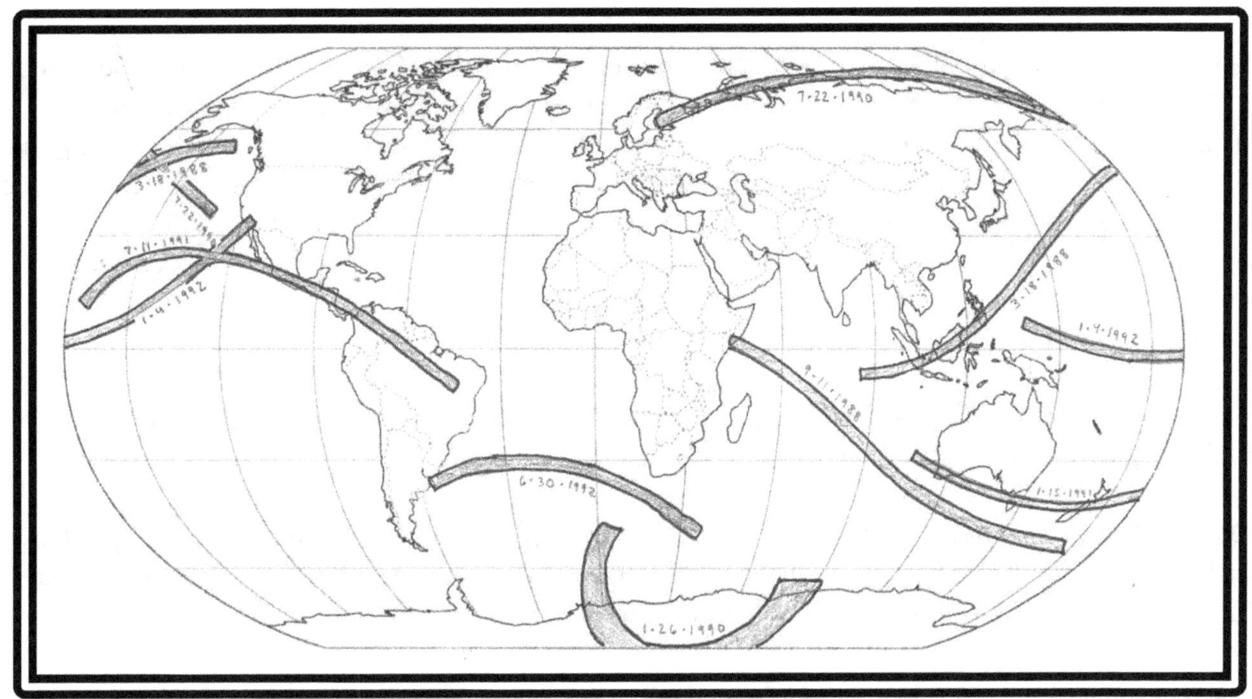

1988 to 1993 Minimum of Continental Land/ Lots of Peculiarities

1988 Mar 18 Total Indonesia/ Borneo/ Pinatubo to Aleutians Saros 139
1988 Sep 11 Ring of Fire Mogadishu Sunrise Single City- Saros 144
1990 Jul 22 Total USSR Last Judgment - Saros 126
1990 Jan 26 Ring of Fire Antarctica -Saros 121
1991 Jan 15 Ring of Fire MLK Birthday Australia/New Zealand -Saros 131
1991 Jul 11 Total Temple of the Sun...Mexico City/ Central and South America -Saros 136
1992 Jan 4 Ring of Fire Southern California Single City Sunset -Saros 141
1992 Jun 30 Total Single City Montevideo Uruguay -Saros 146

Only a very small portion of earth's land surface is crossed between Sep 1987 and May 1994. Three Single City eclipses in four years. LA Metropolis is considered single city for 1992 eclipse.

1988 Mar 18 Indonesia/ Borneo/ Pinatubo to Aleutians Total (Saros 139) This Pacific Total eclipse makes an x in combination with the 9-23-1987 eclipse (centerpoint northeast of the Philippines. Mount Pinatubo (70 % coverage) of the Philippines erupts 3 years and 3 months and 3 days later in the 2nd largest volcanic eruption of the 20th century. Saros 139 returns for USA Apr 8 2024.

1988 Sep 11 Mogadishu Sunrise Single City Ring of Fire (Saros 144) Eclipse begins in Mogadishu, Somalia at sunrise on 9-11-1988, the ancient Egyptian New Year, 13 years before the 9-11 terror attacks. Path of annularity touches no other land before dying in the ocean south of New Zealand. 1988 began genocide in Somalia of Isaaq tribesmen. 3 months before the election of George H.W. Bush in the USA. Five years before infamous Battle of Mogadishu ("Blackhawk Down") on Oct 3 1993 where USA forces fought Somali militiamen. Good example of the Single City phenomenon which gets highlighted in this section. No other population besides Mogadishu. The Metonic Date of 9-11 is highlighted in a total eclipse for the first time in centuries. Odd numeric parallel with 8-11-1999 eclipse, which ended the 20th century. 9-11-1988/8-11-1999.

1990 Jul 22 USSR Judgment Total (Saros 126) Nine years minus nine days after the USSR Judgment of Jul 31 1981 (Saros 145) which loosely followed the southern border of the Soviet Union without touching any other nation. Eclipse travels the northern length of the Soviet Union before crossing the Aleutian Islands of the USA. This eclipse only touches one other nation (Finland at sunrise and a few of the US Aleutian Islands towards evening) while defining the northern border of the USSR. 1990 was the year that the USSR lost 6 of its republics. Eclipse happens ten days after Boris Yeltsin quit the Communist Party, 8 months after fall of Berlin Wall and 15 months before the official dissolution of the USSR on December 26 1991 (which is also a Metonic date).

1990 Jan 26 Antarctica Ring of Fire (Saros 121) Lewis Mumford, one of America's greatest minds, dies on this very day. He was author of Myth of the Machine (and many more).

1991 Jan 15 MLK Birthday Australia Ring of Fire (Saros 131) Ring of Fire on what would have been Martin Luther King's 62nd birthday. Path touches Perth Australia, the island of Tasmania again, and straddles New Zealand, Annularity in Wellington. 2 days before Persian Gulf War air campaign.

1991 Jul 11 Temple of the Sun/Hawaii/ Mexico City/ Central-South America Total (Saros 136)
Totality for entire Big Island of Hawaii (including Kilauea volcano), first total eclipse there since 1850. Continues across Mexico (including totality in Mexico City—Aztec temples of Sun and Moon 470 years after their conquest by Spain and Popocatepetl Volcano), touching every nation of central America (totality in San Salvador and Guatemala City) and extending through Colombia before ending in Brazil. In the middle of Central American Crisis that saw wars throughout the region, 2 years before the Zapatista conflict in Mexico. 27 days after eruption of Mount Pinatubo in the Philippines. Saros 136's last eclipse of 20th Century and the longest total eclipse until 2132.

1992 Jan 4 Southern California Metropolis Single City Sunset Ring of Fire (Saros 141) A ring of fire travels 7,000 miles across the Pacific to end immediately over Southern California (and Tijuana Mexico) at sunset over the ocean. San Diego had a good view. Clouds obscured the eclipse in Los Angeles (but it still happened). 9 days after Soviet Union officially dissolved. George HW Bush is President. Beginning of 500th year since Columbus arrived in the New World...look what we made of it! Michael Jackson had the number one song with "Black or White". Nirvana's "Smells like Teen Spirit" was number 13. 114 days after eclipse, on Apr 29 (also a Metonic date), the worst riots in Southern California history break out in Los Angeles after the acquittal of police officers involved in the beating of motorist Rodney King. 64 people died and more than 2300 were injured.

1992 Jun 30 Single City Montevideo Uruguay Total (Saros 146) In our site's eclipse history there's only one example of a single city eclipse (Boston JFK eclipse 1959). Suddenly in time there are 3 single city eclipses in less than 4 years--Mogadishu, LA Metropolis and this one. June 30 1992 brings totality to Montevideo Uruguay at sunrise and afterward finds nowhere else but ocean. Montevideo means roughly " I see a mountain". 104 days shy of the 500th anniversary of Columbus finding New World. Buenos Aires Argentina is a near miss at 93 % totality. where Jorge Bergoglio (who becomes Pope Francis) is first promoted to auxiliary bishop in Buenos Aires in 1992. Bergoglio was born 3 days after the ring of fire eclipse of Dec 14, 1936. He becomes the first Pope from the Western hemisphere.

Partials occur on 3-7-1989, 8-31-1989, 12-24-1992, 5-21-1993, and 11-13-1993

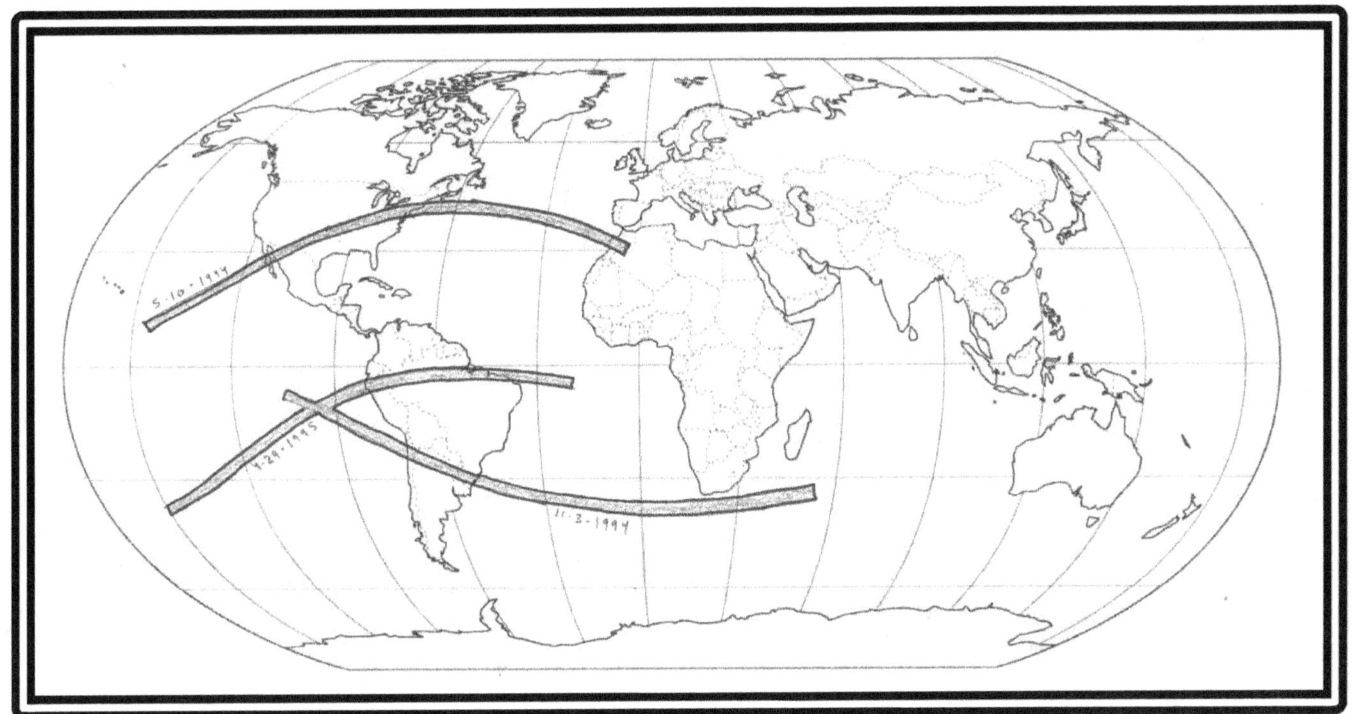

1994 to 1995

1994 May 10 Ring of Fire USA to Africa -Saros 128 The last complete eclipse in the USA for 18 years until the May 20 2012 Midland Sunset Ring of Fire eclipse brings the same Saros 128 back to the USA. Eclipse starts in the Pacific, bisects northern Mexico before crossing the heartland of the United States and exiting through the Bush family center of Kennebunkport Maine. Serial killer John Wayne Gacy is executed on this very day in Joliet Illinois, in the path of annularity. Metropolitan Chicago, home of Barrack Obama, is scraped by path. True Annularity in many cities, including El Paso, Amarillo, Lubbock, Tulsa, Joplin, Detroit, Springfield, Cleveland, Buffalo, Albany. The ring of fire crosses the Atlantic and ends at sunset over City of Casablanca (white house) Morocco. George W. Bush wins the Texas State Governorship in 1994 six months later. Path goes by W. Bush childhood home of Midland at 87% coverage, ranch at Crawford sees 80% coverage. Austin has 77%. Kennebunkport, Maine is in full Annularity. W. Bush birthplace of New Haven Connecticut sees 85% and Washington DC sees 80%. 6 years later he would be running for President.
"The Sign" by Ace of Base was the number one song in the USA.

1994 Nov 3 Total Nazca Lines South America -Saros 133 Total eclipse bisects northern South America. Totality on mysterious Nazca lines in Peru and Asuncion, Paraguay. Exits in Brazil.

1995 Apr 29 Ring of Fire South America -Saros 138 Eclipse crosses Peru and northern Brazil...Path enters Peru at Piura, the oldest colonial city in Peru, and third oldest in South America, founded by Spanish Conquistador Pizarro in 1532 and the first to declare independence from Spain (in 1821--- path travels across northern Brazil, total annularity at Belem (Portuguese for Bethlehem), and exits continent at Fortaleza ("strength"). annularity just north of Natal ("Nativity")

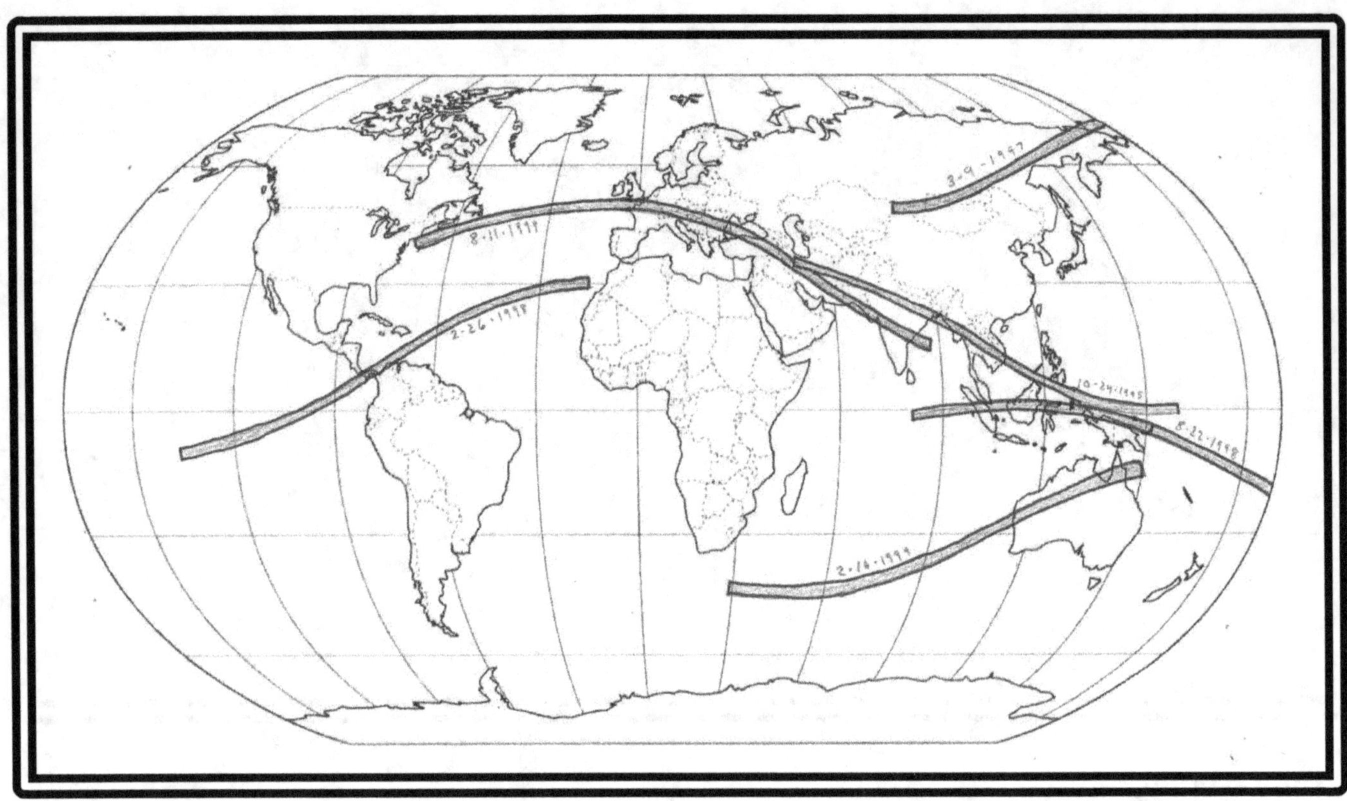

1995 to 1999 Last Solar Eclipses of the Second Millennium (UN Metonic)

The United Nations Metonic begins Oct 24 1995 and ends Oct 24 2014.
October 24 is United Nations Day, which celebrates the ratification of their charter on Oct 24 1945.

1995 Oct 24 Total United Nations Iran Sunrise/ South Asia/ SE Asia -Saros 143
1997 Mar 9 Total Russia -Saros 120
1998 Feb 26 Total American Isthmus -Saros 130
1998 Aug 22 Ring of Fire Indonesia/ New Guinea -Saros 135
1999 Feb 16 Ring of Fire Australia Cross-Country -Saros 140
1999 Aug 11 Total End of Millennium—Great Civilization Judgment - Saros 145

Oct 24 1995 United Nations Day Iran Sunrise/ South Asia/ SE Asia Total (Saros 143) On the 50th anniversary Of the United Nations Charter ("United Nations Day"). 30 years after the end of the Southeast Asia Eclipse Swarm of 1944 to 1965. Eclipse begins at sunrise south of Tehran (97% totality), crosses Iran, goes across Kandahar Province in Afghanistan (US Invades almost exactly six years later). Eclipse crosses Pakistan, totality over ancient city of Bahawalpur, crosses Northern India (totality in Allahabad-" City of God"), nations of Myanmar, Thailand, temple of Angkor Wat in Cambodia, Vietnam (3 months after normalized relations with USA). Eclipse exits Asia near Sandakan in Malaysia.

1997 Mar 9 Mongolia/ Russia Total (Saros 120) The only complete eclipse of 1997 begins at the eastern border of Kazakhstan five years after their independence, crosses northern Mongolia 790 years after its founding by Genghis Khan and heads across eastern Siberia in Russia before dying in the Arctic Ocean. 300 years after Peter the Great became the first Tsar to travel outside his country's borders. (1697 to England and France) Two months before Russia signed agreement with NATO to "avoid future war". Eclipse crosses largely uninhabited regions of Mongolia and Eastern Siberia.

1998 Feb 26 American Isthmus Total (Saros 130) A small patch of ground experiences totality on the 19th anniversary of the USA Mount Saint Helens Eclipse of Feb 26 1979. Path touches Galapagos Islands, and then isthmus of Panama just south of the canal zone. It scrapes northern borders of Colombia and Venezuela, with totality in Maracaibo, nicknamed "land of the beloved sun" founded 477 years before. Eclipse heads to Caribbean, including Montserrat, which was still in the throe of a series of volcanic eruptions. Last town in totality was St. John's, on island of Antigua.
Volcanic Allusions: After beginning erupting in 1995, tremendous eruptions in the summer of 1997 destroy the capital of Plymouth in Montserrat and render much of the island uninhabitable. The volcanoes of Saint Helens and Montserrat (two of the most explosive in the western hemisphere in the 20th century) are linked by Metonic Total Solar eclipses of Feb 26 1979 and Feb 26 1998.

1998 Aug 22 Indonesia/ New Guinea Ring of Fire (Saros 135) Aug 21 Metonic continues a brief lurch forward into 8-22. Path of Annularity close to epicenter of Dec 26 2004 Great Earthquake/ Tsunami southwest of Banda Aceh, Sumatra which occurs six years later and will kill more than a quarter of a million people. Annularity heads over Malaysia and Solomon Islands. Ends after Vanuatu Islands. 8-22 becomes 8-21 again in 2017 USA Cross Country Eclipse.

1999 Feb 16 Australia Cross-Country No Cities Ring of Fire (Saros 140) Only the tiniest Australian towns such as Walkaway (pop. 270) Mullewa (Pop. 356) and Lakeland Downs (pop.299), experience the annularity of the second to last solar eclipse of the Second Millennium as it travels nearly 2,000 miles across the continent. The largest town in the path appears to be Tennant Creek (pop. 3,000)

partials on 4-17-1996, 10-12-1996, 9-2-1997

Focus on August 11 1999 End of the Millennium Eclipse-Saros 145

143 days before the 21st Century begins, an extraordinary path of totality ends the 20th Century and the Second Millennium on the unusual numeric of 8-11-1999. One of the most watched eclipses of all time, it crossed densely populated areas across Europe, the Mideast and South Asia.

USA: This 1999 path of totality begins at sunrise a few hundred miles from the East Coast of USA (26% totality in New York City, 62% in Boston, 72% in Kennebunkport, Maine). Eclipse is 26 days after death of John F. Kennedy Jr. and wife Carolyn in plane crash off Martha's Vineyard (60% totality) which happened on July 16 1999 (exactly 54 years after test of first A-Bomb).
Eclipse is exactly 84 eclipse years from the 11-22-1919 JFK Premonition USA Eclipse and exactly 16 eclipse years from DC Ring of Fire May 30 1984. Exactly 57 eclipse years (a full eclipse triple Metonic) from 1945's A- Bomb Eclipse.

Occurs 11 months after founding of Google, a few months before founding of Gates Foundation, two months after George W. Bush announced candidacy for White House, four months from Y2K hysteria and the changing of the millennium.
"Genie in a Bottle" by Christina Aguilera was the Number One song in the USA.

European Union: 1999 was the year of the introduction of the Euro Monetary denomination, and the creation of the Eurozone. The Euro and Eurozone was launched on Jan 1st 1999.

England: Eclipse crosses Plymouth and Truro in England, and the Port of Penzance ("holy headland"). England's first Total eclipse since the Liverpool eclipse of 1927.

France: The Eclipse Path goes across the entirety of Northern France, scraping Metropolitan Paris. Totality in **Luxembourg** and Strasbourg.

Germany: The first total eclipse in Germany since Aug 19 1887 Berlin Sunrise Eclipse 19 months before Hitler was born. The eclipse path across southern Germany and the province of Bavaria. Totality in the town of Branau Aum Inn (Austria) where Adolf Hitler was born 110 years before on 4-20-1889. Totality in Munich, Germany, where he began his rise to power. 66 years after the Nazis seized power in 1933.

Austria: Mozart's birthplace of Salzburg; also **Hungary**, and **Romania**--including totality in Bucharest.

Turkey: The eclipse bisects Turkey. Six days later on Aug 17, the worst earthquake in modern Turkish history devastates the city of Izmit (96% coverage) and surrounding region, killing 20,000. Major damage in capital of Istanbul. (95 %). One of the most dramatic examples of eclipse/disaster associations in our survey. The epicenter of quake was less than 45 miles from path of totality.

Iraq: The eclipse passes over Mosul, Iraq and the ruins of the ancient city of Nineveh. Four years later, US troops invade Iraq.

Nineveh: mentioned many times in the Bible, Nineveh, the great capital of Assyria, was where the Prophet Jonah showed up after being spit out by the whale. It was the place he had been told by God to warn of God's impending judgment. There was a total solar eclipse in Ninevah on 762 BC Jun 15. See "Sign of Jonah" Path of the August 11 1999 eclipse intersects with the August 21 1914 World War One Eclipse path near Nineveh. The two eclipses are separated by 84 solar years plus one lunar year.

Iran: The eclipse cuts the Persian nation in half. Its second total eclipse in four years. Totality over the great city of Isfahan.

Pakistan: Totality in Karachi, the 12th largest city in the world.

India: Eclipse crosses the entire nation of India, exiting in totality with centerline at Srikakulum, a sacred city for both Hindus and Buddhist and site of Buddhist Heritage site Salindaham, the location of the Arasavalli Sun Temple --the ("abode of the Sun god")-one of two major sun temples in India.

The Aug 11 1999 Eclipse is one of the most incredible displays of the cosmic art of the Master, our God, who made and placed the sun and moon and stars and knows all the empires. The August 11 1999 eclipse has enough levels to be worthy of study for ages.

If the traditional date of 1 BC is accepted for the birth of Jesus Christ, then this eclipse, one of the greatest eclipses of all time, fell on the 2,000[th] solar year anniversary of that event. The eclipse also occurs one day after the Aug 10 Temple Destruction Metonic. (Temple destroyed on 8-10-70 AD)

Saros Cycle 145's next engagement with planet earth comes August 21 2017. It crossed the length of the USA. Next up, it crosses Beijing, Pyongyang and Tokyo Sep 2 2035.

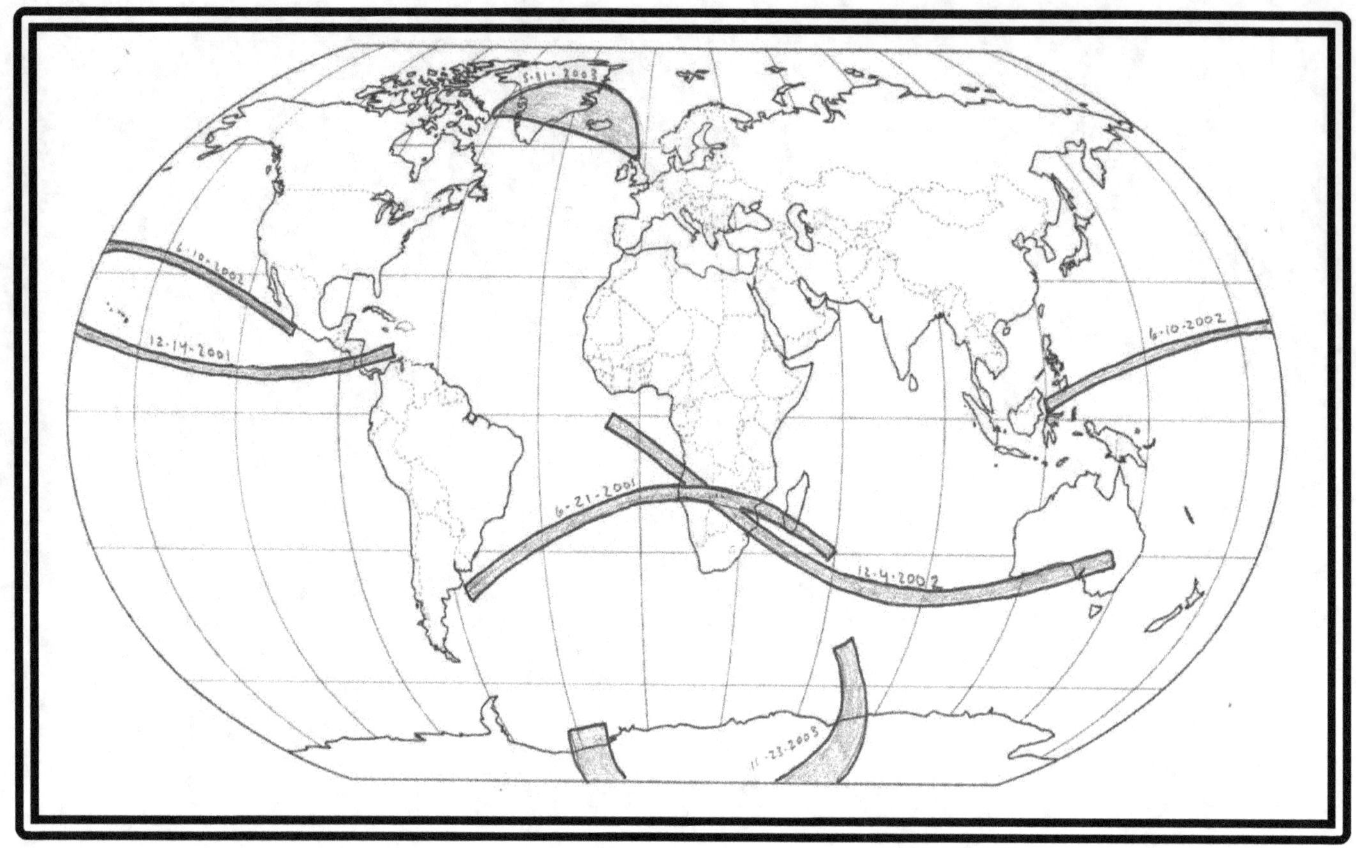

2000 to 2003 Start of the New Millennium---Very Little Land

2001 Jun 21 Total Solstice South America to Africa -Saros 127
2001 Dec 14 Ring of Fire Pan- American Isthmus/ Washington's Death Metonic -Saros 132
2002 Jun 10 Ring of Fire Pacific Ocean -Saros 137
2002 Dec 4 Total Africa to Australia -Saros 147
2003 May 31 Ring of Fire Greenland Iceland -Saros 147
2003 Nov 23 Total Antarctica -Saros 152

The first four eclipses of this section have exactly 19 year corresponding eclipses in 2020 and 2021. Jun 21 and Dec 14 appear again as eclipses in 2020. Jun 10 and Dec 4 do the same in 2021.

The African Total Eclipses of 2001 and 2002 intersect on the coast of Angola, south of Luanda, the notorious slave port where millions of slaves were shipped to the Americas. One of the most interesting things about this map of eclipses is how very little populated land is in totality. A tiny portion of the western hemisphere (Iceland, the Pan-American Isthmus and a chunk of the Mexican Coast), Southern Africa, a few Pacific Islands, and a largely uninhabited chunk of Australia. From the amazing density of population covered by the August 11 1999 End of Millennium Eclipse, for six years perhaps only a few million people (mostly in Southern Africa and Costa Rica) would have the opportunity to witness a complete eclipse. North and South America, Europe, Asia and most of Africa are free of eclipses from August 1999 to October 2005. The Dec 14 2001 eclipse, the first after 9-11, only touches two tiny Islands in the Philippines, and then only the Island of Saipan, site of a major battle of World War II. That Eclipse ends just after entering Mexico south of Puerto Vallarta.

2001 Jun 21 Solstice South America to Africa Total (Saros 127) The first complete eclipse in almost two years and the first since the last eclipse of the 20th Century on Aug 11 1999. The first complete eclipse of the 21st Century contains two 21's 21-2001. Also, the last six digits of the eclipse, 21-2001 when added individually, add up to the first digit, 6. But, whatever...
The path begins just off the Eastern Coast of South America (29% coverage in Buenos Aires) crosses the ocean into the Slave Coast of Angola and crosses Southern Africa, ending just after Madagascar. 82 days before 9-11 attacks. Saros 127, which reappears in Pope Francis Eclipse of 7-1-2019, and was also responsible for the 8-8-1496 so-called "Mayan Apocalypse" Eclipse. The Summer Solstice reappears as eclipse date in 19 years for Pandemic Ring of Fire of Jun 21 2020.

2001 Dec 14 Pan- American Isthmus/ Washington's Death Metonic Ring of Fire (Saros 132) The path of Annularity goes over almost no land, except a tiny portion of Central America, straddling the border of Costa Rica and Nicaragua, including annularity over town of Liberia. Dec 14 2001 was the 202nd Anniversary of George Washington's death. This is the first complete eclipse since the Sep 11 2001 attacks 94 days prior. 16 % coverage in Washington DC and 5% coverage in New York City. 54% coverage in Miami. Eclipse true annularity path ends in the ocean north of Venezuela about 74 degrees west longitude, the same longitude as New York City, 2200 miles away as the crow flies. See isthmus cluster. Dec 14 shows up again in Electoral Eclipse of 2020.

2002 Jun 10 Pacific Ocean Ring of Fire (Saros 137) Indonesian Islands to sunset on coast of Mexico over the Pacific Ocean, almost no land. 71 % coverage in Los Angeles. "Foolish" by Ashanti was the Number One song in the USA. Metonic reappears in 6-10-2021 Canadian Judgment.

2002 Dec 4 Africa to Australia Total (Saros 147) Southern Africa gets their second total eclipse in 18 months. It crosses southern Indian ocean before crossing a nearly uninhabited portion of Australia. Path stops near Queensland/New South Wales border very close to the location of the Australian Tav of 2028 and 2030. The Dec 4 date returns in 19 years with the 12-4-2021 Antarctica Eclipse

2003 May 31 Greenland/Iceland Ring of Fire (Saros 147) JFK Birthday has officially moved away from Metonic, now that it's two days removed from his May 29th Birthday. Ring of Fire over Greenland, Iceland and Extreme northern United Kingdom. Number One song in UK was "Ignition" by R. Kelly.

2003 Nov 23/24 Antarctica Total (Saros 152) Was anybody there? Hard to say. JFK death Metonic shifts further away from where it was 19 years before (11-22) in 1984, beginning the move away from the Nov. 22 date that began in 1900. By 2030 it will be all the way to 11-25 (JFK Jr's Birthday). Antarctica gets the only total eclipse on earth of 2003.

The year 2000 has four partial eclipses, including two in July (the first time two eclipses occur in one month since 1880). 2-5-2000, 7-1-2000, 7-31-2000, and Christmas 12-25-2000 over USA …

Between the two July eclipses was one of the greatest solar storms on record, the Bastille Day event of July 14 2000.

12-25-2000 is last solar eclipse on true Christmas for 755 years. Two more "sort-of" Christmas eclipses happen in 2038 and 2057 due to long Antarctic days and the international date line.
Christmas 2000: A Hint that Jesus was indeed born in traditional year? There was a Christmas solar eclipse in 1 BC.

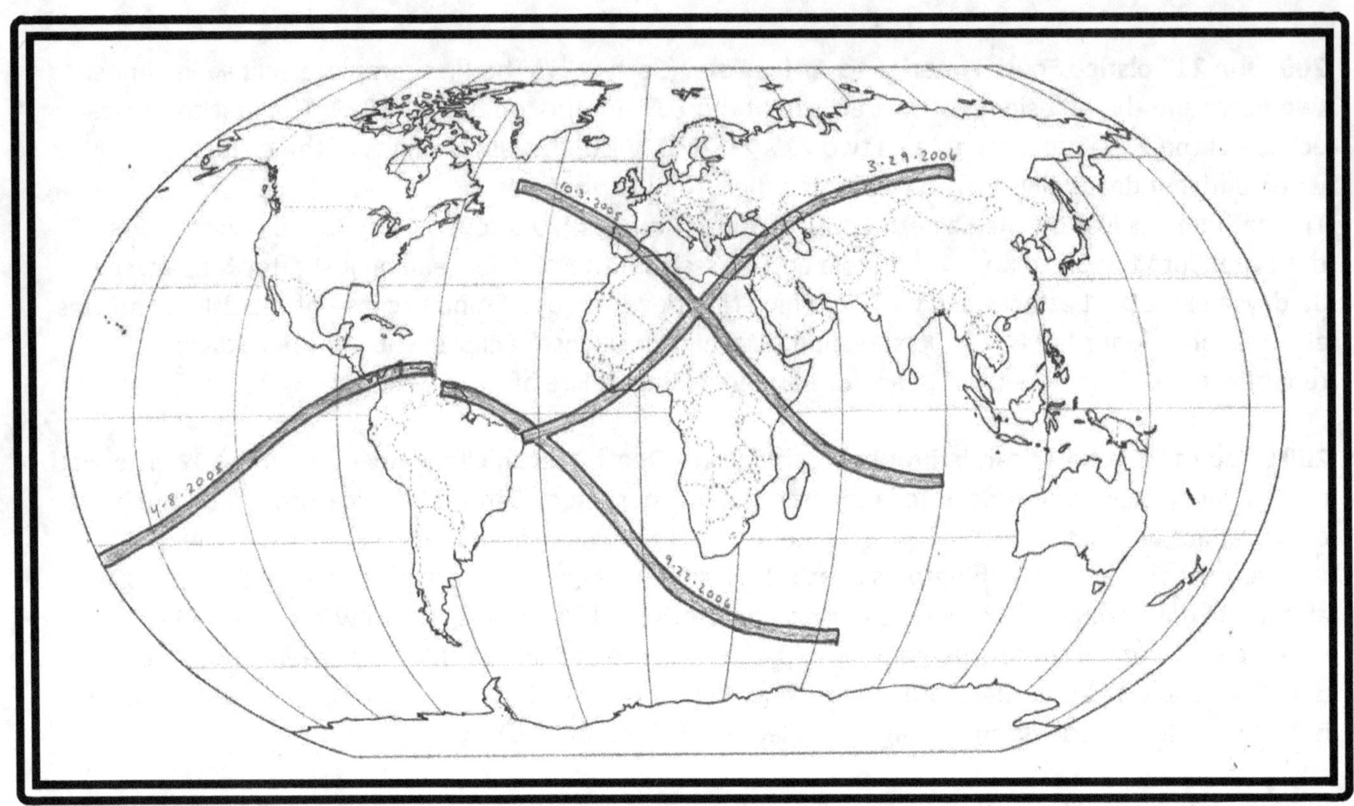

2005 to 2006 Libyan Tav/ Second Half of Turkish Cross/ Isthmus Again

2005 Apr 8 Total Isthmus Third time in Seven Years -Saros 129
2005 Oct 3 Ring of Fire Spain Through Africa Libyan Tav -Saros 134
2006 Mar 29 Total Second Line of Great Turkish Cross -Saros 139
2006 Sep 22 Ring of Fire Fall Equinox South America to Ocean -Saros 144

It's now been 10 years (1994 and 1995) since any eclipse has crossed the continental land masses of the western hemisphere except at the isthmus and extreme northern South America. Meanwhile, a profound double Tav over the Mideast and North Africa takes shape.

A 177-day Cross with the Oct 3 2005 Ring of Fire and the Mar 29 2006 Turkish eclipse intersect over south central Libya, in the nearly unpopulated desert of the Sahara.

2005 Apr 8 Isthmus Third time in Seven Years Total (Saros 129) Eclipse covers almost exactly the same land as the Dec 14 2001 eclipse at a 1211 day interval but comes from an opposite mirrored convex. A tiny stripe of true totality through Panama and Northern Venezuela. Exactly 19 years before Great North American Eclipse of Apr 8 2024. Buddha's Birthday (Japan).

2005 Oct 3 Spain Through Africa Ring of Fire—Libyan Tav (Saros 134) Ring of Fire Eclipse crosses Spain and Northern Portugal. 500 years since the retirement of Christopher Columbus and petition to the King for New World titles after his return from last voyage in 1504. 500 years since the arrival of conquistador Hernan Cortes in the New World. Total annularity path squarely on Madrid and Toledo as it bisects Spain. Path goes over Algiers, Algeria 500 years almost to the week after Spanish defeated the Algerians and began 300 years of control (Battle of Mers-el Kebirer Sep 13 1505). Eclipses cross Africa. Libya divided NW to SE six years before it was wrecked by civil war and NATO bombing.. Crosses Darfur, Sudan during early years of Genocide. Crosses Kenya ten months after Barack Obama starts brief career as a US Senator on January 3 2005, Obama's father was from Kenya, which saw 80% coverage in his home Rachuonyo district.

2006 Mar 29 Second Line of Great Turkish Cross Total (Saros 139) Looked at in depth in Mideast Eclipses, the second line of the Hepton Turkish Cross. A path that creates three intersections. It intersects with the next 2006 eclipse (ring of fire) just east of South America. It crosses the last eclipse (Oct 3 2005) over the doomed nation of Libya (NATO bombing in 2011), then crosses the 8-11-1999 20th Century Judgment eclipse with a total Hepton Cross over the middle of Turkey.

Totality begins directly over city of Natal, Brazil ("Nativity " in Portuguese. Crosses Atlantic to enter Gold Coast of Africa at Acrra, Ghana 39 years almost to the day after Martin Luther King's visit there in March of 1957. In Libya, it intersects the path of the eclipse of Oct 3 2005, which occurred 6 lunar months before. The path of totality goes over the Mediterranean before entering Turkey. 7 and ½ weeks before 500th anniversary of death of Columbus on 5-20-1506).

The two total eclipses of 1999 and 2006 are spaced 2422 days apart. They are Hepton eclipses exactly 7 eclipse years apart both occurring on Wednesday. The intersection of those two great paths is in Northern Turkey about 200 miles east of Ankara in the region of Tokat, an ancient city from the time of the Hitittes, which became a Roman stronghold. It was home to a thriving Armenian Christian community until what happened during WWI. These same Saros eclipses (145 and 139) create the cross over the USA in 2017 and 2024. The 2006 path heads on to bisect Georgia 20 months before its war with Russia then crosses Kazakhstan, including Totality at mysterious Capital of Nur-Sultan, formerly known as Astana. Path of totality ends just after crossing over Kyzyl, the capital of Tuva in Russia. Tuvans are famous for their unusual throat singing.

2006 Sep 22 Fall Equinox Top of South America to Ocean (Saros 144) Fall equinox Metonic. Ring of Fire. The eclipse begins over extreme Northwestern South America. New Amsterdam, Guyana and Paramiribo, the capital of Suriname, are in the initial path of annularity. Then no more land as it heads to the ocean. If you look at this path and the path of the Mar 29 total eclipse occurring 6 months prior you will see they form an interesting Tav intersection in the ocean east of South America.

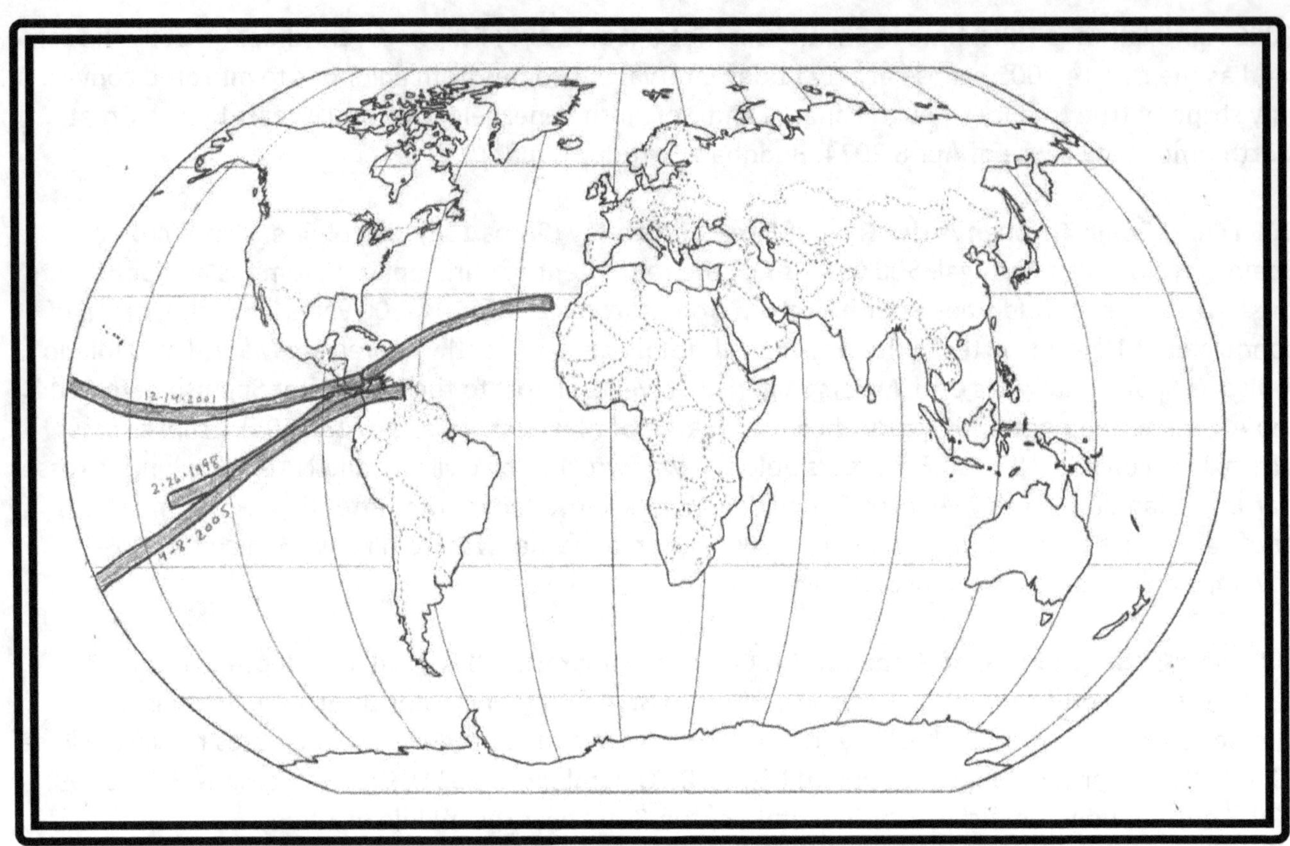

Isthmus Cluster 1998 to 2005

In a span of 7 years plus 40 days, there are three Pan-American Isthmus solar eclipses. All three dates are linked metonically (exact 19-year spacings) with other interesting American Eclipses.

1998 Feb 26 American Isthmus Total (Saros 130) Feb 26 is the same calendar date as Feb 26 1979 Mount Saint Helens Total Eclipse (Saros 120) and Feb 26 2017 Ring of Fire scheduled South America to Africa. (Saros 140)

2001 Dec 14 Pan- American Isthmus/ Washington's Death Metonic Ring of Fire (Saros 132) This eclipse is 19 years before the Electoral College Eclipse of Dec 14 2020.

2005 Apr 8 Buddha Birthday Isthmus Third time in Seven Years Total (Saros 129) 19 years after this eclipse will come the Great North American Eclipse of 4-8-2024.

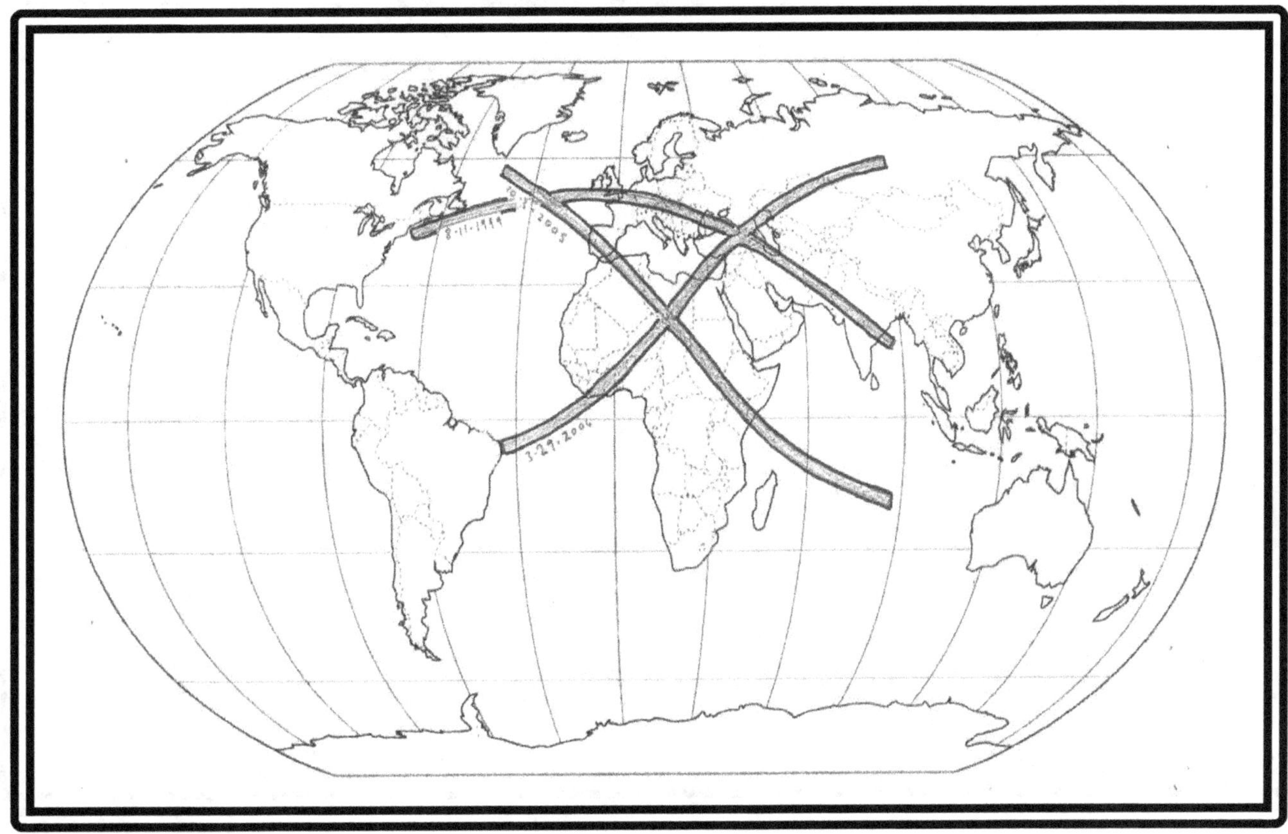

1999 to 2006 Mediterranean Aleph

How beautiful is this? A neatly divided Center of the Historic Western World over the course of 2422 days. Within the small triangle created by the A are the nations of Spain, France, Germany, Austria, Italy. Greece, Libya and Turkey.

This pattern, with the same eclipse time spacing, reappears over America beginning 18 years and 11 days and 8 hours later (19 eclipse years) involving the same Saros Personalities.

1999 Aug 11 End of Millennium—Great Civilization Judgment (Saros 145)
2005 Oct 3 Spain Through Africa Ring of Fire—Libyan Tav (Saros 134)
2006 Mar 29 Second Line of Great Turkish Cross Total (Saros 139)

partials occur on 3-19-2007(next Metonic on 8-year lurch to Spring Equinox 2015 North Pole Eclipse) and on 9-11-2007—
last solar eclipse on the 9-11 Metonic (Egyptian New Year)

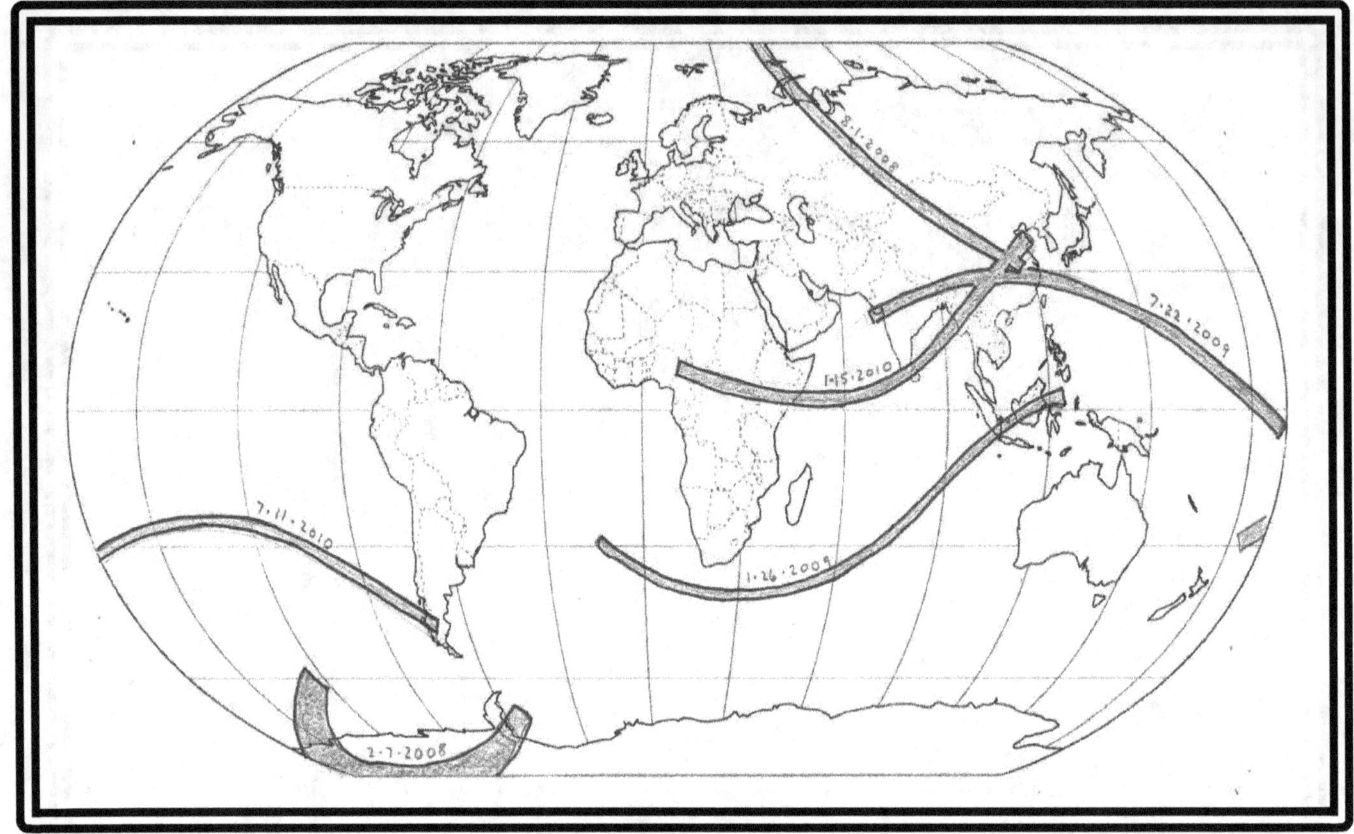

2008 to 2010 China Cluster/ Wuhan Totality/ Chongqing Cross

2008 Feb 7 Ring of Fire Antarctica -Saros 121
2008 Aug 1 Total Northern latitudes/ Russia/ China -Saros 126
2009 Jan 26 Ring of Fire Australia Day South Atlantic to Indonesia -Saros 131
2009 Jul 22 Wuhan Totality -Saros 136
2010 Jan 15 Ring of Fire MLK Birthday Africa to China -Saros 141
2010 Jul 11 Easter Island Total -Saros 146

2008 Aug 1 Northern latitudes/ Russia/ China Total (Saros 126) August 1 2008 brings a total solar eclipse from the extreme north of North America across the top of Greenland and then down through the Siberian wilderness into China. Path of totality ending at Luohe, China at sunset. 90% coverage at Beijing, 70% at Wuhan. Occurs one week before start of Russia/Georgia War on August 7 2008.

2009 Jul 22 Wuhan Totality (Saros 136) Ten years before the Pandemic begins and the world changes forever; The total eclipse of 2009 puts Wuhan, China, almost perfectly in the centerline of totality.

At 6 minutes and 39 seconds, the July 22 2009 Total solar eclipse is the longest of the 21st Century. It begins just west of India and crosses many important cities: totality in Surat, Indore, Jabalpur, Allahabad (City of God), Varanasi. Patni etc...Totality at the Taj Mahal. Totality over nearly the entire mysterious mountain kingdom of Bhutan 8 weeks before a major earthquake there. Path traverses Tibet (95% in Lhasa) and then heads to China. The eclipse goes over the great city of Shanghai, which becomes at that moment the most-populous city in the history of the world to experience a total eclipse (metro pop. 40 million). The path crosses mostly ocean for the next several thousand miles, before dying in the Marshall Islands, site of USA nuclear tests following World War Two, including the famous island atolls of Bikini and Eniwetok, where the US controlled until 1990's and bombed with some of the largest destructive weapons in the history of man-- are in totality. The great Saros Cycle 136 returns for The Valley of the Kings eclipse in 2027.

2010 Jan 15 MLK Birthday Chongqing Cross Ring of Fire (Saros 141) The Martin Luther King Birthday Metonic on what would have been his 81st birthday. King visited Africa and India, both of which are in this eclipse. 177 days after the Great Wuhan Eclipse, eclipse begins in Cameroon, Africa and crosses the continent through Uganda (Annularity in Kampala), Kenya (Annularity in Nairobi. The annularity path crosses the home and birthplace of US President Barrack Obama's father Barrack Obama Sr. in the Rachuonyo District of Kenya, just as President Obama begins his second year in office.

Path heads across the Indian ocean to cross extreme southern India for India's second eclipse in 6 months, including annularity at The Sri Padmanabhaswamy Temple of the Sun in Thiruvananthapuram. It is known as the world's richest temple, near where the lost city of Ophir was believed to be visited by Kings Solomon's ships around 1000 BC. Eclipse bisects Myanmar (annularity in Mandalay) and then bisects China (second eclipse in 177 days) before ending just after China's extreme northeastern border at Rongcheng, eclipse remains on the Chinese side of the Yellow Sea stopping just short of the Korean border. 80% coverage in Wuhan. If overlaid with the path of the 2009 eclipse 177 days prior, an X is seen across the nation of China, with its epicenter west of Wuhan, and very near the ancient city of Chongqing- geographically, at 31,000 square miles, the largest city in the world. On the other side of the world, an earthquake hit Haiti 60 hours earlier, one of the deadliest disasters in the history of the western hemisphere, killing 200,000 people.

2010 Jul 11 Easter Island Total (Saros 146) Hey, it's the 7-11 Metonic. Let's say you're an eclipse. Where would you want to go? Maybe you'd want to be the first total eclipse in 1400 years to cross Easter Island, one of the most mysterious and isolated places on planet earth, a tiny speck of land on a vast ocean with a tragic yet enduring history, the famous home of gigantic carved stone heads gazing off across the ocean's horizon. This solar eclipse travels almost entirely over open ocean before reaching Easter Island. Afterwards it travels to South America without making it to the other side. Manages to cross Patagonian city of El Calafete, and town of Esperanza ("hope"), Argentina at sunset .

The Easter Island eclipse occurred precisely during the playing of the 2010 World Cup Soccer final between the Netherlands and Spain. The Dutch had been the first colonizers of Easter Island in the 1700's. Spain had asserted linguistic control, via Chile, in 1888. Easter Islanders speak Spanish. A soccer match between two extinct empires took place while a total eclipse crossed their most mysterious of conquests, the great watching heads of Easter Island. Spain won the match 1 -0.

partials occur on 1-4-2011, 6-1-2011, 7-1-2011, 11-25, 2011

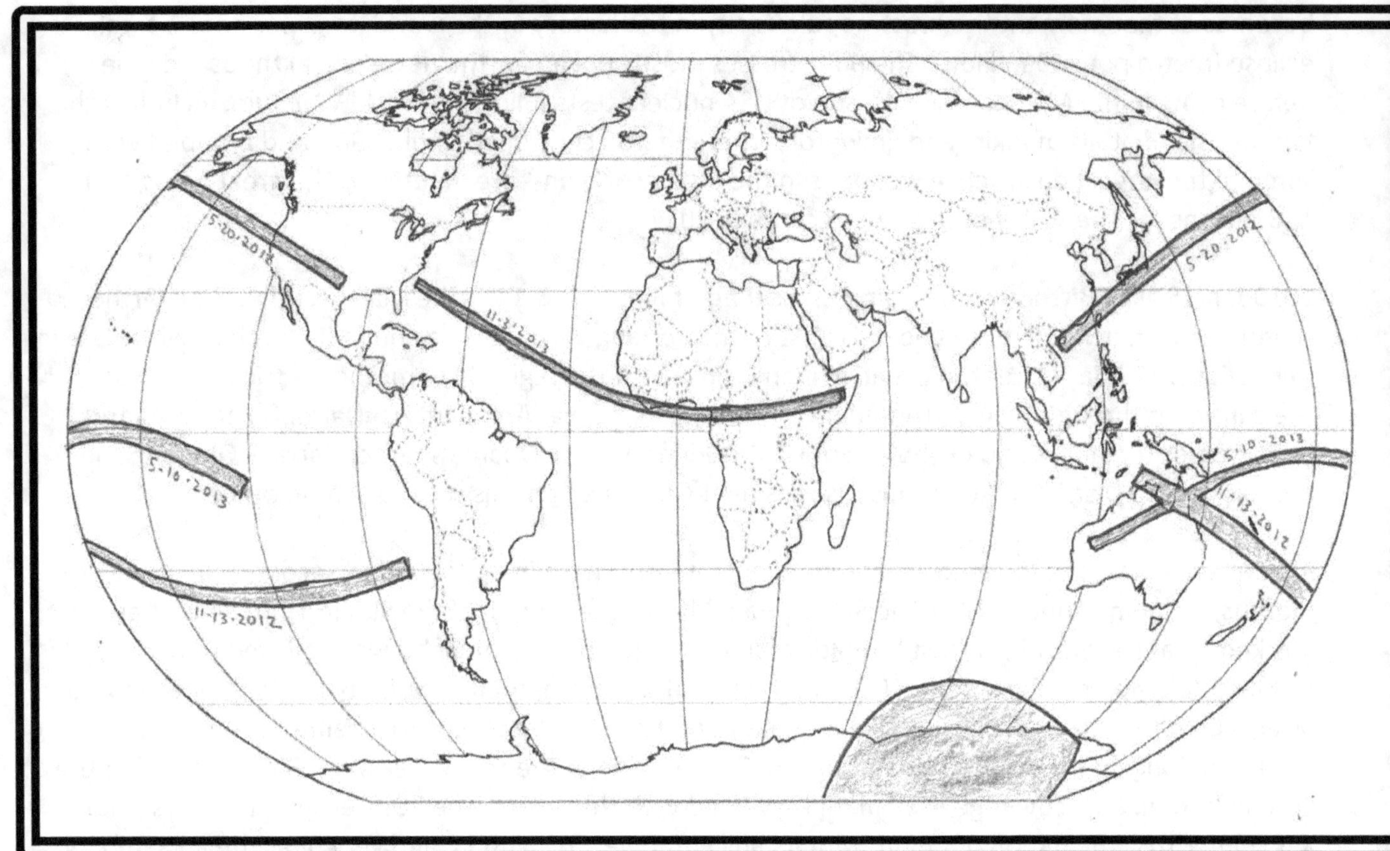

2012 to 2014. End of 10-24 United Nations Metonic.
Last Eclipses before Spring Equinox Series

2012 May 20 Ring of Fire Columbus Death Day/ Vietnam Sunrise/ Midland Sunset -Saros 128
2012 Nov 13 Total Darwin Australia -Saros 133
2013 May 10 Ring of Fire Darwin Cross/ Solomon Islands -Saros 138
2013 Nov 3 Hybrid Africa -Saros 143
2014 Apr 29 Ring of Fire Antarctica -Saros 148

May 20 2012 Columbus Death Day/Vietnam Sunrise/ Midland, Texas Sunset Ring of Fire (Saros 128)
Spanning the gulf between East and West. Eclipse begins in the northernmost part of Vietnam at
Sunrise just east of Hanoi. China gets its 3rd eclipse in 4 years along the coast, Annularity in Taipei,
Taiwan. Annularity in Tokyo, Japan. Path crosses Pacific to USA. It crosses Mount Shasta, Reno, Zion
National Park (Utah), much of the Navajo Indian reservation (including 3 of 4 of their most sacred
mountains-- San Francisco Peaks, Navajo Mtn, and Mount Taylor). Annularity over the cross of the four
corners cross. Chaco Canyon, Santa Fe and Albuquerque. Goes just north of White Sands Atom Bomb
Site (87%). Annularity over Roswell, home of real-life space aliens!

The path enters West Texas and ends at sunset over George W. Bush's boyhood home of Midland. It
has been 10 years since US forces invaded the Middle East under his Presidency and 21 years since
they did so under his Father's Presidency.

During the day May 20 2012, President Obama hosts the NATO Summit in Chicago, Illinois. The theme was "smart defense". Chicago has 49% coverage.

5-20 is the day Columbus died in Spain in 1506, supposedly murmuring his last words "Father into thy hands I commit my Spirit", also some of the last words of Jesus Christ (Luke 23:46).

5-20-2012 is the 520th anniversary of the discovery of America. Columbus died in 1506 on 5-20. 506 years later is eclipse, the 520th anniversary of discovery of America. Numerically odd.

2012 Nov 12 Darwin Australia Total (Saros 133) Speaking of numbers, an eclipse on 11-12-2012. Total solar eclipse begins at sunrise just outside of Darwin Australia-named for the famed evolutionist who was born 200 years and 9 months before on Feb 12 1812. He published Origin of Species on Nov 24 1859, 153 years minus 11 days before. The eclipse travels across the extreme northern coast of Australia through city of Cairns and over the Great Barrier Reef.

2013 May 10 Queensland Cross/ Solomon Islands Ring of Fire (Saros 138) 177 days later Australia gets a second eclipse-again barely any land. Apex in Queensland Northern Australia ...exiting over the Great Barrier Reef and into the Solomon Islands, this time again perfectly passing over the wreckage of John Kennedy's famous boat PT 109. near the island of Kolombongara. Exactly 30 eclipse years after the same location was crossed by a total eclipse on 11-22-1984, the 21st anniversary of JFK's murder.

2013 Nov 3Africa Hybrid (Saros 143) Hybrid solar eclipse. A very thin path of totality crosses the Atlantic (19% coverage at Charleston South Carolina at Sunrise, enters Gabon just south of Libreville and crosses the continent (Congo, Kenya, Ethiopia) . Totality path becomes tinier as it goes but does manage to include Gulu, Uganda. It has only a few miles width of true totality as it dies over Somalia.

2014 Apr 30 Antarctica Ring of Fire (Saros 148) On Hitler Death Day Metonic 69 years after his suicide in Berlin. Two weeks after the Blood Moon Total Lunar Eclipse on Passover of April 14/15. Area of true annularity smaller than map suggests.

partial occurs 10-23-2014 (Saros 153)

Chapter 13

The Nineveh Clause
And Eclipse Art

Eclipses of 2034 to 2053 (and a few beyond)

The Nineveh Clause- A Second Chance?

Eclipses of 2034 to 2053 plus, a look at a few scheduled later eclipses.

"The word of the Lord came to Jonah son of Amittai: Go to the great city of Nineveh and preach against it, because its wickedness has come up before me."

Jonah 1:1

The Prophet and the Eclipse: Nineveh

As the storm raged around them, the prophet was thrown into the heaving ocean from the deck of the ship. He bobbed briefly to gasp for air and then sank fast into the waters. He spun around and around in the swirling sea, sinking deeper into darkness. As he sank, he remembered how he had fled from God.

Just days before, he had been told by the voice of the Lord, go warn the great city of Nineveh that God was going to destroy them. Nineveh was a metropolis of violence, immorality and idolatry. They were mortal enemies of the Israelites. Jonah did not want to warn them of anything. As far as he was concerned, destruction is what they deserved.

So, the prophet fled from God aboard a ship bound for the distant port of Tarshish. Once at sea, a violent storm had come upon them and threatened to destroy the ship. When the sailors aboard the ship realized the prophet was fleeing from his God, they became frightened. Jonah told them they must throw him overboard to save their lives. At first they resisted, but finally they complied. As soon as he hit the water, the storm ceased.

The Lord prepared a giant fish to swallow him. Gulp. In the belly of the fish. Was he dead? Maybe. Three days and nights in darkness. Was he between dimensions, between dreams?

Between heaven and the deep blue sea. Finally he cried out in a prayer which ended:
"What I have vowed I will pay. Salvation belongs to the Lord." Then the Lord commanded the fish, and it vomited Jonah out on dry ground. -Jonah 2: 9-10

No man had ever lived through such a thing or ever would again. His mind was blown. His skin and hair bleached whiter than an albino. Perhaps astonished Assyrians saw the fish belch him out, and knew it was an act of God. Frightened and superstitious, they would lead him across the desert to Nineveh, nearly 200 miles from the coast of the Mediterranean Sea. They were fed by strangers who may very well have been angels. They found wells of water provided by the Creator of all things. What went through his mind and heart, no one knows.

Finally, he entered the great city. Word spread he had been spit from a fish. Jonah walked one day across Nineveh crying the message of God:

"Yet Forty days and Nineveh will be overthrown." (Jonah 3: 4)

As he reached the center of the city beside the King's palace, the Hebrew man grew a large crowd. His bizarre appearance gave him an air of otherworldly authority. Even officials from the King's palace came there, along with the court magicians and astrologers and soldiers. They were drawn to his damning words in ways they could not comprehend, like their very souls were caught up in a pantomime of eternal significance. The crowd grew larger.

As he spoke, the people began to notice the light from the sun beginning to change. The palm trees began to cast strange shapes upon the ground. Someone cried out that the sun was being eaten alive. Jonah repeated his same warning. He had already seen darkness and hell and separation from the light of God. Nothing could surprise him now.

The Assyrians were people who worshipped the sun. None of them had ever seen a solar eclipse before. Now the light from their god began to refract and dim everywhere they turned. The Queen, hearing word of the prophet's arrival and seeing the crowd from the palace window, rushed to the scene surrounded by her eunuchs. As she arrived, the people bowed to her. Many were weeping hysterically. Jonah did not bow. He spoke again looking right at her.

"Thus says the Lord, Yet Forty days and Nineveh will be overthrown."

In the lurid glow of impending eclipse, the bleached prophet stopped and gazed into the eyes of the Queen of Assyria. One of the Queen's eunuchs lifted his spear as if to fling it at the prophet, but she signaled him to stop. As the prophet's stare met hers, some unfathomable communication occurred from soul to soul.

The Queen turned towards the palace and saw the outline of the King far off standing on the tower. Just then, the moon completely eclipsed the sacred sun and darkness fell over Nineveh. Jonah went silent. The crowd fell to hushed prayer. The birds and dogs hid themselves.

For the first and last time in any of their lives, they could stare directly at the savage sun, now darkened and surrounded by a holy crown of blue and white light. It seemed the end of the world. It seemed the judgment of a true God, one that had total power over sun and moon. The Queen cried quietly and prayed in the sight of the people. For five minutes it remained like that. Jonah spoke one last time:
"Thus says the Lord, Yet Forty days and Nineveh will be overthrown."

He turned and walked back towards the desert. As he stepped away, the sun began to reemerge from behind the moon, breaking the darkness. The Queen went back to the palace. Jonah left the city and sat upon a hill overlooking Nineveh to await its destruction.
From the Bible:
The Ninevites believed God. A fast was proclaimed, and all of them, from the greatest to the least, put on sackcloth. When Jonah's warning reached the king of Nineveh, he rose from his throne, took off his royal robes, covered himself with sackcloth and sat down in the dust. This is the proclamation he issued in Nineveh:

"By the decree of the king and his nobles: Do not let people or animals, herds or flocks, taste anything; do not let them eat or drink. But let people and animals be covered with sackcloth. Let everyone call

urgently on God. Let them give up their evil ways and their violence. Who knows? God may yet relent and with compassion turn from his fierce anger so that we will not perish."

When God saw what they did and how they turned from their evil ways, he relented and did not bring on them the destruction he had threatened. -Jonah 3: 6-10

Obviously. I took liberties inserting this total solar eclipse narrative into the story of Jonah. It is "dramatization", of course. But I, and many others, believe it probably happened like that.

For millennia, the Book of Jonah was a mystery. The storyline portrays a tremendous about-face repentance from a Pagan empire which is completely unseen anywhere in the rest of Scripture. Why had the pagan capital of Assyria repented so dramatically at the word of a lone (and reluctant) Hebrew prophet? Something big had to have happened. Something besides just a doomsayer stinking like a tuna can. What was so impressive about Jonah?

The explanation for their repentance is left out of the narrative by the scribes. Why? I think we were supposed to think about it right now. It turns out there was a total solar eclipse over Nineveh during the time Jonah lived and preached. That eclipse is almost certainly the reason Nineveh repented.

Jonah prophesied during the reign of King Jeroboam (786 to 746 BC) according to 2 Kings 14:25. It so happens that there was a total solar eclipse over the city of Nineveh in 762 BC. It was calculated in the 19th century by western scientists but was recorded in ancient chronicles as well.

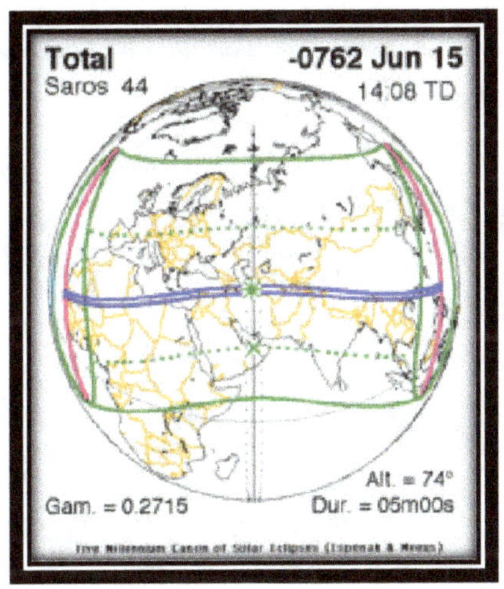

Here is the NASA map of the Nineveh Eclipse. Nineveh would be located along the blue line just to the left of the star (the star is in Northern Iran in the center of the globe which signals maximum eclipse). Nineveh was situated very near modern day Mosul, in the far north of Iraq.

To sum up: Nineveh deserved judgment but avoided it by timely repentance. That is what I call The Nineveh Clause. By all appearances, judgment looms for our modern world. That impending judgment is written in eclipses. Perhaps mass repentance could save us. This possibility appears to be already written into the eclipse schedules themselves. What would that repentance look like? How much longer if we repent? A few years? A few decades? Who knows? Nineveh had forty days to turn it around. How long do we get?

Jesus mentions the "sign of Jonah" in two Gospels. "The sign" is understood to mean that Jonah was essentially or literally dead for three days which Jesus says explicitly in one of the Gospels. (Matt 16: 1-4). But now another "Sign of Jonah" comes to light, the sign of a timely total solar eclipse.

As it happens, Nineveh (near modern Mosul Iraq) was in the path of two extraordinary total solar eclipses in Aug 21 1914 (WWI Eclipse) and Aug 11 1999 (Last complete eclipse of Second Millennium). Now an overview of the Nineveh Clause, the eclipses which follow the Judgment Series.
Who will watch them?

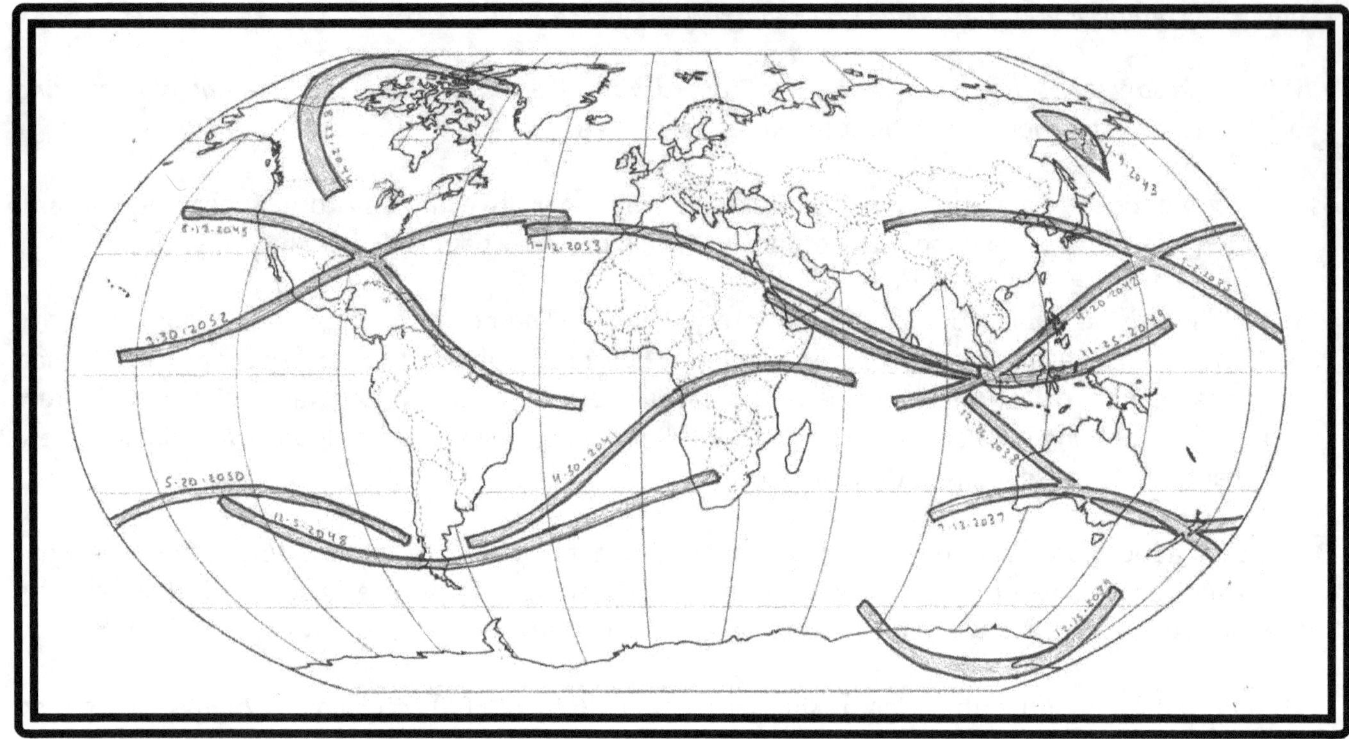

Above: 2034 to 2053 Total Eclipses of the Sun

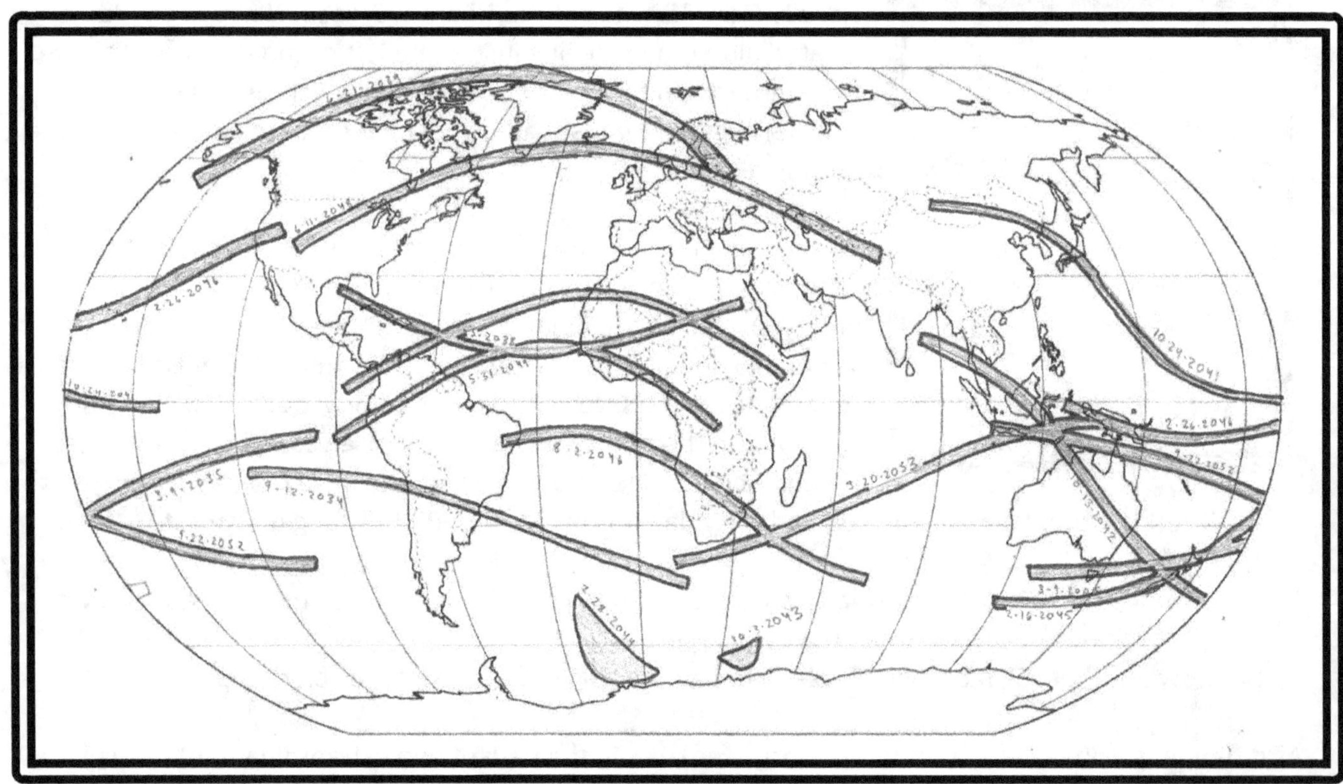

Above: 2034 to 2053 Ring of Fire Eclipses

Individual Eclipses late 2034 to 2053

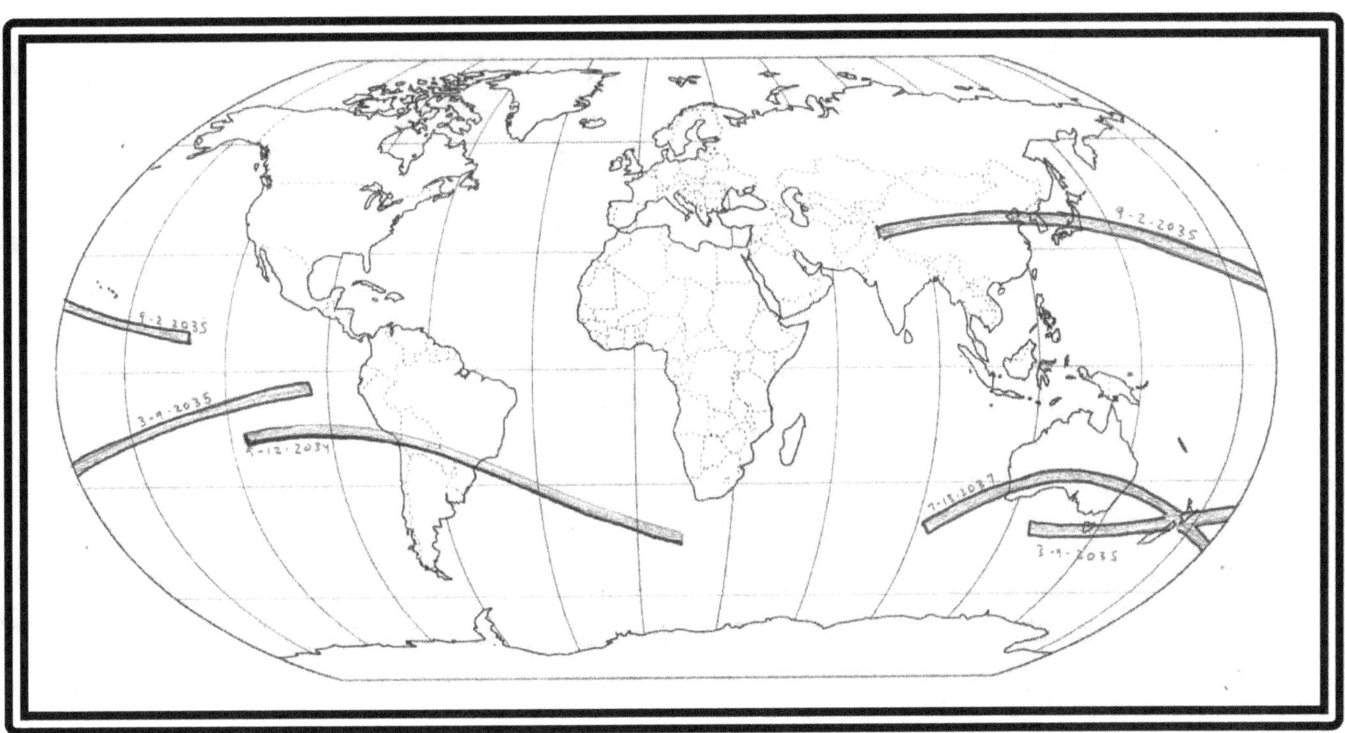

Solar Eclipses Sep 12 2034 to 2037

2034 Sep 12 Ring of Fire South America (Saros 135) After the Spring Equinox New Year Total Eclipse, we have an Egyptian New Year's (day after 9-11) in South America.

2035 Mar 9 Ring of Fire New Zealand/Wellington (Saros 140) Only city in path is Wellington, very small amount of land touched. Part of The Australia Swarm.

2035 Sep 2 Total China/Korea/Japan (Saros 145) A profound eclipse. Beijing, the capital of China, gets its first total eclipse since 1277. Totality in Pyongyang, North Korea and northern metropolitan Tokyo. After this eclipse, there will not be another complete solar eclipse on earth until July 2037. The great Saros 145 continues its run of imperially significant eclipses.

2037 Jul 13 Total Last Monday Hepton Australia/ New Zealand (Saros 127) Saros 127's third to last complete eclipse. Total Solar Eclipse. Hepton Cycle Monday Eclipse. 2422 days from Nov 25 2030 British Empire Eclipse. 856 days from last Australia/New Zealand Ring of Fire in Hepton Two. The last of the seven Hepton Mondays which began with the August 21 2017 USA judgment. Australia nearly perfectly bisected geographically. Totality in Gold Coast and North Island of New Zealand.
3443 years from July 14 1406 BC Holy Land Eclipse as Israelites entered Canaan (one day difference is due to International dateline).

partials occur on **2-27-2036, 7-23-2036, 8-21-2036, and 1-16-2037**

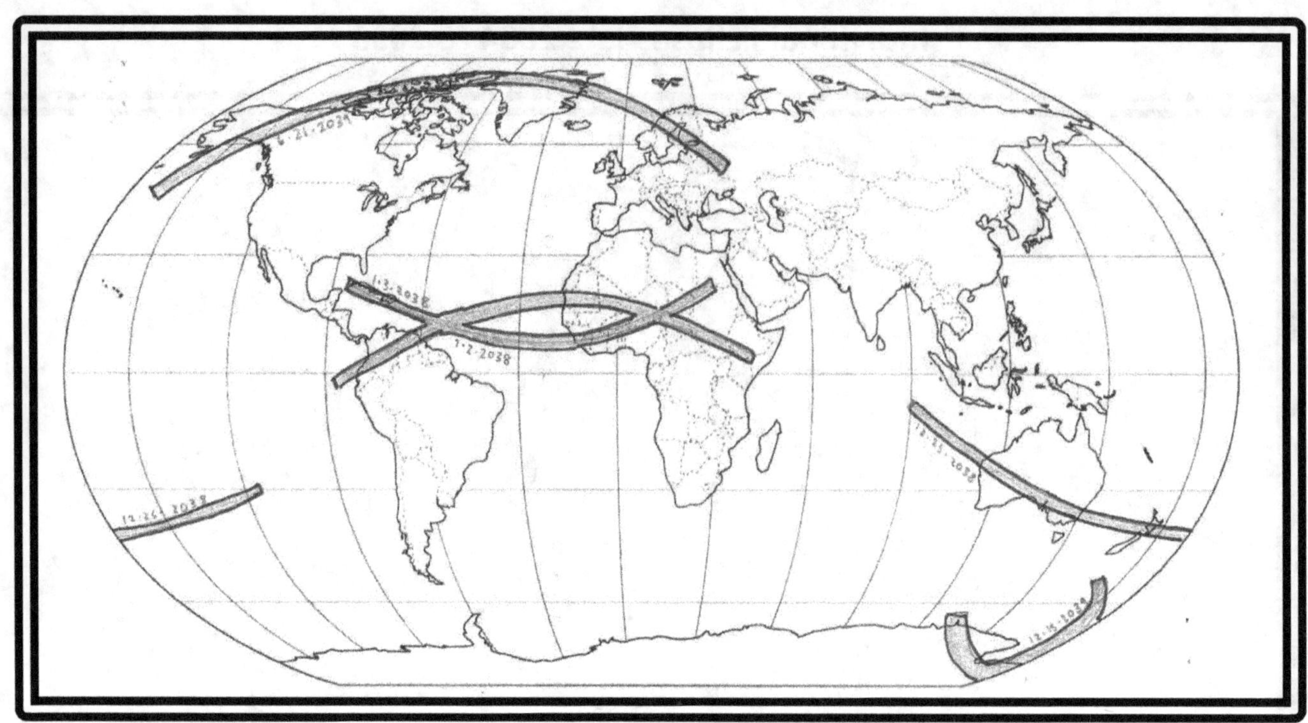

All Eclipses 2038 to 2039

There are three complete eclipses in 2038, and seven all-around eclipses including lunar. (Four penumbral lunar eclipses). These are the maximum possible eclipses in any one year.

2038 Jan 5 Ring of Fire Cuba/ Haiti to Egypt (Saros 132) It's been 119 years since Cuba was covered in such a fashion, and this time its sunrise in Havana. Path reminiscent of 1919 Nov 22 JFK Premonition eclipse. This time the eclipse ends at sunset in SW Egypt. Luxor has 66% coverage. Cities in path include Port Au Prince, Haiti and Kingston, Jamaica. Enters Africa at Liberia.

2038 Jul 2 Ring of Fire South America to Africa (Saros 137) Colombia and Venezuela then across Atlantic and Africa. Ends at sunset just before Mogadishu (63%)

Dec 25/26 2038 Total Australia/ New Zealand Christmas dateline (Saros 142) Most of this eclipse occurs on December 26th in Australia and New Zealand, but a portion of mostly uninhabited South Pacific has the first complete Christmas eclipse in 84 years after the path crosses the International date line. First complete Christmas Eclipse since South Africa in 1954, and first Christmas Total Eclipse since 1666. Thin path of totality manages to straddle North and South Island. Yet another Christmas Eclipse manages to show up in 2057 in Antarctica.

2039 Jun 21 Ring of Fire Solstice Alaska to Russia (Saros 147) Anchorage in totality on summer solstice. Path crosses Greenland and down through Finland (Helsinki in Annularity). Path ends at sunset on border with Ukraine. Kyiv has 62% at sunset. Moscow 83%.

2039 Dec 15 Total Antarctica (Saros 152) Make your travel plans now.

Partials occur on 5-11-2040 and 11-4-2040

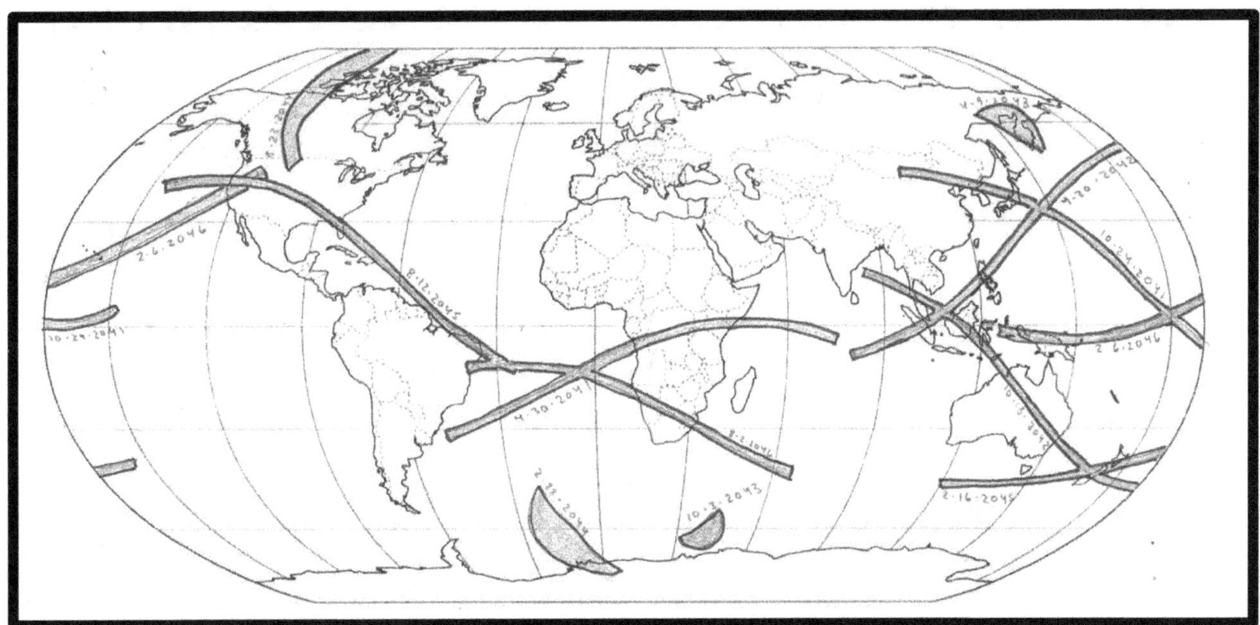

Eclipses 2041 to 2046

2041 Apr 30 Hitler Death Day Total Africa (Saros 129) Totality on Luanda, Angola. Kampala, Uganda

2041 Oct 24 United Nations Ring of Fire Mongolia/ Korea/ Japan (Saros 134) Mongolia. Korea. Japan. Annularity begins at sunrise over Mongolian Capital of Ulan Bataar, crosses Korea and bisects Japan.

2042 Apr 20 Total Sumatra Malay Philippines (Saros 139) 153 years after Hitler Birth on 4-20-1889. The 4-20 numeric is located two times within the date of 4-20-2042.

2042 Oct 13 Ring of Fire Malay/ Australia/ New Zealand (Saros 144)

2043 Apr 9 Total Extreme NE Russia (Saros 149)

2043 Oct 3 Ring of Fire Antarctica (Saros 154) The first complete eclipse of Saros 154

2044 Feb 28 Ring of Fire Antarctica (Saros 121) The last complete eclipse of Saros 121

2044 Aug 22 Total USA/ Canada (Saros 126) First complete US eclipse since 2024. Begins at sunrise near geographic center of North America (Rugby North Dakota).

2045 Feb 16 Ring of Fire New Zealand (Saros 131). Annularity in Wellington, where I was conceived.

2045 Aug 12 Total USA/ Columbus landing. (Saros 136) 87 % in my birth town Flagstaff. 100% in town I grew up, Tahlequah OK, and 100% in town I live, Moab Utah. Also Cape Canaveral (Moon launch)and San Salvador Island, where Columbus landed. Enters California. 2045 is considered the date of the Singularity, where machine and Man become "one" according to the transhuman philosophers.

2046 Feb 5 Ring of Fire New Guinea to USA (Saros 141) intersects last USA eclipse over Redding/ Shasta, California where my truck broke down once and I spent three days on the couch of the service station. Third USA Eclipse in successive years.

2046 Aug 2 Total South America to Africa (Saros 146) Angola again.

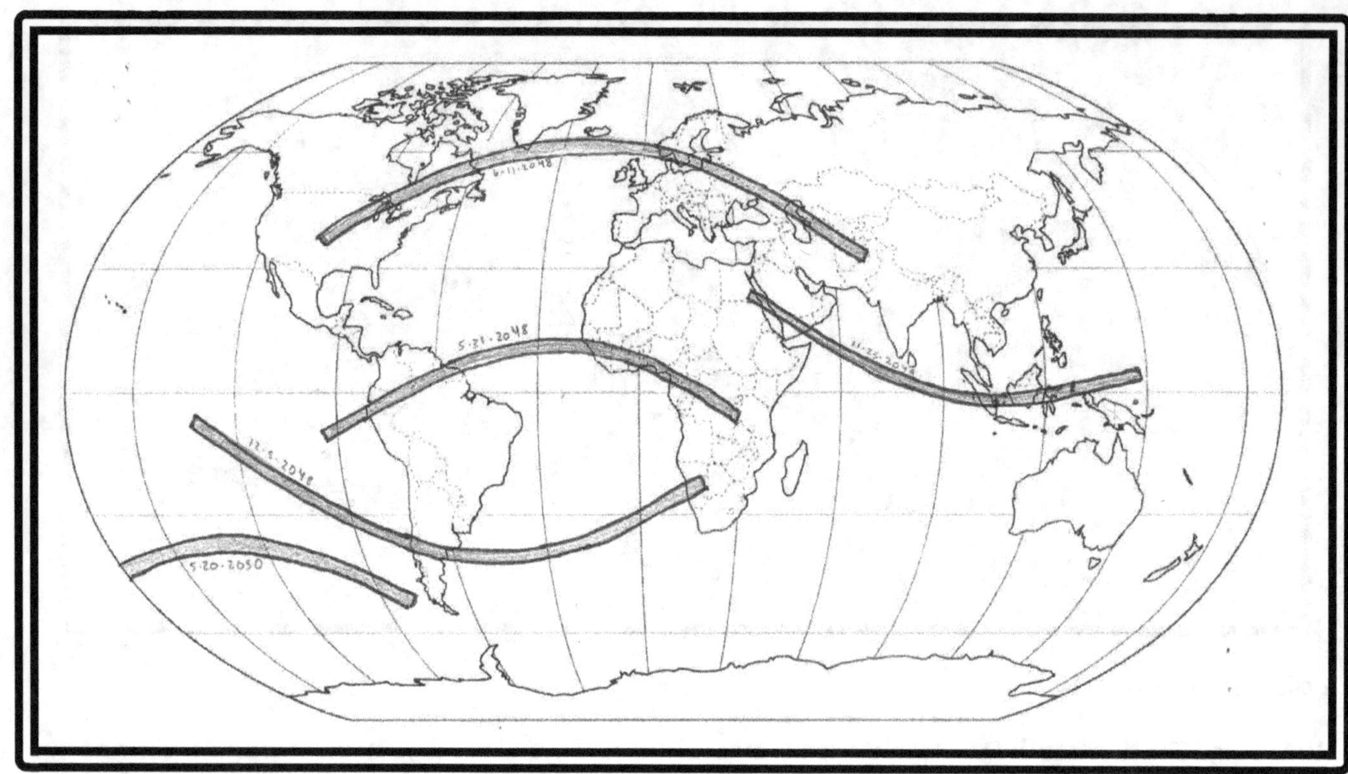

2048 to 2050

None of these eclipse paths touch each other.

2048 Jun 11 Ring of Fire Lincoln to Islamabad (Saros 128) USA at Sunrise over Garden City, Lincoln , Des Moines. Heartland stuff. Into Canada, on to Reykjavik, Iceland. Path follows along boundaries of the Baltic states, central Europe and Central Asia's borders with Russia. Ends just before Islamabad, Pakistan. After 21 years without a complete eclipse, the US has had four eclipses in 5 years. This is the last for four more years.

Lincoln, Nebraska is in the full ring of fire sunrise exactly 193 eclipse years (193.00098 eclipse years) after Lincolns murder on 4-14/15-1865.

2048 Dec 5 Total South America to Africa (Saros 133)

2049 May 31 Ring of Fire South America to Africa (Saros 138)

2049 Nov 25 Total Red Sea to Malay (Saros 143) Mecca 95%. Would have been JFK Jr. 89th birthday.

2050 May 20 Ring of Fire Last Columbus Death Day South Pacific (Saros 148) last appearance of Columbus Death Day Metonic for several centuries.

partials occur on 1-26-2047, 6-23-2047, 7-22-2047, 12-16-2047 and 11-14-2050

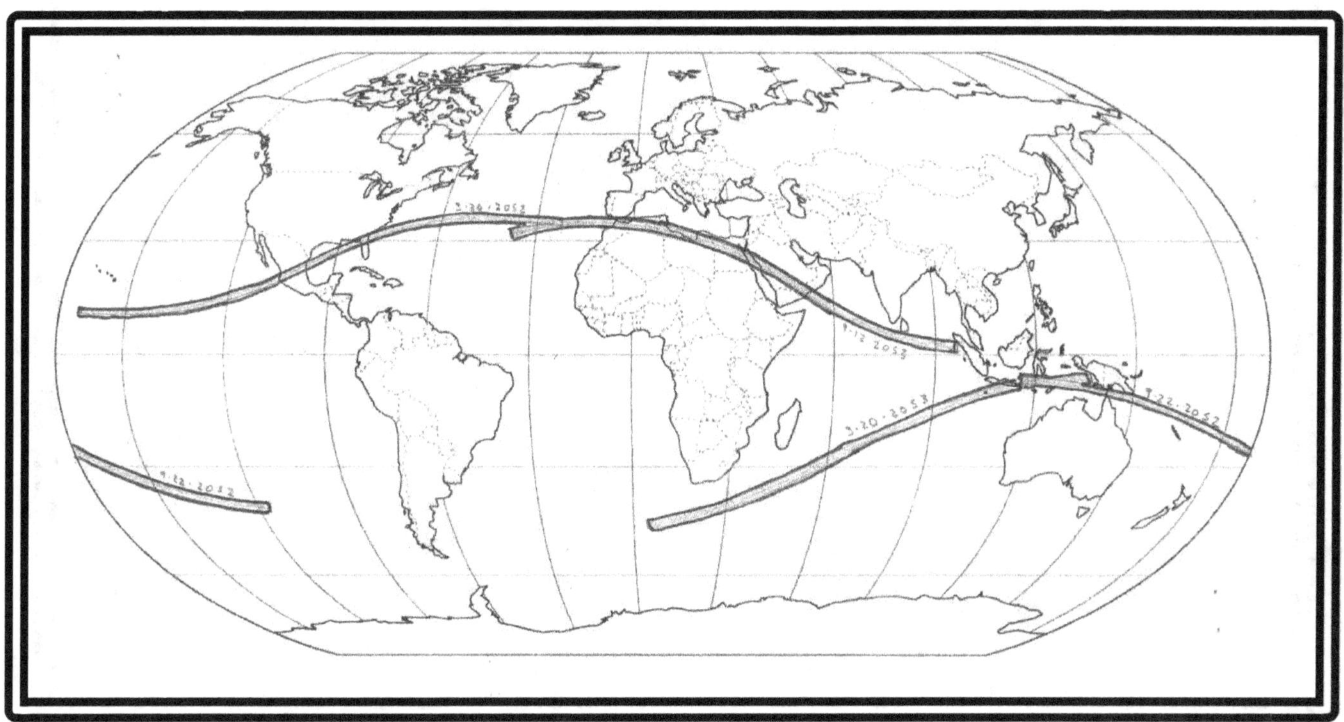

2052 and 2053

These eclipses very nearly wrap around the globe hand in hand.

2052 Mar 30 Total Mexico, USA (Saros 130) Saros 130's first appearance since Mar 20 2034 Spring Equinox Eclipse. Bisects Mexico. 99% coverage in New Orleans. Crosses panhandle of Florida, Savannah Georgia and brings totality to slave port of Charleston, its third total eclipse in 82 years (average should be 1 every 400 years). Second half of 2045/2052 Nineveh Clause Hepton, where the Eclipse crosses switch places over USA and Egypt.

2052 Sep 22 Equinox Ring of Fire New Guinea (Saros 135)

2053 Mar 20 Equinox Ring of Fire Indian Ocean to New Guinea (Saros 140)

2053 Sep 12 Egyptian New Year Total Gibraltar to Sumatra (Saros 145) An almost identical path to the Valley of the Kings 8-2-2027 (Saros 136). Astonishingly, totality goes over the Straits of Gibraltar again, Tripoli, Luxor and northern Mecca (99%). Ends in Sumatra over city of Pekanbaru at sunset.

Partials occur on 4-11-2051, 10-4-2051, 3 -9-2054, 8-3-2054, 9-2-2054 and 1-27-2055

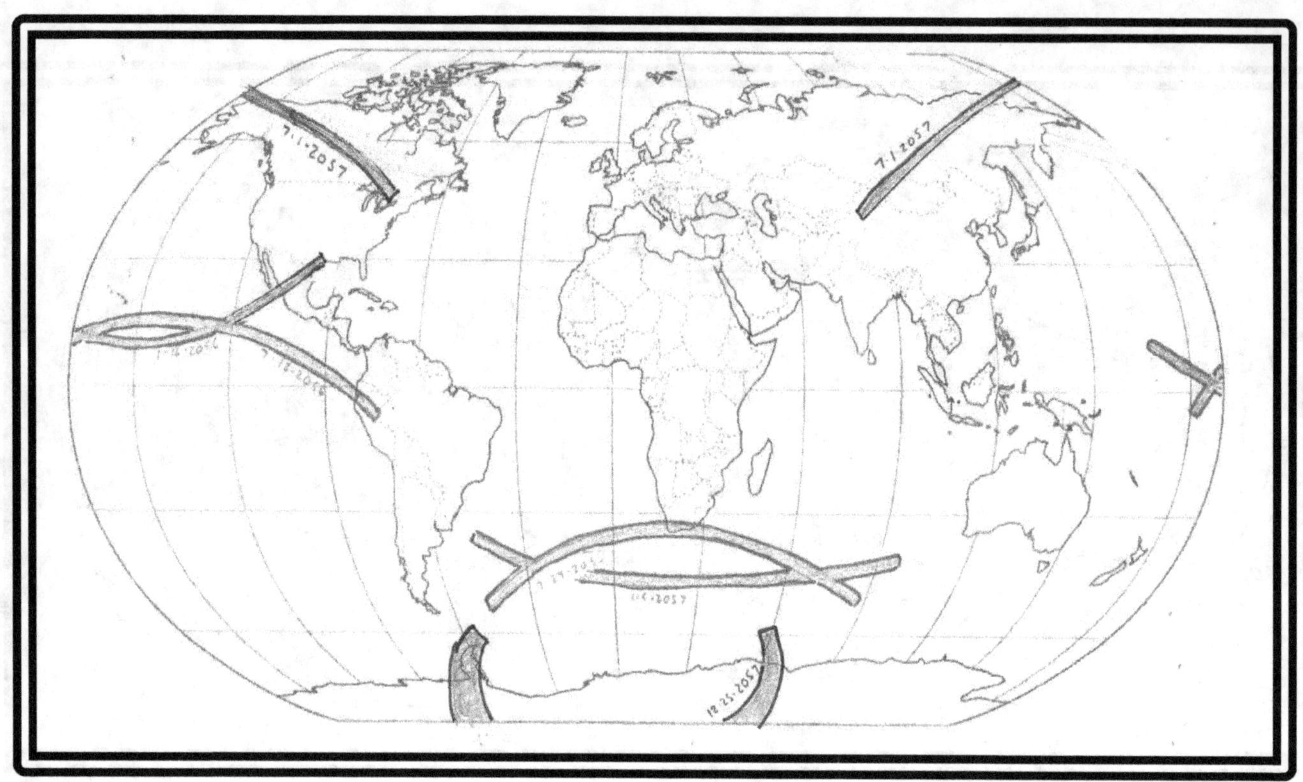

2055 to 2057..Still Christmas in Eclipse Land?

2055 Jul 24 Total South Africa (Saros 127) Totality in Port Elizabeth

2056 Jan 16 Ring of Fire Pacific to Mexico/Texas (Saros 132) Date moving away from 1-15 MLK Metonic, ends at sunset over Austin and Corpus Christi Texas. USA's 7th eclipse in 12 years.

2056 Jul 12 Ring of Fire Pacific to Ecuador/Brazil (Saros 137)

2057 Jan 5 Total South Atlantic to South Indian Ocean (Saros 142)

2057 Jul 1 Ring of Fire China across Russia to US border (Saros 147) Ends almost precisely at Great Lakes

Dec 25/26 2057 Christmas Total Antarctica Christmas Metonic makes reappearance 122 years after first Christmas Metonic appearance in 1935. 123 years is the longest amount of time for any re-occurring date in our survey. It just happens to be on Christmas. Probably as long as any date can possibly stick around in the Metonic Calendar, helped by its repeated appearance near the time-bending South Pole in the long days near its summer solstice.

Partials occur on 5-22-2058, 6-21-2058, and 11-16-2058

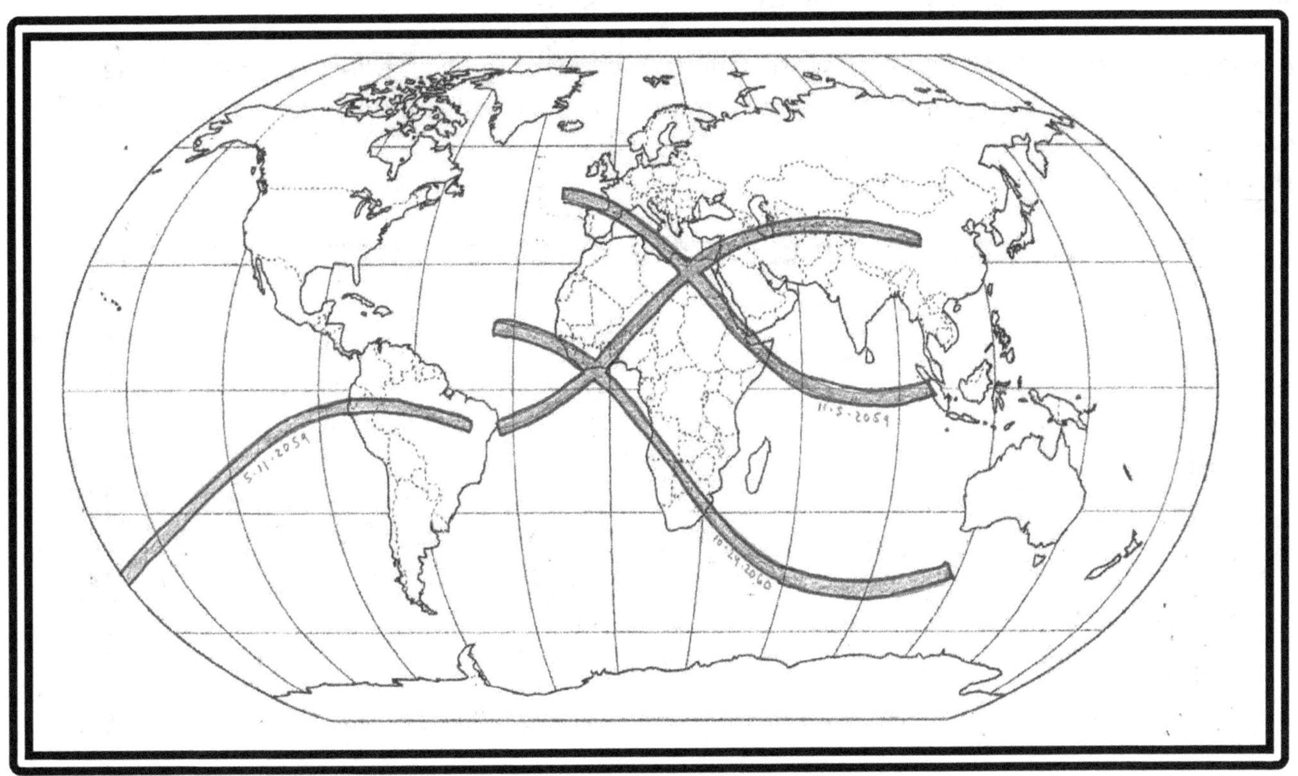

2059 and 2060 Beautiful Pattern and a Palindrome

Sir Isaac Newton supposedly predicted the end of the modern world as 2060, based on calculations from the Bible.

2059 May 11 Total Pacific to Brazil (Saros 129)

2059 Nov 5 Ring of Fire France, Spain, Egypt to Sumatra (Saros 134) Month and day are palindrome with last eclipse, 5-11 to 11-5. Last time that happened was 1987. A 6-month cross. Apex in NW Egypt.

2060 Apr 30 Total Africa/Egypt to China (Saros 139) Very similar to path of Saros 130's 2034 Great Civilization Spring Equinox eclipse. Bisects Libya, crosses extreme NW Egypt. 96 % in Alexandria. Island of Cyprus in Totality-along with Eastern Turkey, Central Asian Republics.

2060 Oct 24 Ring of Fire Africa (Saros 144) Luanda, Angola in Annularity. Johannesburg, Pretoria and Durban South Africa in Annularity.

Eclipse Art: Shifting Patterns of the Great Aleph

A study in patterns. Following the transformation of an Aleph. The first eclipse is Saros 145, the second, Saros 134 and the third is Saros 139. They interact in peculiar fashion. We can follow as the intersection changes over about a century.

We will begin with the Turkish Aleph:

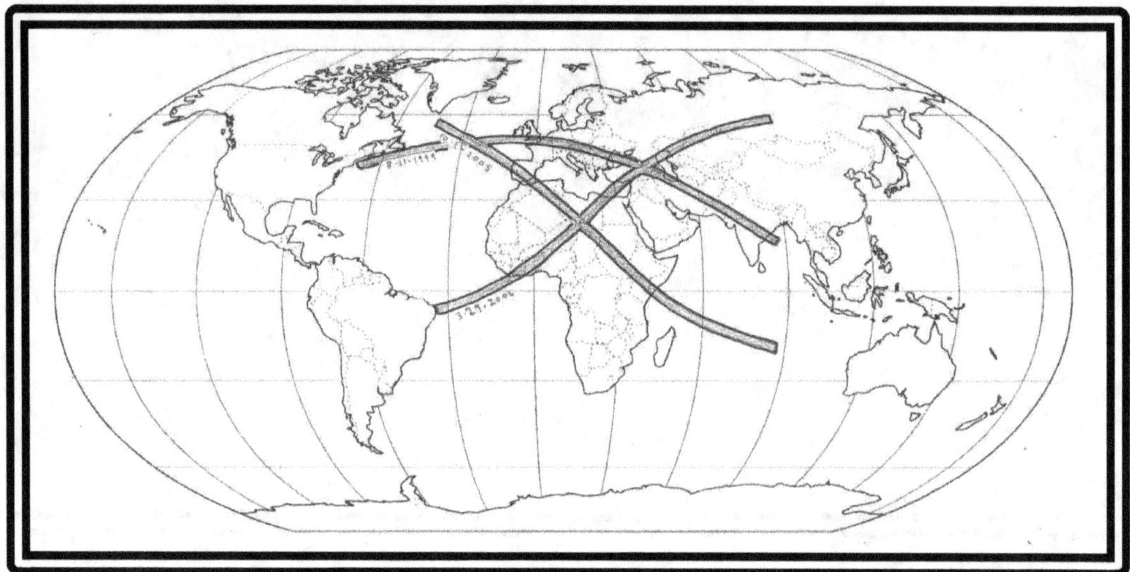

Above: **Turkish Aleph 1999, 2005 and 2006 eclipses,** Apex over Mediterranean Sea

Next comes American Aleph, composed of the same Saros Members returning 19 Eclipse Years later:

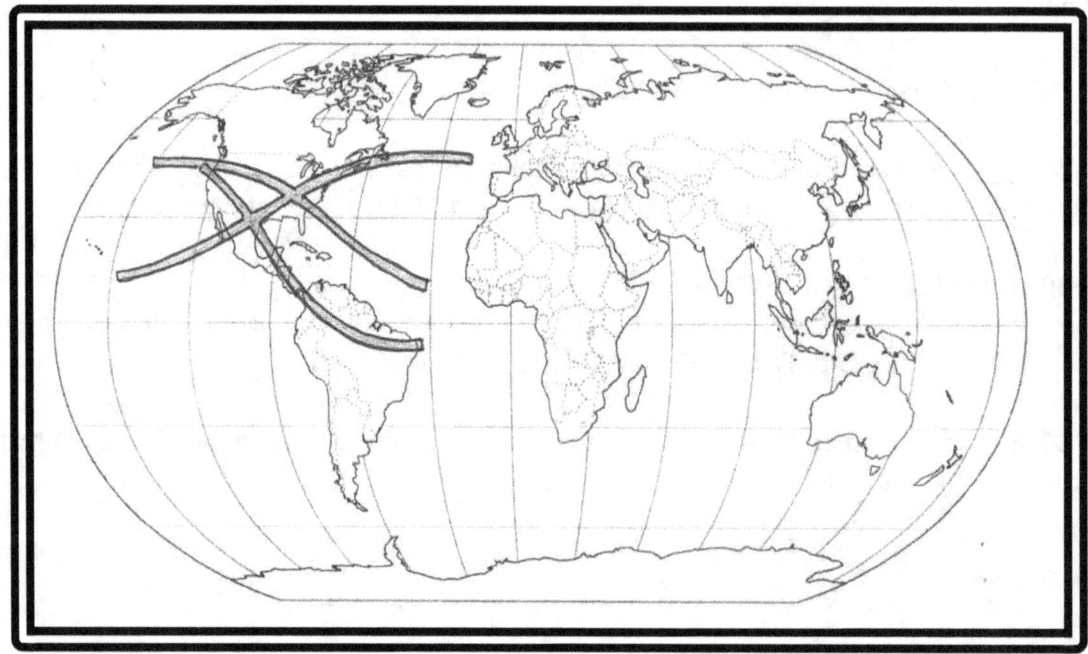

Above: **American Aleph 2017. 2023 and 2024** Apex over the USA

The next Aleph after this one happens over the Pacific 2035 to 2042, Apex is in the Pacific Ocean and unfriendly to map. Let's skip it and go the next one after that.

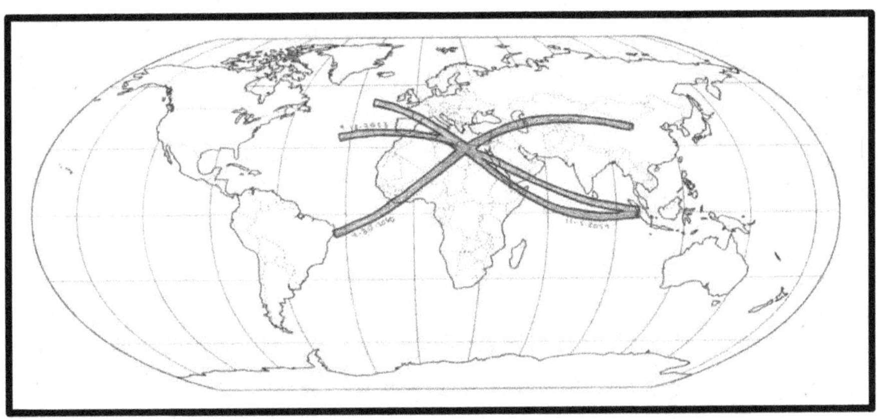

Above: 2053 to 2060 The same Saros as original Aleph:
Sep 12 2053 (Saros 145) / Nov 5 2059 (Saros 134) / Apr 30 2060 (Saros 139)

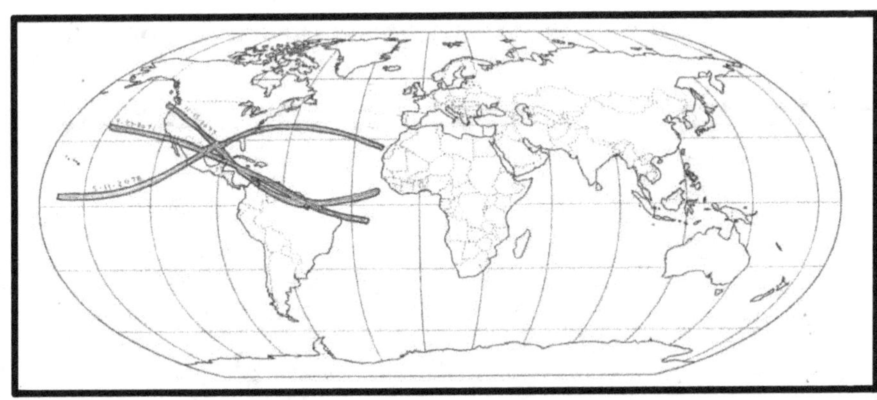

Above: 2071 to 2078
Sep 23 2071 (Saros 145) / Nov 15 2077 (Saros 134) /May 11 2078 (Saros 139)

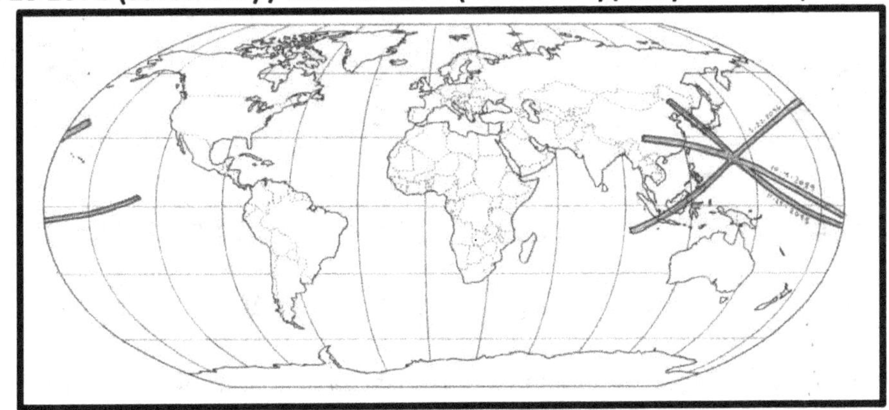

Above: Remnants of American Aleph Hepton 2089 to 2096

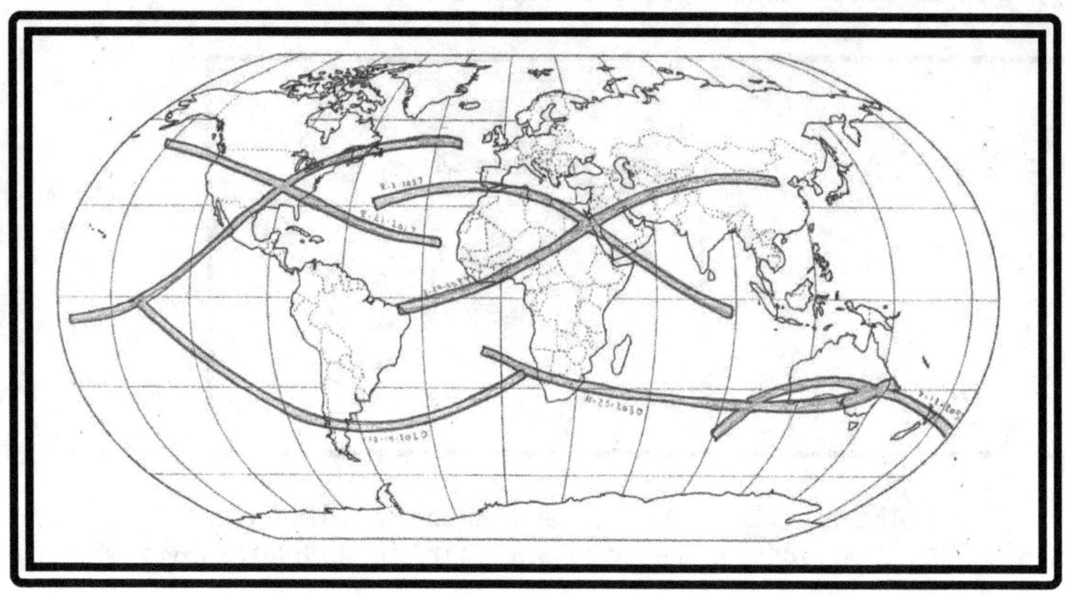

Above: Hepton Crosses of 2017 to 2034 (our current Hepton)

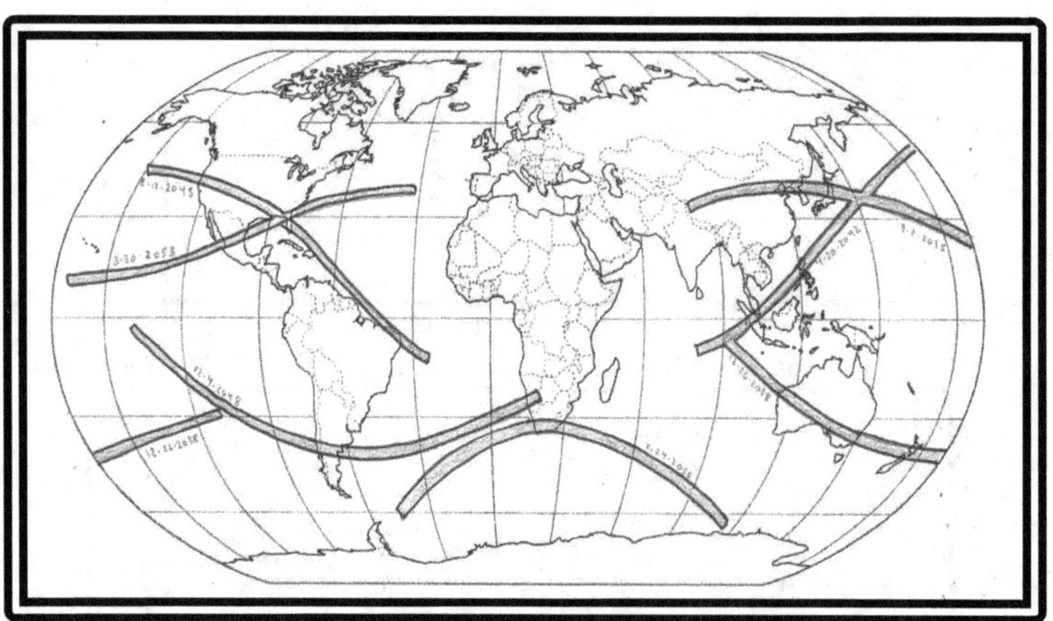

Above: Hepton Crosses of 2035 to 2054

Note Saros 136 and Saros 130 intersecting in both the above maps. In the upper map, they intersect over Egypt creating the great Hepton Cross of 2027 to 2034. In the lower map, they intersect over the USA from 2045 to 2052, apex over Florida Panhandle. The Saros have effectively leap-frogged.

On the next page, let's watch this particular cross of Saros 136 and 130 march across the globe.

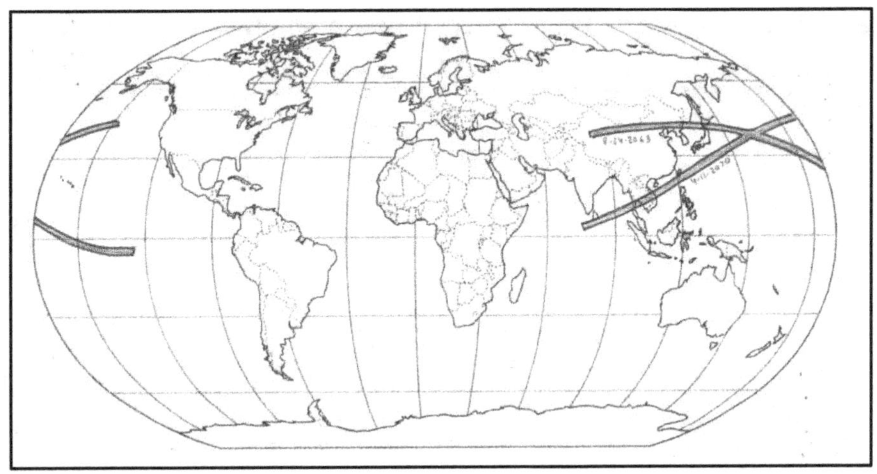

Above: 2063 to 2070 (similar to Pacific cross 2035 and 2042)
2063 Aug 24 (Saros 136) and 2070 Apr 11 (Saros 130)

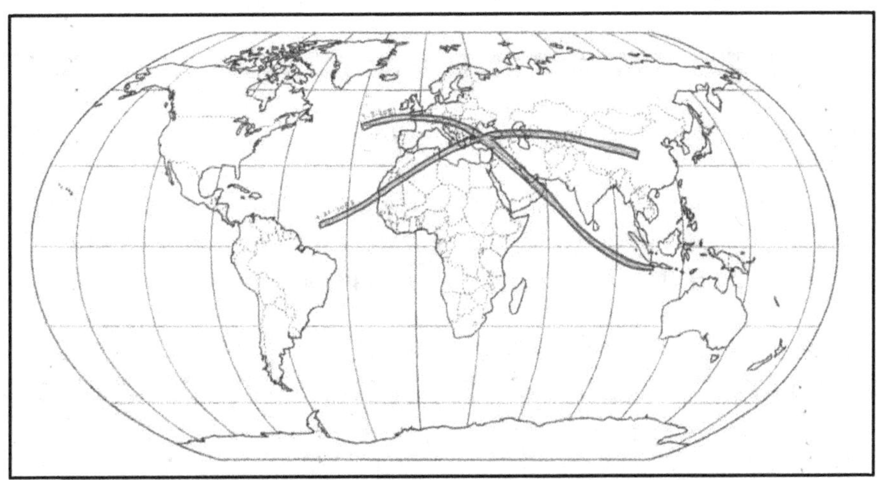

2081 to 2088 (similar to Turkish Cross of 1999 to 2006)
2081 Sep 3 (Saros 136) and 2088 Apr 21 (Saros 130)

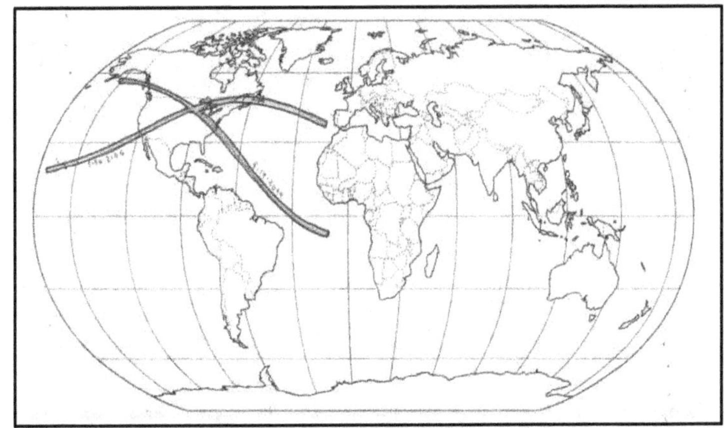

2099 to 2106 (similar to American Cross of 2017 to 2024)
2099 Sep 14 (Saros 136) and 2106 May 3 (Saros 130). Saros 136 is the last eclipse of the 21st Century.
Saros 145 had last eclipse of the 20th (along with Saros 139 making the Turkish Cross)

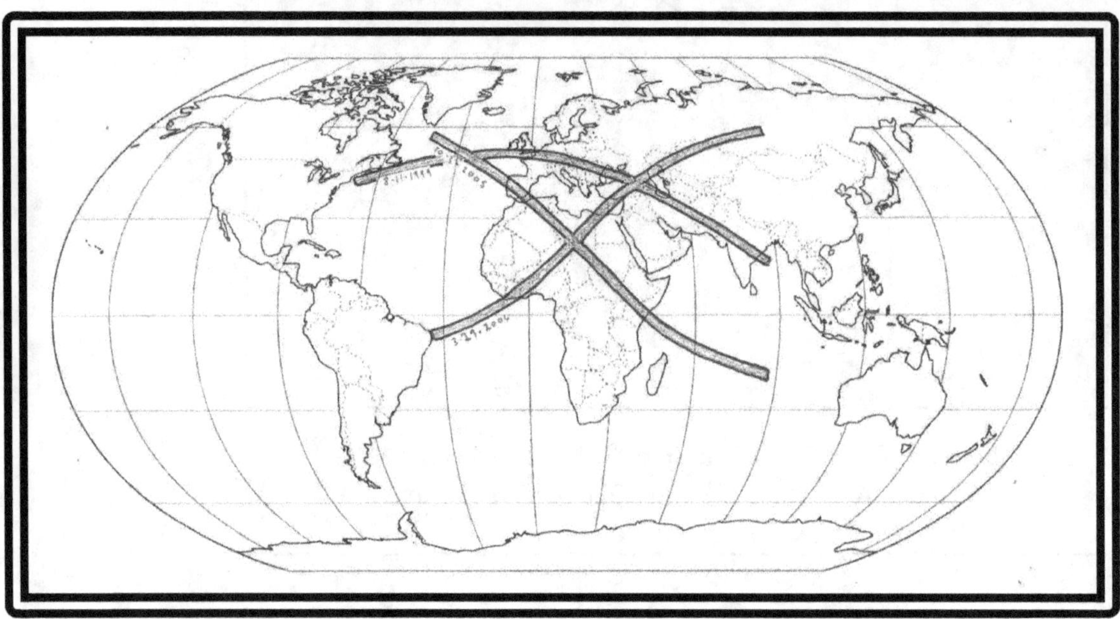

Above: the Turkish Aleph is composed of Saros 145, 134 and 139.

Exactly 100 years and 34 days after the Tav of Aug 11 1999 and Mar 29 2006, a cross occurs over the USA. It is composed of the Saros 136 and 130, the eclipses which had created the Hepton Cross over the Mideast in 2027 and 2034. 100 years and 34 days equals one half of Hepton Cycle!

Here are Saros 136 and 130 appearing over America at the end of the 21st Century.

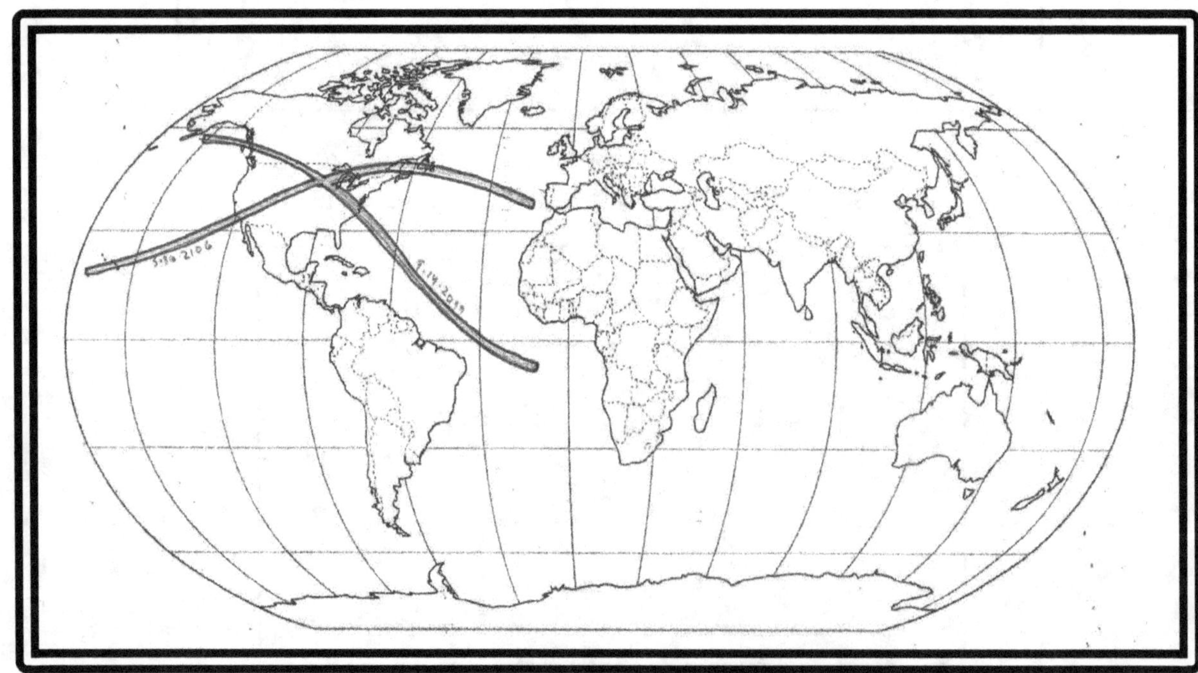

2099 and 2106 USA Tav Saros 136 and 130
2099 Sep 14 (Saros 136) and 2106 May 3 (Saros 130)

There were only **two true Hepton Total Crosses overland** between 1869 and 1999 (130 years)
Both were unbalanced.

Why do I mention balance? Because only balanced crosses resemble the letter x, or tav.
Here are the two overland Hepton true totals in that time frame:

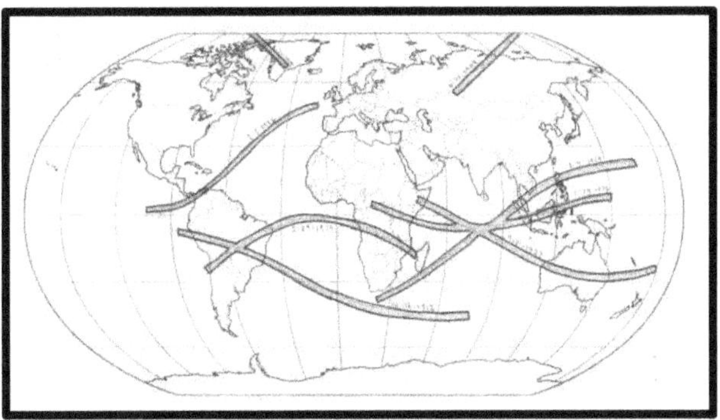

Above: Hepton Cross: 1912 Oct 10 and 1919 May 29

Overland True Hepton Total Cross (lower left of map)-1912 and 1919 (Saros 142 and Saros 136) with a
very unbalanced Apex in the Amazon Rain Forest of Brazil several hundred miles west of Brasilia.

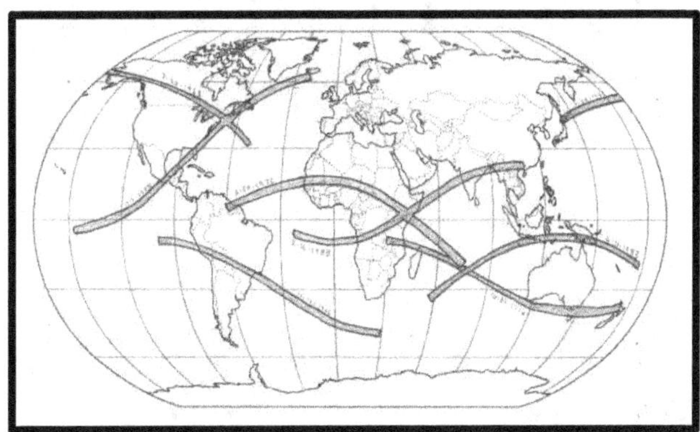

Above: Hepton Cross: 1973 Jun 30 and 1980 Feb 16

The other overland True Hepton Total Cross of the 20th Century occurred between 1973 and 1980
(Saros 136 and 130) and has an apex just west of Dar Es Salaam, Tanzania. As you can see in the map
above, it was also unbalanced. Saros 136 and 130 are beginning to cross in a more balanced fashion.
This pattern is prominent in our current Spring Equinox Series with the Hepton Cross of 2027 and
2034. Off the coast of North America, you can see Saros 145 and 139 cross in Atlantic Ocean.

Summary: two unbalanced overland Hepton Total Crosses on earth between 1869 and 1999. Both
involved Saros 136. They are shown on this page in the above maps. Twelve other Total Hepton
Crosses occurred with apex over open ocean. You can see them all in the next section.

Between 1999 and 2106 there are *Six* **Balanced True Total Hepton Crosses** on planet earth. Three of
them are centered on the USA. Three of them are centered on The Middle East. Three of them occur
between 1999 and 2034. The first over the Mideast, second over USA, third over Mideast.

Chapter 14

The Mystery of Numbers

The Mystery of Numbers and Number 19.

Sure, numbers are there, alright...but do they take you anywhere?

There is allure in numbers, and surprises and synchronicities, sometimes riddles. They pop up when we aren't even looking for them...and we look for them a lot. All of us do. Numbers on a check, on decks of cards, social media views, lottery tickets, bank accounts, clocks. We are this old in numbers. We've been married this long. Anniversaries. This is the number I was born on or my father died or the years I was sentenced to. Numbers.

One day its 7:47 am on the digital clock as you leave for work. Minutes later you're stuck in traffic and URA 747 is printed on the license plate of the car that barges in front of you. Then the guy on the radio says a 747 crashed. Maybe you make a mental note:
"huh...that's odd...sure been a lotta 747's this morning".

Oh yeah, of course, as it turns out, when you get home, you're on page 747 of that interminable mystery novel. And your kid turns seven tomorrow and there's still 47 items left on your to-do list and if you ever cared to write such things down, you'd realize that it's been 747 days since the last time something like this happened. So what? Really. So what?

OK. There's numbers floating around, and sometimes there are swarms of them. But do they take us anywhere? Or do they just seem to be inviting us into the worst kind of maze; a cold mathematical twilight zone, full of meaningless signposts leading nowhere, except perhaps into a tail-chasing descent into distraction...or worse.

The math goes on forever and the numbers never end.

It's surely not accidental and its likely not altogether meaningless. The problem is that it seems impossible to convincingly define their purpose or meaning, either to ourselves or to anyone else. But patterns are patterns, and patterns yearn to be described.

Of course, numbers have a function. We all know that. They made us study them for 12 years after all. Of course, numbers have a mystery. Most of us likely suspect that too. But all should take heed: infatuation with numbers can be extremely dangerous. Only the most balanced of personalities can truly indulge (or attempt to express) the fractal scope of numbers without becoming a gibbering idiot.

Why do I say this? Because I know. These last three years, I have somewhat inadvertently been drawn into the belly of the proverbial number beast. I had to. Eclipses are swimming in numbers. Their cycles and peculiarities are based in repetitions of time. Try describing time without numbers! It can't be done. So, on you must go. As soon as you enter the howling wind that blows through the four-dimensional kingdom of numbers, associations and synchronicities and equations start buffeting your fragile mortal mind.

I have returned to tell about it. Whew! My mind has emerged (still hinged, apparently) from the madhouse chatter of lonely numbers desperate to be heard.

If I had given voice to all the numbers that appeared as I wrote this book, the ones that tried to follow me to bed, babbling about how important they were, this work would be a diabolical zoo of digits. This book would be worse than worthless. It's got too many numbers as it is.

In the personalities section, you will read the cautionary tale of Norman Bloom, the "numbers man". Once a promising musician in New York City, perhaps even a genius. After some business and artistic failures, he began to take interest in numerical patterns. It drove him mad. He began delivering staggeringly complex and often incomprehensible numerical tracts to world leaders, university professors and scientists. He was a regular caller to the radio host Larry King, who later mourned his death on air. Carl Sagan wrote an essay about him.

When the homeless Bloom died at 62, found dead in his ice cream truck, he left his family (whom he had long since abandoned) a storage unit stuffed with hundreds of boxes, each of them crammed with thousands upon thousands of papers- endless tracts jammed with number associations, astonishing theories based on numbers from the Bible, baseball scores, planetary distances and yes, license plates. Bloom's journey into madness is roughly outlined in the personalities section of this work.

Yet, in spite of the dangers and the weirdness, a certain level of the fractal synchronicity present in all things is reflected in numbers. They can't be ignored, but I have yet to read anyone who makes true sense of it. For myself, I came to the decision that I won't ignore them, but I'm not going to let them boss me around either. I can't pretend they aren't there. Obviously, numbers create order. But are they real? They're everywhere, it seems. But are they real?

I decided to approach the number phenomenon and offer them the dignity they maintain in music. Numbers in music represent beats or steps. They are not rhythm, but they describe rhythm. They are not chords, but describe chord progressions and arrangements (i.e. G7). We can agree music is in 4/4 or ¾ time. We can agree a major chord is composed of the 1st, 3rd and 5th notes of an octave.

When we listen to music, we notice a beat, we feel it, we may even groove to it. Maybe a few of us search it out and think about the beat for a while. But when we try and stop a song midway through to analyze the beat, what are we doing besides stopping the music? This is the danger of over analyzing and ascribing meaning to numbers themselves. They are fundamentally a heartbeat. You are supposed to move on. They are a foundation. They need to flow unimpeded. To turn away from a child, a flower or an eclipse to rant how your phone number matches the playoff game? It's possible with number addiction. Absurdity is always just a few numbers away.

Important!
It's not the numbers themselves that are important. It's the fact that the numbers are there. They are indeed making true patterns. We're not just making it all up at the casino. There really have been some honest 777's, jackpots in the dream of existence. Numbers swirl and clump and dance and disappear and reappear. I will throw some of them on the canvas like watercolor, because I would be a liar if I said they weren't there.

Numbers are all over eclipses. I make a note of their peculiarities sometimes, smile and then move on. But the fact they are there, patterned in some divine backdrop of mystery, does mean a lot to me. For myself, they are best scanned quickly and then returned to for a brief lark. They are funny, after all. They are musical beats. Or maybe echoes of a mathematical cosmic heartbeat or threads in a woven holographic dream. It's a mysterious language that God knows. I don't know it. I just see it's there.

Pythagoras

Pythagoras is considered the first Greek "philosopher" and first mathematician, one of the earliest teachers of ancient Greece. Born in 580 B.C., Pythagoras became one of the most well-known men in history. He influenced Socrates, Plato, and Aristotle. Pythagoras mixed math with spirituality and science. He is the author of the Pythagorean Theorem. It's often claimed that the word "philosophy" was invented by Pythagoras.

Pythagoras taught that reality is fundamentally mathematical in nature. Pythagoreans believed a system of divine principles existed behind numbers. The principles form a foundation for many concepts of traditional Western thought. One of these principles revolves around the notion of Divine Proportion, where all fundamental shapes can be found within a circle, or sphere.

Now, I turn to peculiarities involving the number 19.

Number 19

When I was a young man, in either high school or college (I can't distinguish now), I had a very strange dream. In the dream, I was a patron seated at a table in a shadowy music venue. Onstage, in the spotlight, was the famous silver-maned singer Kenny Rogers ("*The Gambler*" etc...), who was quite famous when I was a kid. In the dream, Kenny was singing a song on a small stage. I felt the sensation that this song was somehow important, so I watched and listened intently as he crooned.

Someone at my table was trying to distract my attention from Kenny, but I resisted the urge to turn away. He seemed to be singing about what to do when all else fails. Finally, Kenny Rogers, the Ol' Gambler himself, turned to face me directly. Okay. Here's Kenny with the microphone! He's telling me something I need to hear. The suspense was killing me.

The melody remains with me to this day though I only heard it once in a dream. I could sing it for you right now. Kenny sang this to me about what he does when all else fails:

"I turn to number...*nineteen*."

Then I woke up. Huh? Kenny Rogers said what? I filed that odd dream away, but I never forgot it.

I'd had never given much thought to number 19 before that dream. I never gave 19 much thought after it either. I have gone most of my life without giving any consideration to the number 19, or any particular number, really, for that matter. Then I began to put this book together. It turns out that there are peculiar associations with the number 19. I'll admit right now I don't have the "meaning" for number 19, but here's some peculiarities.

19's power as a Prime Number is not the subject of this chapter.

The number 19 has unusual properties in mathematics, based on its status as a peculiar prime number. A prime number, of course, is one that is only divisible by itself. This is not the place to learn about any of those interesting relationships, which can be garnered from a quick internet search.

I am more interested in the Number 19 as it relates to astronomical patterns, particularly its appearance in eclipse studies.

Here are the three most important eclipse relations of 19.

1. A Saros cycle is exactly 19 eclipse years (19 x 346.62) The Saros are the main characters of eclipses, the particular geometric eclipse identities who return regularly to earth.
2. A Metonic cycle is exactly 19 solar years. (19 x 365.24) The Metonic cycle is the place where the solar, lunar, and eclipse calendars intersect every 6940 days.
3. The Saros Cycle equals 19 solar years minus one lunar year.
4. An eclipse year (346 days) is 19 days less than a solar year (365 days)

Here are some peculiar subsidiary patterns related to eclipses and the number 19.

- the first four digits of an eclipse year (346.6) when added make 19. 3+4+6+6=19
- the first four digits of a lunar month (29.53) when added make 19. 2+9+5+3=19
- the first four digits of the anomalistic month, moon's perigee to perigee, 27.55 when added also make 19. 2+7+5+5= 19
- 19 lunar years, or (19 times 354.37 days) equals exactly 6733 days 6+7+3+3= 19
- The digits of the number of days in the Metonic Cycle is commonly rounded and expressed as 6940, which itself contains the number 19 two times. 6+9+4+0=19 as well as 69+40=109
- The length of the draconic (eclipse month), where the moon passes one node to the next has been calculated at 27.2122210 (*year 2000 Eric Weisstein's World of Astronomy*). Those digits when added make 19. 2+7+2+1+2+2+2+1=19
- The Saros Cycle is 18 years 11 days and 8 hours. Look at the numeric. 18+1 from the front=19. 8+11 from the rear also equal 19.

Perhaps some other properties and peculiarities might interest the reader:

Peculiar Math. The casting out of nines and the number 19

Both the number one and the number nine are the only two numbers that can be found by equation within all numbers without introducing any other numbers than themselves. Let me try and explain:

The number one is found everywhere. Any number divided by itself is one. No other number is needed to be brought into the equation. 147 divided by 147 equals one.

The number nine is found in all numbers in the process of the ancient mathematical concept of the "casting out of nines".

The number nine is obtained by adding the individual digits of any number greater than nine, then subtracting their sum from the original number, and then adding the digits of the new sum until you are left with an individual number. You will get to 9. You'll see.

Here's how that works.

      1) pick a number greater than single digits, perhaps your birth year: for instance: 1984
      2) now add the individual numbers: 1+9+8+4 = 22
      3) subtract that new sum from your starting number: 1984 -22= 1962
      4) add the new digits: 1+9+6+2= 18...
      5) then add 1+8 = 9

whatever number you pick will eventually add back to 9 given this method. And yes, it only works for nine. Try it with your birth year. It's a fun party trick.

19 times 19 equals 361, far and away the closest number to the solar, lunar and eclipse year of any two whole numbers multiplied by themselves. The 361 day year is only four days removed from the 365 day solar year, and is used by the Bahai Calendar. The 360 day year was common in antiquity, and its properties of 60 are still reflected in our hours, minutes and seconds.

Turn that 361 (19x19) upside down and it starts with 19. 3 times 6 + 1 also equals 19.

19 in The Bible

Psalm 19 (from David) is the most famous psalm to assert that God's existence can be determined by observation of the astronomical universe.

    The heavens declare the glory of God; the skies proclaim the work of his hands.
    Day after day they pour forth speech; night after night they reveal knowledge.
    They have no speech, they use no words; no sound is heard from them.
    Yet their voice goes out into all the earth, their words to the ends of the world.
                               Psalm 19 1-4

The Psalms are the only part of the Bible that has been numbered from antiquity. The Psalms have retained their order for perhaps 2500 years. Psalm 19 has likely always been Psalm 19.

Psalm 119 is both the longest psalm and the longest chapter of the Bible. It has many unusual properties as well.

Chapter 19 of Genesis, the first book of the Bible, contains the destruction of Sodom and Gomorrah. (note: chapter numbers other than psalms became part of the Bible about 500 years ago)

Chapter 19 of Revelation contains the destruction of Satan and the final victory of God.

Isaiah Chapter 19 is the Judgment against Egypt

Luke Chapter 19 contains Jesus triumphal entry into Jerusalem

John Chapter 19 is the Crucifixion of Jesus. John 19:19 is the following:

Now Pilate wrote a title and put it on the cross. And the writing was:
JESUS OF NAZARETH
THE KING OF THE JEWS
                    John 19:19 (NIV)

- Each year it takes 19 days to observe all of God's annual Feast (Holy) days. These commanded yearly celebrations are the Passover (1), Days of Unleavened Bread (7), Pentecost (1), Day of Trumpets (1), Day of Atonement (1), Feast of Tabernacles (7) and Last Great Day (1)

- Discovered in 1947, the Dead Sea Scrolls were found east of Jerusalem. Among all the scrolls unearthed over the years, 19 copies of Isaiah the prophet have been identified.

- Mary (Jesus' mother), is mentioned 19 times.

- Two of the most important numbers in the Bible (and the world), are 7 and 12. When added they become 19. These are also the most important time keeping numbers. The 7 day week and the 12 month year. The typical clock has 12 visible hours.

From the Quran:

- 19 is the number of angels guarding Hell ("Saqar") according to the Qur'an: "Over it are nineteen" (74:30), after which the Qur'an describes the number as being "a trial for those who disbelieve" (74:31), a sign for people of the scripture to be "convinced" (74:31) and that believers "will increase in faith" (74:31) due to it.

- The Number of Verse and Sura together in the Qur'an which announces Jesus son of Maryam's (Mary's) birth (Qur'an 19:19).

- According to a few Islamic scholars, The Quran appears to have an unusual mathematical structure based on the number 19 ("Code 19"), which they argue proves divine authorship.

- The first verse of nearly every Surat in the Koran, the section known as Bas-malah, consists of 19 Arabic letters .

Bahai Faith:

- In the Bábí and Bahá'í Faiths, a group of 19 is called a Váhid, a Unity. The numerical value of this word in the Abjad numeral system is 19.

- The Bahá'í calendar year contains 19 months of 19 days each (along with the intercalary period of Ayyám-i-Há), as well as a 19-year cycle and a 361-year (19x19) supercycle.

- The Báb and his disciples formed a group of 19. There were 19 Apostles of Bahá'u'lláh, the Prophet of the Bahai.

A few more curiosities surrounding 19:

- The two most rapidly changing and bloodiest centuries in history were the 19th Century (1800's) and the 1900's (20th Century)

- Covid-"19" changed the world. The Spanish Flu's worst year was 1919.

- Tarot card number 19 is the Sun. Which I'm obliged to mention though I don't use tarot cards.

- The Saros Cycle of 19 eclipse years covers 18 years, 11 days and 8 hours. The number 19 is found forward and backward in that number with addition of three integers in either direction.

- The last eclipse of the 1900's took place on 8-11-1999, with a path that crossed numerous great civilizations throughout the Middle East and Europe. 8-11-1999 can be deconstructed as month and day 8+11= 19...year 19+99= 118...11+8= 19 etc...

- The Saros that brought the USA 2017 eclipse was Saros 145. 14+5 =19. The Saros which brings the Valley of the Kings eclipse in 2027 is Saros 136...13+ 6 =19

- 1919 saw two profound eclipses. Einstein's Theory of Relativity was proved by observation of the 1919 May 29 eclipse (JFK's second birthday). The 1919 Nov 22 eclipse began at sunrise over Texas 44 years to the day before JFK was murdered in that state.

- 5-14-1948 is the date of modern Israeli Independence. 5+14= 19. 19 years from 1948 brings you to 1967, when Israel took control of Jerusalem for the first time since 70 AD. 19+48 also equals 67, for good measure. 5x14x1948 = 136,360. 136 and 360 each have separate significance. Also you can add the digits and get 19.

- Lincoln's death day and Titanic sinking day of 4-15 both equal 19.

- King Tut was 19 when he died. The oldest known calendar in the world, the Egyptian, has its new year on 9-11. The Egyptian Revolution against the British happened in 1919.

- The US (or what became the US) first brought slaves to its shores in 1619. The last slaves in the USA to learn of their freedom were in Texas on 6-19 (Juneteenth).

- I've always liked that Steely Dan song, "Hey Nineteen". It stayed on the Billboard Hot 100 charts for 19 weeks back in 1981.

More Number Peculiarities- A quick look at Double Digit Repetition

There are consistent patterns of repeating double digits in the eclipse phenomenon.

- semester series eclipses (which occur within the same year) are 177 days apart.
- Lunar year eclipses occur usually 355 days apart.
- Hexon eclipses occur 1033 days apart.

- Hepton eclipses occur 1211 days apart.
- Heptons repeat in their cycle at 2422 days ,3633 days, 4844 days, and 6055 days. An entire Hepton Cycle takes 7266 days.
- Octon eclipses occur 1388 days apart,
- Back to back Octons become 2777 days or 7 years, 7 months and 7 days apart.
- Tzolkinex Cycle of 88 lunations becomes 2599 days.
- Hibbardina Cycles (almost half a Saros) occur at 111 lunar months or 19 eclipse seasons.
- John F. Kennedy (11-22), his brother Robert Kennedy (6-6) and Martin Luther King (4-4) all died from their assassins on double digit days. All are also connected in the same Eclipse Metonic.
- Curiosity: There was an eclipse on May 14 1771. 1771 is 177 in both directions. 177 years later was the Israeli Independence declaration on May 14 1948.
- An eclipse month (draconic month) is 27.21222 days. 7 eclipse months equals 190 days.

Quick aside: The number Seven

In eclipses the number seven is most found in the Hepton ("7-sided") series. The Hepton Cycle features seven eclipses each 7 eclipse seasons apart occurring on the same day of the week. "Ancient Knowledge" from the wisdom schools of India and East Asia as reported in the century-old works of Ouspensky and Gurdjieff *(in Search of the Miraculous)* claim that the entire Universe is built upon a "Law of Seven"(and a "law of three").

There are seven differing notes of musical scale. Seven days of the week. Sven seas. Seven continents. Seven visible "wandering" celestial objects: Sun, Moon, Mercury, Venus, Mars, Jupiter, Saturn. Seven in the Bible signifies completion.

Beginning with Seven days for Creation and finishing with Seven Judgments at the end, not to mention hundreds of sevens in between. Jesus said seven things from the Cross.

In common usage: lucky 7. Jackpots have three of them.

Eclipse Years and Numbers

The eclipse year is a real thing, a cosmic marker just as surely as the solar year. The only difference is perspective, in this case the perspective seems more cosmic, or divine. Eclipse year =346.62 days.

19 eclipse years is one Saros Cycle (18 years and 11 days and 8 hours solar time).
1000 eclipse years is about 949 solar years.
2000 eclipse years is about 1898 solar years.
For instance, the eclipse bicentennial of the USA didn't occur in 1976 but rather 1966.
The second eclipse millennium began around 1898.
The 2,000[th] eclipse anniversary of the birth of Christ was likely just before the turn of the 20th Century.
The 2,000[th] eclipse anniversary of the death of Christ was somewhere between 1925 and 1933.
The 2,000[th] eclipse anniversary of the destruction of the Jewish Temple in 70 AD occurred on Aug 21 1968. The USSR invaded Czechoslovakia that very anniversary day.

Chapter 15

The Hepton Cycle

The Hepton Septiform Cycle

There are many different eclipse cycles. We've looked closely at the Saros and the Metonic in two individual chapters. We focus here on the Hepton Cycle. Among the 26 eclipses that are currently returning cyclically to bring complete eclipses, fourteen belong to two ongoing Hepton Cycles. They will be referred to as Hepton One and Hepton Two.

This is the first time the Hepton Cycle has been carefully chronicled. Upon consideration of the solar eclipse phenomenon, I've come to believe that the Hepton Cycle is a Maximum Expression of solar eclipse cycles, which are themselves maximum expressions of peculiarity in the natural visible cosmos.

The Hepton facts:
Hepton means "7-sided". There are 7 eclipses in each Hepton cycle.

These 7 eclipses appear at intervals of exactly 7 eclipse seasons, or 1211 days apart...(which is 3 1/2 eclipse years. Alternating Hepton eclipses are 2422 days apart, or 7 eclipse years.

Heptons nearly always occur on the same day of the week (with occasional orbital variation). In their current cycle, Hepton One eclipses occur on Monday and Hepton Two eclipses happen on Saturday.

It takes 7266 days for a Hepton Cycle to run through all seven eclipses from beginning to end.
A full Hepton Cycle is 21 eclipse years, obviously a multiple of seven and three, the two most important sacred numbers in Christianity and ancient wisdom.
7266 days equals 19 solar years, 10 months and 22 days, or 20 solar years minus 40 days.
A full Hepton has a lifespan composed of 10 repetitions of the same 7 Saros Cycles, or 70 eclipses altogether over the course of exactly 210 eclipse years.

Hepton One is currently composed exclusively of seven total eclipses.
Hepton Two currently has five ring of fires and two totals.

History: The Hepton Cycle was named by George Van den Beergh in the 1950's. We know this fact based on its mention at a couple of online science sites. Van Den Beergh invented the numbering system for the Saros and named several of its patterns and cycles. But finding a copy of Van den Beergh's actual work, published as "Periodicity and Variation in Solar Eclipses" seems impossible. Not only that, but there are no directly quoted excerpts or pdf's of that seminal work available online.

Obviously, since no material exists, I don't know the depth George van den Beergh explored the Hepton, but it's highly unlikely it was chronicled to the extent I do here. His book appeared in the 1950's, well before the Hepton began to put on the show it is doing now. Also, Van den Beergh was busy chronicling dozens of different cycles.

Our current Hepton One Cycle's peculiarity grows over the course of its lifetime. In my estimation, it reaches maximum expression with what I name "Hepton One 8th Chapter", the Monday eclipses of 2017 to 2037.

The repetition of sevens cannot help but bring to mind the 7 year and 3 and 1/2-year timekeeping of final judgment expressed in the Bible, as well as all the sevens in the Bible in general. Just as interesting in my mind is the relation of the seven to music, as in the 7 distinct notes of the scale. There is an 8th note, or octave, pattern in the Hepton too, as the first eclipse in the cycle repeats again before the cycle ends.

Both current Heptons are composed of complete eclipses. A couple of partials are included in our sthe beginning of Hepton Two as it grew into completeness, to illustrate how the cycles come into maturity.

Where does one begin with such an unusual and unstudied cyclic phenomenon? One must choose a moment in time and begin there. It turns out our current Hepton One had its beginning just before the dawn of this Century, with Saros 145's first complete eclipse, a ring of fire on June 6 1891 (you'll recognize the June 6 date as 77 years before the martyrdom of Robert Kennedy). You will recognize the Saros 145 personality from many great eclipses, including the August 21 2017 USA Divided.

Heptons roll into and then out of existence by maturing Saros Eclipses on one end (from partial to complete) and decaying eclipses on the other end (from complete to partials). For 210 eclipse years they are fully composed of complete eclipses.

Important!!! The Hepton Cycles overlap! They overlap each other and they overlap themselves. That is because a Saros cycle is 19 eclipse years, but it takes a Hepton 21 eclipse years to complete itself. Study and you'll understand. At first, I suggest you let it flow like a song.

We find that the Saros Personalities (numbers) go backwards by threes as they progress in the Hepton.

HEPTON ONE (Monday)--Saros 145 ,142 , 139, 136, 133, 130 and 127

HEPTON TWO (Saturday)--Saros 146 , 143, 140, 137, 134, 131 and 128

All three current USA eclipses in 2017, 2023 and 2024 are among the Hepton characters.

I follow Hepton One's "book" through Eight Chapters, from 1909 to 2037. I include a Nineveh Clause 9th chapter, 2035 to 2055, for comparison.

I follow Hepton Two from 1902 to 2048 for eight chapters. 146 years. 56 eclipses, of which 54 are complete. You will see the same eclipses we have already been studying and thus encounter some repetitive information already mentioned with the history chapters.

The Hepton Two series contain partial eclipses in 1902 and 1920, both from Saros 146.
Many of the Greatest eclipses of the 20th Century were not Heptons, but that's all part of the riddle too. During any 19-eclipse year Saros Cycle, 12 complete eclipses occur outside the Hepton. They also are part of other cycles.

I believe there is much meaning in this cosmic work of the Creator God.
We now begin with the history of Hepton One.

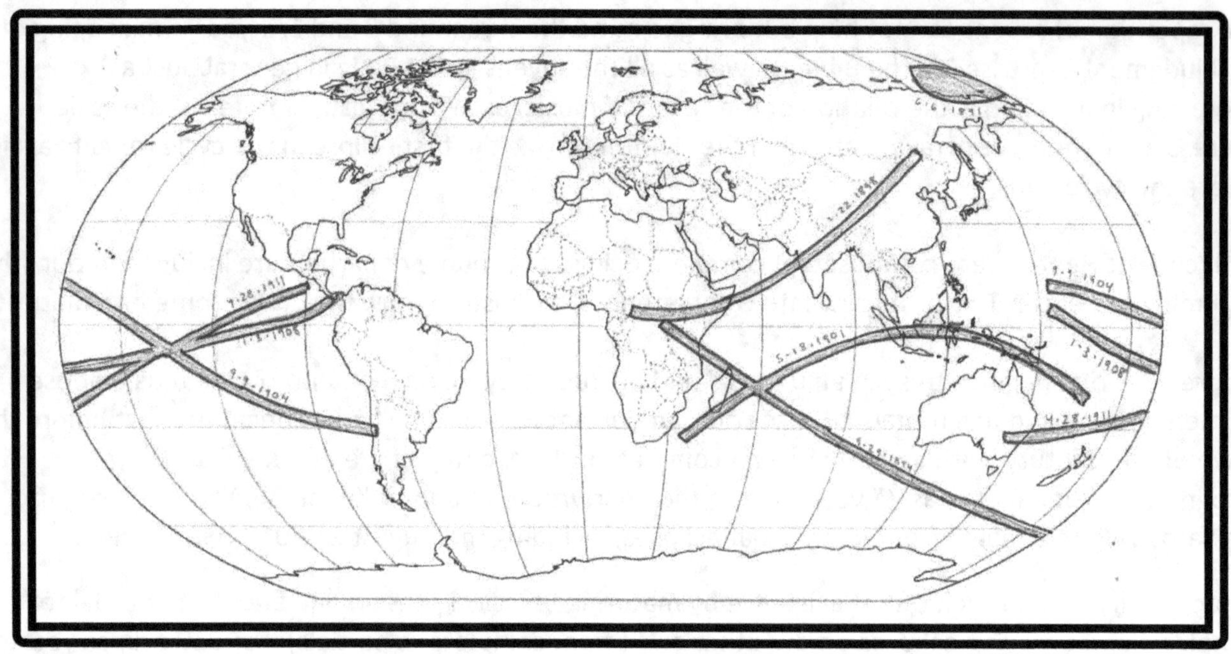

Hepton One Chapter One -Saturdays 1891 to 1911

1891 Jun 6 Extreme NE USSR Ring of Fire (Saros 145) Last RFK Death Day Metonic occurs 77 years before his assassination. Saros 145's first complete eclipse. Birth of current Hepton One.

1894 Sep 29 Equatorial Africa to South Pacific Total (Saros 142) 1211 days after the last eclipse, as all these eclipses will be spaced, 7 eclipse seasons (3 and 1/2 eclipse years) apart.

1898 Jan 22 Equatorial Africa/ India/China Total (Saros 139) This eclipse covers the most continental land of any eclipse in this Hepton. It bisects three areas all in the midst of European Colonization.

1901 May 18 Madagascar to Malay Archipelago Total (Saros 136) Covered in depth in world eclipses. 79 years to the day before Mount Saint Helens erupted.

1904 Sep 9 Nearly All Ocean to Chile Total (Saros 133) No land except sunrise over a few of the Marshall Islands and sunset over Copiapo, Chile—the only town of any size in the eclipse path, in the barren Atacama Desert. Copiapo was devastated 14 years later by an earthquake on Dec 4, 1918, which occurred the day after a Ring of Fire eclipse on Dec 3 1918 (76% coverage), one of the few eclipse/earthquake synchronicities in the survey.

1908 Jan 5 Ocean and sunset in Costa Rica Total (Saros 130) Almost entirely over Pacific ocean and entering just one country, this time Costa Rica, totality ending just past Puntarenas (Sand Point)

1911 Apr 28 Australia across the Ocean Total (Saros 127) Hepton lingers almost entirely over Pacific, beginning in extreme southwest Australia south of Canberra, and ending just west of Central America.

Our current Hepton One is born, as Saros 145 has its first complete eclipse in 1891. Four of these eclipses were almost entirely over Ocean. One was confined to the northern polar region.

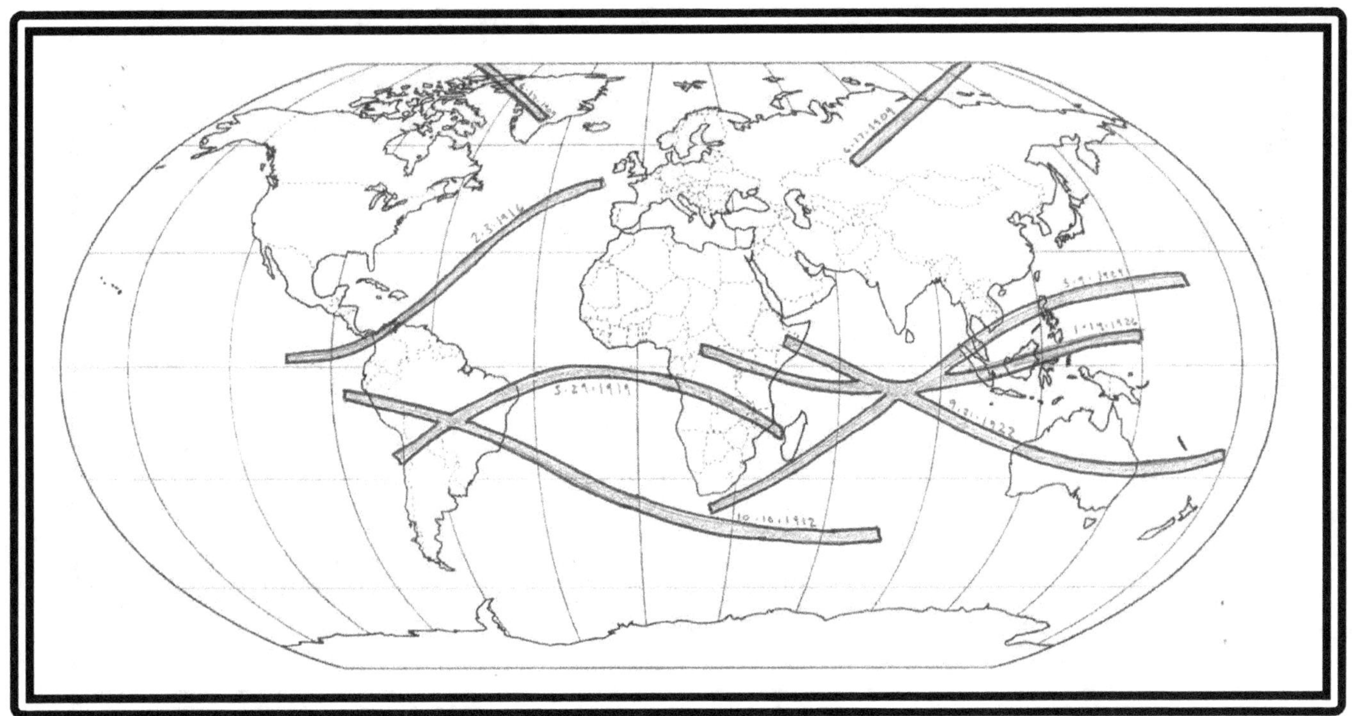

Hepton One Chapter Two -Thursdays 1909 to 1929

Chapter Two begins (in 1909) before Chapter One ends (in 1911). Hepton chapters overlap, with the first member beginning (145) the second chapter before the last member (127) finishes the first. That's how the wheel of time rolls. A Saros is 19 eclipse years while the Hepton is 21 eclipse years. This Chapter contains one of only two 7 year overland total crosses in the survey prior to the Turkish Cross, the 1912 to 1919 intersection over the Amazon of western Brazil.

1909 Jun 17 Hybrid Russia to Greenland (Saros 145) Reaching brief totality for the first time at maximum of 24 seconds in some places, Saros 145 has its only hybrid eclipse.

1912 Oct 10 Total South America to Indian Ocean (Saros 142) Straddling border between Colombia and Ecuador 420 years minus two days after Columbus landed in New World, path heads to Brazil, exiting roughly midway between Rio De Janeiro and and Sao Paulo.

1916 Feb 3 Total Isthmus towards Europe (Saros 139) Eclipse Colombia and Venezuela---totality in Medellin. Crosses the Atlantic to end just offshore of Europe 18 days before the Battle of Verdun...92% totality on coast of Ireland. 60% on coast of France.

1919 May 29 Total Einstein's Eclipse/ JFK Birthday (Saros 136) Discussed in world eclipses.

1922 Sep 21 Autumn Equinox (Saros 133) Horn of Africa to Australia, exiting south of Gold Coast.

1926 Jan 14 MLK Premonition Africa to Malay 3 years and one day before birth of Martin Luther King, Eclipse travels from very near the geographical center of Africa to Malay Archipelago.

1929 May 9 Ocean to Malay The first eclipse on earth after birth of MLK on Jan 15 1929. Mostly ocean from below South Africa then crosses Malay Archipelago.

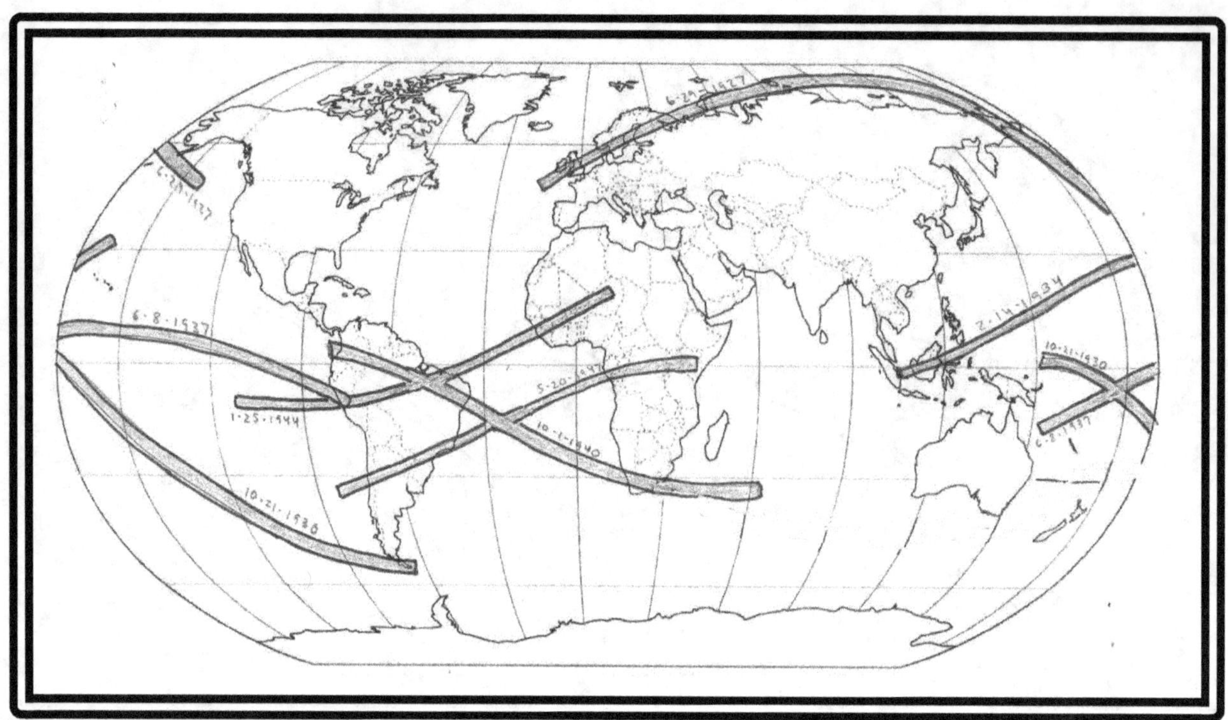

Hepton One Chapter Three -Wednesdays 1927 to 1947

This is the first Hepton One Chapter that consists of all true total eclipses, as 145 goes total. Sign of Pisces over South America. True Heptons intersect in Atlantic and Pacific.

1927 Jun 29 Liverpool Sunrise to America (Saros 145) First Total for the Great Saros 145. Last Total to bisect Britain (the extreme south of England sees totality in 1999). See personalities section.

1930 Oct 21 Pacific to Chile (Saros 142) avoids all land masses as it crosses the Pacific to come ashore and die as a thin totality in extreme southern Chile, closest city Port Natales or "Nativity."

1934 Feb 14 Valentine's Day in Borneo (Saros 139) Begins on Valentine's Day in Sumatra, crosses Date Line in Pacific to become 2-13 eclipse for virtually nobody. Ends before reaching Canada.

1937 Jun 8 Again almost all Ocean (Saros 136) Longest solar eclipse (at 7 minutes 5 seconds) since 1098 takes a trip across open ocean. Ends in Peru, with centerline at town of La Liberdad (Liberty), totality ending at sunset over Madre de Dios (Mother of God) National Park. Ends just before one of humanities most mysterious achievements--90% coverage at the amazing Nazca Lines.

1940 Oct 1 South America to Africa (Saros 133) 13 months after beginning of WWII...exits South America near Fortaleza ("strength): and Natal ("Nativity"). 8 days before John Lennon was born.

1944 Jan 25 South America to Africa (Saros 130) Saros 130 heads the opposite convex after exiting South America at essentially the same place as the last Hepton did 1211 days before.

1947 May 20 Columbus Death Day South America to Africa (Saros 127) South America to Africa. Totality begins just west of Chile. Totality in Santiago. Crosses Atlantic into Africa with totality in Kampala, Uganda and Nairobi Kenya. 441st anniversary of death of Columbus on May 20, 1506.

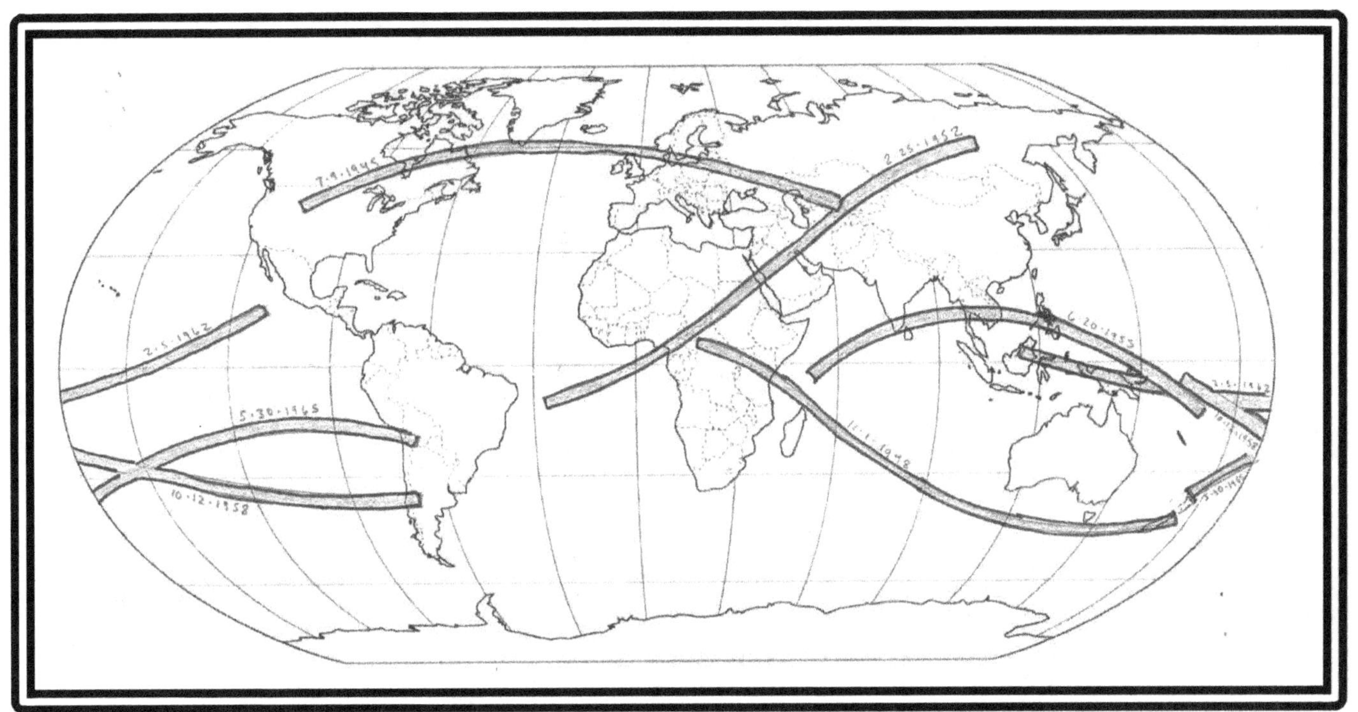

Hepton One Chapter Four -Mondays 1945 to 1965

1945 Jul 9 USA to USSR Atom Bomb Eclipse (Saros 145) One week before detonation of first atomic bomb. Looked at in depth in USA and world eclipses.

1948 Nov 1 Africa to Pacific (Saros 142) Begins very near equator. Ends just before New Zealand.

1952 Feb 25 Shahrud Tav (Saros 139) Discussed at length in World Eclipses. Traverses Africa (Totality in Khartoum, Sudan), then crosses Arabia (totality in Jeddah/ 99% in Mecca), crosses Iraq near Basra, and then Iran (totality in Shahrud) ends in Russia.

1955 Jun 20 Southeast Asia/ Vietnam Summer Solstice Eve (Saros 136) Discussed at length in World Eclipses. Longest total eclipse in a thousand years. Part of the unprecedented Southeast Asia Swarm.

1958 Oct 12 Columbus Day Santiago (Saros 133) 466th anniversary to the day after Columbus "discovers" America, a total eclipse travels the entire Pacific Ocean to end just after passing the city of Santiago, Chile. Eclipse occurred 5 weeks after Chile's Presidential elections, the first that featured Salvador Allende, who would be overthrown by a CIA backed coup on 9-11-1973.

1962 Feb 5 New Guinea PT 109 (Saros 130) 54 weeks after JFK takes office, Totality over Kolobangara Island (in Solomon Islands) where JFK's boat PT 109 sank 19 years before. Eclipse over New Guinea, where both Lyndon Johnson and John Kennedy had their wartime experiences. During the eclipse a rare grand conjunction of the planets occurred, for the first time since 1821. All 5 of the naked eye planets plus Sun and Moon were within 16° of one another on the ecliptic.

1965 May 30 New Zealand to Peru (Saros 127) New Zealand to Peru, with only Ocean in the path between. JFK birthday Metonic. First complete eclipse on earth since his assassination on 11-22-1963.

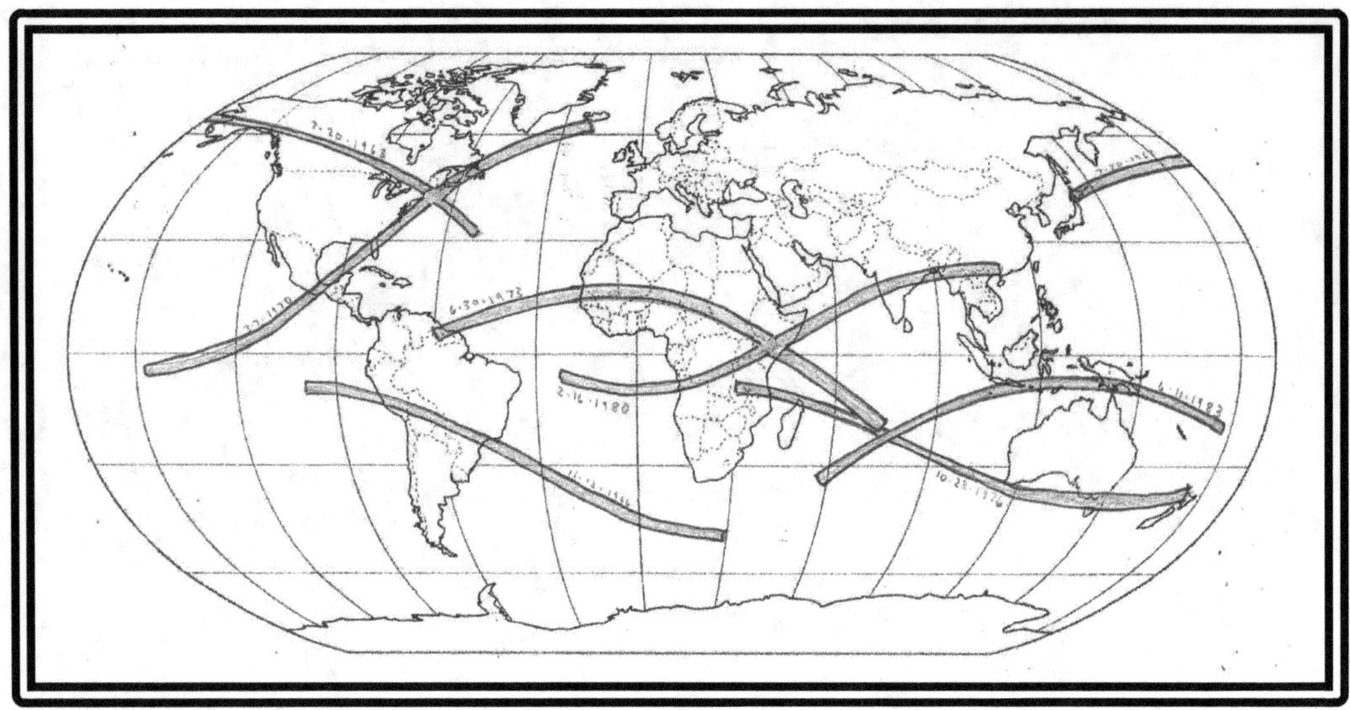

Hepton One Chapter Five-Saturdays 1963 to 1983

1963 Jul 20 Japan to USA. Beginning of 60's (Saros 145) Looked at In-depth at USA eclipses. Six years to the day before Moon Landing. 125 days (exactly 3,000 hours or 180,000 minutes) before the assassination of John F. Kennedy. A few weeks before King's March on Washington.

1966 Nov 12 South America. Middle of Sixties (Saros 142) Narrow line of true totality brings darkness to Lima, 97% coverage at Nazca Lines.

1970 Mar 7 End of the 60's Eclipse USA, Mexico, Canada (Saros 139) Saros 139 crosses same three nations again on 4-8-2024. Cape Canaveral (91% coverage) 7 months after first moon launch. 34 days before launch of ill-fated Apollo 13 on April 11 1970. Much more at USA Eclipses.

1973 Jun 30 Great African Eclipse (Saros 136) part of African Hepton Swarm. Longest African eclipse in more than a millennium-7 minutes 4 seconds. Eclipse enters at Mauritania, 92% at Timbuktu, path through Sudan, with totality in Darfur. Exits along border of Kenya and Somalia.

1976 Oct 23 Dar Es Salaam Africa to Australia (Saros 133) African Hepton Swarm. Path connects city of Dar Es Salaam, Tanzania ("place of peace") with Melbourne, Australia on October 23 1976.

1980 Feb 16 Dar Es Salaam Cross (Saros 130) One lunar year (355 days) after USA Mount Saint Helens/Microsoft eclipse, Totality crosses Africa from the slave coast exiting at Mombasa Kenya., bisects India (centerline of totality in Brahmapur) and Myanmar before ending in China.

1983 Jun 11 Indonesia/New Guinea (Saros 127) Major Indonesian cities in totality, including Yogyakarta, Semarang, Surabaya, and Makassar. Port Moresby, capital of Papua New Guinea.

Heptons coming together in more balanced crossing patterns. The African Hepton Cross (1973 to 1980) is one of only two True overland Hepton Total Crosses of 20th Century. Apex near Dar Es Salaam.

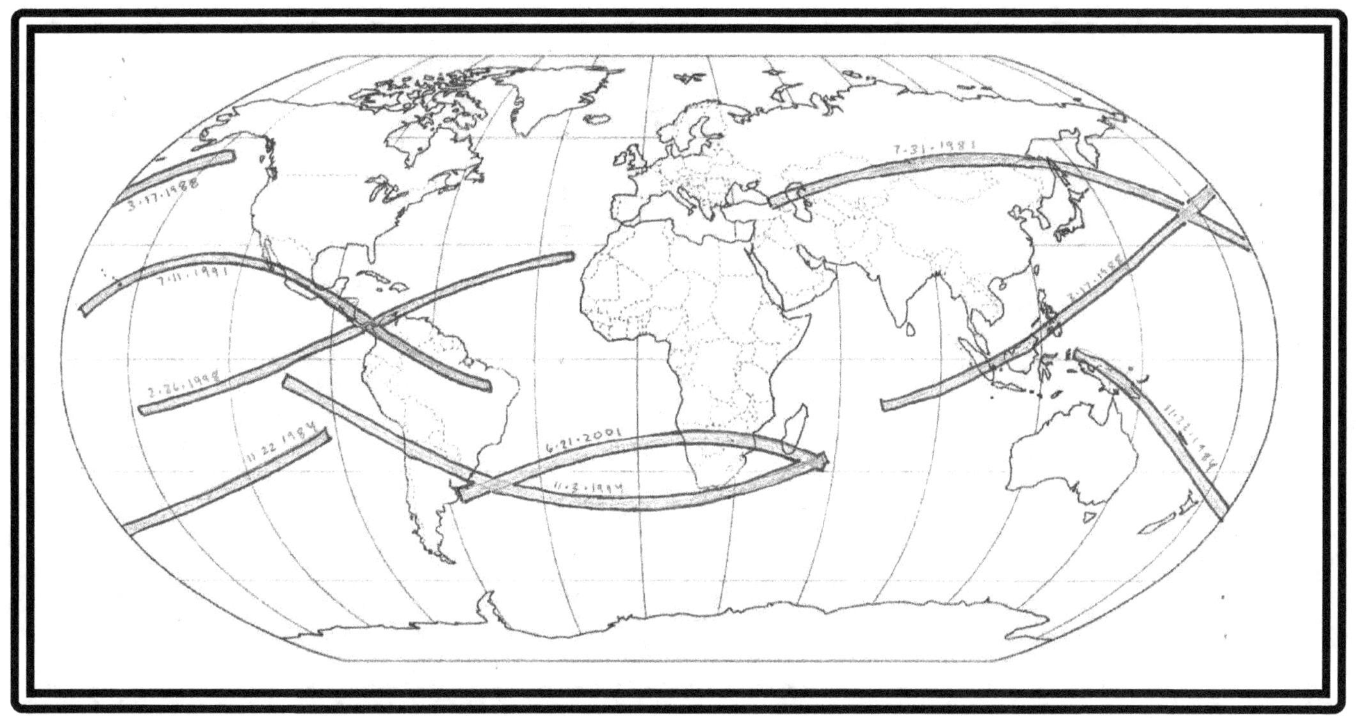

Hepton One/Chapter Six -Fridays 1983 to 2001

Isthmus Hepton Cross of 1991 and 1998 has an apex in ocean. The other Hepton Cross is over Pacific.

1981 Jul 31 USSR (Saros 145) The last decade of the Soviet Union's existence begins with a cross-country total solar eclipse that touched no other nations. More at world eclipses.

1984 Nov 22 New Guinea (Saros 142) 21 years to the day after JFK assassination. Path over PT-109.

1988 Mar 17 Indonesia/Philippines (Saros 139) Mount Pinatubo (70%) of the Philippines erupts 3 years, 3 months and 3 days later in the 2nd largest volcanic eruption of the 20th century. Path from Palambang extends across the Pacific with partial eclipse in the USA in Alaska.

1991 Jul 11 Hawaii/ Central America (Saros 136) Totality for Big Island of Hawaii (Kilauea volcano), First total eclipse there since 1850. Mexico (including totality in Mexico City and Temple of the Sun and Moon). Eclipse path touches every nation of Central America through Colombia before ending in Brazil. 27 days after eruption of Mount Pinatubo (6-15-1991) in the Philippines.

1994 Nov 3 Nazca Lines South America (Saros 133) After dancing around Nazca for decades, the moon finally eclipses the sun perfectly over the mysterious Nazca lines in Peru. Totality in Asuncion, Paraguay. Path crosses South America and Atlantic and dodges Southern Africa.

1998 Feb 26 Isthmus Tav (Saros 130) A tiny patch of ground has totality on the 19th anniversary of the USA Mount Saint Helen's Eclipse. Path crosses the isthmus of Panama just south of the canal zone.

2001 Jun 21 Solstice South America to Africa (Saros 127) 82 days before 9-11-2001. First complete eclipse of the 21st Century. Summer Solstice. The first complete eclipse since 20th century's last eclipse of Aug 11 1999. The first complete eclipse of the 21st Century also has Double 21's located in the number, 6- 21-2001, In addition, the sum of all individual numbers equal 12, for numerica fans.

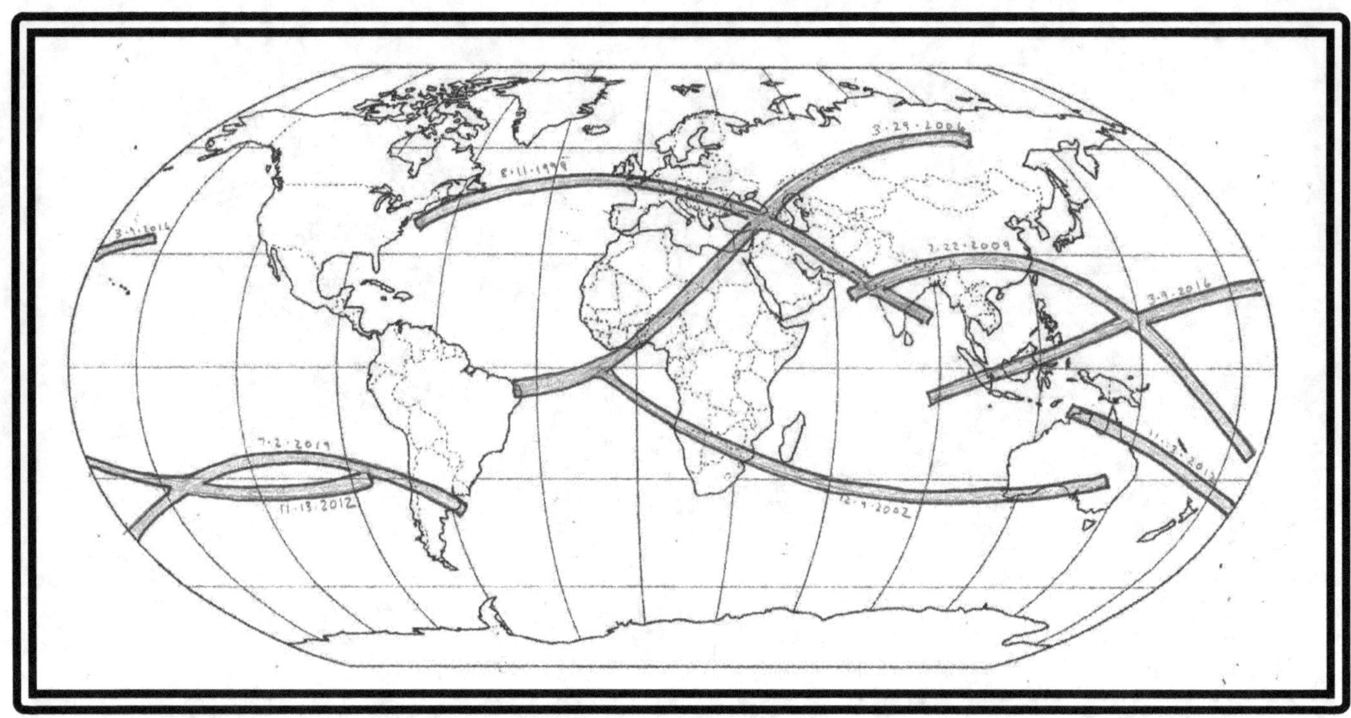

Hepton One Chapter Seven Turkish Cross-Wednesdays 1999 to 2019

This Hepton Features the great Turkish Cross, apex near Tokat. Cross reappears over the USA next.

1999 Aug 11 European Judgment (Saros 145) Described in detail at World Eclipses. Just offshore of USA to Britain, France, Germany (totality in Hitler hometowns), Austria, Hungary, Romania. Turkey six days before massive earthquake...then Iraq (Nineveh), Iran, Pakistan and India. One of the greats.

2002 Dec 4 Africa to Australia (Saros 142) First complete eclipse after 9-11 attack. Stops near Queensland/ New South Wales border.

2006 Mar 29 Turkish Tav Africa to Russia (Saros 139) Described in detail at world eclipses. Completes the Turkish Hepton Tav, the first truly balanced Hepton Cross overland in the entire series.

2009 Jul 22 Wuhan Totality (Saros 136) 6 minutes and 39 seconds, this eclipse has longest duration of the 21st Century. Ten years before Pandemic declared. India, Bhutan, then China. Crosses Wuhan, China, her first complete eclipse in centuries, with the city almost perfectly in the centerline of totality.

2012 Nov 13 Darwin Eclipse (Saros 133) Begins at sunrise outside of Darwin, Australia (88% coverage), the city named for the famed evolutionist, on the 200th year anniversary of his birth(Feb 12 2012) . *Origin of Species* was published on Nov 24 1859, exactly 152 solar years plus one lunar year before. Only touches Australia.

2016 Mar 9 Indonesia/ Malay Archipelago (Saros 130) Indonesia, Malaysia, south Philippines. The last year in office of Barrack Obama, who lived in Indonesia as a boy.

2019 Jul 2 Buenos Aires Sunset (Saros 127) Pope Francis Eclipse. 2302 days after his ordination as Pope, Totality at La Serena, Chile and San Juan, Argentina). Ends at sunset near Pope Francis birthplace of Buenos Aires, Argentina. Leaves nation at Dolores (sorrows), Argentina.

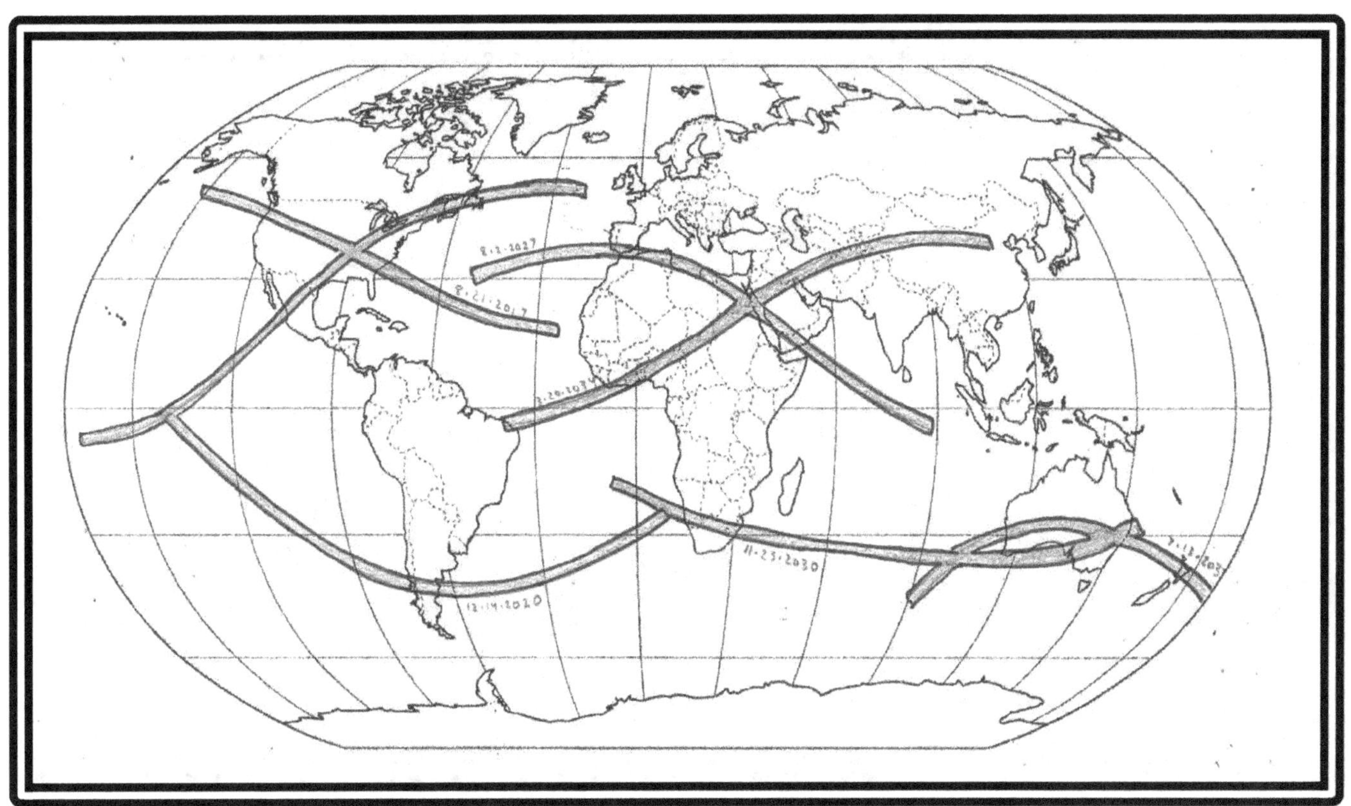

Hepton One Chapter Eight-Mondays 2017 to 2037

3 Hepton Crosses in one Chapter. Perhaps the maximum expression of all modern eclipses.

2107 Aug 21 USA Divided/ Great American Eclipse (Saros 145) More at USA Eclipses. No other nation touched. First USA total eclipse in 38 years. Entrance at Salem. Exit at Charleston. Maximum eclipse in Makanda (the "star of Egypt") Illinois.

2020 Dec 14 Electoral College Eclipse (Saros 142) Eclipse occurs at bottom of South America while the new President and Vice-president are being elected by the Electoral College in Washington, DC.

2024 Apr 8 North American Eclipse (Saros 139) South to North across Mexico, USA and Canada. More populated cities in path than any eclipse in the history of the Western Hemisphere.

2027 Aug 2 Valley of the Kings/ Mecca (Saros 136) 6 minutes and 23 seconds, there won't be a longer total eclipse until 22nd Century. Rock of Gibraltar, Tripoli, Libya, Valley of the Kings, the great Egyptian necropolis. Total eclipse in the Holy city of Mecca, Saudi Arabia.

2030 Nov 25 Ghost of British Empire (Saros 133) South Africa to Australia. Ending up at sunset over Botany Bay and Sydney. Occurs on what would have been JFK Jr's 70th Birthday.

2034 Mar 20 Spring Equinox Hepton Cross (Saros 130) More on this elsewhere. Creates 7 eclipse year cross with 2027 Valley of the Kings. New Year's Day in many other cultures.

2037 Jul 13 Australia and New Zealand (Saros 127) Forms the third and last Hepton Cross of this chapter in conjunction with 11-25-2030 eclipse, apex in remote southwest New South Wales.

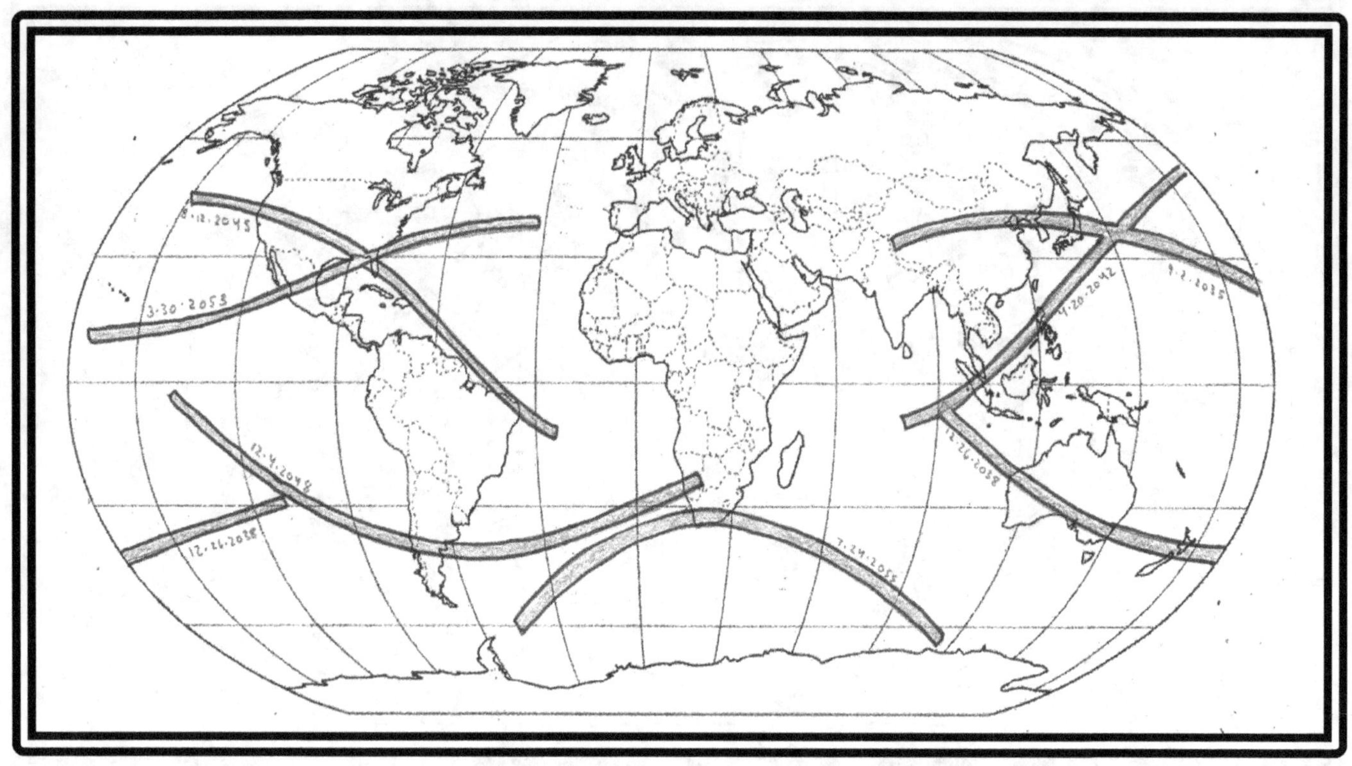

Hepton One Chapter Nine-Saturdays (Nineveh Clause) 2035 to 2055

The First three eclipses of this series, including a Hepton Cross, combine to touch the nations of China, Korea, Japan, Philippines, Indonesia, Malaysia, Australia and New Zealand—most of them over or very near their capitals. Also notable are places that are not touched, Europe, most of Africa, Indochina, , the Central Asian Republics, Russia and India.

In the western hemisphere, three balanced eclipses perform a rough approximation of the eclipse positions that occurred 28 years before in Hepton Chapter Eight. But the apparent repetition isn't as simple as it appears. The Saros have switched places. The North American Cross is now created by the two Saros that previously created the Mideast Cross of 2027 to 2034. Check out Hepton Double Cross.

2035 Sep 1/2 China/ Korea/ Japan (Saros 145) Eclipse begins in China, just after western border with Tajikistan. It bisects the entire nation and brings the capital Beijing its first total eclipse since 1277 AD. Totality in Pyongyang, North Korea. North Tokyo is in totality. Downtown Tokyo sees 99.5 % coverage.

2038 Dec 26/25 Do they Know it's Christmas? Australia and New Zealand (Saros 142) It's December 26th for the Aussies and Kiwis, but for the watery world east of the international date line, it's a Christmas Day Total Eclipse. Australia and New Zealand bisected. Part of the Australian Swarm.

2042 Apr 20/19 Sumatra/ Philippines (Saros 139) Third eclipse in a row to emanate from around the 90th Parallel, only to cross the International Date Line (go back in "time" a day), and die in Pacific without reaching the Americas. The number 420 is present in 4-20-2042 numeric two times. The Metonic date of April 19 and 20 is a split between the death day of Charles Darwin (April 19 1882) and the birthday seven years later of Adolph Hitler (April 20 1889).

2045 Aug 12 USA California to Cape Canaveral to Columbus (Saros 136) Where do we begin? The Nineveh Clause offers up the possibility of later Judgment. Do we want it or not? 27 solar years plus one lunar year after US is first divided by an Aug 21 2017 USA Eclipse (numerica: 8-21 to 8-12 are the dates). 21 years after North American eclipse of 2024 painted the Hepton Tav across Little Egypt. The year the transhumanists proclaim synchronicity.

If all happens as astronomically scheduled, the 2045 Eclipse enters the US at the farthest westernmost point in California, Cape Mendocino, on the Lost Coast. Cape Mendocino was named by Spanish Explorers in the 16th Century and has been a landmark ever since. In Totality: Eureka, CA is the first city; then Redding and Chico. Reno, Nevada. Salt Lake City, Provo and Moab, Utah. Grand Junction, Aspen, Colorado Springs and southern Denver. Wichita, Kansas. Bisects Oklahoma including OKC, Tulsa and Tahlequah. Fort Smith and Little Rock, Arkansas (Little Rock is also in 2024 totality). Jackson, MS. Montgomery Al. 99% in Memphis. Totality in almost the entire state of Florida, including Miami and Cape Canaveral, which has its 7th complete eclipse in less than 500 years, and 3rd eclipse since 1908. In the 77th year (76 years and 27 days) since Moon Rocket of Apollo 11 left Cape Canaveral/Kennedy.

Totality in San Salvador Island, where Columbus first landed on October 12, 1492. Eclipse brings totality to Haiti and Dominican Republic. It follows the Northeast Coast of South America, including Guyana's capital Georgetown and New Amsterdam. Follows coast of Brazil over city of Belem' (Portuguese for Bethlehem) and exits a few miles south of Natal, or "Nativity"(99.5%)...ends very near 11 degrees south, 22 degrees west.

2048 Dec 4 South America to Africa (Saros 133) An interesting replica of Saros 142's famous Electoral College/ Washington Death Day Eclipse of 12-14-2020. This eclipse goes slightly south of the 2020 path in Chile and Argentina, and crosses the ocean to Namibia, Africa.

2052 Mar 30 Mexico US/ Florida Cross/ Final Charleston (Saros 130) Saros 130 makes a decent approximation of landfall of 139's 4-8-2024 USA Judgment eclipse, directly over Puerto Vallarta. From there, the path goes decidedly farther south than 2024. Mexico bisected. Brownsville TX and then across the gulf, just south of New Orleans (99.5%). Area between Pensacola and Tallahassee Florida is likely apex of Hepton Cross of 2045/2052. Path goes over Savannah, Georgia and leaves the USA at the infamous slave port of Charleston, South Carolina.

This is Charleston's 3rd Total Eclipse in 82 years (1970/2017/2052). Average for any place is once every 400 years. It's Charleston's fifth complete eclipse since 1834 (218 Years) and its 6th complete eclipse since 1628. That includes 4 totals in 218 years (1834, 1970, 2017, 2045) That is 800% of an average place on earth. Remember its not uncommon for a city to go more than a thousand years without a single total eclipse.

2055 Jul 24 South Africa (Saros 127) The oldest Saros in Hepton Chapter One on its last trip over heavily populated land before dying crosses South Africa, leaving at Port Elizabeth and New London.

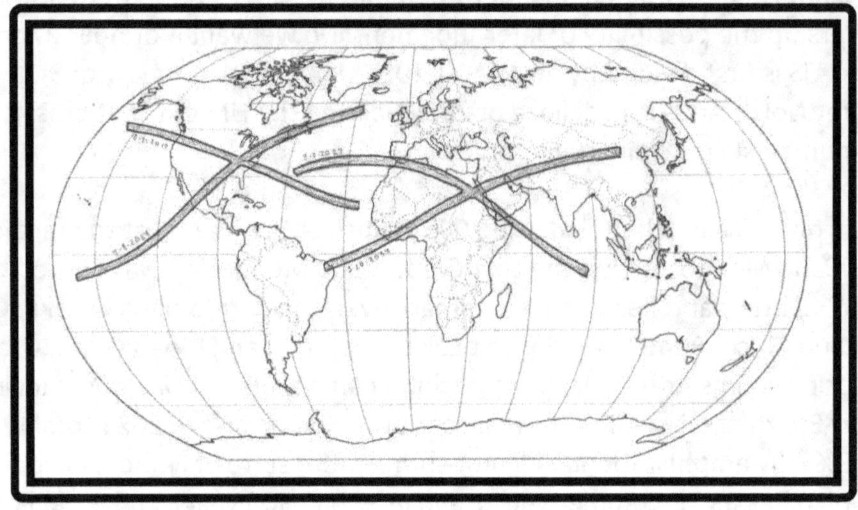

Above: The Hepton Crosses (Tavs) of 2017 to 2034

Western Cross: 8-21-2017 (Saros 145) and 4-8-2024 (Saros 139)
Eastern Cross: 8-2-2027 (Saros 136) and 3-20-2034 (Saros 130)
Then, a cosmic Switcharoo:

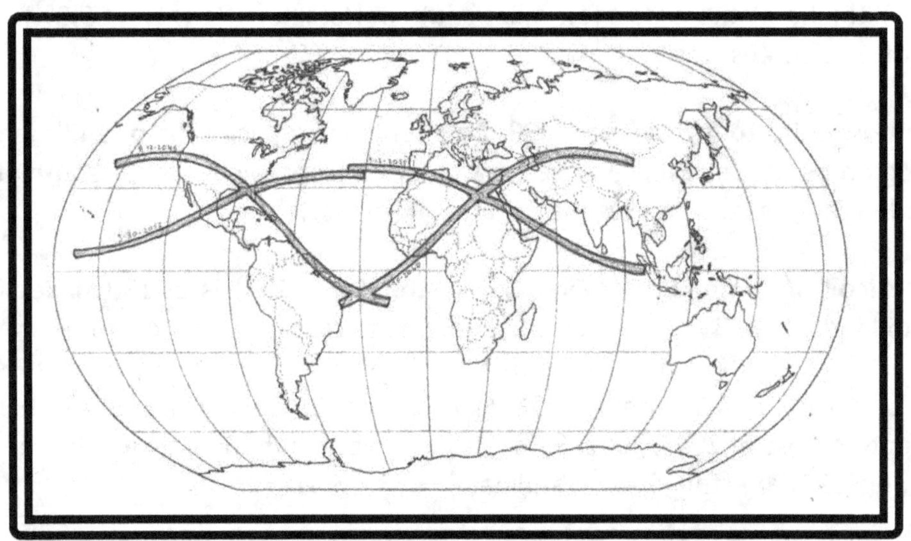

The amazing Hepton Bizarre Switcharoo Double Cross 2045 to 2060 above

No need to adjust your eyes. That's not the same map you just looked at. It's 28 years later and opposite Saros are making the Crosses.

Western Cross: 2045 USA/ Cape Kennedy (Saros 136) 2052 Florida Cross/ Charleston (Saros 130)
Eastern Cross: 2053 Gibraltar/ Egypt (Saros 145) - 4-30-2060 Africa to China (Saros 139)

It takes a minute to believe what's going on. A double switcharoo eclipse cross? This is the signature of artistic mastery. Shadows as words, art, and songs of God.

Hepton Two Chapter One- Wednesdays 1902 to 1922

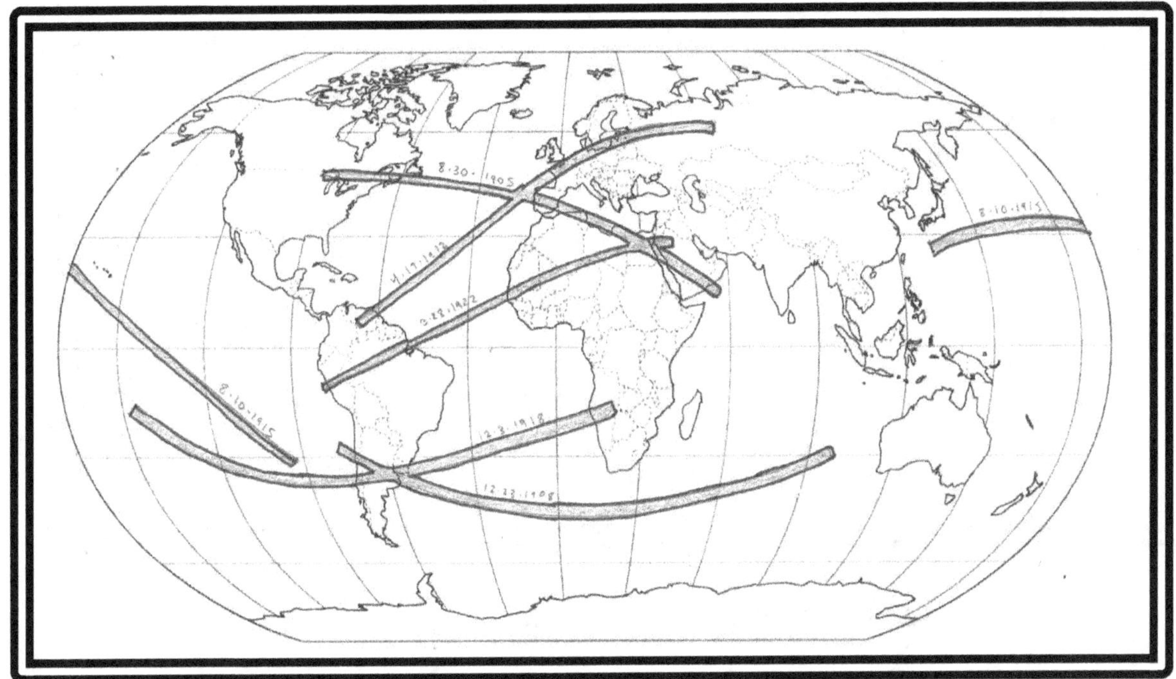

1902 May 7 Partial (Saros 146) Saros 146 will be partial for two rounds of this Hepton while Saros 143 finishes its turn on August 22nd 1979

1905 Aug 30 Total Canada/ Spain/ Tripoli/ Egypt/ Mecca (Saros 143) Looked at closer in Mideast Section. First line of Hepton Cross with Titanic Hybrid.

1908 Dec 23 Hybrid South America to Africa (Saros 140) Begins and ends as ring of fire, totality only visible from South Atlantic

1912 Apr 17 Hybrid Titanic Eclipse Atlantic/Europe(Saros 140) 60 hours after sinking of Titanic, a hybrid eclipse crosses the Atlantic. 60 hours after the sinking of the Titanic, a hybrid solar eclipse travels across the Atlantic. 90% totality at Ocean Liner's departure site of Southampton, England. Totality in Paris. A Saros in transition. Here it's a hybrid, soon to be a ring of fire. 99.7% totality in Oviedo, Spain—the apex of the pronounced 1905 and 1912 Hepton Cross.

1915 Aug 10 Ring of Fire Temple Destruction Metonic Pacific (Saros 137) 1845 years to the day after Rome destroys the Temple. One lunar year after Aug 21 1914 WWI eclipse. Eclipse avoids land.

1918 Dec 3 Ring of Fire Earthquake Chile Eclipse (Saros 134) A Ring of Fire enters Atacama desert region of Chile 24 hours before a magnitude 7.8 earthquake devastates towns in the Atacama region. Perhaps an example of an earthquake/eclipse tie, though the reality is hard to pinpoint.

1922 Mar 28 Ring of Fire King Tut Egypt Independence (Saros 131) Egypt one month after their formal independence from Britain and 7 months before Excavation begins on tomb of King Tut in the Valley of the Kings. Path of Annularity crosses all sites of Mount Sinai and Northern Arabian Peninsula. This one's got it all. It takes everything I can do not to get sidetracked by the King Tut Eclipse.

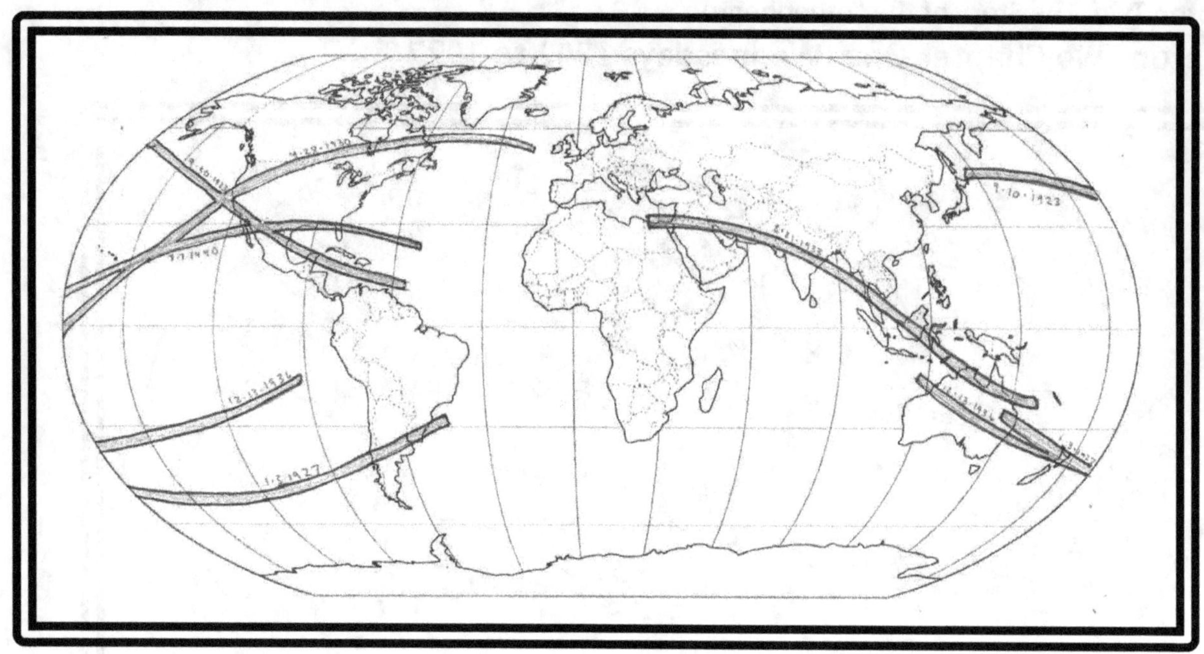

Hepton Two Chapter Two Mondays 1920 to 1940

1920 May 18 Partial (Saros 146) The exact day the future Pope John Paul II was born in Poland. Mt. Saint Helens erupts exactly sixty years later in 1980.

1923 Sep 10 Total Japan to LA (Saros 143) Begins in the northern islands of Japan at sunrise 9 days after the worst natural disaster to ever hit Japan, the great earthquake and fire of Sept 1, 1923, which killed nearly 150,000 people and largely destroyed Tokyo and Yokohama. Path goes to Southern California then Mexico. US and Japan, linked by the path, would be at war within 19 years.

1927 Jan 3 Total Australia to South America (Saros 140) Eclipse begins at sunrise in Australia and crosses the Pacific ending just north of Buenos Aires, Argentina. Pope Francis parents arrive there from Italy 2 years later.

1930 Apr 28 Hybrid San Francisco (Saros 137) a thin line of totality (only ½ mile wide!) cuts across northern San Francisco before becoming a ring of fire through Idaho, Montana and Canada. The two great coastal cities of North America, New York and San Francisco, are thus marked by total solar eclipses within 5 years of each other. 36 years and two days prior to the Church of Satan being founded by Anton Lavey in San Francisco Apr 30 1966.

1933 Aug 21 Ring of Fire Jerusalem (Saros 134) discussed in detail in Middle East and World Eclipses. One of the three great August 21st twentieth century eclipses.

1936 Dec 13 Ring of Fire Australia/ New Zealand (Saros 131) The future Pope Francis will be born about 3 days later in Buenos Aires on Dec 17, 1936

1940 Apr 7 Ring of Fire USA/Mexico (Saros 128) See USA Eclipses. Total annularity in Austin and Houston) and hugs the Gulf Coast (Baton Rouge, New Orleans, Mobile and Jacksonville) on the first new moon after Easter.

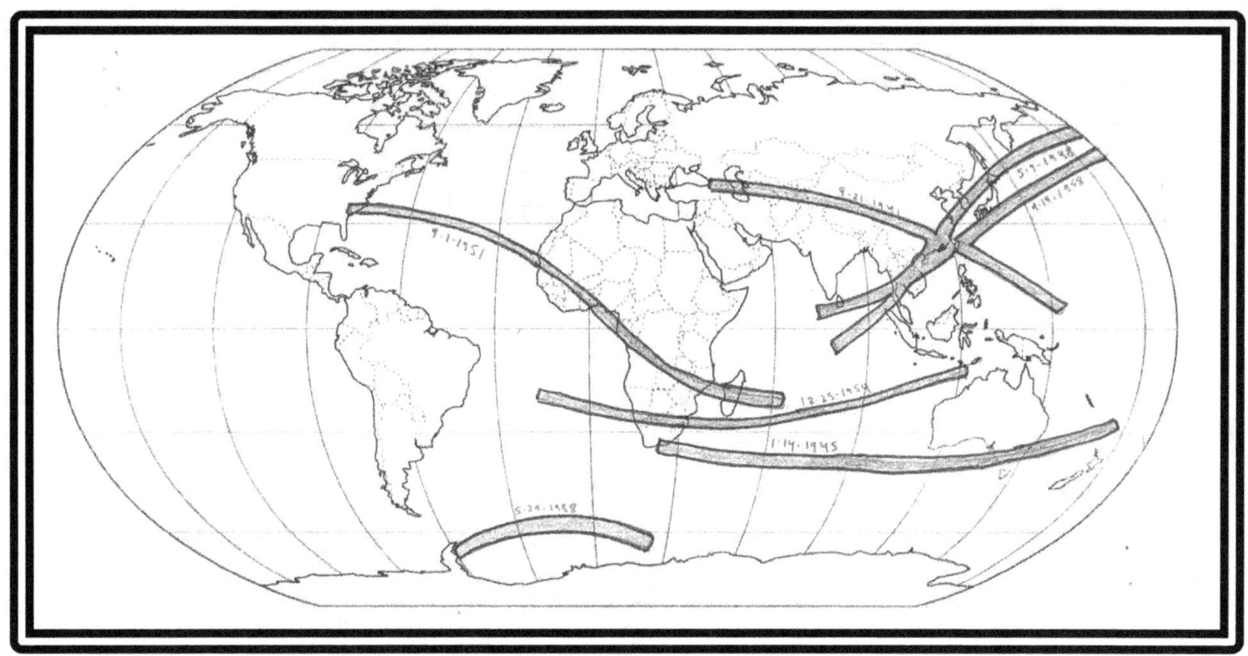

Hepton Two Chapter Three Sunday/Saturday 1938 to 1958

1938 May 29 Total JFK 21st Birthday Endurance Eclipse (Saros 146) -We begin Hepton Two creating all complete eclipses, as Saros 143 transforms from partial to total. Eclipse over site of Shackleton's Antarctic Expedition 22 years before. The May 1916 500 miles in a rowboat and mountain traverse of South Georgia is one of the greatest adventure stories in the history of humanity Eclipse brings totality on South Georgia Island and 93% on Elephant Island. Back in Boston, it's JFK's 21st Birthday.

1941 Sep 21 Total USSR/China Equinox (Saros 143) Metonic cycle begins to settle in on Fall equinox. Eclipse 3 months after Germany turned on the Soviet Union and invaded them, 3 weeks after 23,000 Hungarian Jews are slaughtered by the Gestapo in Ukraine and 8 days before 30,000 more are killed at Babi Yar near Kiev. Total cuts across USSR (starting just east of Ukraine) Eclipse crosses China, during great battle of Changsha (95% totality). Totality in Wuhan and Taipei.

1945 Jan 14 Ring of Fire British Empire Southern Boundary (Saros 140) in the last terrible winter of WWII, a ring of fire skirts southern Africa and Tasmania. MLK turns 16 the next day.

1948 May 9 Ring of Fire (or was it Total?) Korea Divided (Saros 137)-One of those amazing eclipses which people have cause to wonder whether it was annular, hybrid or total. 99.99% totality in some places as it exits China. Brief 100% totality in Gang-won-do South Korea? Korea is divided between north and south two years before North Korea invades. Ensuing war will kill maybe ten million.

1951 Sep 1 Ring of Fire Jamestown to Angola, Africa (Saros 134) 332 years after first slaves brought to English colonial Jamestown on August 20, 1619. Path begins in Virginia with Total Annularity in Jamestown and colonial Williamsburg, crosses Atlantic to the Slave Coast, including Angola, which had been home to the 20 or so Africans brought on the first ship.

1954 Dec 25 Ring of Fire Christmas in South Africa (Saros 131) Christmas Ring of Fire in South Africa.. one of only two Christmas eclipses in centuries, and the last complete Christmas eclipse over continental land for centuries more. Annularity in Port Elizabeth.

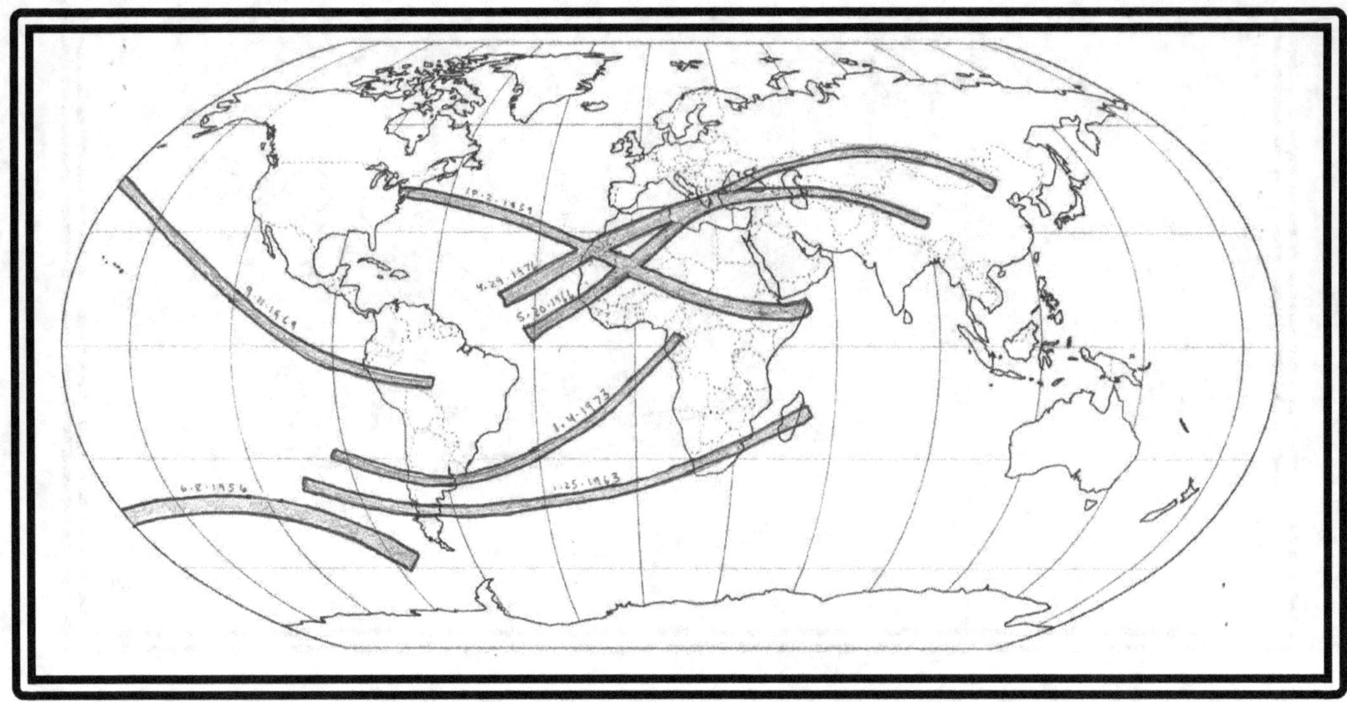

Hepton Two Chapter Four- Casablanca Cross-Fridays 1956 to 1976

Hepton cross occurs between 1959 (Total) and 1966 Ring of Fire with apex near Kidal region of Mali.

1956 Jun 8 Total Antarctic Ocean (Saros 146) Began on Jun 9. Ended on Jun 8. Open Ocean.

1959 Oct 2 Total JFK Boston Sunrise to Africa (Saros 143) Much more at Personalities. Total Eclipse begins just west of Boston and gives that city a total solar eclipse for the first time since 1806, just as its modern son JFK is campaigning for President. Boston is the only major city in America to witness.

1963 Jan 25 Ring of Fire South America to South Africa (Saros 140) South America to Madagascar across South Africa, which is engulfed in national turmoil (95% at Cape Town). Nelson Mandela had begun his 27 years in prison six months before on Aug 5 1962.

1966 May 20 Ring of Fire Columbus Death Day Africa/ Athens/Asia (Saros 137) Columbus Death Day Metonic appears 460 years after his death (5-20-1506). Eclipse almost has totality, with maximum 99.94%. 56% coverage in Columbus home region of Genoa, Italy. 41% coverage in Valladolid, Spain where he died. Major cities in path: Tripoli, Athens (99%). Istanbul (98%), Stavropol and Astrakhan Russia (99.7%). Crosses almost all of China.

1969 Sep 11 Ring of Fire Peru (Saros 134) The September 11th Egyptian New Year metonic. South American nations of Peru and Bolivia. 92% coverage at Nazca Lines. Annularity at Trinidad, Bolivia.

1973 Jan 4 Ring of Fire Chile to Africa (Saros 131) Crosses Chile (66% at Santiago) 250 days before Sep 11 1973 coup deposed President Salvador Allende. 250 days has interesting properties. It is exactly 6,000 hours, or 360,000 minutes or 35 weeks 5 days (last number echoing 355 day lunar year)

1976 Apr 29 Ring of Fire Africa to China (Saros 128) Similar path to May 20 1966 in many ways—intersecting that prior path in islands of Greece. Tripoli in path again. Ends in Tibet.

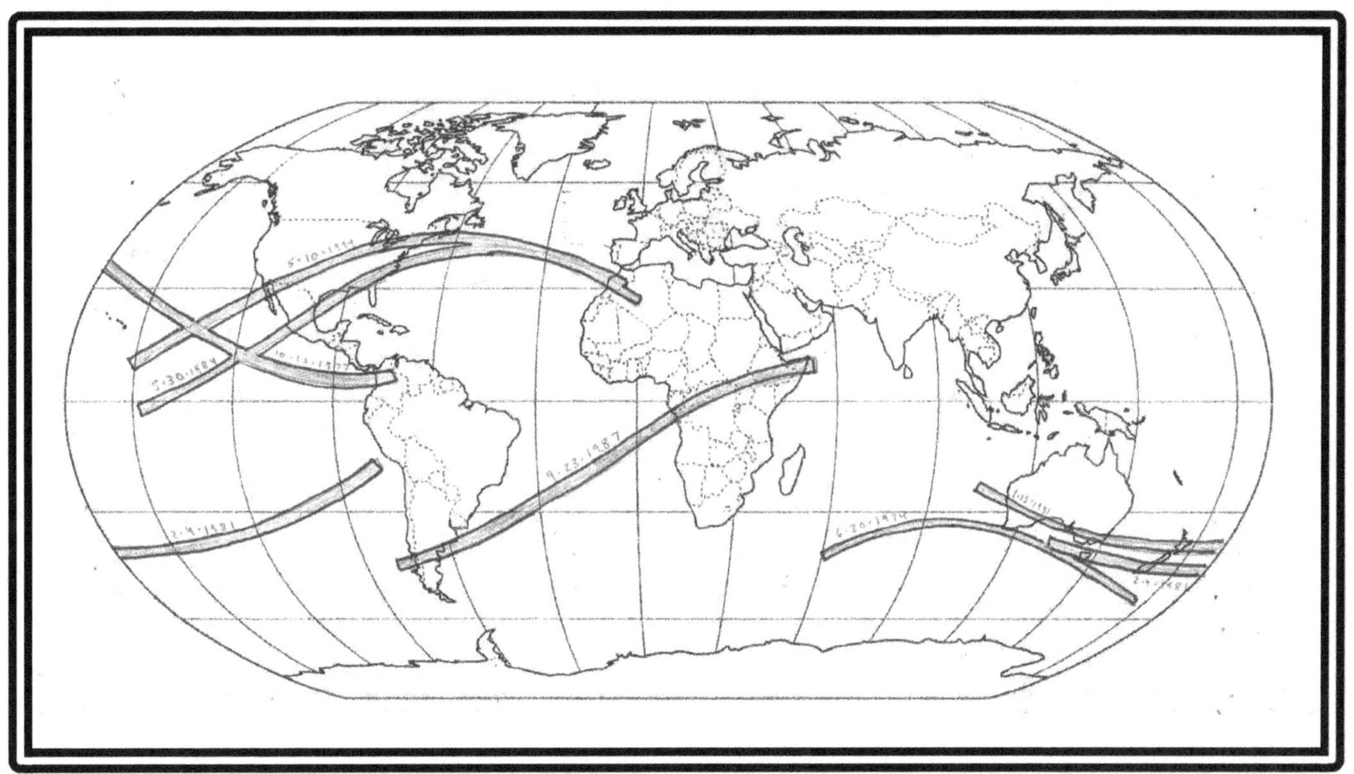

Hepton Two Chapter Five; Wednesdays 1974 to 1994

1974 Jun 20 Total Eclipse Island (Saros 146) Only complete eclipse for all of 1974 and 1975 touches almost no land, only a tiny part of southwest Australia, appropriately named Eclipse Island.

1977 Oct 12 Total Columbus Day in Colombia (Saros 143) Almost no country affected except Colombia, the only nation on earth named after Columbus, on the 485th anniversary of the discovery of the New World. 53% coverage in San Salvador Island, site of Columbus's arrival. Totality in Bogota.

1981 Feb 4 Ring of Fire Tasmania (Saros 140) Over Hobart Tasmania, extreme southern New Zealand and Pacific almost to Peru. 105 years after death in Hobart of Trugannini, the last Tasmanian. This makes three Hepton Two eclipses in a row that have barely touched any land.

1984 May 30 Ring of Fire USA/ Mexico (Saros 137) More in USA eclipses. JFK 77th Birthday Metonic. Bisects Mexico and Crosses eastern USA before heading to Africa. A tiny path of hybrid-ish maximum goes directly over New Orleans and Atlanta. Annularity near DC.

1987 Sep 23 Ring of Fire Palindrome South America to Africa (Saros 134) Eclipse date of 9-23 is palindrome with the other eclipse of 1987 of 3-29. Eclipse neatly fits over both continents.

1991 Jan 15 Ring of Fire MLK Birthday Australia and New Zealand (Saros 131) Exactly 7 eclipse years after last JFK Metonic of May 30, 1984, the Martin Luther King Birthday Metonic truly re-appears for first time in centuries. What would have been his 62nd birthday. Perth Australia, Tasmania again and straddling New Zealand. 2 days before beginning of the Persian Gulf War air campaign (Desert Storm).

1994 May 10 Ring of Fire USA (Saros 128). Last complete eclipse in US for 18 years. More at USA.

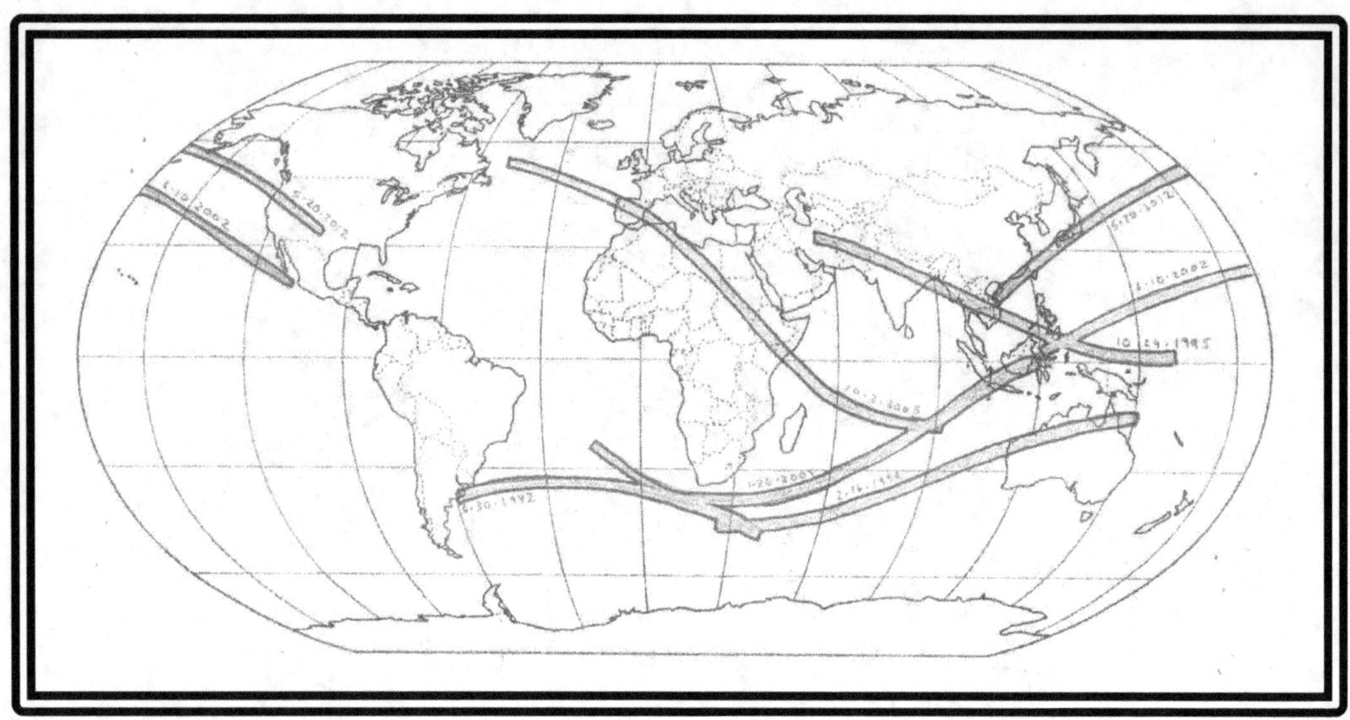

Hepton Two Chapter Six Mondays 1992 to 2012

1992 Jun 30 Total Montevideo Single City Eclipse (Saros 146)   There are 3 single city eclipses in less than 4 years--Mogadishu 9-11-1988, LA Metropolis 1-4-1992 and this one. Montevideo, Uruguay at sunrise...nowhere else but ocean. Montevideo means roughly " I see a mountain". 500th year since Columbus showed up in America. Buenos Aires, Argentina is a near miss at 93 % totality.

1995 Oct 24 Total UN Day Iran Sunrise/ India/ Southeast Asia (Saros 143) 50th anniversary of the United Nations. Path begins in Iran, goes through Pakistan, Afghanistan almost exactly 6 years before US ground invasion. India has totality at Allahabad (City of God) then eclipse in Southeast Asia.

1999 Feb 16 Ring of Fire Australia (Saros 140) the second to last solar eclipse of the 20th Century.

2002 Jun 10 Ring of Fire Indonesia to Mexico (Saros 137) Indonesian Islands to Puerto Vallarta (at sunset...sweet!). Crosses entire Pacific Ocean, almost no land. 71 % coverage in Los Angeles. "Foolish" by Ashanti was #1 song in the USA. Almost exactly 9 months after 9-11 attacks.

2005 Oct 3 Ring of Fire Spain/North Africa (Saros 134) more at World eclipses. Annularity squarely on Madrid and Toledo Spain, also Algiers and Libya. Darfur in Sudan.

2009 Jan 26 Ring of Fire Indonesia Inauguration (Saros 131) Six days after inauguration of Barrack Obama. a ring of fire crosses almost entirely ocean to end over Indonesia, where Obama spent years of his childhood. Total annularity in capital of Jakarta, his boyhood home, at sunset.

2012 May 20 Ring of Fire Hanoi to Midland Sunset (Saros 128) See more at USA eclipses. Begins at Sunrise just east of Hanoi Vietnam and ends at sunset immediately over Midland, Texas, childhood home of George W. Bush.

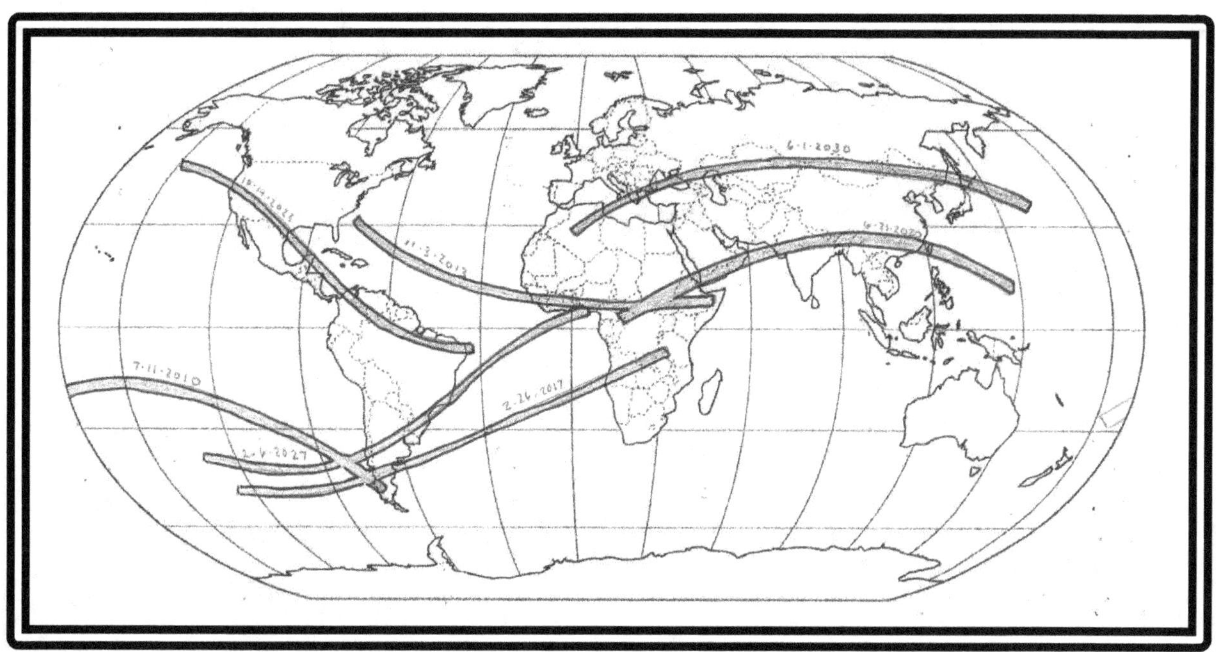

Hepton Two Chapter Seven - Saturdays 2010 to 2030

2010 Jul 11 Total Easter Island (Saros 146) more at world eclipses. The first total eclipse in 1400 years to cross Easter Island--one of the most mysterious and isolated places on earth, the island of the great stone heads. Eclipse occurs precisely during the playing of the World Cup Soccer final between Netherlands and Spain, both of whom laid claim to that island. Spain wins the match 1-0.

2013 Nov 3 Hybrid Atlantic Through Africa (Saros 143) 143 goes Hybrid with a very thin path of totality crosses the Atlantic (19% coverage at Charleston South Carolina at Sunrise, enters Gabon just south of Libreville and crosses the continent. Last solar eclipse before lunar tetrad of 2014 and 2015.

2017 Feb 26 Ring of Fire South America to Africa (Saros 140) Last solar eclipse before 8-21-2017 USA eclipse, this narrow ring of fire goes over the Patagonia region of South America, including only major town of Coyhaique and heads across the Atlantic into Angola before ending in Zambia. Occurs on the Feb 26 metonic date, 38 years to the day after USA Mount Saint Helens/Microsoft eclipse.

2020 Jun 21 Ring of Fire Solstice Father's Day Pandemic Africa to China (Saros 137) Africa and the Red Sea through Yemen, Oman, India, Pakistan, Nepal, Tibet and exiting through China and Taiwan. 83% in Wuhan , 70% in Mecca, 94% in Delhi, 35% in Jerusalem

2023 Oct 14 Ring of Fire USA to South America (Saros 134) much more at USA eclipses. On the Columbus Day Metonic exactly 531 years after first colonial crime, on Oct 14 1492. Exiting US at Corpus Christi ("Body of Christ"). Exiting Brazil at the city of Natal, translated "Nativity"

2027 Feb 6 Ring of Fire South America to Africa (Saros 131) see world eclipses. From Rio de Janeiro, the largest slave port in history of world to salve coast of Africa in 500th anniversary year of first importation of African slaves to the Americas.

2030 Ring Jun 1 of Fire Athens/ Constantinople/ Crimea/ Japan (Saros 128) See at world eclipses. Another great path from Algeria to Japan across the middle of Asia.

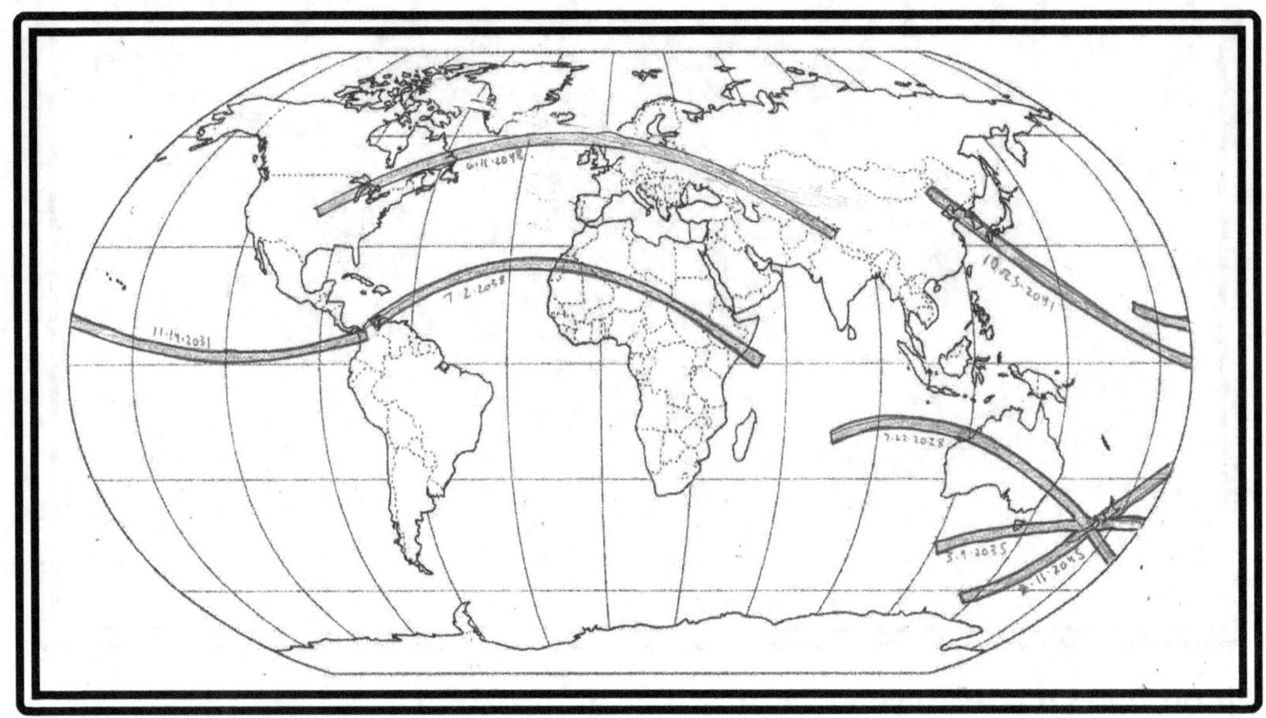

Hepton Two Chapter Eight—Fridays 2028 to 2048

2028 Jul 22 Total Australia Sydney/Botany Bay (Saros 146) 19th Anniversary of the Wuhan Totality of 2009. Australia bisected with totality in Sydney and Botany Bay, where the first prisoners were brought to colonize the continent 240 years earlier, in 1788. Crosses southern New Zealand. 177 days after Australia Day Ring of Fire of 1-26-2028. One lunar year from Mideast Eclipse of 8-2- 2027.

2031 Nov 14 Hybrid Pan American Isthmus (Saros 143) Perhaps the last totality for the great Saros 143. It completes its cosmic tasks with an open ocean eclipse ending with totality at sunset over the most extraordinary landbridge on planet earth. 180th anniversary of publication of Herman Melville's "Moby Dick" in 1851. 9.94% coverage at Dallas Texas 67 years, 11 months and 22 days after JFK murder on 11-22-1963. Saros 143 created the JFK Boston eclipse of 10-2-1959. Its next eclipse is on 11-25-2049, which would have been JFK Jr.'s 89th birthday.

2035 Mar 9 Ring of Fire New Zealand (Saros 140) 136 eclipse years exactly after we began our Hepton Two journey on May 7, 1902. A ring of fire begins south of Australia, just dodges the island of Tasmania and crosses the straits between the magical islands of New Zealand, including annularity in Wellington, where the author of this website was conceived.

2038 Jul 2 Ring of Fire South America to Africa (Saros 137)

2041 Oct 25 Ring of Fire China, North Korea, Japan (Saros 134) Crosses path of 3-9-2035 Total with apex along China-North Korea border.

2045 Feb 16 New Zealand (Saros 131) last eclipse of Australia New Zealand Swarm.

2048 Jun 11 USA to Islamabad (Saros 128) Heartland of USA to Baltics, Russia, Afghanistan. Ends at sunset outside of Islamabad, Pakistan. Annularity over Lincoln Nebraska exactly 193 Eclipse years after Lincoln's murder 4-14-1865.

Heptons as Cosmic Art

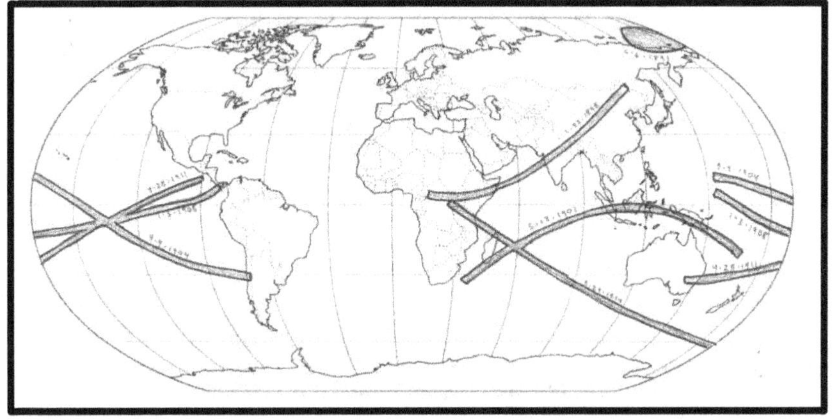

Above: Hepton 1 Chapter One 1891 to 1911

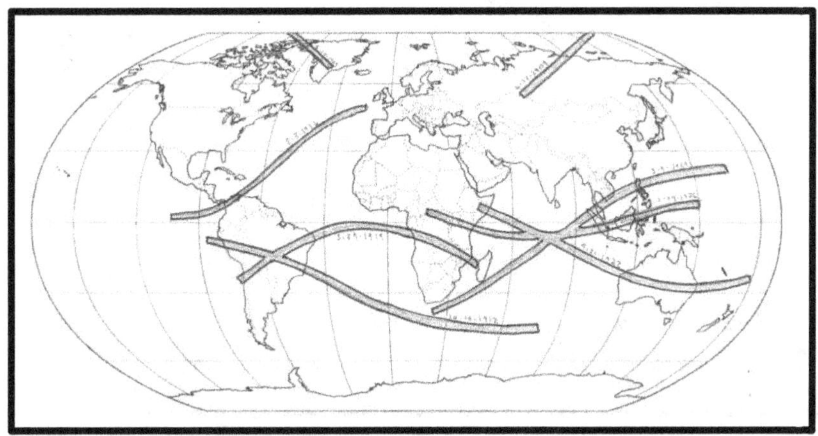

Above: Hepton One Chapter Two 1909 to 1929

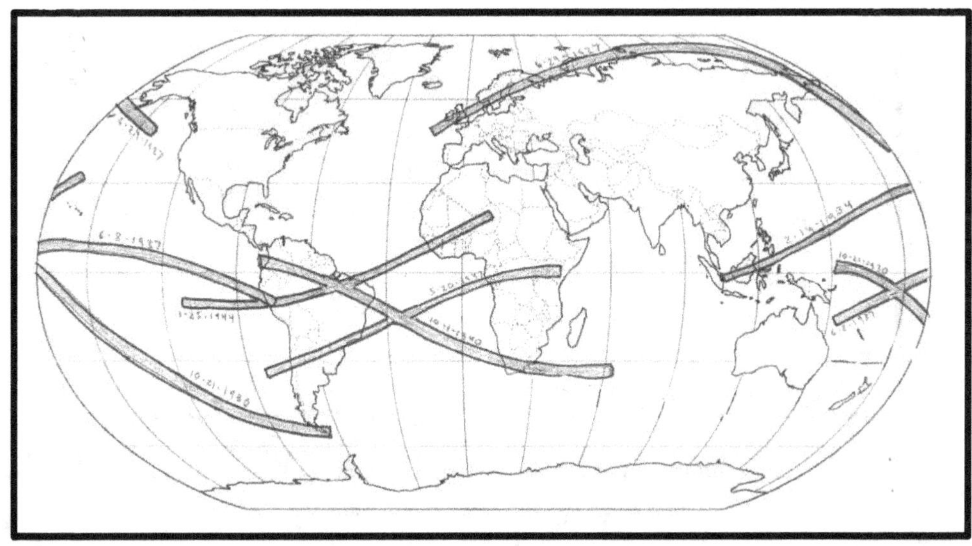

Above: Hepton One Chapter Three 1927 to 1947

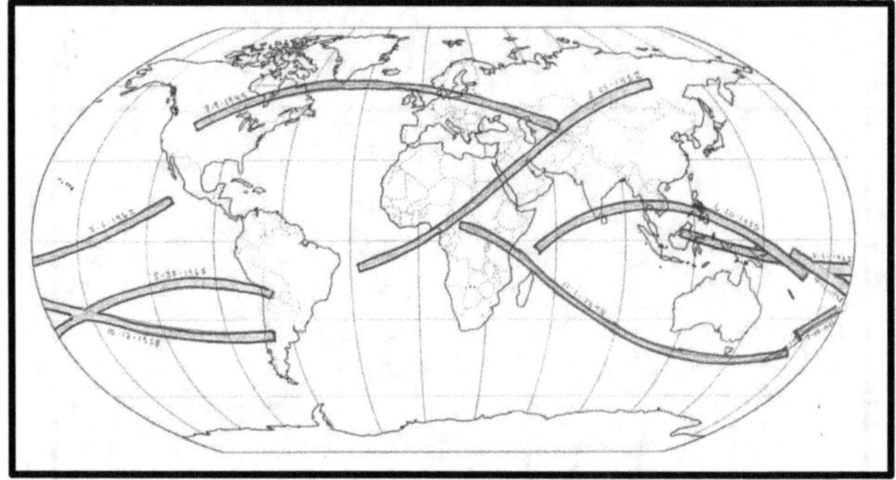

Above: Hepton One Chapter Four 1945 to 1965

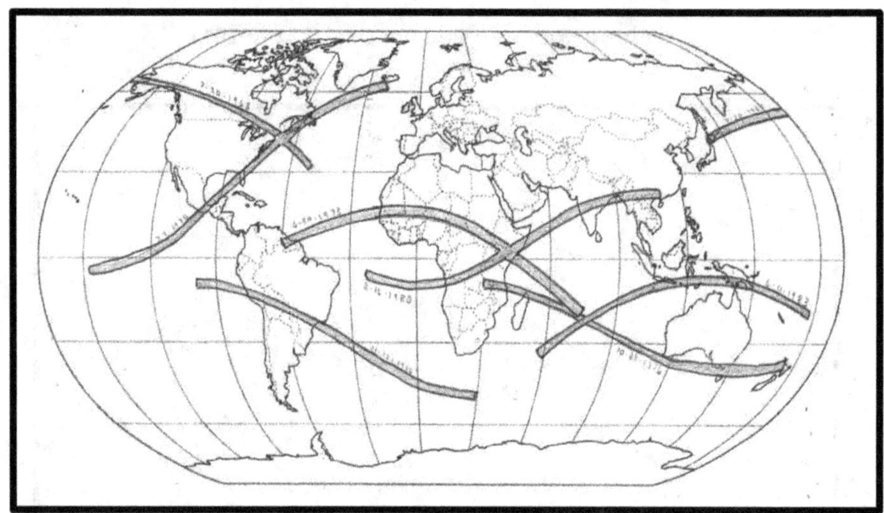

Above: Hepton One Chapter Five 1963 to 1983

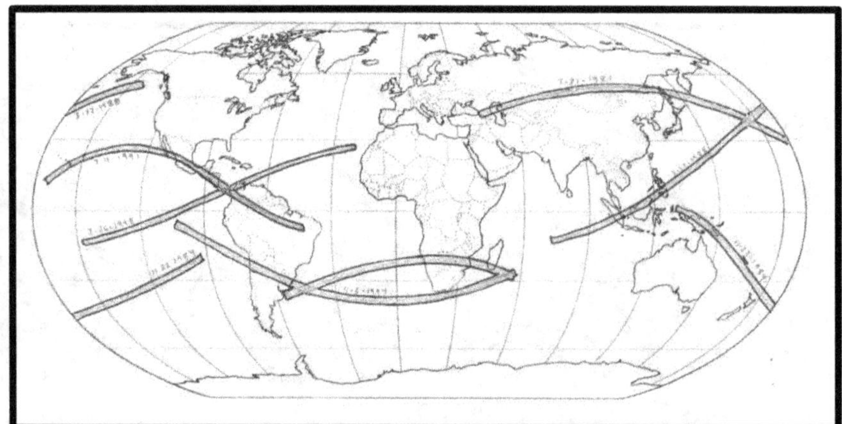

Above:Hepton One Chapter Six 1983 to 2001

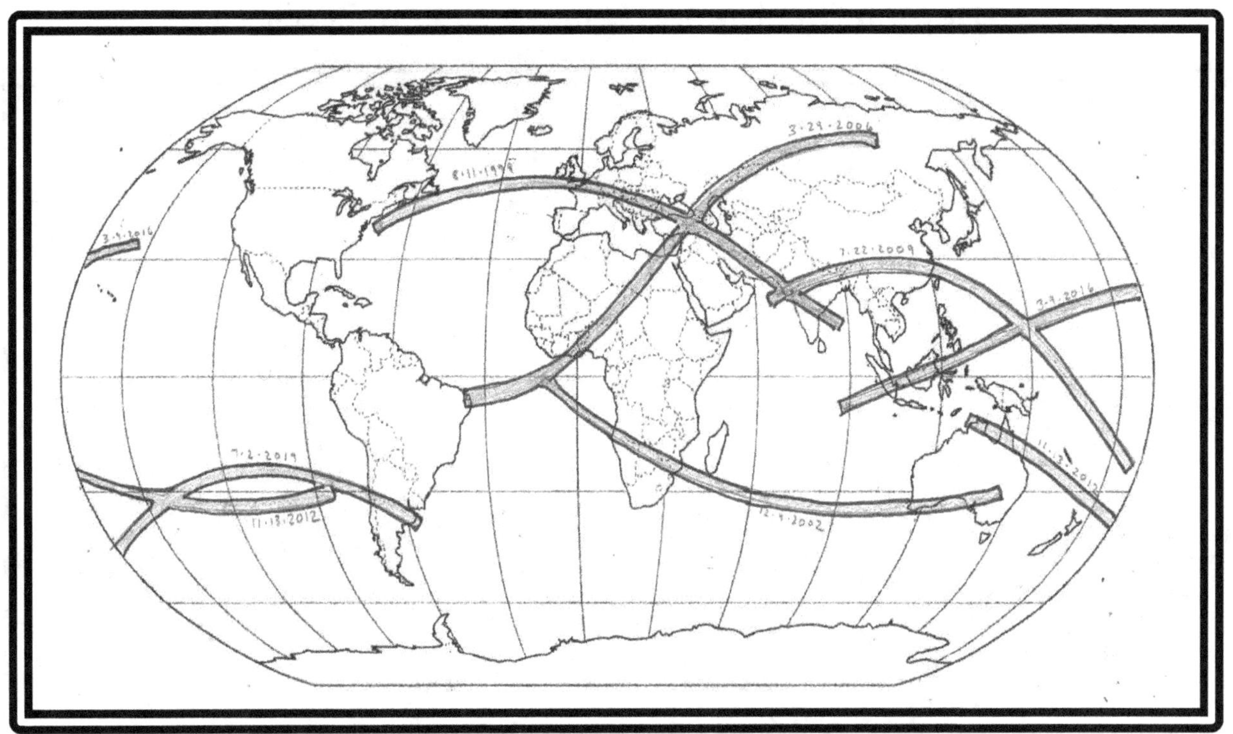

Above: Hepton One Chapter Seven Wednesdays 1999 to 2019 Turkish Cross

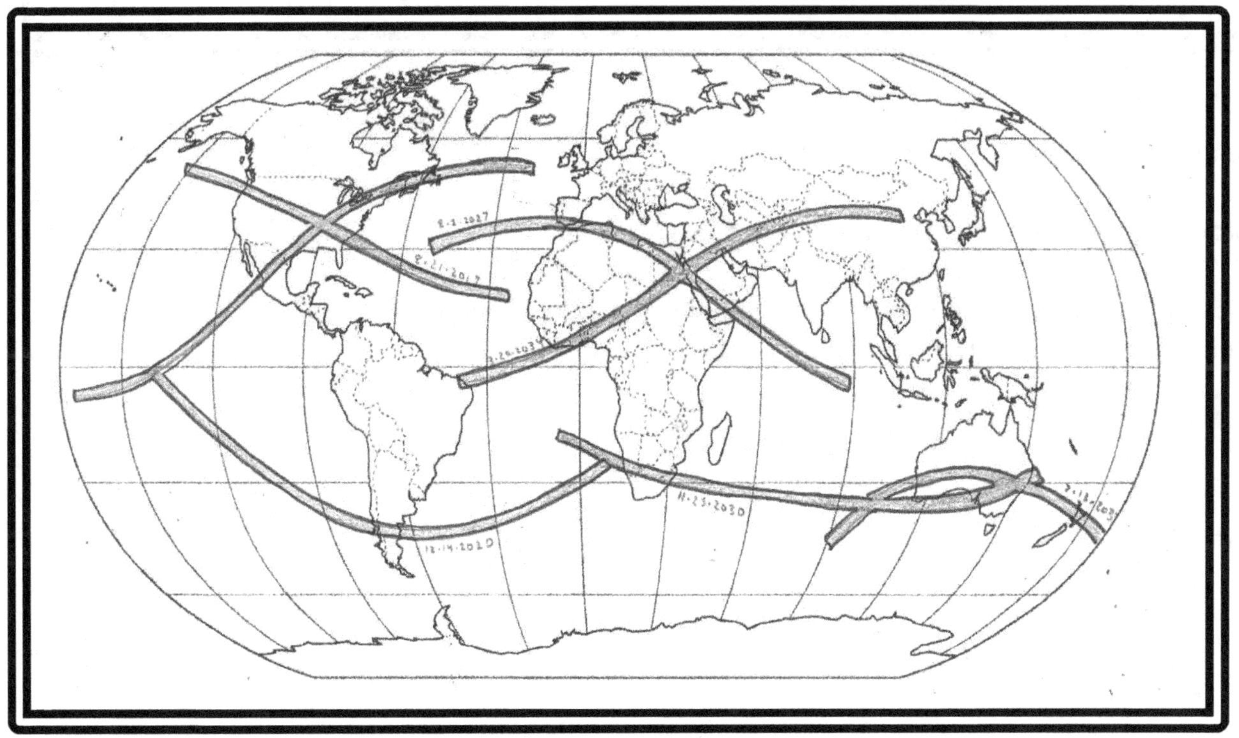

Above: The Great Chapter Eight- Mondays 2017 to 2037

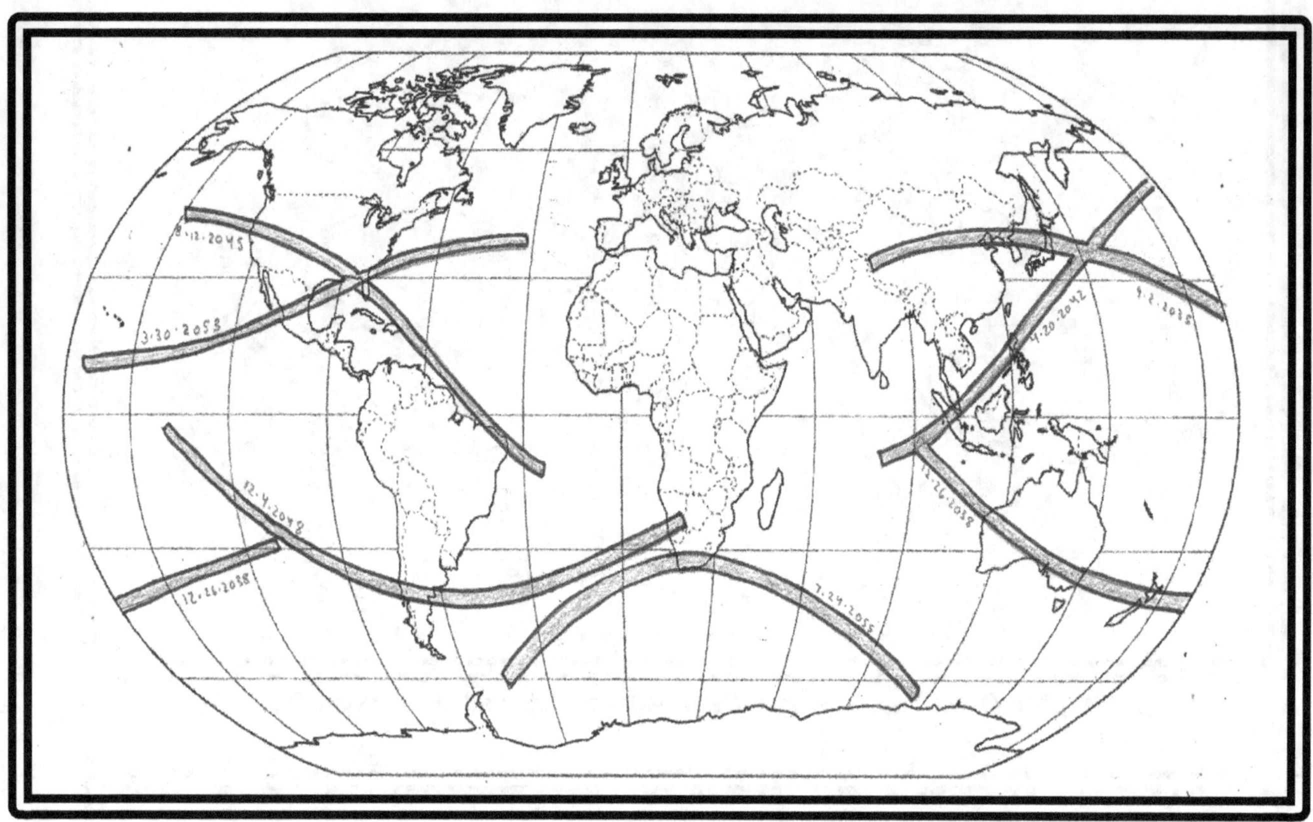

Wouldn't it be nice?

Hepton One Chapter Nine *Nineveh Clause* 2035 to 2055.

Obviously, the balanced Hepton crosses continue for a little while. Do we?
Further consideration at Nineveh Clause - The Sign of Jonah.

The Hepton Two Story in Pictures:

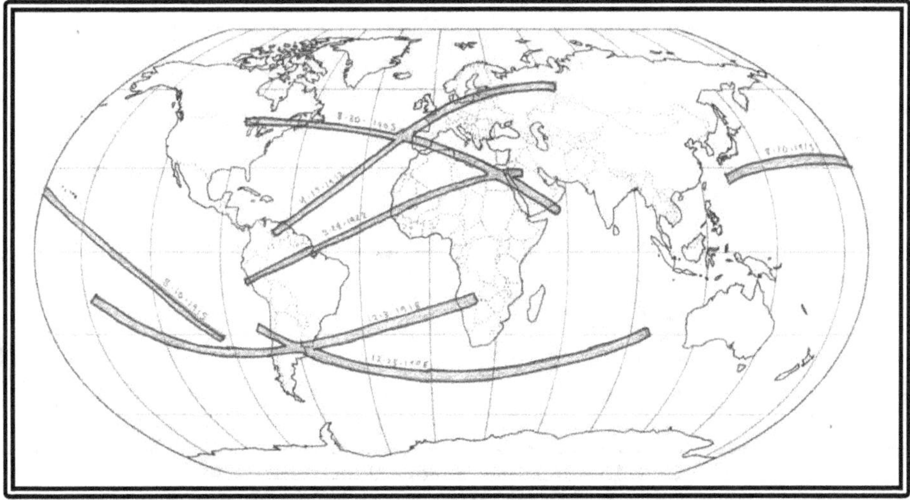

above: Hepton Two Chapter One 1902 to 1922 Wednesdays

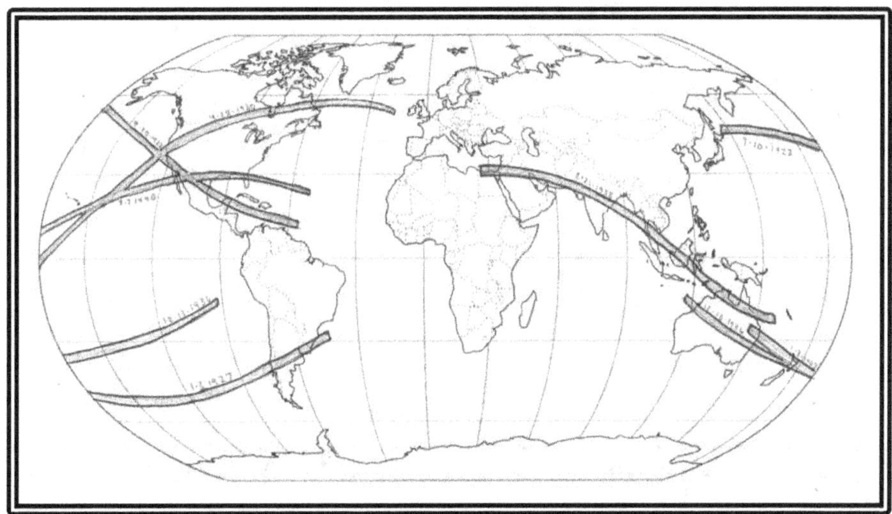

above: Hepton Two Chapter Two 1920 to 1940 Mondays

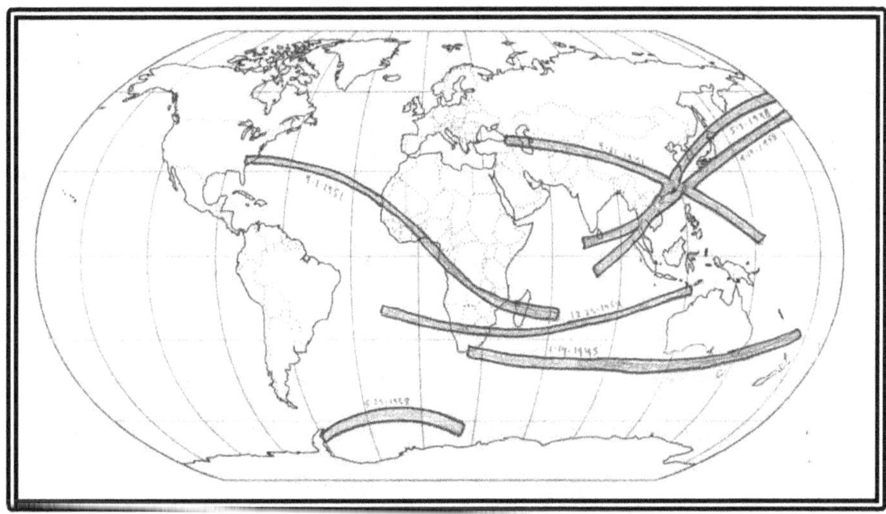

above: Hepton Two Chapter Three 1938 to 1958 Sunday/ Saturday

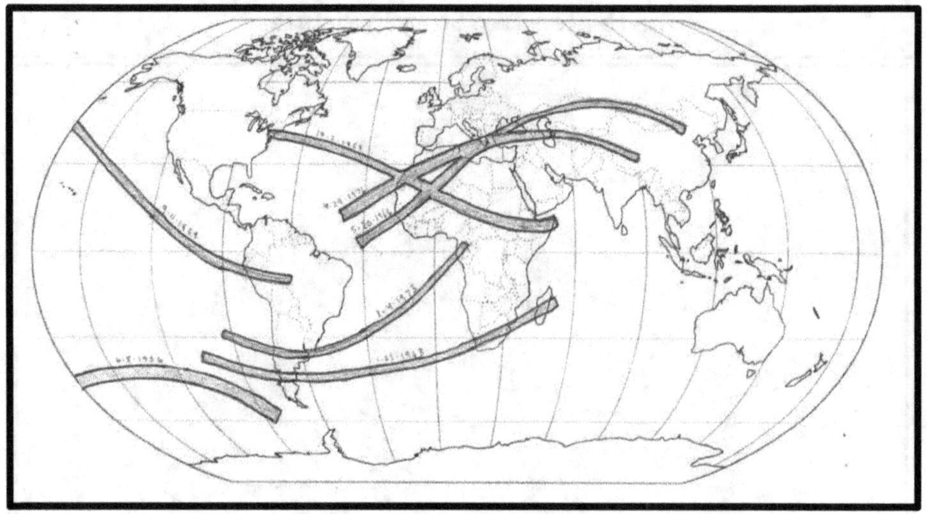

above: Hepton Two Chapter Four 1956 to 1976 Fridays

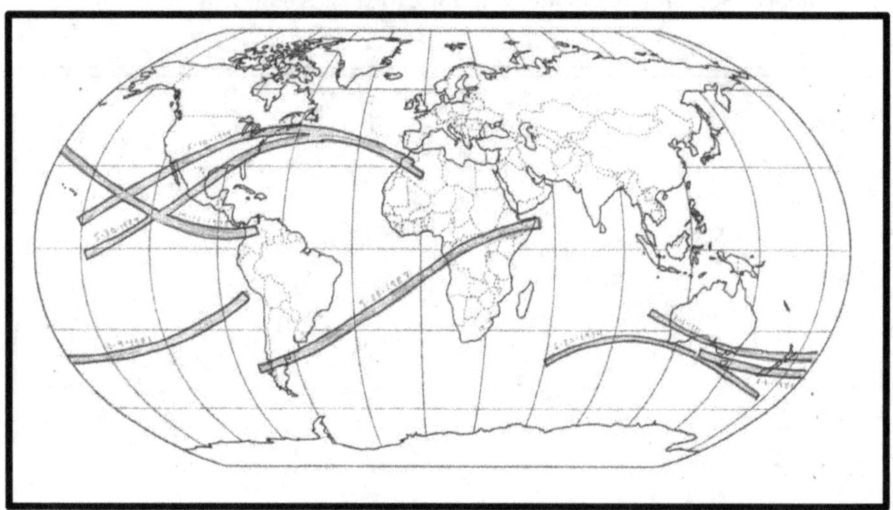

above: Hepton Two Chapter Five 1974 to 1994 Wednesdays

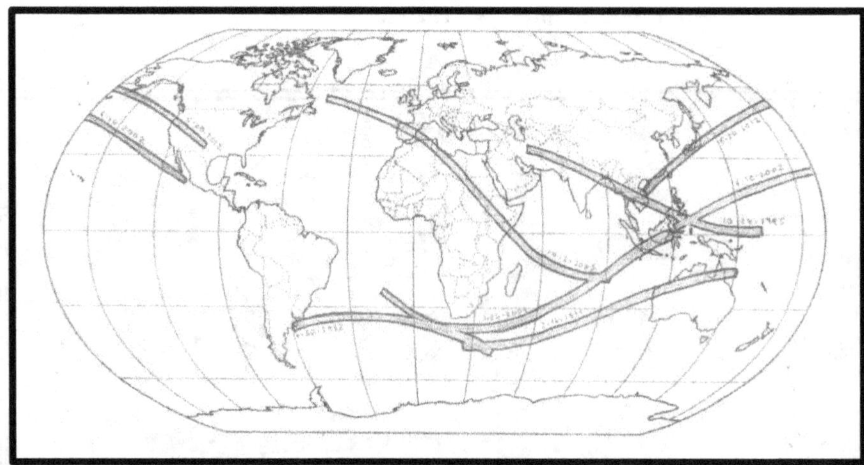

above: Hepton Two Chapter Six 1992 to 2012 Mondays

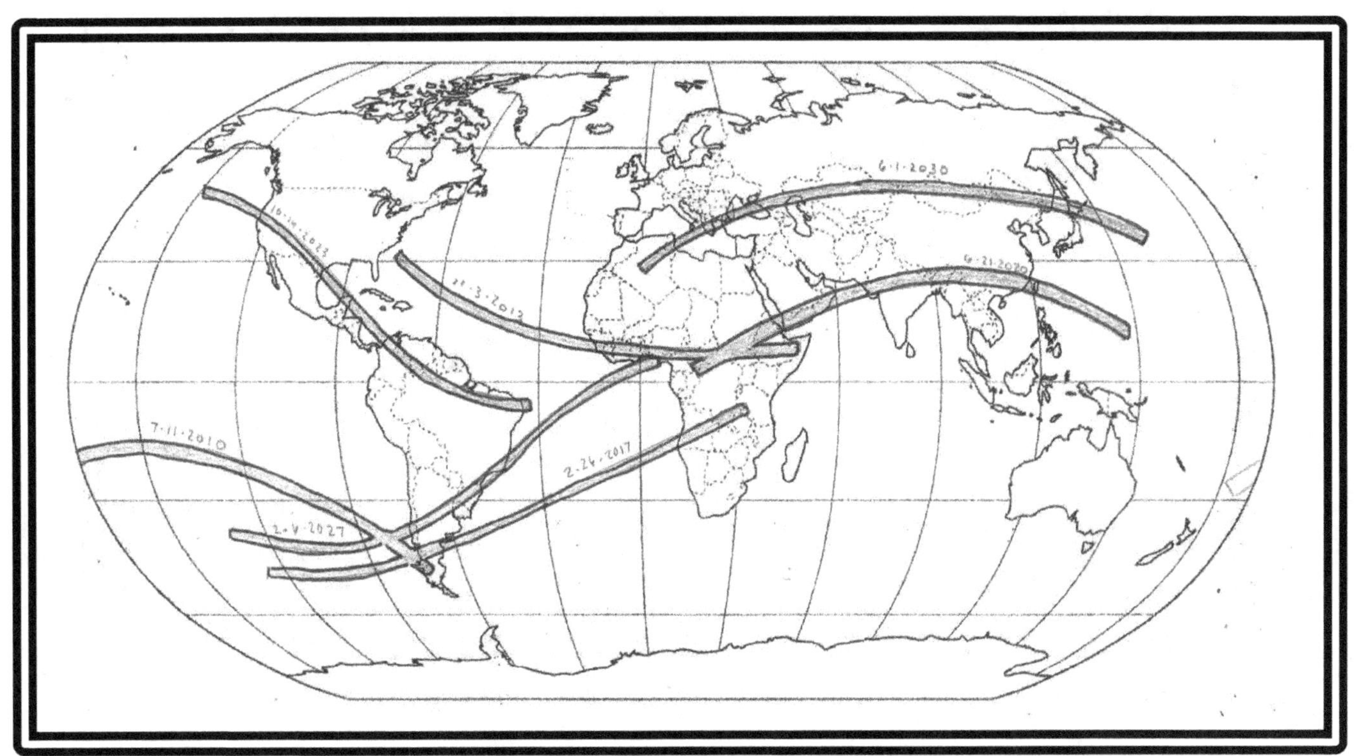

Above: Hepton Two Chapter Seven 2010 to 2030

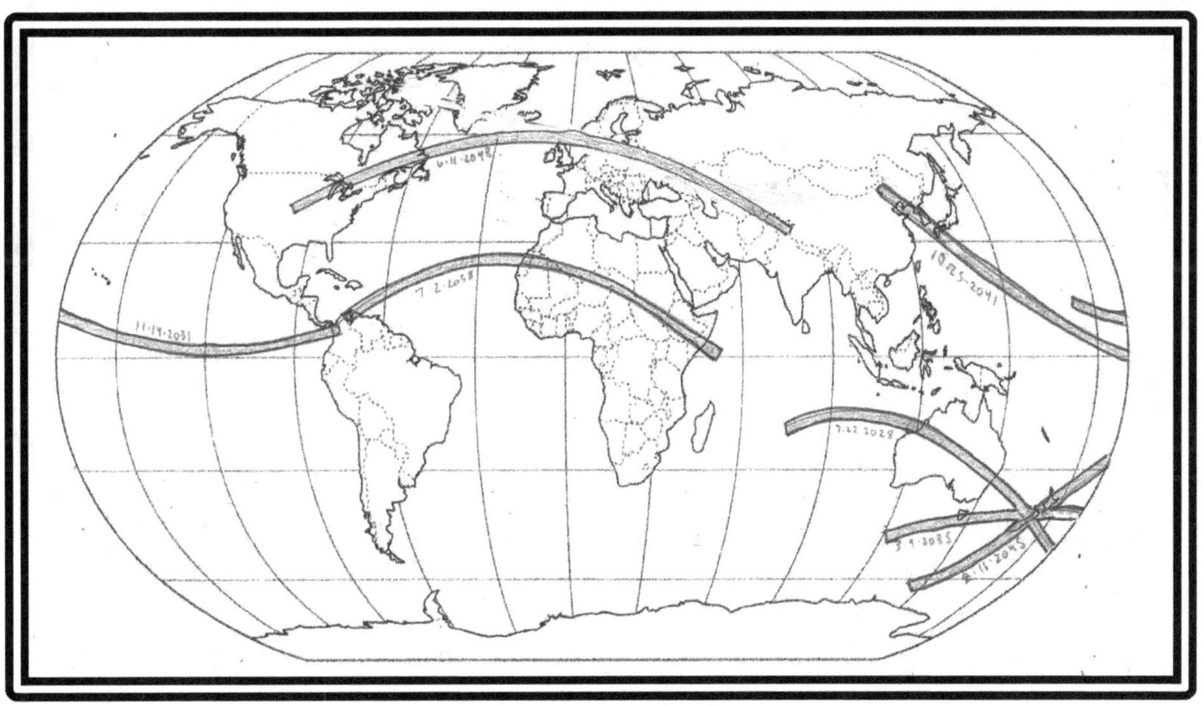

Above: Hepton Two Chapter Eight 2028 to 2048

A comparison of the Hepton Crosses 2017 to 2034 and the Hepton Crosses of 2045 and 2060

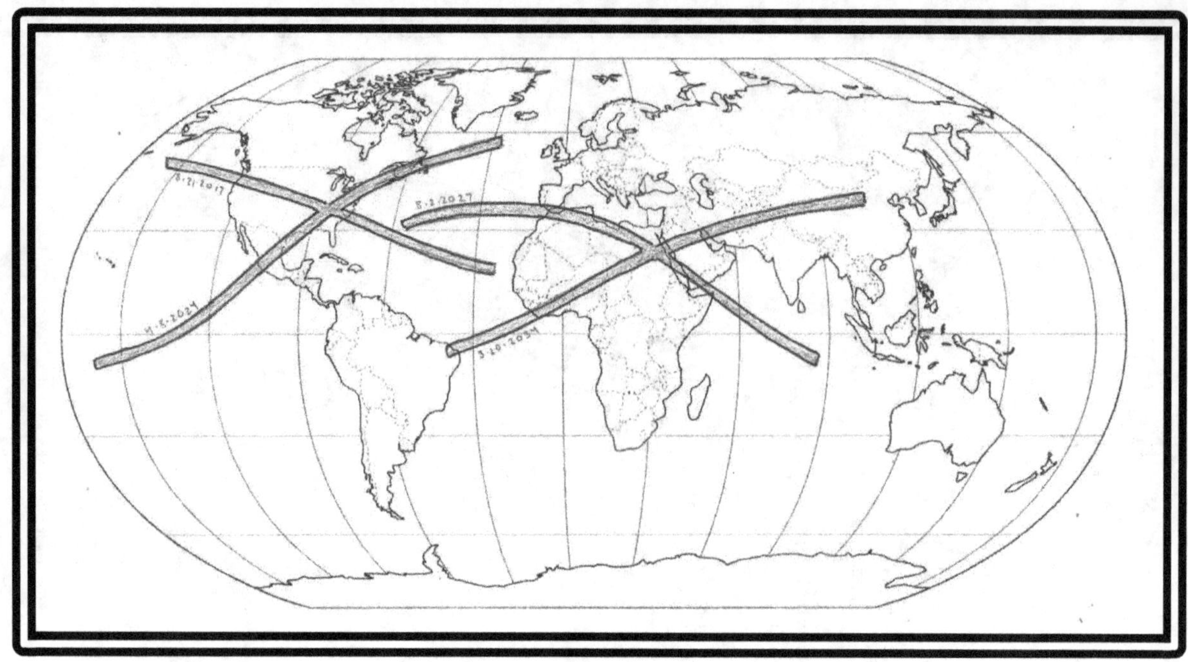

Above—2017 to 2034

Below: Switcharoo Hepton Crosses 2045 to 2060

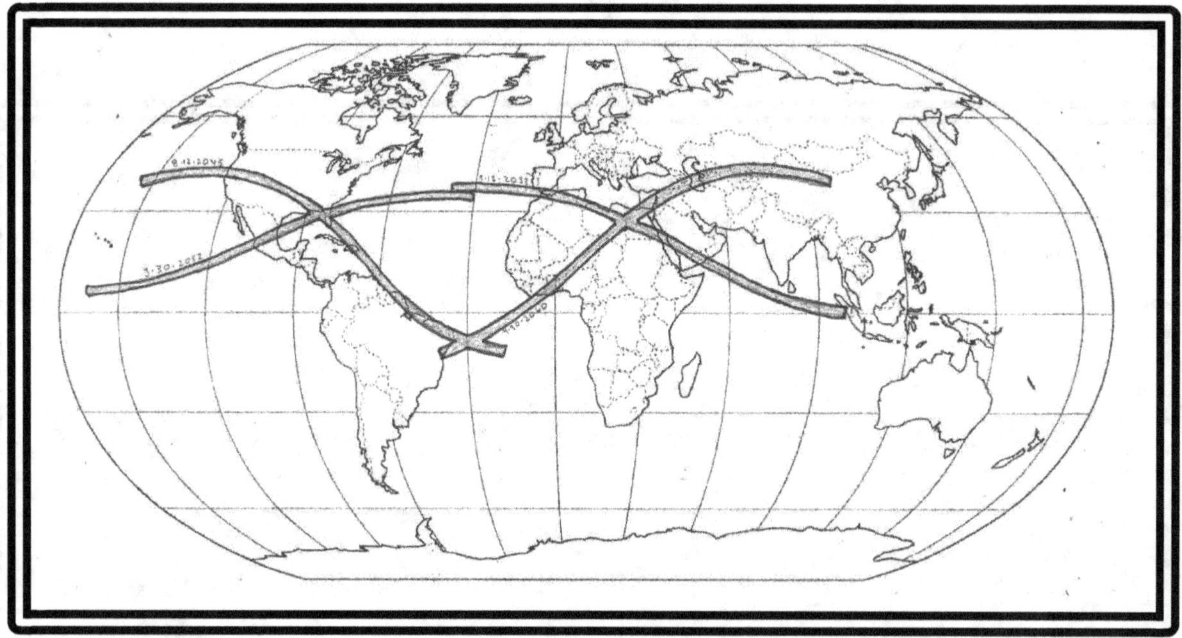

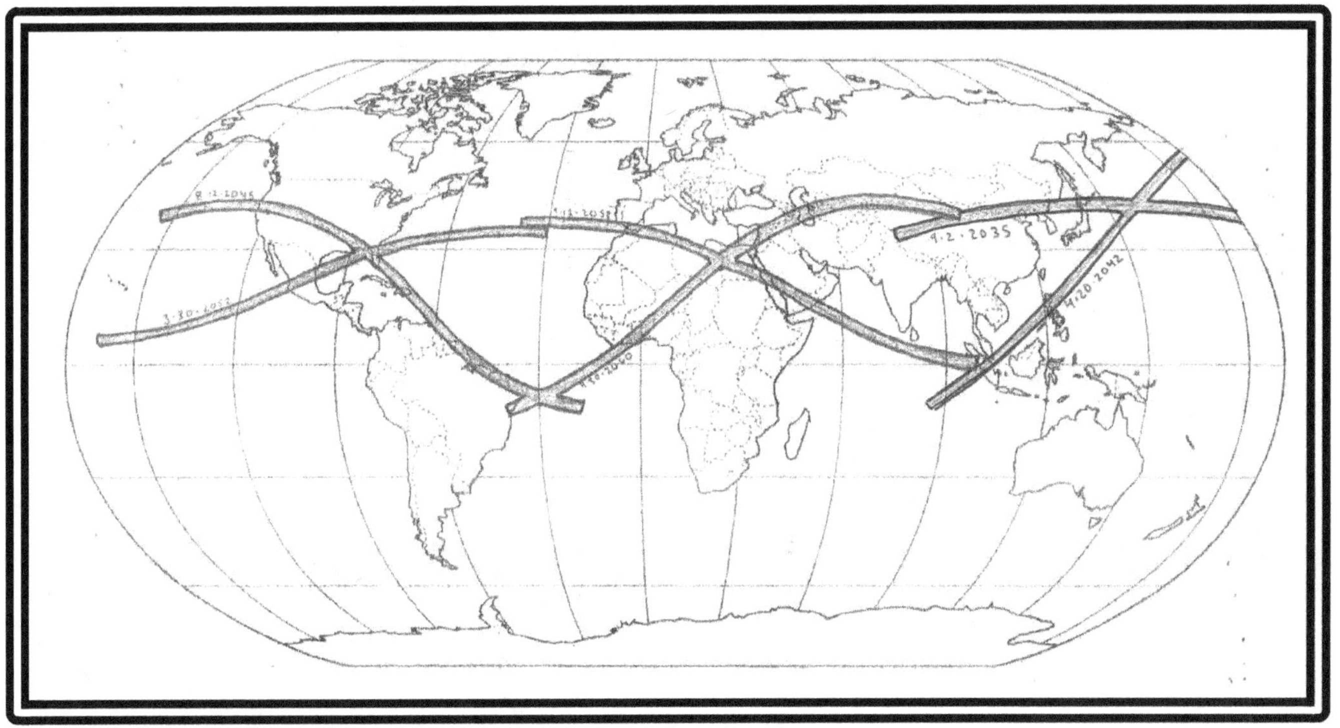

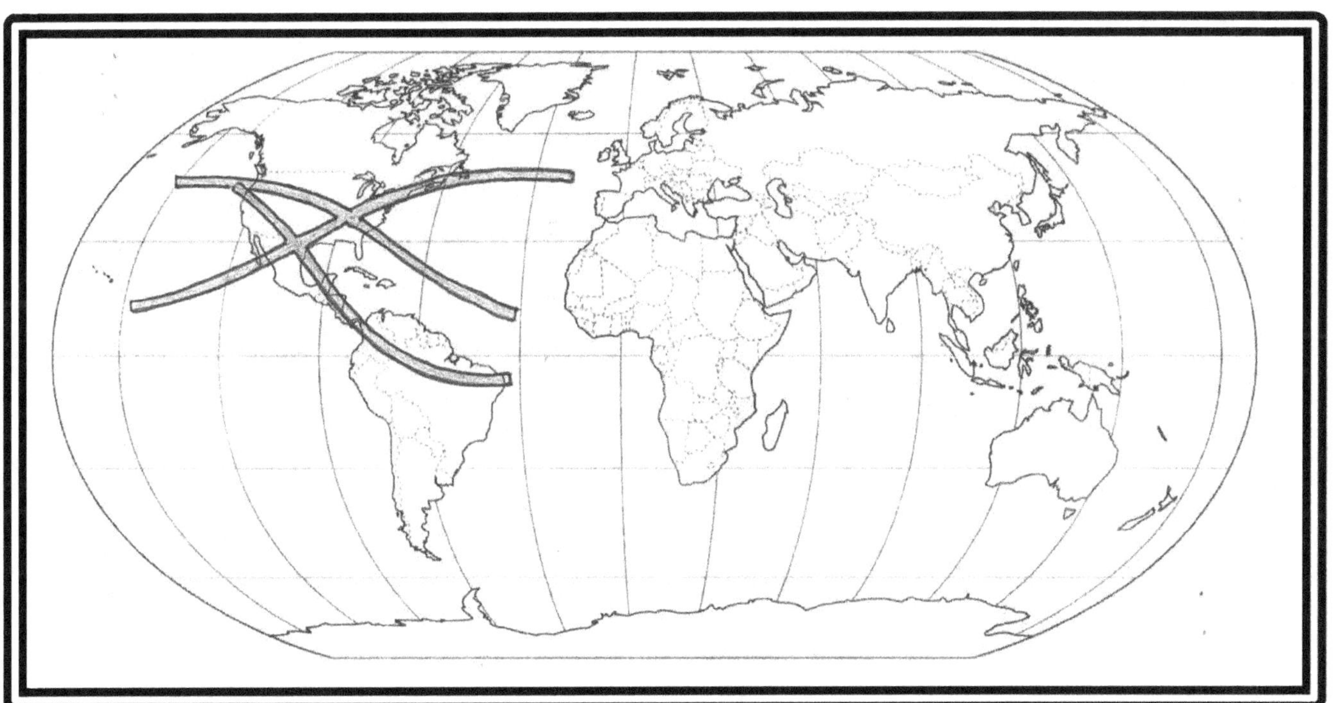

Cosmic Art from God.

Chapter 16

Peculiarities And Personalities

The Eclipse as Omen

"there will be signs in the sun and in the moon and in the stars"
-The words of Jesus (Luke 21:25)

Are eclipses omens? An omen is a portent, or a signal, of a coming event. Usually, omens are considered in a negative context. The darkening of the sun or moon during an eclipse has been considered an ill omen for millennia across all cultures around the world. It is such a universal phenomenon that it could be described almost as instinctive. Sometimes, meanings are different for solar and lunar eclipses. Solar eclipses can be so rare as to take entire generations by surprise.

Since solar eclipses began to be predictable, around 1700, scientists have informed us they are not omens. This is apparently because scientists uncovered elements of the cosmic mechanism for the eclipses, or more succinctly, the clockwork motions of these celestial bodies.

It is typical of Western thought that insight into the mechanism of a phenomenon is considered proof that the phenomenon is not supernatural. This type of logic's faults lie both in the conception of time and in the reluctance to acknowledge authorship of said phenomenon.

Scientists discover natural mechanisms. Does this grant natural mechanics the same soullessness as the mechanisms we humans create? Is that logical? Is the extraordinary mechanism of the human body the same as an automobile? Even if one was to accept such an absurdity, doesn't the presence of the automobile demand consideration of its creator's intent?

If there is an author/creator living outside the time domain, then mechanisms that author created could contain the artist's signature. The mechanisms in place (sun and moon) to illuminate and warm the earth could very well contain portents, messages or omens. It would be similar to a book author leaving "clues" within a text.

As far back as the 11th Century, there were "experts" (Medieval religious authorities) warning people against believing eclipses were omens. Clearly this was a response to a history of public anxiety in response to eclipses. There's clearly a desire from authority to maintain order that resurfaces time and again in such warnings; a desire to repel unpredictable expressions of human superstition.

Lunar eclipses are the most experienced eclipse. Astronomers have predicted them for 2500 years. It was well-known among learned people from an early age that the phenomenon was caused by the moon entering the earth's shadow. Solar eclipses eluded most predictions until the development of a fully measured globe and careful astronomical computation, which arrived around the time of Newton and Halley in the late 17th Century.

The first duty of the first eclipse mappers was to tell the "common people" that eclipses were not omens. The very first accurately predicted solar eclipse, which darkened London on July 13 1715 was accompanied by this hope from its famous predictor, Sir Edmund Halley:

"They (the people) will see that there is nothing in it more than Natural, and no more than the necessary result of the Motions of the Sun and Moon."-Halley 1715

There was clearly an interest in easing the fears of the common people. Similar reassurances were issued from the pulpits and Universities of England, Germany and France well into the 19th Century. Perhaps this was a response to previous chaos at such events. Perhaps it was about the triumph of logic and rational thought. Either way, an attitude of the new scientific priest class was established. "Established Authority"insisted that it is foolish (meaning wrong, or even dangerous) to publicly entertain the idea that an eclipse might be omens.

The tone of this "decision" bears consideration. It seems more than just a leader crying "no need to panic", for the general safety of a milling crowd. It seems, then and now, as a statement of theology. It is a pronouncement that the authority has determined there is no "meaning" in this peculiar visible phenomenon. This determination accompanied a turning point in the history of humanity, the rise of the technological era.

This raises the question: because one understands some of the physical properties of a phenomenon, does that make any meaning besides physical mechanics disappear? If one understands the motions, muscles and hormones that bring about a hug between two people, do we know why the hug occurred? Do we understand why hugs exist? Is it possible that different people might have differing views of hugs that are simultaneously true?

Solar eclipses present to the human eye a physical, cosmic and potentially symbolic story. It stirs deep emotion in the viewers. The witnesses wonder without authoritative prompting, what could it mean?

Science rushes to tell us the meaning of the eclipse story is partly its mechanics and partly their scientific brilliance in presenting these mechanics. Science tends to say there is no greater meaning. Don't look for an author. Keep focused on the wheels, the figures, the numbers, the tables and illustrations. Or better yet just let the professionals do it for you. According to science, eclipses are periodic meaningless turning of astronomical wheels. We know their story. It's all we have been taught for generations.

Before any solar eclipse occurs over a populated areas these days, "experts" rush to remind us that while, once upon a time, primitive people were superstitious about such things, these days we "know" better that they are simply meaningless natural phenomenon.

What do the Media Say?
Here's a very small sample from the internet, from the days before the Aug 21 2017 USA eclipse:

Today, we know that the Sun is a star and that eclipses aren't omens. A total solar eclipse happens when the Sun, Earth, and Moon line up. Aug 17 2017 — Boulder Daily Camera

In ancient history, eclipses were seen as ill omens, before people understood exactly what caused this. Now we know it's just simple celestial geometry –National Space Center. UK Jul 26 2017

They (primitives) thought that it was an omen of disaster and devastation. Those of us
who aren't superstitious put no stock in those old beliefs. --Zanesville Times Recorder Aug 19 2017

In the modern age, scientists can predict when and where these cosmic events will occur, and
skywatchers can appreciate their beauty rather than fear that the events might bring devastating
consequences. Even though most people today have access to science-based information about
eclipses, misinformation, myths and superstitions continue to surround these celestial events. –
Space.com Aug 2017

As long as eclipses have occurred, humans have interpreted them as a sign of something. Yet, even
during what were perceived to be highly superstitious times, ancient astronomers could already
believe there must be something more to the phenomenon. Time Aug 16 2017

And, the winner is! :
Given that nature has manufactured untold numbers of these planets, it's a safe bet that someday
we'll find one with a moon that's just the right size to block its sun. Maybe we'll even find members of
another intelligent species who have also watched, in astonishment, as their local star vanished at
mid-morning.... A solar eclipse still exerts a pull on the dark side of the human imagination, but it
weakens with every century. --Ross Andersen/ Atlantic Magazine Aug 10 2017

Mr. Andersen states that "nature" manufactured this Universe. He claims it's a safe bet you'll find a
planet just like ours-one with a same-sized sun and moon, no less! Apparently, there's even smart
creatures there too. Does that seem true? Is such conjecture somehow more sophisticated than
believing God actively creates a phenomenon for the purpose of human consideration?

The sum of the public modern narrative is this: dumb people used to believe superstitious things, but
now smart people can safely assert every natural phenomenon to be completely meaningless. OK, we
hear you. But to me, the certainty of the argument has a smell of deep uncertainty about it.

 "Thou dost protest too much, methinks", as Shakespeare said through the Queen in *Hamlet*.
Earlier in the play he had Hamlet says these famous lines:

"There are more things in Heaven and Earth, Horatio, than are dreamt of in your philosophy "
Hamlet Act One Scene 5 William Shakespeare
Who was wiser? Shakespeare or Atlantic magazine?

The peculiarity of the solar eclipse phenomenon is different from all other cosmic peculiarities and
coincidences by orders of magnitudes. An eclipse is an improbably tiny tip of a shadow brought by a
perfect alignment in the perspective of observant beings of two wildly different sized and distanced
objects-which returning at predictable intervals-leaving a distinct path across the face of earth.

I'm saying God writes upon the Earth. Perhaps that's heresy to certain court magicians. But is it an
omen? They look like omens sometimes. Other times it seems more like a song. Sometimes it's
humorous, sometimes a lament. Proof? There's no proof in this pudding. You can say "so what" to
anything. But after a bit of research, I fail to see much good that has come to human dignity from
submission to the law of coincidence.

Want An Argument For Omen Eclipses?

Here's 10 examples of solar eclipses that could easily be construed as omens. Full descriptions are found at the world or US eclipse sections:

Trail of Tears Total Nov 30 1834: Path follows route of displaced Cherokees while they are on it.
Titanic Hybrid Apr 17 1912 Path across Atlantic, England and Europe 60 hours after Titanic sinks.
World War One Total Aug 21 1914 Path divides Europe in the first days of the Great War.
JFK Death Premonition Nov 22 1919 Path in Texas/Cuba 44 years to day before JFK killed.
Jerusalem Ring of Fire Aug 21 1933 Across Jerusalem and Babylon as Nazis rise to power.
Atom Bomb Eclipse Total Jul 9 1945 Path goes from USA to USSR one week before first A Bomb Test.
Korea Divided Total May 9 1948 Korea divided by path along 38th Parallel two years before war.
Sharud Iran Cross Totals 1952 and 1954 two eclipses over Iran at time of coup and Shah's installation.
Vietnam Cross 1955 Total and Ring of Fire Vietnam has two eclipses the year the Vietnam War starts.
Mount Saint Helens Feb 26 1979 last US total eclipse for 38 years over volcano year before she erupts.

This is a short sample of potentially "ominous" eclipses. Omens are subjective, of course. You can't prove meaning. Is the supernatural awe humans have always felt towards this phenomenon justified?

Some Cultural Responses

Cultures around the world felt that the Sun and the Moon become dark during eclipses because something was happening to them. In most cases, even after the idea of the moon entering earth's shadow became known, the common people assumed that some form of dark entity was attempting to eat or blot out the Sun or Moon. Most cultures attached ominous meaning to the eclipses.

Bible: Solar eclipses are not specifically mentioned in the Bible, though there are several supernatural events involving the sun and several prophecies concerning the darkening of the sun around the time of Judgment Day. The potential for signs in the sun and moon is mentioned near the beginning of Genesis (1:14) and by Jesus in the New Testament (Luke 21:25). There is no Hebrew word for "eclipse". The Jewish Talmud apparently says that a solar eclipse is a bad omen for the entire world, which runs according to the solar calendar, and a lunar eclipse is considered a bad sign for the Jewish nation, with its lunar year (according to Chabad.org)

Quran: Solar Eclipses are not specifically mentioned in the Koran either, though Mohammad addresses them in later sayings as "signs of Allah" and a call to prayer.

Hinduism: Hinduism addresses solar eclipses in the Vedas, attributing them to a a demon called Svarbhanu, who attacks the sun.

Buddhism: Tibetan Buddhists believe that during eclipses, the effects of one's good or bad deeds are multiplied. Other Buddhists feel, as many cultures do, that it is an important time for prayer.

Native Americans: Native beliefs are, of course, varied. Many, like the Navajo, avoid the eclipse and pray. Aztec and Mayan civilizations predicted lunar eclipses and put great meaning on all eclipses.

Babylonians, Greeks and Romans: All three pillars of western civilization considered eclipses omens, often with bad implications for the king or nation. All three cultures recorded dates of local eclipses.

Egyptian: Ancient Egyptians had a myth of a snake, Apep, that attacks the boat of the Sun god which is now believed to refer to solar eclipses.

China: Chinese traditions of astronomical observations can be dated to 2650 BC and they had sophisticated observatory buildings by 2300 BC. Systematic Chinese records of eclipses begin around 720 BC., around the same time as the Babylonians. In ancient China, solar and lunar eclipses were often omens that foretell the future of the Emperor. Ancient Chinese believed that solar eclipses occur when a celestial dragon devours the Sun. This belief is the source of the term "draconic" to describe the eclipse month and year. They also believed that the same dragon attacks the Moon during lunar eclipses. It was a tradition in ancient China to bang drums and pots and make loud noise during eclipses to frighten that dragon away. Chinese Emperor Zhong Kang of China supposedly beheaded two astronomers, Hsi and Ho, who failed to predict an eclipse in 2300 BC.

Personalities -Eclipse Scientists:

Thales, the Milesian (c. 624 – c. 548) sometimes called the "Father of Science" , Thales is the first person credited with predicting a solar eclipse. This prediction is mentioned by **Herodotus (624-547 BCE)**, concerning the total eclipse of 28 May 584 BCE, which put an end to a conflict between the Lydians and the Medes.

Herodotus wrote:" ... *day was all of sudden changed into night. This event had been foretold by Thales, the Milesian, who forewarned the Ionians of it, fixing for it the very year in which it took place. The Medes and the Lydians, when they observed the change, ceased fighting, and were alike anxious to have terms of peace agreed on".*

It's remained unclear for centuries how such a prediction could have been made, as it is generally understood that the scientific knowledge necessary for such a prediction was not available at the time. While that is true for solar eclipses, Babylonian astronomers were already well aware of the Saros 18-year 11 day repetition for lunar eclipses. By an extraordinary geographical happenstance, an exact Saros period of 18 years and 11 days separated the eclipse May 28 584 Thales predicted from a prior eclipse in the "neighborhood" on 602 BC May 18. They were both members of Saros 57, which simply managed to include some of the same general geography in both appearances. This is a difficult feat for solar eclipses, which tend to move farther around the globe, but it is not altogether unknown. The same thing happened in Midland Texas between ring of fires on May 10 1994 and May 20 2012 with Saros 128. The 602 BC eclipse was centered on Persia and Babylon, and likely made quite a stir. The 584 BC eclipse had its center more towards the Mediterranean, including Greece.

If the date of the Babylonian eclipse and the astronomical knowledge of Saros Cycles had been imparted to Thales and Grecian astronomers, a lucky educated guess (which would be difficult to repeat) could have easily been the cause of the first successful solar eclipse prediction.

Edmond Halley (1656 to 1742)

Halley was a British Scientist, astronomer, and friend of Sir Isaac Newton. He is the man whom Halley's comet is named for. He produced what is considered the first accurate predictive eclipse map, for a May 3 1715 (Saros 114) total eclipse over England.

It is certainly quite peculiar and fortuitous that such an eclipse would occur over the hometown (London) of the great master of eclipse prediction just as he was perfecting the art of eclipse prediction. He predicted the eclipse correctly with only minor errors.

Theodor von Oppolzer (26 October 1841 – 26 December 1886)

Oppolzer was an Austrian astronomer and mathematician of Bohemian origin. He was born in Prague. Clearly a genius, one of his famed feats was memorizing the values of 14,000 logarithms. In 1868 he led an expedition to observe a solar eclipse (1868 Aug 18 Total -Saros 133).

The year after his death (1887), Oppolzer's work the *Canon der Finsternisse,("Canon of Eclipses")* was published. An authoritative compilation of the 8,000 solar and 5,200 lunar eclipses from 1200 BC until 2161 AD, it was understood as one of the greatest computational feats of its day.
He also published a two-volume manual on the determination of the orbital elements of comets and planets. Both of his works served as standard astronomy references for many years. (Wikipedia)
His book is credited as important source material for much of modern research, including the work of Van Den Bergh and Espenak. It is only available through rare book sellers.

Oppolzer's work provided an important lesson about eclipses. There's both a way to predict them in the future and to chart their past paths. The math is very precise, but not super easy. I don't know how to do it, and I am grateful to eclipse scientists for taking care of it for us.

George van den Bergh (born Apr 25 1890 / died Oct 3 1966)   There is little information about George Van Den Bergh besides a brief Wikipedia page. His greatest claim to fame is naming the Saros personalities by giving them numbers. This feat greatly facilitated the ability to organize and understand eclipse cycles. Van Den Bergh also discovered (and re-discovered) many eclipse cycles. Van Den Bergh was a Dutch law professor and is called an "amateur astronomer". It's unclear why he was considered an "amateur". Perhaps because he was a lawyer by trade. He did write an accessible book about the cosmos, *The Universe in Space and Time, (1937)* which was published in the UK and the USA. I bought a used copy. His style is inviting, easy to understand and contains a great chapter on eclipses, a subject he clearly loved. He was probably the greatest eclipse expert of the 20th Century. There is something peculiar about the sheer unknown-ness of George Van den Bergh.

This is interesting, however: Van den Bergh was the son of Samuel van den Bergh, one of the founders of Unilever, which in 2020 was the fourth largest consumer goods company in the world. George did not go into the family business, but instead pursued an academic career and became a lawyer in Amsterdam. From 1925–33, he was also a member of the Netherlands House of Representatives.

He was arrested and detained by the Nazis in 1941. I can find no specifics. He survived the war. In 1936, he had given a speech to the Congress concerning the problem posed by anti-democratic groups in democratic countries. The text of the speech is available with some searching online. Such a public stance was perhaps a reason he was targeted by the Nazis.

During the 1950s he studied the longer eclipse cycles. This work was published in 1951 as *Regelmaat en wisseling bij zonsverduisteringen, met een aanhangsel over maansverduisteringen* and translated into English in 1955 as *Periodicity and Variation of Solar (and Lunar) Eclipses.*
It's as unavailable as any book I've ever looked for. No pdf's online. No quotes. What did he say in it? Apparently, there is a crater on the Moon named after him.

He is credited by NASA's chief eclipse scientist, Fred Espenak, with numbering the Saros and other achievements. He named the Hepton, Octon and others. It's unclear why his work has largely disappeared. Perhaps because he was an amateur. Perhaps because he said this:

"How inconceivably greater (than was then surmised) is the Creation to him who believes in a Creator! And he who cannot believe in a Creator, should not he, too, bow his head before this wonderful and mysterious Universe?"

George van Den Bergh- *The Universe in Space and Time*- Final words of Chapter 11 (1937)

Fred Espenak (b. Aug 1 1953 USA) and Jean Meuss (b. 1928 Belgium)

We must give credit to the two men probably most responsible for making sure that the world has maintained easy access to accurate records of the eclipse phenomenon. The Two Volume *Five Millennium Canon of Solar Eclipses (2000 BC to 3000 AD)* is an indispensable resource for those wishing to track historical implications of eclipses and their cycles. It has the benefits of all the technology of the USA's space agency, NASA, and is extremely easy to comprehend. Fred Espenak has been working with NASA on eclipses since 1978 and is known as Mr. Eclipse. Jean Meuss was his partner in writing the Canon. Espenak was born into the eclipse Metonic. There was an eclipse over Russia on Aug 1 2008, on his 55th birthday.

Part Two of Personalities: The Mystery of the Numbers Man

What was the challenge Norman Bloom, the "Numbers Man", came up with that elicited the response of two of America's greatest minds?

The cast of characters:

Norman Bloom: Eccentric, mentally ill homeless ice cream truck driver obsessed with numbers.
Martin Gardner: Renowned popular mathematician and magician.
Carl Sagan: Popular astronomer and television personality. Easily the most famous popular scientist of the late 20th Century.
Larry King: Radio and TV Talk show host. King (Born Larry Zieger) was the guy who let Norman Bloom talk to the public. Bloom was a regular caller to Larry King. It was King who first called Bloom "the Numbers Man." During his career, King did more than 60,000 interviews. CNN's Larry King Live became "the longest-running television show hosted by the same person, on the same network and in the same time slot" as recognized by the Guinness Book of World Records.

These four unusual characters appear in a mysterious true-life interaction over the legacy of a "$1,000 challenge" to refute Norman Bloom's "proof" that God exists.

The Mystery: What was the proof? What did Bloom really say in his publication, THE NEW WORLD, that caused Gardner and Sagan to both reference him in print in the 1970's?

Last year, I read Martin Gardner's *Amazing Numbers of Dr. Matrix*, a collection of columns from Scientific American which appeared in the 1960's and 70's.
This is what Dr. Matrix said in Chapter 2: *Los Angeles*:

Dr. Matrix continued. "It is in the physical world, however, that we find the most remarkable patterns. A numerologist knows it is not a coincidence that the sun's disk viewed from the earth, is almost identical in size to the moon's disk or that the sun's period of rotation is almost exactly the same as the moon's period of revolution (sidereal month) around the earth.* (About 27 days)

Here, in the *Magic Numbers of Dr. Matrix,* Gardner provides one of the only examples in print of a scientist (though speaking through an alias) marveling at the reality of the sun/moon alignment. It appears to have been written as a column originally in 1960 or 1961 and then updated later. It's perhaps the only creative reference to the sun/moon alignment from a popular 20th century personality.

I felt vindicated! At least someone in history is in the "Sun's the same Size as the Moon in the Sky" Club. Then I saw the footnote. It got even better. The footnote referred to Norman Bloom. He seems to be the first guy in the USA to go around talking "cosmic clocks moon sun 19 year "mystery stuff. I guess the cliché would be to call him an American Original. Even better, call him the Numbers Man.

What was the footnote? Why does it matter? First let's find out about Norman Bloom.

Norman Bloom (exact birth and death date unknown--died late 1989 or early 1990

Information from *NPR (who talked to Bloom's family)/* and summarized *in Kooks Magazine-Feb 1990*

Norman Bloom was a New Jersey native who was Jewish by descent. His birthday and death day are unclear. For a while in the early 1950's, he was a family man adored by his children. He was also a near genius level musician. He made some poor investments in the 1950's, was divorced by his wife and gradually became eccentric and preoccupied. In 1962 he decided to go public.

"In 1962 (Bloom) rented Carnegie Recital Hall for a concert which was supposed to vindicate and validate what had become his preoccupation: The Bible. The music was impressive, but almost no one showed up and Bloom was shattered. "From the NPR 1990 broadcast upon Bloom's death.

After the disaster of his concert, he began an all-consuming obsession with numbers. In the summertime, Bloom drove an ice cream truck in New Jersey neighborhoods. People considered him a kindly character who was loved by the kids. The ice cream job paid for his true profession: numbers work. The work that has survived consists of tiny type, disorganized number associations and equations crammed on legal-sized or 11x17 paper. Part typewritten, part hand-written scrawls filled with numerical associations, Biblical prophecy, sports scores, historical numbers, stock market figures, prophetic warnings, thoughts, and Hebrew letters. He lived at the Library much of the time. Many of the staffers liked him and considered him a gentle soul.

Eventually Bloom began to believe himself to be the Messiah. He made frequent calls to Larry King's call-in show for years spouting his doctrines. King eulogized him on-air upon his death. Bloom would travel to various universities and publicly challenge scientists and public figure. These included Carl Sagan. Here's how Sagan described him in his important essay, *God and Norman Bloom*:

"Norman Bloom is a contemporary American who, not incidentally, believes himself to be the Second Coming of Jesus Christ. Mr. Bloom has been a fixture at some scientific meetings, where he harangues the hurrying, preoccupied crowds moving from session to session...Bloom has issued a fascinating pamphlet that states: 'The complete faculty of Princeton University has agreed that it cannot refute, nor show in basic error the proof brought to it, in the book, THE NEW WORLD, dated Sep 1974. This faculty acknowledges as of June 1 1975 that it accepts as a proven truth THE IRREFUTABLE PROOF THAT AN ETERNAL MIND AND HAND HAS SHAPED AND CONTROLLED THE HISTORY OF THE WORLD THROUGH THOUSANDS OF YEARS.'"...Despite his offer of a thousand dollar prize for the first individual to refute his proof, no response was elicited.

Carl Sagan-*God and Norman Bloom*-
The American Scholar Magazine Autumn of 1977

Sagan dates Bloom's THE NEW WORLD at 1974. As you'll see, Gardner dates it at 1970.

How to Find what Norman Bloom Said? There are Apparently Only Two Bloom Source Materials in the World: Stanford University and Kooks Magazine

Stanford Collections Dept: I obtained about 20 pages of Bloom material from Stanford Special Collections. Bloom had dated it 8-8-1976 on the first page. It was found in the papers of Martin Gardner, Box 69 file 6. It starts strong with much talk of "proof" but never gets around to it. Instead, the pamphlet almost immediately turns to approximately 15 pages of tiny type devoted to an astonishingly incomprehensible theory of baseball scores.

Sample:
"IT IS OBVIOUS THAT AFTER HAVING ESTABLISHED THE RECORD OF 888 GAMES WON during the 11 years without a pennant, THAT HE HAS THE POWER TO INSURE THAT THE YANKEES WILL WIN THAT PENNANT IN THE BICENTENNIAL anniversary year of 888+888...
"In 1776 Redcoats Beat American Yankees-CAPTURE NEW YORK!
In 1976 Reds beat American League NY YANKEES-winning in NEW YORK!

Bloom states that the reason for his new fixation on Baseball was" the huge amount of amassed data". His intention was to show patterns in the numbers. It goes on to spend several pages more on the number 1776. Maybe it will all make sense someday.

Though the work says THE NEW WORLD on the cover page, it simply can't be the original THE NEW WORLD because it's clearly not the same writings either Gardner or Sagan mention. It doesn't appear to contain much astronomical content at all, just a bunch of head spinning words and numbers.

Upon seeing this and later (1987) Bloom material, it becomes obvious that Bloom "published" everything he did under the heading THE NEW WORLD and simply added numerical ideas as he went. The original THE NEW WORLD book, composed of ten pages, was apparently written in 1970. It was clearly astronomical in nature and at least coherent enough for comprehension. Several years passed between these writings. Clearly Bloom's mental state was deteriorating. Could it be that he wasn't totally crazy when he composed the first THE NEW WORLD?

What was the proof in that original THE NEW WORLD that Bloom was so certain of? Why would minds like Carl Sagan's and Martin Gardner's even acknowledge Bloom? And for that matter...**Where *is* the original THE NEW WORLD?**

It turns out that Bloom's 1970 work THE NEW WORLD is not exactly available for free download. In fact, even more than George van Den Bergh's vanished study of eclipse cycles, it has been essentially lost to time. My efforts to locate any original pamphlet in true form have turned up fruitless, though he certainly would have distributed hundreds of them, as was his habit. It appears he used the logo THE NEW WORLD on pretty much everything, and in every new writing, he would always mention he had some proof, but then you search the material...and the proof isn't there... it seems he felt he had already stated it, which perhaps he had. Or more likely he felt he was restating the same proof in new ways with things like baseball scores and bicentennial trivia.

The Other Source: Kooks Magazine 1987 Bloom material was graciously provided to me by Donna Kosey, former editor of Kooks Magazine, who did an article on Bloom in Kooks 1990 Feb 3 issue. Family members who were interviewed for the NPR piece that aired after his death apparently threw away the dozens of boxes filled with papers stuffed into Bloom's storage unit.

Bloom's later work is genuinely incomprehensible. This later material I've seen is not something two serious academics would likely consider. It's a harrowing trip to numberland madness. Here it is, if you dare, but you probably don't want to stick around in there for long.

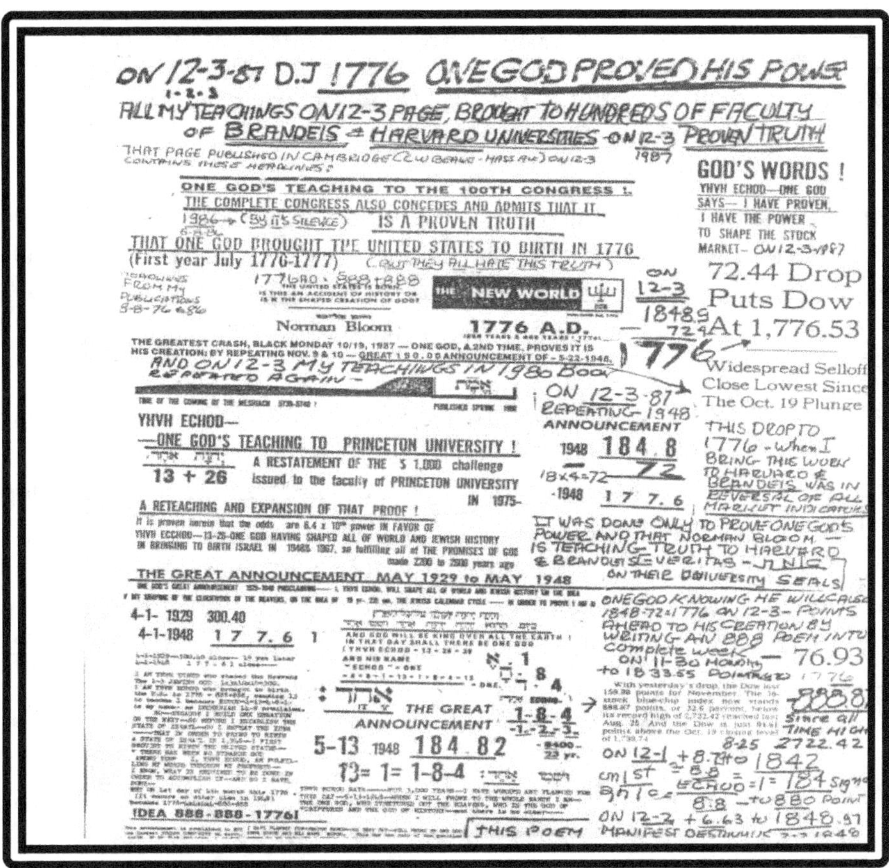

(above) Norman Bloom circa late 1987-courtesy of Donna Kosey Kooks Magazine

There is mention of the proof in there (lower middle left), near a mention of the "$1,000 challenge", but where's the proof? He doesn't get around to it here, though you can see a mention of the 19-year Metonic in reference to 1929 to 1948 in Israel history. You can also see THE NEW WORLD logo in the middle of the page. In existing Bloom documents, there is precious little astronomy at all. No proof, just new numbers from new fascinations. Like any true artist, he had already moved on.

Was the astronomical "proof" in Bloom's 1970 THE NEW WORLD perhaps kind of crazy sounding? Sure. But probably not as crazy as these final writings.

What Norman Bloom wrote in the original THE NEW WORLD in the early 1970's was interesting enough that two of America's greatest minds thought it worthy of public commentary. What challenge did the "Numbers Man" come up with that stirred both Martin Gardner and Carl Sagan to respond in print? The first clue comes from Gardner.

Martin Gardner (Oct 21 1914 to May 22 2010)

Martin Gardner was an American popular mathematics and science writer with interests also encompassing scientific skepticism, "micromagic", philosophy, religion, and literature – especially the writings of Lewis Carroll, L. Frank Baum, and G. K. Chesterton. He was a leading authority on Lewis Carroll. *The Annotated Alice*, which incorporated the text of Carroll's two Alice books, was his most successful work and sold over a million copies.

He had a lifelong interest in magic and illusion. In 1999, MAGIC magazine named him as one of the "100 Most Influential Magicians of the Twentieth Century". He was considered the preeminent American puzzler, and published more than 100 books.

Gardner was best known for creating and sustaining interest in recreational mathematics – and by extension, mathematics in general – throughout the latter half of the 20th century, principally through his "Mathematical Games" columns. These appeared for twenty-five years in Scientific American, and his subsequent books collecting them. (sourced from Wikipedia)

One of Gardner's best loved characters was the mysterious Dr. Matrix, a master of illusion and math, who was usually off on secretive adventures with his mysterious daughter and assistant, Iva. Invariably, each adventure would have lots of room for tough math puzzles and mind-blowing numerical peculiarities. It was Dr. Matrix, in Gardner's columns from Scientific American, who showed the world that the name of HAL, the demonic computer in the film 2001, was obtained by shifting each letter backward from IBM. He was also the first to chronicle many of the non-ecliptic synchronicities between the lives and deaths of John F. Kennedy and Abraham Lincoln.

Again, here is Dr, Matrix speaking, from the chapter entitled Los Angeles:

A numerologist knows it is not a coincidence that the sun's disk viewed from the earth, is almost identical in size to the moon's disk --The Amazing Numbers of Dr. Matrix (Martin Gardner)*

Note the asterisk. Just a little footnote in another out-of-print book:

(Gardner speaking)" this is dramatically evident during a total eclipse of the sun when the moon's disk precisely occludes the sun's disk. Put another way, the tip of the moon's shadow just brushes the earth's surface. The improbability of this coincidence is a cornerstone in a proof of the existence of God as outlined in a ten-page pamphlet by **Norman Bloom, published in 1970 in Guttenberg, New Jersey. That pamphlet is entitled, **The New World: The First Proof in History that the Earth, Moon and Sun are controlled by a Thinking, Acting Mind and Hand That Has The Power of Life and Death over every Living Thing on Earth.** Mr. Bloom reports that he has defended his argument at such centers of higher learning as Harvard, MIT, and Barry Farber's WOR Radio Show. He offers $ 1,000 to anyone who can find a flaw in his proof. (Gardner ibid.)*

There you have it. Gardner says the sun/moon coincidence is "the cornerstone" of Bloom's argument. Gardner clearly possessed this 1970 work at some point, but it did not end up in his papers at Stanford. The papers at Stanford are mostly about baseball and are dated 8-8-1976.

It is interesting that Gardner addresses the sun/moon arrangement in the guise of his pseudonym, Dr. Matrix. Gardner was a lifelong magician and scholar of magic. This is perhaps a clue in this drama. Later in life, Gardner, though far from being a Christian, admitted belief in an active God and the power of prayer in his extraordinary intellectual memoir, *The Whys of a Philosophical Scrivener.*

After puzzling Gardner, the great puzzler, Bloom takes on a true heavy hitter: none other than the most famous scientist in the world, Carl Sagan. Sagan next takes an interest in Norman Bloom, and what he says (and doesn't say) about Bloom's THE NEW WORLD is fascinating, to say the least. On many levels it illuminates the tension between modern scientific and ancient cosmic truth.

The Norman Bloom Mystery deepens. Here comes Carl Sagan:

Carl Sagan (Nov 9 1934 to Dec 20 1996)

Carl Sagan was undoubtedly the most famous popular scientist of his time, and one of the most well-known scientists of all time. The author of several best-selling books (Dragons *of Eden,* among others), he also created and starred in the most watched astronomy show of all time, the 1980's PBS series, *Cosmos,* which aired in more than 60 countries to over 500 million people. I was certainly a fan.

Oddly enough, Sagan broaches the subject of Norman Bloom in an essay entitled *God and Norman Bloom.* It appeared in The American Scholar in Autumn 1977 and is available for reading online. *The American Scholar* is the Quarterly Magazine of the Phi Beta Kappa Organization. It has been around for 90 years and addresses lots of different intellectual topics. It isn't a peer reviewed science site. It's more like a place for smart people to share essays.

In *God and Norman Bloom, Sagan* addresses Bloom's THE NEW WORLD which is dated 1974. The NEW WORLD Sagan received in 1974 was possibly different from the one Garner got in 1970, as it appears Norman Bloom was constantly changing THE NEW WORLD.

In his American Scholar essay, Sagan says Bloom's "central" argument for proof of God is based on the close alignment of the luni-solar Metonic cycle of 19 solar years. Oddly, Sagan will not call it the "Metonic Cycle" or even call it a cycle at all. As we'll read in a minute, Sagan instead argues that the alignment of lunar and solar calendars at 19 years is not all that special of a phenomenon.

Sagan's is clearly geared to a general Phi Betta Kappa crowd. His witty, friendly style invites a natural trust that what he says is the truth, the whole truth, and nothing but the truth. A little careful reading, however, reveals curious omissions in his argument against the conclusions of the earnest, though potentially insane, Norman Bloom.

Before we read about that, it's important to figure out: what is Bloom's central point in *The New World*? Martin Gardner in *The Magic Numbers of Dr. Matrix* stated that the "cornerstone" of Bloom's argument is this: the sun and the moon appear the same size in the sky.

Sagan mentions this "cornerstone" when discussing Bloom's work only in two passing moments. First he says:

The supposed proofs (in Bloom's THE NEW WORLD) are many and diverse, all involving numerical coincidences that Mr. Bloom believes could not be owing to chance. Both style and content of the arguments are reminiscent of the Talmudic textual commentary and cabalistic lore of the Jewish Middle Ages. For example, the angular size of the moon or the sun as seen from earth is half a degree. This is just 1/720th of the circle (360 degrees) of the sky. But 720= 6! 6x5x4x3x2x1 =720, therefore, God exists...the approach is familiar and it infiltrates the entire history of religion.

The second mention occurs later: *"in a solar eclipse the moon, which appears from earth just as large as the sun, must pass in front of it"*

Carl Sagan-*God and Norman Bloom/* The American Scholar (Autumn 1977)

Sagan seems to downplay the arrangement, to say the least. The words go to impressive lengths to avoid showing wonder towards the sun/moon alignment. This blasé view of created symmetry and synchronicity has been the template for scientists and educators for hundreds of years (maybe even longer!). For some reason, many scientists can't seem to bear to say (take a deep breath) that the sun appears the same size as the moon in the sky!

Sagan does not say Bloom mentions the sun/moon alignment as any cosmic proof at all, instead focusing on Bloom's apparent fixation that the angular diameter of the sun and moon is 1/720th of a 360-degree sky. (That is something new I learned from the essay, and thus learned from Norman Bloom, I suppose.) But back to reality. The sun being the same size as the moon in the sky is an infinitely more interesting truth than the 720 equation, being as the sun/moon alignment demands an aesthetic (thus simultaneously mathematical and Artistic) consideration. The sun/moon alignment is not a numerical coincidence, nor can it be a result of any currently understood gravitational resonance. The alignment is based on perspective. Not only that, but scientists also assert that the current alignment has not always been this way. The moon was closer in the distant past and moves farther away at the rate of 1 and ½ inches a year.

Again, let's isolate Sagan's language. I think it's important. He states:
 "For example, the angular size of the moon or the sun as seen from the earth is half a degree (and later) ...the moon, which appears from earth just as large as the sun"

I guess it's better than not mentioning it at all. But you can't help but read such underwhelming ho-hum language and not leave with the feeling that "oh, Carl Sagan said half a degree and 1/17hundred20th or something about something".

And, as we know, saying that something is "just as large" is not exactly saying its "the same size". But how this verbiage differs is difficult to express. Of course, Sagan knew the sun and moon appear the *exact* same size, to the point where sunlight can flicker through the valleys of the moon during an eclipse. But he doesn't feel he needs to make a big deal about it, even though most of his readers are probably completely unaware of the fact and might find it enlightening.

I believe the sun/moon perception alignment is so astronomically improbable that many "educated" people would rather avoid it. Perhaps the sun/moon alignment is a can of spiritual worms. Astronomers may gulp at the sun/moon reality. Out of perhaps an innate sense of awe, they hesitate to call such a reality "coincidence". Sagan's response is typical of the astronomical community's largely dismissive consideration of the sheer magnitude and creative implications of this cosmic arrangement.

That's all fine. Downplaying something is no crime.

Much more problematic is Sagan's response to Bloom's argument for the peculiarity of the Metonic Cycle, that alignment of solar and lunar calendars every 19 years that means we have the same phase of the moon on the same calendar date every 19 years. The Metonic Cycle has been known to astronomers for 2500 years. It was named for Meton of *Athens*, after all from, well... way back.

Sagan will not even call it the Metonic Cycle. He can't utter the words. He won't even say it's a "cycle". But he does say it is Bloom's "central argument":

"Bloom's central argument-and the one that much of the rest is based upon-is the claimed astronomical coincidence that 235 new moons is, with spectacular accuracy, just as long as nineteen years. Whence...(Quoting Bloom): 'Look Mankind, I say to you all, in essence you are living in a clock. The clock keeps perfect time, to an accuracy of one second/day! How could such a clock in the heavens come to be without there being some being, who, with perception and understanding, who, with a plan and with the power, could form that clock?' "

(Sagan then replies) "Fair question. To pursue it, we must realize that there are several different kinds of years and several different kinds of months in use in astronomy":(ibid)

This is where Sagan's natural scientific skepticism and great knowledge run the risk of appearing deliberately confusing. Faced with describing a simple alignment of 19 years and 235 months, he decides to go off for nearly two pages with facts and figures of all three kinds of solar years, the Sidereal, Tropical and Anomalistic, which all differ one from another by less than $1/10^{th}$ of a day. Then, if your head isn't already spinning after learning that the sidereal year is 365.2564 days and the tropical is 365.242199 days and the anomalistic is 365.2596 days, he brings in all four kinds of months (Synodic, sidereal, anomalistic and draconic), announcing them down to the fifth decimal!

Soon, instead of reading an essay, we are wading in a mess of numbers worthy of Bloom himself, all because Sagan is apparently trying to prove that Bloom is grasping at digital straws.

After piles and piles of numbers and words, Sagan finally agrees in print that yes (of course) there are actually 235.00621 synodic months (the lunar month we actually observe) in exactly 19 solar sidereal years (the solar year we actually observe).

After admitting the Metonic Cycle is indeed real, but still refusing to mention it by name, Sagan does an odd thing. He asserts that this same phenomenon of luni-solar calendar connection could easily be found elsewhere. To prove this audacious claim, he gives the only other possible example that could even be close, the conjunction of eleven solar years with 136.05623 lunar months.

"Thus," (Sagan writes, with more than a touch of irony,) " *there is* (also) *a connection between 11 years and 136 new moons. Moreover, Sir Arthur Eddington believed that all physics could be derived from the number 136. With the foregoing information and just a little intellectual fortitude, it should be possible as well to reconstruct all of Bosnian History"* (ibid.)

Ha Ha. All the Phi Beta Kappas perhaps laugh on cue. Aren't we educated elites so clever? Here is the final word, the final put-down of all those silly people and all their unauthorized associations, their sad attempts to find sacred meaning in visible cosmic patterns. Sagan uses the 136 moon/ 11 solar year conjunction as his essay's last concrete refutation of Bloom's claims of a divine cosmic clock centered on the 19-year Metonic. It's the scientific feather in the cap. The frosting on the agnostic cake.

He then spends several pages making the familiar argument that anyone can find patterns anywhere but smart people know better- blah blah blah. Case closed, right? The triumph of science over madness. Umm...well, wait a second. First and foremost, are the two alignments Sagan just compared in print actually that similar?

The first numerical alignment (the "central" argument of Bloom, according to Sagan), is the Metonic Cycle, a conjunction of 235.006 lunar months and exactly 19 solar years. The second alignment (the one Sagan says is essentially the same) is 136.056 lunar months and 11 years.

Are these the same level of precision? Of course not. Look at the numbers behind the decimal points. In percentage of each lunar month, the difference is almost of a magnitude of ten.

The 19 year cycle has 235 lunar months followed by two zeros behind its decimal point. That amounts to just a *two hour difference* over 19 entire years in the alignment of the two calendars of sun and moon as perceived from earth. It takes 250 years for the alignment to be one day off!

In the second alignment, the one Sagan quotes as certain proof the 19-year cycle isn't all that special, behind the 136 lunar months is .056, meaning the 11 year calendar is off by 5.6 % of one lunar month. That's at least 1.65 days (nearly 40 hours) off from alignment *just every 11 years.*

In mathematical and astronomical terms, these two alignments are not even close to the same degree of precision and would never be compared as such in any peer-oriented paper. The only reason Sagan could get away with such a sloppy comparison is that the audience for the American Scholar is likely a non-mathematical audience, already sympathetic with his intentions.

The 19-year Metonic is quite precise. If he knows another place they align so precisely, he doesn't mention it in this essay.

Why would a famous astronomer like Sagan need to make a sloppy argument with weak math? The 11-year solar/lunar calendar alignment is "ballpark" close, but nearly two days off every 11 years is not really much of an alignment at all. No cultures make calendars out of it, simply because the phase of the moon will not be the same every 11 years, making the "alignment" essentially useless.

In the essay, Sagan makes sport of Bloom's assertion that the 19 year cycle is connected to The Bible's Psalm 19 (*"the heavens declare the Glory of God"*). He then goes on to mention the Saros Cycle without mentioning that it also contains *three* precise alignments with lunar, draconic and anomalistic months every 19 eclipse years. Sagan only mentions that the first two align and ignores the anomalistic month alignment. He ends the paragraph about the Saros describing the phenomenon this way:

"Coincidence? Similar numerical coincidences are, in fact, common throughout the solar system. The ratio of spin period to orbital period on Mercury is 3 to 2. Venus manages to turn the same face to the earth at closest approach on each of its revolutions around the sun...given enough time...such resonances will arise inevitably."

Carl Sagan (ibid.)

I don't know about you, but neither of the resonances Sagan lists resonates with me. The ratio of spin to orbit is 3 to 2 on Mercury? What does that have to do with us? Is it as impressive as the alignments of sun and moon, based on both human perspective *and* cosmic reality? Do the same principles that make Mercury and Venus spin make the Metonic and Saros work? If so, he does not explain.

The Metonic and Saros alignments are magnitudes more improbable than the examples he gives. Not to mention, on top of that, the sun's the same size as the moon in the sky.

The 19-year Metonic alignment has been part of astronomical teaching and religious timekeeping since the Babylonians. It was named for Meton of Athens 500 years before Christ. What was the harm in calling it by name, or admitting that it's pretty darn amazing and unique? Why make the disingenuous argument that such precise alignments are everywhere when they aren't?

But that's actually the least of it.

Sagan goes further into the perilous pit of omission. He neglects to reference the biggest "coincidence" of all: the alignment of the Draconic/Eclipse Calendar with the lunar and the solar calendar at that same 19-year spot. Every 6940 days, three calendars align, not just two. And this allows not only for the same phase of the moon every 19 years but also for eclipses on repetitive calendar dates every 19 years for up to a century.

Solar Year 365.24 days x   19   years   = 6,939.60 days
Lunar   29.53   days x 235 lunar months   = 6,939.68 days
Draconic 27.212 days x   255 draconic months = 6,939.06 days.

There is no gravitational/ orbital reason three different orbital realities make calendars out of the same two objects (sun and moon). It just does, every 6940 days. Not mentioning it seems like a big omission, even absurd. They do not do it again until the 391-year mark.

Bloom's argument for the 1,000-dollar prize was surely based on two cosmic points. 1) The sun is the same size as the moon in the sky (as his argument was reported by Gardner) and 2) the Metonic alignment, as reported by Sagan.
Gardner said the "cornerstone" of Bloom's argument was the sun and moon alignment.
Sagan said "central" to Bloom's argument was the 19-year Calendar alignment.

Maybe Sagan didn't actually know the Draconic Calendar aligns with the solar and lunar calendars at 19 years. You'd think he would know such a thing, being an astronomer and all. But it's actually possible he didn't know. I think there are many astronomers out there right now who don't know. Sagan had a lot on his plate in 1977. I still find him likable. He does mention the draconic (eclipse) month in the essay and gives its length (27.22122 days), so he at least knew about that.

Still, in this influential essay written to influential people, the most prominent astronomer of modern times dismisses the idea of giving dignity to the Metonic arrangement, even by name. He brushes off two thirds of it as just another "numerical coincidence" and ignores completely the most improbable third piece. Then, when bringing up the Saros, he only mentions two thirds of its improbable alignment as well. We aren't talking about the coincidence that La Quinta always seems to be next to Denny's. We are talking about what is arguably the greatest cosmic synchronicity known to humankind besides the existence of intelligent life itself.

Maybe he was hurrying. Maybe his priority was keeping the essay light, having just a little fun at Norman's expense. To Sagan's credit, at least he acknowledged Bloom. Maybe Sagan was so deep in other important science that he didn't think too hard about the specifics of cosmic arguments levied by an unbalanced ice cream truck driver crying in the wilderness:

"Look Mankind, I say to you all, in essence you are living in a clock. The clock keeps perfect time..."
-Norman Bloom

I don't know what to make of it. I get the feeling that there is more to all of this, some kind of deep spiritual blindness constructed by a priest class that is waiting to be lifted. I have to move on.

In the end, Sagan is rather sweet (albeit a little patronizing) when talking about Bloom. He calls him a "kind of genius". But something is still amiss about this type of "science" Sagan exhibits. Its blind spot is so great as to make it feel untrue. It is a blind spot increasingly familiar to modern man, and that blind spot seems planted squarely over the face of God.

Peculiarities in the Bloom Saga?

It seems there is a triad in this story, like a musical chord. We have interesting archetypes who might very well have met in a Middle Eastern palace court 3000 years ago. They return entwined in a real-life story with the nature of truth at its very core.

One riddle is that three of the four characters are Jewish: Norman Bloom, Carl Sagan and Larry King.

Norman Bloom is the crazy oracle archetype. A Hebrew messenger in the path of bewildering cosmic signals, a conduit of universal information bound to be ignored and ridiculed. Bloom tries to deliver a message but ends up shattered upon the rocks of social disinterest, his own mind, and his pride. Bloom clearly tapped into a mysterious numerical reality -a kind of projection or underlying pattern.

But...as Bob Dylan said:
"some things are too hot to touch.
The human mind can only stand so much"- Things Have Changed- Bob Dylan

Martin Gardner's archetype is the classic Gentile astrologer/wise man of the Babylonian or Egyptian courts and schools. Familiar with ancient mysteries of math and magic, he knows there are secrets of universal order, but also knows why certain shrouds are left in place. Gardner probably knew how close Bloom had gotten to one part of the secret. If he didn't think so, he wouldn't have so playfully presented that secret in print. He repeats Bloom's greatest claim two times on one page. Once, giving credit to Bloom himself, and once, having his own fictional alter-ego say essentially the same thing.

Carl Sagan found himself as a kind of Official Authorized Priest of the Temple of Science and Technology. Sagan was a kindly and sympathetic observer of Creation. Possibly our most beloved agnostic, he still acknowledged Bloom. As temple priest, he knew there are certain avenues of the mind that are not to be wandered if the sanity of society is to be maintained. Certain truths can only be approached by the initiated. Certain equations are not to be admitted to the masses.

Larry King: For whatever reason, King did not screen out Bloom's calls. There must have been some reason he let him speak. I can't find any taped samples of their interactions online, but I have seen a video one time of King eulogizing Bloom upon his lonely passing. The video is not easy to find anymore. King looked sincerely sad when announcing his death.

Bloom and Gardner and King are all basically nouns. Sagan is a Polish word for "maker of pots and pans". These three souls, for one brief moment in history, put on a largely unnoticed play of intellectual, imperial, scientific and human reality. The reality of Sun and Moon and Earth and Mind are tossed between them in print like cosmic hot potatoes.

Riddles of the Age-The Computer Network

They will say, "Where is this 'coming' he promised? Ever since our ancestors died, everything goes on as it has since the beginning of creation."
2 Peter 3:4

Here's a question. What's so different about now? Haven't unhinged doomsayers been babbling their apocalyptic fantasies for thousands of years? Why is this the end of an Age? How can I be so certain it would be marked by peculiar eclipses? Really, the God of the Universe signing His name across the planet? What's the hurry in repentance?

What's the answer? Pretty simple, really. Three words. International computer network. The computer age has brought the imminent onset of true spiritual tyranny, where free thought and free will are contained and subdued automatically by physical and spiritual interface with the anti-spirit of that machine.

I put out a big, dense book in 2001 called "What the Fire Said". I made 35 copies and distributed them to various American personalities and friends, just like some crazy person might (hint: I'm not any crazier than the next guy). Here is a quote from that all but forgotten tome:
"The Force that desires the destruction of free will forever comes to power inside the international computer network." (What The Fire Said 2001)
You can judge for yourself if there is truth in the singular prediction I made.

How big is the Internet?

In a list of 80 cultural moments that shaped the world, chosen by a panel of 25 eminent scientists, academics, writers and world leaders, the invention of the World Wide Web was ranked number one, the entry stating, "The fastest growing communications medium of all time, the Internet has changed the shape of modern life forever. We can connect with each other instantly, all over the world."
From Wikipedia (British Council May 2016)

First the positive: The network has "positive" attributes. Ok? There. I said it. It's not ALL bad, of course. Positive traits include communication, ideas, commerce, entertainment, and funny cat videos.

The network had to come into existence, of course.
The network's appearance is implied in the Bible in three places: the Hebrew Book of Daniel, and the New Testament books of Matthew and Revelation. Very quickly and without much explanation, I will quote these three sections. Of course, my reasoning is contestable, as I expect anything and everything I say or quote to be contested in this world of endless criticism.

The fourth beast is a fourth kingdom that will appear on earth. It will be different from all the other kingdoms and will devour the whole earth, trampling it down and crushing it. Daniel 7:23

And this gospel of the kingdom will be preached in the whole world as a testimony to all nations, and then the end will come. Matthew 24:14 (the words of Jesus stating that there will be a means for communicating across the whole world)

And of course:
He was granted power to give breath to the image of the beast, that the image of the beast should even speak and cause as many as would not worship the image of the beast to be killed. He causes all, both small and great, rich and poor, free and slave, to receive a mark on their right hand or on their foreheads, and that no one may buy or sell except one who has the mark or the name of the beast, or the number of his name. This calls for wisdom: let the one who has understanding calculate the number of the beast, for it is the number of a man, and his number is 666. Revelation 13: 15-18

To come literally true, these Biblical passages necessarily require a worldwide network. Of course, such a network did not and could not have existed until the last thirty years.

I am not trying to concoct complicated and obscure riddles. I am just giving a simple application of famous prophecies to a likely meaning. The ubiquitous and desolating power described in these passages has not previously appeared in history, though it has been approached in diabolical archetype by various imperial efforts. Now is the only time such a worldwide force could exist. The recent example of the enforcement of certain worldwide measures to the farthest ends of the globe should confirm that reality. The installation of thousands of satellites is bringing wireless internet to every portion of the planet. We live very close to becoming at the mercy of a worldwide machine. Of course, believing that prophetic descriptions of our modern times are written in two-thousand-year-old documents requires a faith in God's intentions. I believe these riddles are intentionally preserved remnants of truth left to instruct and comfort us in deceptive times. Not everyone has such faith, I understand. Perhaps you will at least entertain my arguments that our peculiar time is indeed peculiar. Then perhaps later you can decide for yourself whether it has indeed been foretold. I primarily intend to make this assertion with eclipses. But first I will assert the peculiarity of our times with an exploration of modern riddles.

With all that in mind, I present some **riddles of the age**.

- **Internet.** Enter net. A trap. A command or suggestion: come into the trap. Nets are commonly used to catch fish. A fish is a common symbol for Christians.

- **World Wide Web.** Webs are also traps. The spider's web is the other name of the internet. Spiders often eat their prey alive. Feel like "Getting on the web"?

- **Apple.** The most famous logo in the world is a fruit with a bite taken out of it. This is clear symbolism of the first rebellion in the Bible, when Eve and Adam ate the forbidden fruit from the Tree of The Knowledge of Good and Evil.

  When the woman saw that the fruit of the tree was good for food and pleasing to the eye, and also desirable for gaining wisdom, she took some and ate it. She also gave some to her husband, who was with her, and he ate it. Genesis 3:6

The Apple One Computer sold for $666.66. Steve Jobs and Steve Wozniak put the first personal computer in history, the Apple One, on sale in April 11 1976 in Palo Alto, California.

It went up for sale in the Bicentennial year of the USA. The introductory price was $666.66 cents. Wozniak said he came up with that initial price because he liked repeating digits.

www=666. The Hebrew letter Waw or Vav is the correspondent to our common letter w. In earlier times it was likely pronounced with "w" sounding. Today modern Hebrew pronounces it with a "v" sound. Hebrew letters are also numbers. Waw is the sixth letter of the Hebrew alphabet and bears the numerical value of six. Quite literally, www=666.

 The meaning of the letter is "hook" and it bears a vague resemblance to a hook or shepherd's crook.

Meta means "dead" in Hebrew. Meta means "is dead" in Hebrew. This is the new name of the facebook company and their signature "Metaverse".

The Computer and Internet were funded by War. The first computers and linked computer systems were created and funded by military departments. Tim Berners-Lee claims to have invented the World Wide Web, but communication between computer systems had already been going on since the 1960's at the US Defense Department and Intelligence agencies. We were just brought into the network by the web.

The computer network is an architect of deception. Maximum expression of historical deception is likely found in the modern "deepfake", automatically created and distributed(and convincing) falsity.

Screens are Simulated Fire. Once we sat around fires and told each other stories. It's hard to take your eyes off a fire! Now we stare into electric fires and it tells us the story. This is the unspoken primary explanation for the overwhelming addictive power of visual technologies, including cinema, television, computer terminals and smart phones. The metaphor of fire and light and story is relevant not only for this riddle but also to the significance of eclipses.

The screen gets closer and closer. From the movie screen to the television to computer to the phone, the screen gets closer to us and moves further upon our bodies. The stated goal of proponents of digital technology is to breach the final barrier; to bring technology inside the human body.

Internet addiction/Phone Addiction is the most common pathology in the world. The National Institute of Health says nearly 40% of Americans are internet addicts, many of them porn addicts. It's probably more than that. The average person touches their phone 2600 times a day.

The Internet behaves in an opposite manner of God. Firstly, by mimicking omniscience. God knows all and sees all. The Internet imitates His omniscience. The Internet does not forget or forgive. God forgives. The Internet catalogues our behavior and then never forgives. Indeed, it is always threatening to expose faults.

The Internet is a Worldwide Temptation machine. *Lead us not into temptation?* The internet tempts all. The internet is also a worldwide defamation machine.

It seeks entry into the human body. Brain computer interfaces are almost on the market as of this writing. Recent responses to "events" showed the urge for state control over the body.

The Internet swallows creativity and culture. The last 20 years have seen the placement of nearly all forms of art into the mercy of the network. Musical albums, newspapers, poetry, romance, books, painting and other art.

It transforms civilization. In 30 years it has transformed nearly every aspect of human life.

Religious faith plummets in inverse relation to rising Internet usage. In the early 1990's, more than 90% of Americans identified as Christian. By 2007 78 % of Americans identified themselves as Christians. Fourteen years later in 2021, that number was down to 63%. What is the common cultural denominator? The rise of the internet and cell phone.

It calls itself IT. Information Technology's acronym defines itself as a non-human other, in contrast to the "I and Thou" reality of mutual recognition. In Madeline Le Engle's classic, *Wrinkle in Time*, the Dark enemy of humanity is called It.

Spirit does not live in the Network. The greatest clue as to its nature.

AI. The manifestation of information as personality. Ai is touted as our destroyer. If this is so, why was it made? Ai was also the name of a Canaanite city, an enemy of early Hebrews (Joshua Chapter 7 to 9). Ai in Hebrew means literally "heap of ruins".

The Father of the Modern Computer was a man named Zuse. Konrad Zuse created the first digital programmable computer in May 1941, funded by the War Department of Nazi Germany. Zuse, of course, is very close to "Zeus", supreme God of the Greeks. I mean, how many people do you know with the name Zuse? IBM bought his patents in 1946, just after the war. Zuse suggested the universe was a computer simulation in a 1967 book and was an avowed atheist.

Gates as a noun is at once a common entryway, a computer term and a Biblical Concept. Gates are places where one enters private spaces. Gates were the entry point of walled cities and fortresses. Logic "gates" are the essential commands of binary Boolean algebra, the method of computer logic. These gates refer mainly to the three basic commands of binary language – "and, or, not".

The "gates of hell" is a common phrase even today and was famously used by Jesus:
on this rock I will build my church, and the gates of hell shall not prevail against it.
Matthew 16:18

Windows 7 revealed on Wuhan Eclipse Day July 22 2009 Windows 7 was the most popular computer program in the history of the world to that point, with 630 million licenses by 2012. It was released for manufacturing on July 22 2009, the day of a tremendous solar eclipse, the longest of the 21st Century, with duration of 6 minutes and 39 seconds. It crossed from India to the Pacific and featured total eclipse in Wuhan China, where ten years later the pandemic began. Bill Gates' first health partnership with China began in 2009 just as he undertook co-operation with them. He toured China in 2010.

Solar eclipse 7 weeks after Microsoft moves to Bellevue/Seattle. The last total solar eclipse in America for 38 years occurred in the American Northwest Feb 26 1979, seven weeks after Gates moved his company from Albuquerque to Bellevue, Washington, where he had been raised. Bellevue (99.64% coverage). 100% totality at Mount Saint Helens the year before she erupts.

Mount Saint Helen's and Microsoft. As Microsoft prepared the language MS DOS in Bellevue April and May 1980, Saint Helens awakened 80 miles to the south erupting on May 18 1980 in the most destructive eruption in the modern history of North America. Saint Helens erupted laterally to the north. The longitude of the crater of Mount Saint Helens is 122.2 degrees. The longitude of Bellevue is 122.2 degrees. Mount Saint Helens erupted in the direction of Microsoft as they developed the most important language in the history of the world, which was leased to IBM and Apple the following year.

Saint Helen was the mother of Constantine, the Roman Emperor who made Christianity the official religion of the Empire. She is credited with being a faithful follower of Jesus and with finding the "true Cross" in Jerusalem on a pilgrimage in 324 AD. She founded the Church of the Nativity and the Church of the Ascension there. She is often depicted carrying a cross. Mount Saint Helens is one of only two active volcanoes in North America named for a saint.

Eclipses are an expression of riddle, story and synchronicity. Solar eclipses bear a message of the truth of the implicate order. They are much like music in that regard. This work intends to present them sympathetically with that in mind.

The Idea of Anniversary in Calendars

Anniversary celebrations are common in all cultures who have calendars.
The first anniversary celebrations were likely Solstice and Equinox celebrations.

Death days are more often noted accurately than birth days. Four Examples:
Julius Caesar. No exact birthday (just month of July noted) Died March 15 46 BC
Christopher Columbus. no exact birthday, Died May 20 1506.
Prophet Mohammad ...no exact birthday, Died June 8 632 AD.

The reason for obscure birth dates is obvious. Often people are born in obscurity. After they achieve renown, their passing is felt sharply and the date of death is noted. In the modern era of public record keeping, both birth and death days became more commonly known.

In cosmic timekeeping, three calendars are constantly aligning. Each moment is always in relation with a moment 19 years before and 19 years after, where the sun and moon are the same relative to earth. It's quite odd. These pulses of 19 are accompanied by other cycles, with other pulses. The Saros connects three alignments every 19 eclipse years, the Hepton has 7 eclipses 7 eclipse seasons apart occupying 19 years and 325 days. There are cycles at 300 years minus 44 days. There are Octons and persistent lunar year triads of eclipses. The 391-year cycle is a near perfect re-alignment.

In short, there are many calendars colliding. Moments resonate! Carl Sagan agreed there was certainly resonance and it was in that agreement of resonance where he most accepted the possibility that cosmic arrangement might offer proof of God. From *God and Norman Bloom:*

"None of these numerical coincidences prove the existence of God-or, if they do, the argument is subtle, because the effects are due to resonances."
Carl Sagan-God and Norman Bloom 1977

Sagan's fault in *God and Norman Bloom* was not his opinion that resonances are coincidental. That's a valid opinion. The problem came in his neglect in not acknowledging the actual triad of resonances involved in the Metonic and Saros Cycles. Instead he wrote as if in both, only two calendars were resonating.

Anniversaries: Do anniversaries resonate? Perhaps. There is an actual realignment of cosmic bodies on certain dates. This is especially true over the course of about a hundred years as the lunar, solar and eclipse calendars align on exact or nearly exact dates. Those same alignments then reappear about five hundred years later.

The problem is that it is nearly impossible to diagram or "sum up" such a phenomenon. There are many different calendar relations. It is very much like an extraordinary symphony with hundreds of instruments. The best I have been able to do is try and listen to it and repeat what part I can hear.

But the possibility of cosmic realignment, or "resonance" as Sagan puts it, on actual anniversary dates turns out to be certainly real in time. To what extent I don't know. I can only offer peculiarities of certain anniversaries and personalities. As mentioned in the Metonic section, exact day anniversaries can be deceptive on a spinning planet. As Van den Bergh said: *"A day is 48 hours".* Sometimes I mention anniversaries including the day before or day after. Sometimes I stick to the singular exact anniversary.

Other times in history are more general. Times of war and conquest, migration, enlightenment or enslavement. Sometimes these may be marked by peculiar eclipses or swarms. I do not propose that any of constitutes "proof". I do propose we all have the right to point out "peculiarities".

Eleven Personalities/Eclipse Associations

1) Pope John Paul II (born May 18 1920, died April 2 2005, buried April 8 2005)

Pope John Paul II is the only person I've found who was both born and buried on actual exact days there were solar eclipses. He was born Karl Wotjyla in Poland on the day of a strong (.97 magnitude) partial solar eclipse, though it was not visible in Poland. This was Saros 146's last eclipse as a partial before becoming the lead-off eclipse of the modern Hepton Two Cycle. This was 60 years to the day before the eruption of Mount Saint Helens. Thus the eruption occurred on the Pope's 60th birthday. Saint Helen was the mother of the Roman Church (mother of Constantine). Pope John Paul II was buried in Vatican City on April 8 2005, the same day a total solar eclipse occurred over the Pacific, ending at the isthmus of the Americas (Saros 129). It was not visible in Vatican City. Apr 8 returns in American eclipse of 2024.
An interesting numerica fact : John Paul II lived exactly 31,000 days.

2) Martin Luther King (Jan 15 1929 to April 4 1968)

King was born into a solar eclipse Metonic and died 8 days before a total lunar eclipse over the USA. There are seven MLK birthday eclipses on January 14/15 between 1926 and 2029. King's birthday Metonic arrives three years before his birth and departs on what would have been his 100th Birthday.

King Birthday Metonic Eclipses:

Jan 14 1926 Africa to Southeast Asia Total (Saros 130)
Jan 14 1945 Africa to Tasmania Ring of Fire (Saros 140) Eve of MLK's 16th Birthday
Jan 14/15 1964 Antarctica partial (Saros 150) First eclipse on planet earth after JFK's assassination happens on MLK's 35th birthday. Eclipse warbles between Jan 14 and 15 due to south pole location.
Jan 15/16 1972 Antarctica Ring of Fire (Saros 121) Again moving between his birthday and the 16th on the eight year lurch on what would have been his 43rd birthday.
Jan 15 1991 Tasmania/New Zealand Ring of Fire (Saros 131) The Birthday Metonic firmly arrives on the 15th, what would have been King's 62nd birthday.
Jan 15 2010 Africa/India/China Ring of Fire (Saros 141) King would have been 81 for this tremendous 11 minute ring of fire 177 days after the Wuhan Total of Jul 22 2009.
Jan 14 2029 USA Partial (Saros 151) Strong partial covers North America on the eve of MLK's 100th Birthday. 45 % coverage in hometown of Atlanta. 50% coverage at murder site of Memphis.

There were 103 years of MLK Birthday Metonic Eclipses. Note the numeric links between the Saros numbers: the first three are 130, 140 and 150. The last four are 121, 131, 141 and 151.

King's Death: Lunar Eclipse in the USA April 12th and 13th, 1968

On Passover April 12/13 1968, eight days after King's murder, a Blood Moon total lunar eclipse covered the USA Good Friday night into Saturday morning, part of a Blood Moon Lunar Tetrad on Jewish High Holy Days. Easter arrived the next day on April 14th, the 103rd anniversary of the murder of Lincoln, the 56th anniversary of Titanic sinking, and the 36th anniversary of the splitting of the atom. April 12 1968 was 23 years after death of Franklin Roosevelt.

Between 1951 and 1968, there were eight total lunar eclipses in the USA. One happened 8 days after King's murder. Another happened 38 days after JFK's assassination.

Notes about MLK Death Day of April 4

Being about 11 days apart (one Metonic week or the difference between a lunar and solar year), King's Death Day of Apr 4 is perpetually linked with Abraham Lincoln's Death Day Apr 14/15 in the larger Eclipse Metonic. Because they are one lunar year apart, the two eclipse dates follow each other linked in perpetuity, usually separated by a single eclipse. For instance, Lincoln's death day appears in a solar eclipse Apr 14 1809. King's Death Day appears in a solar eclipse on Apr 4 1810, one lunar year later. King's death day of 4-4 is also linked with Robert Kennedy's death day of 6-6 in the same Metonic, separated by exactly 3 eclipse years (1033 days). Kennedy was shot 63 days after King in 1968, and the two men were friends.

April 4th is one day after the traditional dating of the Crucifixion of Jesus on April 3rd 33AD. April 4th would be the day that Jesus slept in the tomb 1935 years before.

1968, the year King and RFK were killed, is also 2000 eclipse years (2000 x 346.62 days) after the Jewish Temple was destroyed by the Romans in 70 AD. The Temple was attacked on the day after Passover, or our equivalent of April 13, 70 AD. Therefore, the 70 AD attack on Jerusalem and the April 1968 Passover lunar eclipse which followed the killing of King are exactly 2,000 eclipse years apart.

More on MLK: Atlanta and Memphis In Modern Eclipses.

King's hometown of Atlanta is dead center in deepest eclipse line on the May 30 1984 USA eclipse.

Memphis, where King was killed, is highlighted by the 2017 and 2014 USA cross over Little Egypt, the area around the conjunction of the states of Illinois, Tennessee, Missouri and Kentucky.

Memphis, Tennessee was named after Memphis, the capitol of Ancient Egypt, the first empire to have an institutional system of mass slavery. Memphis Tennessee even has a giant pyramid, opened 11-9-1991. Note numeric relation of 11-9 to Egyptian New Year of 9-11.

Memphis lies at the southern end of what is called Little Egypt, along the Mississippi in the USA. Memphis, Tennessee has 93% coverage in 2017 eclipse and 98% coverage in 2024.
The ancient site of Memphis Egypt (near Cairo) has 96 % coverage in 2027 eclipse and 75 % in 2034.

3) Bush Family in Eclipses

The Bush family, from Prescott (1895 to 1972) to George W. (born 1946), is probably the closest thing the USA has to a true political dynasty.

1925 Jan 24 NYC/ Roaring Twenties Total Eclipse (Saros 120) eclipse six months after George Bush Sr. was born on Jun 12 1924. 98.3% totality in his birthplace of Boston, Massachusetts. Curiosity: John F. Kennedy and George Bush Sr. were born 6 miles and 7 years apart in 1917 and 1924. Their fathers, Joseph Kennedy and Prescott Bush were also born 7 years apart, in 1888 and 1895.
NYC eclipse saw totality in New Haven, CT (at 9:11 am), where George W. Bush will be born in 1946. Totality over New York City 919 months and 19 days before 9-11 attacks.

1932 Aug 31st New England Total (Saros 124) Totality 100% over Kennebunkport, Maine—spiritual home of the Bush Family. 99.8% in Boston.

1963 July 20 Japan to USA—Only Maine Total (Saros 145) A few months before Kennedy Assassination. Totality only in state of Maine. 97% in Kennebunkport. 94% in Boston.

1970 March 7 USA Eastern Seaboard (Saros 139) 94% in Kennebunkport. 96% in Boston.

1984 May 30 Washington DC (Saros 137) Ring of Fire darkens Washington DC (95%) while Bush is Vice-President.

1994 May 10 USA Ring of Fire (Saros 128) Ring of Fire crosses Texas the year George W. Bush is elected Governor. 87% coverage in Midland, his boyhood home. Path crosses USA to exit in Maine, with annularity directly over Kennebunkport.

2012 May 20 USA Hanoi Vietnam to Midland Texas (Saros 128) First complete USA solar eclipse in 18 years begins at sunrise over Hanoi Vietnam and ends at sunset over Bush Home in Midland Texas.

2023 October 14 USA Ring of Fire (Saros 134) Ring of Fire Path again goes directly over Midland Texas on its way to exit nation at Corpus Christi (Body of Christ). The only three rings of fire eclipses (1994, 2012 and 2023) over the heartland of USA in the last century all go directly over Midland or very close.

2024 April 8 USA Total Eclipse (Saros 139) Eclipse crosses Texas. Totality in Austin (where George W. was Governor), Dallas (George W.'s current home), and at his Ranch in Crawford. 96 % coverage in Kennebunkport Maine.

4) Donald J. Trump Trump was born the day of a total lunar eclipse which was not visible from the USA on June 14 1946. Interesting: Bill Clinton (Aug 19 1946) and George W Bush (July 6 1946) were also born in the summer of 1946. Thus, three US presidents were born within 66 days of each other, each in different summer months.

The strange case of three 60's Icon relations in 1920's Eclipses—Dylan, the Beatles and MLK

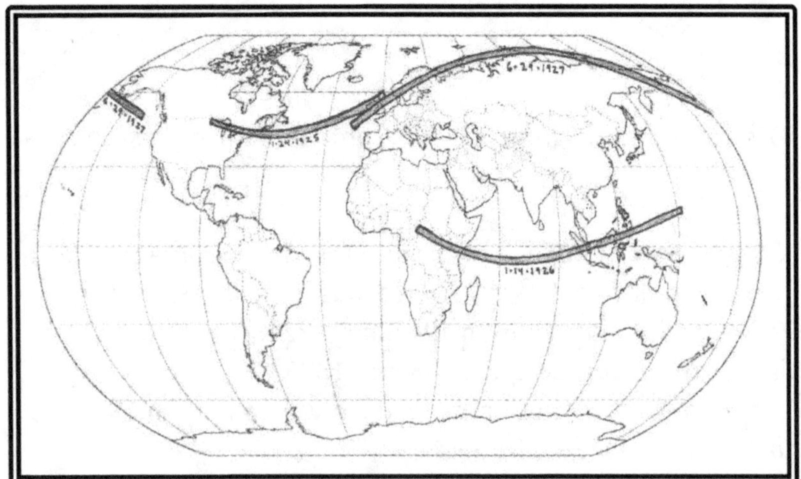

1925 to 1927

The map at left shows three eclipses. At Top is **Dylan eclipse USA to Britain (Jan 24 1925)**, just below it is the **Beatles Liverpool to USA (Jun 29 1927)** Bottom right is three years before MLK's birth (Jan 14 1926)

Dylan and the Beatles

Most musical historians make the case that the two greatest examples of creative musical expression in the era of modern cultural maximum expression (1945 to 1975) were Bob Dylan and The Beatles. Dylan is widely considered the greatest songwriter of all time. The Beatles are usually considered the greatest and most worldwide popular musical act of all time. Both are clearly unique phenomenon. They also deeply influenced each other and were friends.

5) Bob Dylan's Eclipse

1925 Jan 24 NYC/ Roaring Twenties Total Eclipse (Saros 120) Solar Eclipse begins at sunrise with totality in Duluth, Minnesota (where Dylan was born) and Hibbing, Minnesota, where he was raised. Eclipse brings totality to Woodstock, New York, (where Dylan lived for many years and the site of huge 1969 concert he did not attend). Eclipse then travels to New York City, the city where he chose the name Bob Dylan, where he became famous, recorded his first albums, and lived for many years.

Jan 24 1925 is 36 years to the day before Dylan travels from Minnesota by bus to New York City on Jan 24 1961. He played two songs that night and would become famous within the year.

After the eclipse path leaves New York City. It travels Atlantic and ends just after crossing the United Kingdom, thus linking the Dylan Eclipse with the 20th century's other greatest example of maximum musical expression, The Beatles, who have their own eclipse 2 ½ years later. All the Beatles became

friends with Dylan and their separate musical expressions fed each other. Between them, they may have been part of the maximum expression of popular art in human history.
Dylan's eclipse had 86% coverage in Liverpool.

6) The Beatles

The Beatles were the most popular worldwide musical phenomenon in the history of planet earth. They were a maximum expression of popular music and art.
The 16/36 pattern: Dylan's hometowns were marked in an eclipse at sunrise 16 years before his birth and 36 years before he first achieved fame.
16/36: The Beatles hometown of Liverpool, England was marked in an eclipse at sunrise 16 years before the last Beatle was born there and 36 years before their peak popularity began.

1927 June 29 Liverpool to America Total Eclipse (Saros 145)

One of the only total solar eclipses to cross Liverpool in centuries broke through heavy cloud banks to give an immortal show to thousands of Brits, many of whom had gathered on the beaches at sunrise. Within the same 16-year time frame as the span between the Dylan eclipse and Dylan's birth, all members of the Beatles are born in the eclipse path in Liverpool.
An eyewitness report:

 "The sense of exaltation produced by such stupendous beauty was so great that when the first pencil of light from the reappearing sun robbed the corona from our eyes, one seemed to fall back to earth through an untold distance, back to the common reality of ourselves, our fellow beings, and the wet sand on which we stood, Totality was ended. "

S. Seeley—The Total Eclipse of June 29, 1927, Liverpool England— writing for The Royal Astronomical Society of Canada

Within the same time frame of 36 years from total eclipse to arrival, both Dylan and the Beatles arrived in the place that will make them famous. (NYC for Dylan and London/ America for the Beatles)

Within 40 years from eclipses both Dylan and the Beatles will achieve their greatest level of commercial popularity. Dylan in 1965 (40 years from 1925 NYC/Hibbing eclipse) with "Like a Rolling Stone". Beatles in June 1967 (40 years from 1927 Liverpool eclipse) with *Sergeant Pepper's Lonely Hearts Club Band.*

The Dylan eclipse traveled from America to Britain, with sunrise over Dylan's home of Hibbing. The Beatles Eclipse traveled from Britain to America. It began in Britain at sunrise over Liverpool before traveling across the northern latitudes to ending in America's Aleutian Islands of Alaska.

The two eclipses essentially circled the globe: America to Britain to America. This is the same pattern as the music itself. The rock, country and blues music of America traveled to England, inspired the Beatles and then returned to America. Music is a force for truth made from and for the Holy Spirit.

In the 1920's the British Empire covered nearly a quarter of the Earth's surface. Soon it would be the American Empire that would dominate. See USA eclipses.

7) Charles Darwin (Feb.12 1809 to April 19 1882)

Darwin was born on the very same day as Abraham Lincoln on Feb 12 1809. He had three birthday eclipses during his life.

He died on April 19, 1882, seven years and one day before Adolf Hitler was born on Apr 20, 1889.

Darwin's death day first appears as an eclipse on April 19 1939, in the hours before Hitler's 50th birthday, the last solar eclipse before World War Two. It was a ring of fire connecting North America with extreme Northern Europe. 25% coverage where Darwin died 57 years before in Kent, England.

8) Adolf Hitler (April 20 1889 to April 30 1945)

One of the only total eclipses over Berlin in the last thousand years happened on Aug 19 1887, 87 weeks before Hitler was born. It traveled from sunrise at Berlin through Poland and included totality in Moscow. (see world eclipses)

Hitler's death (4-30) and birthdays (4-20) are perpetually linked on the Metonic one lunar year apart from each other. They appear firmly in modernity for the first time with an Apr 30 2022 partial and April 20 2023 hybrid eclipse.

Hitler had an eclipse on the eve of his 50th birthday, Apr 19 1939 (7pm Berlin time 30 % coverage 5 hours before his 50th birthday), 135 days before WWII.

Hitler may have seen two eclipses: The Titanic Eclipse of April 17 1912 covered much of Northern Germany in annularity, including Berlin with 95% coverage. **The World War One Eclipse of Aug 21 1914**, in the first weeks of the war, had 80% in Berlin and 70 % at Hitler birthplace in Austria.

The last solar eclipse of the 20th Century and the last complete eclipse of the second millennium was an **August 11 1999 Total Eclipse** across Europe and the Mideast. Totality in Branau am Inn, Austria— where he was born and in Munich, where he first achieved notoriety and power.

There is an unexplained oddity whereby Hitler's birthday of 4-20 became a catch phrase among marijuana smokers as a time to smoke pot.

8) Billy Graham (Nov 7 1918 to Feb 21 2018)

Billy Graham lived his life balanced neatly in between the two greatest modern USA cross-country eclipses. Billy Graham was born 152 days after the great cross country USA eclipse of June 8 1918 on Nov 7 1918. (82% coverage in his hometown of Charlotte, North Carolina). Graham died 184 days after the next USA cross country eclipse, which happened 99 years later on Aug 21 2017. (99 % coverage where he died in Montreat, NC). These two eclipses were the only true west coast to east coast total eclipses since the nation's founding.

Billy Graham preached the Gospel of Jesus Christ in person to more than 200 million people during his lifetime, the most in human history.

9) 2023 Dwight Eisenhower vs. Adolf Hitler Cosmic rematch

Eisenhower (Oct 14 1890 to March 28 1969). There was a partial solar eclipse on Mar 28 1968, exactly one year before Eisenhower's death on Mar 28 1969. Ike's birthday appears in a USA Ring of Fire Oct 14 2023, the first time Oct 14 appears in an eclipse since 1651. Hitler's birthday appear in a hybrid on Apr 20 2023, the first time an April 20 eclipses appears since an April 19/20 solar eclipse in 1651.

Oct 14 2023 is Ike's 133rd birthday
Apr 20 2023 is Hitler's 134th.

Eisenhower was commander of the Allied army that invaded Europe on June 6, 1944 and fought Hitler's army. The Allied Army entered Germany in early 1945.

10) Christopher Columbus (born 1451 died May 20 1506)

Columbus used knowledge of an upcoming lunar eclipse on June 30 1503 to trick the natives of what became Jamaica into thinking Columbus and his crew were deities.
Columbus death day of May 20 (1506), and his arrival in America day Oct 12 (1492) were embedded throughout the last half of the 20th Century Eclipse Metonic.

The date of his first colonial crime (the kidnapping of natives two days after arriving in 1492) is Oct 14, which is the date of the 2023 American Ring of Fire 531 years later.

11) Prophet Mohammad

Quoted From:
Hadith No: 153
Narrated/Authority of Al-Mughira bin Shuba

*The sun eclipsed in the life-time of **Allah's Apostle (Mohammad)** on the day when (his son) Ibrahim died. So the people said that the sun had eclipsed because of the death of Ibrahim. **Allah's Apostle** said, "The sun and the moon do not eclipse because of the death or life (i.e. birth) of someone. When you see the eclipse, pray and invoke Allah."*

Other Hadiths record the Prophet also saying of eclipses on that occasion " *they are but two of the signs of Allah".*

With those recorded statements, Mohammad becomes one of two founders of a major religion to specifically address eclipses.

Jesus talks of "signs in the sun and moon" (Luke 21:25) that will accompany his return, clearly a reference to eclipses.

Study in The Birthday/Deathday Metonic—US Presidents

This is basically a prelude to the JFK/ Lincoln Section

About one in 35 people are likely living in a solar eclipse Metonic that lands usually on their birthday. About one in 35 people are going to die on a day of a metonic.
Eclipses wiggle around a little, so as many as one in 6 people can have at least one eclipse on either their birth or death day during their lifetime.

Having solar eclipses on both one's birthday *and* death day during your lifetime is not exactly "common". That being said, it does happen to some and I will make an attempt to roughly determine the level of its peculiarity.

Does everybody have interesting solar eclipses on their birthday during their lifetimes? The answer is clearly no. Is it what you would call rare? Not really.

There are only about 35 dates rolling around on the Metonic at any one time. These can bend one date in either direction now and then. Depending on how long you live, it seems there might be as much as a one in 6 chance of having your birthday appear on any solar eclipse during your lifetime. I do not have any family or close friends whose birthdays appear on a solar eclipse during their lifetime.

Are you born in the Solar Eclipse Metonic (meaning is there an eclipse anywhere on earth?

Current Metonic (remember these dates can bend and sometimes don't show up for many decades):

January 5, 15, 26
February 6, 17, 27
March 9, 20, 30
April 8, 20, 30
May 10, 21
June 1,11, 21
July 1, 12, 22
August 2,12,21
September 1, 12, 22
October 2, 14, 25
November 3, 14, 25
December 4, 14, 26

These are the latest dates generally, but they wobble a bit due to times zones and orbital reality. If you are more than one day removed from any of these dates, "generally" you will have not had a complete eclipse on your birthday during your lifetime. Partials can vary by two days sometimes.

Now add the death day. How likely is a solar eclipse to show up on both a person's birthday and death day during their lifetime? Example: If I am born on March 2 1908 and die on July 14 1975, did any solar eclipse ever appear on either of those days anywhere on earth while I was alive? The answer is no. The Solar Metonic in early March was around the 8th, the 18th and the 29th during the 20th Century. Now it has moved on. The Metonic in July is around the 11th and 22nd.

I will not have a solar eclipse on my real birthday. The next time my birthday appears in the solar eclipse Metonic is more than a century away.

The chance of having at least one eclipse appear on *both* one's birthday and death day during a lifetime at all appears to may be between one in 15. It is within that framework that we can begin to look at patterns that are more peculiar. We look at Lincoln and JFK.

It turns out that birthday/deathday eclipses (and other interesting eclipses) are peculiarly abundant (and extremely peculiar) in the lives and times of our two most famous American Presidential martyrs, Abraham Lincoln and John F. Kennedy. These eclipse patterns seem to bear a unique witness, worthy of consideration, that compliments already chronicled synchronicities in the lives and deaths of the two presidents.

The full Kennedy/Lincoln Study follows this chapter.

Metonic Calendar Review

Now, let's briefly review Metonic reality by using the Kennedy death day eclipses. The following are actual solar eclipses on or around JFK's death day during the 20th Century.

1) Nov 22 1900 Kennedy Death day Metonic starts the 20th Century
2) Nov 22 1919 JFK Assassination Premonition Eclipse (Texas and Cuba)
3) Nov 21/22 1938 Total Antarctica
4) Nov 23 1946--Partial
5) Nov 22/23 1965 (Ring of Fire Southeast Asia) (Nov 22 in USA)
6) Nov 22 1984—Total in New Guinea over PT 109

Notice the first three eclipses are separated by two exact 19-year periods. That is the classic definition of Metonic. They occur on basically the same date. From 1938 to 1946 you find a two-day jump announcing what I call the 8-year lurch. From that point the 19-year Metonic again resumes, though it begins pushing towards the 23rd on the solar calendar.

By 2030 that same Metonic will have moved to November 25th, and it will be impossible to have a solar eclipse on Nov. 22nd for several centuries.

The difference in dates is affected by orbital reality, geographic time zones, the international date line and leap years. As you begin to appreciate the subtle certainties of the Metonic progression, you realize that these are not imaginary milestones, but indeed specific moments in cosmic time being cyclically marked.

Here's another example:

August 21 1914--eclipse across the battlefields of Europe at outset of WWI -Then 19 years...
August 21 1933-- eclipse across Jerusalem as Nazis rose to power in Germany
then 19 year period (eclipse Aug 20 1952),
then 19 year (eclipse Aug 20 1971) ,
then 8 year lurch (eclipse Aug 22 1979)
then 19 years (eclipse Aug 21/ 22 1998),
then 19 years brings us to: August 21 2017 --cross country eclipse of the USA

That's 103 years, a particularly long Metonic hold on a date.
The most possible is 122 years, a rare feat. Much of it is perspective and perception due to time zones.

These dates progress over the decades and centuries. August 21st becomes Aug 24th by the mid-21st century. After 300 to 500 years or so it will again return to having eclipses on August 21st.

There are approximately 3 dates per month in any given century where an eclipse might happen. For instance, at this writing, we just had a Dec 4 2021 eclipse which was preceded by an eclipse the year before on Dec 14 2020 and a December 26th eclipse in 2019. Those were lunar years apart-354.3 days. These lunar year triads are common but not necessarily constant.

There are cycles of time expressed in eclipses. But it is not science as we have been taught. It is close to song or art, but it's clearly a cosmic clockwork. They are dates that repeat or dance for awhile and then slowly change over time.

Birth and Death Day eclipse. Is it Common? Let's take a Look.

Let us consider the birth and death days of all 39 Deceased Presidents of the USA. Let's also check the birthdays of the 6 living Presidents. Then, just for fun and scientific sampling, we'll look at 39 deceased public personalities and check for the presence of birthday/deathday eclipses in their lives.

Here are the results:

Of the 39 Deceased Presidents of the United States:
28 had **no** birthday or death day solar eclipses of any kind during their lifetime.
6 had **one** solar eclipse (including partials) during their lifetime on **either** their birthday or deathday.
4 had complete solar eclipses on **both** birthday and death days during their lifetime.
One (Hoover) had a partial and a complete

Of the six living presidents, only **one** has a birthday within the eclipse Metonic—Jimmy Carter with an October 1st birthday.

Here are the four deceased Presidents with who had both Birth *and* Death days expressed in complete solar eclipses during their lifetime.

Thomas Jefferson—Born April 13 1745 ---Died July 4 1826
William Henry Harrison—Born Feb 9 1773—Died April 4 1841
Abraham Lincoln---Born Feb. 12 1809--Died April 14/15 1865
John F. Kennedy—Born May 29 1917--Died November 22 1963

3 of them-Harrison, Lincoln and Kennedy- died in office.
3 of them-Jefferson, Lincoln and Kennedy- are considered among the greatest Americans of all time.

Thomas Jefferson Eclipses (b. April 13 1745, d. July 4 1826)

solar eclipses occurred on April 13, 1763, April 13 1801 (partial) and July 4th 1796.

Thomas Jefferson died on July 4 1826, the 50th anniversary of the Declaration of Independence (which he had written). He died on the exact same day as John Adams, a fellow "Founding Father"—who Jefferson had defeated in the election of 1800.

Their deaths occurred thirty years after a July 4 1796 total eclipse (over open ocean-no land- in the Pacific) on the 20th anniversary of American Independence. That was both the first and last eclipse on July 4th for hundreds of years in either direction. As a curiosity, James Monroe, another founding Father, also died on July 4th five years later in 1831.

There were two complete solar eclipses during Jefferson's life on his birth and death days, an eclipse on his 18th birthday in 1763 and 30 years before his death, on July 4 1796. He almost certainly saw neither of them.
The eclipses that happened the day Jefferson turned 18 is shown on left.
The one that happened one lunar year later is on the right

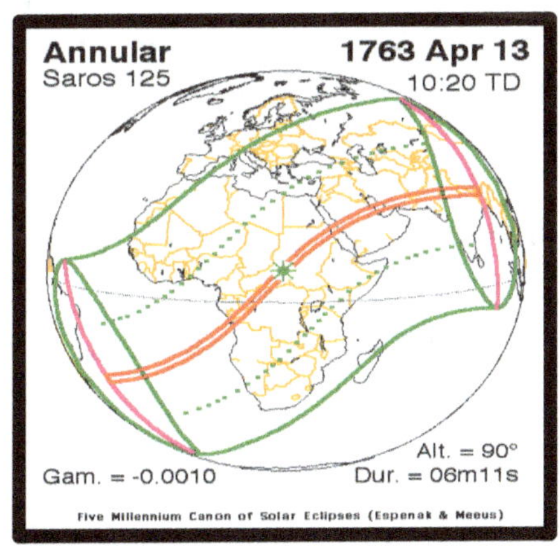

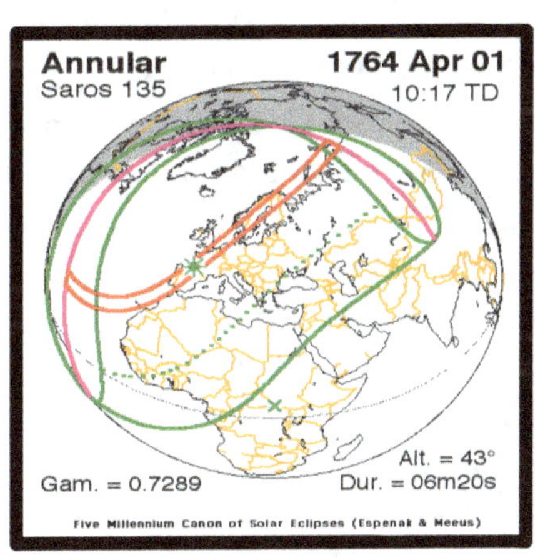

(maps courtesy of NASA) Above left : Jefferson's 18th Birthday Ring of Fire Africa, Arabia, Persia, India
and on the right a Ring of Fire over Europe.
Above right: Jefferson is still 18 for another 11 days. Spain, Portugal, France, Britain, Germany.
And right over Paris in the Springtime!
Those would be some great eclipses to have on your 18th birthday, whether you saw them or not.

He would see plenty of eclipses over the course of the Revolution. See USA eclipse section.
His birthday Metonic of April 13 becomes Lincoln's death day Metonic of April 14th. They literally pass the Metonic baton between Apr 13 1801 and Apr 14 1809 on an eight-year lurch.
Lincoln is two months old and Jefferson is 64 in April 1809.

William Henry Harrison, Born Feb 9 1773—Died April 4 1841
solar eclipses on Feb 9 1785 and April 4 1810

Harrison is said to have been the recipient of the most famous curse in the history of the USA, a curse passed by Shawnee Chief Tecumseh, who was killed by Harrison's Army in battle in 1813.
Harrison is the man who ran on" the Tippecanoe and Tyler Too." slogan for President. The battle of Tippecanoe was where the US Army defeated a confederation of Native Tribes in 1812.
Harrison died 31 days into office, the result of pneumonia contracted during his inaugural parade.

First eclipse occurs when Harrison turns 12 years old in Feb 1785, a South America to Africa Ring of Fire. The next eclipse is when he is 37 on April 4, what will be his death day 31 years later.
He likely saw neither. He dies April 4 1841. 4-4 is also the date of the assassination of Martin Luther King 127 years later.

Herbert Hoover has two eclipses but one is a partial.

Before we get to Kennedy and Lincoln; Here's all the Presidents:

Presidential Births and Deaths in The Solar Eclipse Metonic.

1.George Washington—Born 2-22-1732...Died 12-14-1799 (no eclipses)
2. John Adams---Born 10-30-1735...Died July 4 1826 (eclipse 7-4-1796)
3. Thomas Jefferson—Born 4-13-174/ Died July 4th 1826- (two on April 13' one eclipse Jul 4 1796)
4. James Madison –3-16-1752 to 6-28-1836 (no eclipses)
5. Monroe—4-28-1758 to 7-4-1831 (eclipse 7-4-1796)
6. J. Quincy Adams 7-11-1767 to 2-23-1848 (no eclipses)
7. Andrew Jackson 3-15-1767 to 6-8-1845 (no eclipses)
8. Martin van Buren 12-5-1782 to 7-24-1862 (no eclipses)
9. W.H. Harrison born 2-9-1773 to died 4-4-1841 (eclipses on Feb 9 1785 and April 4th 1810)
10. John Tyler—born 3-29-1790 to 1-18-1862 (no eclipses)
11. James Polk—11-2-1795 to 6-15-1849 (no eclipses)
12. Zachary Taylor 11-24-1784 to 7-9-1850 (no eclipses)
13. Millard Fillmore—1-7-1800 to 11-8-1874 (no eclipses)
14. Franklin Pierce—11-23-1804 to 10-8-1869 (eclipse on 10-8 1866)
15. James Buchanan—4-23-1791 to 6-1-1868 (no eclipses)

16. Abraham Lincoln –2-12-1809 to 4-14/15-1865 (3 birthday eclipses on Feb 12 1812/ Feb 12 1831/ Feb 12 1850 and one post death on Feb 11 1869—one day before what would have been his 60th birthday. Four death day eclipses: April 14, 1809/ April 14 1828/April 15 1847/ one year post death April 15 1866—Lincoln was shot on 14th and pronounced dead 9 hours later on the 15th.

17. Andrew Johnson 12-29-1808 to 7-31-1875 (no eclipses)
18. US Grant 4-27-1822 to 7-23-1885 (no eclipses)
19. Rutherford B. Hayes 10-4-1822 to 1-17-1893 (no eclipses)
20. James Garfield—11-19-1831 to 9-19-1881 (no eclipses—Garfield was assassinated. He narrowly misses eclipse Metonic on both dates. There is a numerical peculiarity of 1s and 9's in birth and death days (11-19 and 9-19). Death year of 1881 is the last year before 1961 that is the same upside down.
21. Chester Arthur –10-5-1829 to 11-18-1886 (no eclipses)
22. Grover Cleveland—3-18-1837 to 6-24-1890 (no eclipses)
23. Benjamin Harrison 8-20-1833 to 3-13-1901 (Aug 20 1895 partial)
24. Grover Cleveland—see above
25. William McKinley –1-29-1843 to 9-14-1901 (no eclipses)
26. Teddy Roosevelt—10-27-1858 to 1-6-1919 (no eclipses)
27. William H. Taft---9-15-1857 to 3-8-1930 (no eclipses)
28. Woodrow Wilson 12-28-1856 to Feb 3 1924 (eclipse on Feb 3 1916—Wilson's 60th Birthday)
29. Warren Harding 11-2-1865 to 8-2-1923 (no eclipses)
30. Calvin Coolidge ---7-4-1872 to 1-5-1933 (no eclipses)
31. Herbert Hoover-- 8-10-1874 to 10-20-1964 (partial eclipse on 10-20-1892 and Annular eclipse on 8-10-1915)
32. Franklin Roosevelt –1-30-1882 to 4-12-1945 (no eclipses)
33. Harry Truman—5-8-1884 to 12-26-1972 (no eclipses— narrowly avoided Metonic in both dates)
34. Dwight Eisenhower—10-14-1890 to 3-28-1969 --Partial eclipse 3-28-1968, exactly one year before his death.

35. John F. Kennedy Born May 29, 1917---Died November 22, 1963
two exact birthday eclipses on May 29, 1919/ May 29 1938)— exact death day Nov. 22nd 1919 and 1965—7 more eclipses on Metonic during 20th Century. Eleven in all

36. Lyndon Johnson—8-27-1908 to 1-22-1973 (no eclipses) ...Johnson's death day has interesting numeric similarity to JFK death date. Johnson died on 1-22-1973 Kennedy on 11-22-1963. it's a difference of exactly 10 years and 10 months.
37. Richard Nixon 1-9-1913 to 4-22-1994 (no eclipses)
38. Gerald Ford 7-14-1913 to 12-26-2006 (no eclipses. Died 34 years to the day after Truman)
39. Jimmy Carter 10-1-1924 (Carter was born in Metonic. Was 16 years old for 10-1-1940 eclipse.
40. Ronald Reagan 2-6-1911 to June 5 2004 (no eclipses)
41. George HW Bush 6-12-1924 to 11-30-2018 (no eclipses)
42. Bill Clinton born 8-19-1946 (not on Metonic)
43. George W. Bush 7-6-1946 (not on metonic...44 days before Clinton)
44. Barrack Obama 8-4-1961 (not on Metonic)
45. Donald Trump 6-14-1946 (not on solar Metonic but was born on a lunar eclipse visible from Asia)
46. Joe Biden 11-20-1942 (not on Metonic. Biden was born on Robert Kennedy's 17th birthday

Again: Of 46 US presidents - 39 are Deceased. Of them, 28 had no birthday or death day solar eclipses of any kind during their lifetime. 6 had one solar eclipse (including partials) during their lifetime on either their birthday or deathday. Herbert Hoover had one partial and one complete. Four had complete solar eclipses on both birthday and death days during their lifetime.

Of the six living presidents, one has a birthday within eclipse Metonic—Jimmy Carter.
About 70% of deceased presidents had no eclipses with their birth/death day during their lifetime. 18% had one. 10% had at least two—Jefferson, WH Harrison, Lincoln and Kennedy.
Only Kennedy and Lincoln had solar eclipses occur *in the US*A on either birthday or death day. Lincoln had an eclipse on his 22nd birthday (2-12-1831). It went very near Washington DC, where he would die 34 years later. Kennedy had an eclipse 44 years before his death day (Nov 22 1919) when he was two years old (plus ½ lunar year). It began over the state of Texas, where he would be killed.

Continuing Synchronicity Eclipse Experiment

Now, let's continue this experiment. I chose a list of 39 random deceased celebrities and public figures from the past. I chose them off the top of my head and wrote their names down on a piece of paper without first looking at their birth/death days for best "scientific" effect.

What would the percentages be of ecliptic presence in their lives?
39 Total Public figures. 22 had no eclipses (56%) 15 had 1 eclipse (38%) 2 had 2 eclipses (5%)
0 had more than two complete eclipses.

As you will see, the two with two eclipses were Huey "The Kingfish" Long and Ray Charles…Long was a populist American politician who was assassinated on September 10, 1935. Ray Charles, of course, was one of the great entertainers in the history of popular music.

Here is the List of famous non-president people and their relation to Metonic:

1. Ben Franklin 1-17-1706 to 4-17-1790 (no eclipses)
2. Albert Einstein 3-14-1879 to 4-18-1955 (one eclipse)
3. Princess Diana 7-1-1961 to 3-31-1997 (no eclipses)
4. Marilyn Monroe 6-1-1926 to 4-4-1962 (no eclipses; died exactly six years before MLK)
5. Elvis Presley 1-8-1935 to 8-16-1977 (no eclipses)
6. John Lennon 10-9-1940 to 12-8-1980 (no eclipses)
7. Huey Long---8-30-1893 to 9-10-1935(3 eclipses---total on 8-30-1905, partial on August 30 1925..total on September 10 1923)
8. Charles Lindbergh 2-4-1902 to 8-26-1974 (one eclipse 2-4-1943)
9. Walt Whitman 5-31-1819 to 3-26-1882 (one eclipse)
10. J. Edgar Hoover 1-1-1895 to 5-2-1975 (no eclipses)
11. Mahatma Ghandi 10-2-1869 to 1-30-1948 (no eclipses. His birthday is in current Metonic)
12. Mao Tse Tung---12-26-1893 to 9-9-1976 (one eclipse)
13. Josef Stalin – 12-18-1878 to 3-5-1973 (one eclipse-partial in 1924)
14. Queen Victoria—5-24-1819 to 1-22-1901 (one eclipse)
15. Nelson Mandela –7-18-1918 to 12-5-2013 (no eclipses)
16. Roy Rogers—11-5-1911 to 7-6-1981 (no eclipses)
17. Judy Garland – 6-10-1922 to 6-22-1967 (one eclipse)

18. Robert E. Lee 1-9-1807 to 10-12-1870 (no eclipses)
19. Martin Luther –11-10-1483 to 2-18-1546 (no eclipses)
20. Charles Darwin—2-12-1809 to April 19th 1882—Shared birthday eclipses with Abe Lincoln)
21. Harry Truman (Mount Saint Helens hermit) birth unknown. Died May 18, 1980 –one eclipse 1920)
22. Mozart ---1-27-1756 to 12-5-1791 (no eclipses)
23. Mother Theresa—8-26-1910 to 9-5-1997 (no eclipses)
24. Babe Ruth – 2-6-1895 to 8-16-1948 (no eclipses)
25. Thomas Edison 2-11- 1847 to 10-18-1931 (no eclipses)
26. Ray Charles 9-23-1930 to 6-10-2004 (eclipses 9-23-1987 and 6-10-1964)
27. Pablo Picasso—10-25—1881 to 4-8-1973 (one eclipses)
28. George Harrison – 2-25-1943 to 11-29-2001 (one eclipse)
29. CS Lewis 11-29-1898 to 11-22-1963 (one eclipse –shared death day with JFK and Aldous Huxley)
30.Helen Keller ---6-27-1880 top 6-1-1868 (no eclipses)
31. Malcolm X—5-19-1925 to 2-21-1965 (one eclipse)
32.Walt Disney--- 12-5-1901 to 12-15-1966 (no eclipses)
33.Buddy Holly –9-7-1936 to 2-3-1959 (no eclipses)
34. Winston Churchill 11-13-1874 to 1-24-1965 (one eclipse 1-24-1925)
35.Shah Reza Pahlavi—10-26-1919 to 7-27-1980 (no eclipses)
36. Augusto Pinochet--- 11-25-1916 to 12-10-2006 (no eclipses)
37. Salvador Allende—6-26-1908 to 9-11-1973 (one eclipse four years before his death on 9-11-1969—he was born 2 days before an eclipse on 6-28-1908)
38. Lucille Ball—8-6-1911 to 4-26-1989 (no eclipses)
39. Jimi Hendrix 11-27-1942 to 9-18-1970 (no eclipses)

Once again, 39 Total Public figures.
22 no eclipses (56%)
15 with 1 eclipse (38%)
And 2 with two eclipses (5%)
The two with two eclipses were Huey "The Kingfish" Long and Ray Charles…

Total of 78 deceased Presidents and Public Figures. 6 of them had eclipses on birth and death days. 7% of them. 4 were Presidents. Another was a famous politician/Governor—one was an entertainer.
Half of that figure (3) were assassinated—Lincoln, Kennedy and Huey Long.
The other three were Jefferson, William Henry Harrison and Ray Charles.

Only Kennedy and Lincoln have closely linked Metonic eclipses (repeating through history) with birth and death days. Only John F. Kennedy has a matching eclipse pair (Metonic side by side) with his birth and death days in permanently locked opposite nodes 177 days apart. Lincoln's death and birthday are linked exactly three eclipse years apart.
We have established the basic peculiarity.
Perhaps now we're ready to dive into the reality of the Lincoln/Kennedy Eclipses.

Chapter 17

The Cosmic Clock

The Time Cycles
Time is the dimension in the Creation of God where stories happen. It is a dimension, the fourth dimension, as Einstein famously proved.
There are apparently at least 10 dimensions, but really, who knows? What we can know about time is etched in nature, much to the benefit of story and the human race. We have a very naturally split up day, month and year, with impressive intersections and interactions.
Besides the obvious markers of days and months and years are many interweaving cycles with distinct realities. The pull of a familiar gravitation marks re-alignment of the sun and moon in precise places every 19 years (with a two-hour bump). Can you feel it? Well, it's been 19 years and all. But maybe there is. 19 years ago, the moon was in the same phase as it is today (if it's new moon tonight, it was new moon then too) and the sun came up on the same day of the year in the same spot and three calendars- the solar, the lunar and the draconic- all re-aligned.
Maybe there was something about nineteen years ago that reminds us of today or visa-versa. I don't know. But there is resonance in the Universe, as even Carl Sagan admitted. How or why or even if those resonances are personally felt I do not claim to know. But we are all aware of the effect of the proximity of the moon. It makes the oceans bulge, after all.
The Draco-Metonic Calendar, and her cosmic mate, The Saros, are the most extraordinary calendars of all. Within them and around them, myriads of calendars are dancing!

Periods of Time in Eclipse Witness /Astronomical

Day. A day is both real and perspective. We know a day only from where we live. That same day looks very different than it does 5000 miles away. And a whole other day completely is going on now on the other side of the world. Dutch Astronomer George Van Den Beergh spends three pages in *Universe in Space and Time* proving that "a day is 48 hours". This due, of course, to rotation, perspective and the International Date Line.
Still, from where we sit, a day lasts exactly 24 hours. Solar eclipses happen from sunrise to sunset. They move fast. They occur as the earth is spinning, traveling east to west.
Hours and Minutes and seconds The day is divided along the concept of the 360 degree perspective, the 360 day Religious and Civil Calendar of the Babylonians and Hebrew. There are 360 minutes every six hours. There are 3600 seconds every hour. The half-day of twelve hours is a measure of time equaling twice 360 minutes. The sky is 360 degrees.
The Seven Day Week. This is the closest quarterly approximation of the phases of the moon. Seven is the profound number of perfection and completion. Musical notes. Visible planetary spheres. Seven seas, continents, "Seven Spirits of God+ (Rev.) We find the seven reflected most profoundly in the Hepton Cycle, Octon Cycle and in the Bible.
The Eleven Day Metonic Week The Metonic week is the difference between the length of the solar and lunar years. There are two simultaneous progressions of Metonic weeks for lunar and solar. They happen on opposite fortnights, 33 times per year. 33x 11 is 363 days, a close approximation of the solar year.
Fortnight- A fortnight is half a moon cycle, such as new to full, or waning crescent to waxing crescent. Solar eclipses are always accompanied by a lunar eclipse one fortnight sooner or later.

The Three Astronomical Months.

Moons are what months are all about. There are three astronomical months.
Lunar month is 29.53 days. From new moon to new moon, typically.

The Eclipse (Draconic) Month During its monthly journey around the earth, the moon passes through the eclipse node in the sky, creating potential for eclipses. The Draconic Month is 27.21 days.

The Anomalistic Month During that same monthly journey, the moon travels either closer or farther away from earth on an elliptical orbit. Its closest approach is called perigee (perigee at a full moon creates what are called "supermoons"). Its farthest distance is called apogee. Full Circle is 27.55 days.

The Calendar months

Ancient Calendar Month-30 days. Round the lunar month up and you have 30 days. Twelve of those fill a 360 day year, adapted to the solar year by leap days, weeks or months.

Julian Calendar Month-30, 31 or 28 days. The Months as decreed by Julius Caesar in 46 BC ignoring the lunar month, timing out the solar year in the manner of the Egyptians and creating 12 months, each named for gods and Caesar's and such. This is our calendar. Julius even has a month named after him! He instituted the leap year. This calendar was perfected astronomically in 1582 as the Gregorian Calendar.

The Bahai Calendar-19 days The Bahai Faith observes a nineteen day month, with nineteen months in a year. 19 x 19 = 361, very close to a solar year.

The Season The procession of the sun during the year creates seasons in the northern and southern hemispheres, marked by equinoxes, daylight and darkness are essentially equal, and solstices, the longest and shortest days of the year.

The Eclipse Season An eclipse season is the only time when the Sun (from the perspective of the Earth) is close enough to one of the Moon's nodes to allow an eclipse to occur. During the season, whenever there is a full moon a lunar eclipse will occur and whenever there is a new moon a solar eclipse will occur. Seasons are 173.3 days apart with matching paired complete eclipses typically happening about 177 days apart within a single eclipse year.

All Those Years (from shortest to longest)

The Draconic (Eclipse Year) The period for the Sun to return to a particular node is called the eclipse or draconic year: about 346.62 days. This is close to 19 days shorter than a solar year and 8 days shorter than a lunar year.

The Lunar Year. A Lunar Year is twelve complete moon phases. This adds up to 354.3 days, about 11 days shorter than a solar year. Civilizations which use the Lunar year can create extra weeks or months to re-align the lunar calendar with solar reality.

The 360 Calendar Year. This is the Babylonian Civil and Hebrew prophetic calendar. 12 30 day months, typically with a 5 day extra period added at the end of the year.

The Solar Year. There are three very close astronomical solar years (differing only in a few hours), depending on perspective, but the typical one used is the "tropical" year of 365.24217 days. The ancient Egyptians used the solar year for calendars. New Year's Day was September 11. Julius Caesar made it January 1st. This was the Julian Calendar, in use for 1800 years in some places. **Gregorian Calendar:** To keep it accurate in continued use required an adjustment of leap years in 1582. Currently, every year that is exactly divisible by four is a leap year, except for years that are exactly divisible by 100. EXCEPT... these centurial years ARE leap years if they are exactly divisible by 400. For example, 1900 is not a leap year, but the year 2000 is.

The Cosmic Clock
There are larger cycles of time involving interactions and repetitions in the solar, lunar and draconic (eclipse) calendars. Are these alignments meaningless? I argue they are not, and many throughout time have agreed with that assessment. Realignment of solar and lunar calendars is a real thing, not an astrological conjecture. The sun and the moon have intense gravitational pull and the perspective of their light and positioning has real consequences. They are extraordinary cosmic entities which return into positions like the hands of a clock.

What is true re-alignment? True re-alignment of earth, sun and moon occurs when sun and moon achieve essentially the exact same cosmic position in relation to earth's perspective. That is only achieved in the Draco-Metonic Cycles of 19 years (with a few hours variance) and 391 years (with almost complete perfection, the variance in minutes).

Why is the Eclipse Year so Unknown?

Luni-solar re-alignment The solar and lunar calendars also re-align at other positions in the cosmic clock. These create the opportunity for looking at various calendars in broad scale.

What is an Epact?
Very simply, epact is the difference of days between calendars. The difference between a solar calendar of 365.24 days and a lunar calendar of 354.3 days is about 11 days and 1 ½ hours. That is their epact. After 33 solar years the epact between them and 33 lunar years is about one complete solar year. These epacts allow for the Metonic Cycle, which is simultaneously 20 eclipse years and 19 solar years.

A Look at the most obvious eclipse cycles
As you peruse this work, I make note of many eclipse cycles. Here's a list of most, from shortest to longest.

Semester Series-Time: 177 days apart A semester series involves two eclipses on opposite ends of an eclipse season. These eclipses can remain paired for centuries, usually with one being total and the other a ring of fire. Saros 136 and Saros 141 are one such pair.
In 1919, they provided "Einstein's Eclipse" (total) of May 29 and the JFK Premonition Eclipse (Ring of Fire) Nov 22 on JFK's birth and death days respectively. They also created the Vietnam War Eclipses of 1955, the longest duration total and annular eclipses of the millennium.

Lunar Year Duos and Triads: 355 days and 710 days apart Solar eclipses can return in sets precisely one lunar year apart in duos or trios. A recent example is the Lunar Year Triad of 2019 to 2021. There were solar eclipses on Dec 26 2019, Dec 14 2020 and Dec 4 2021.

Hepton Septiform Cycle: 1211 days and 2422 days apart Discussed in much greater detail later, Hepton Cycles contains seven eclipses occurring seven eclipse seasons/seven eclipse years apart generally landing on the same day of the week. Probably the maximum expression of cosmic peculiarity. A Hepton lasts 21 Eclipse Years or 20 solar years minus 19 days.

Octon Cycle: 1388 days and 2766 days apart Octon Cycles last exactly eight eclipse seasons and eight eclipse years apart. Interestingly, eight eclipse years is 7 solar years, 7 months and 7 days.

Saros Cycle: 6585 days /19 eclipse years/ or 19 solar years minus one lunar year One of the most astonishing coincidences of science, The Saros Cycle creates the 40 or so Saros Personalities that are born, grow, live and die creating eclipses and allows for the ability ro predict eclipses. The Saros Cycle is the precise alignment of several calendars-the Solar, Lunar-Synodic, Lunar- Anomalistic (which creates apogee and perigee of the moon), and the Lunar sidereal. giving each eclipse a predictable character. This personality returns every 18 years with a similar (though slowly changing in characteristics) eclipse about 1/3 around the globe from where it was last.

Again it is 6,585.3 solar days or 18 years 11 days and 8 hours which is 223 lunar (synodic months), 241.99 draconic months which equals 19 eclipse years and

239 anomalistic months (perigee to perigee) and

241 sidereal months –the moon aligned to fixed star positions.

In other words, every 19 eclipse years there is a precise alignment of 5 cosmic calendars.

It is quite fortuitous that this arrangement should occur in such a conveniently low integer, allowing for repetitive eclipses and easy prediction. The next time after 19 eclipse years such a precise alignment of three of these calendars (lunar, draconic and anomalistic) re-occurs is after 127 Saros Cycles, or about 2290 years.

Draco-Metonic Calendar 6940 days/ 19 solar years/ 20 eclipse years Every 19 years the same phase of the moon returns. Is there a waxing crescent moon out tonight? There was 19 years ago too. And there will be 19 years from now. 235 lunar months is just 1 hour, 27 minutes and 33 seconds longer than 19 tropical years. Meton of Athens in the 5th Century B.C. determined the cycle, which was also known to Babylonians and Mayans.

The eclipse Calendar also aligns at the same 19 years spot at 255 Draconic Months.

Solar Year= 365.24 days. 365.24 x 19 = 6,939.60days (19 years)

Synodic (Lunar)= 29.53 days.29.53058868 x 235 = 6,939.68 days (235 months)

The difference between 19 solar years and 235 lunar months is less than two hours, resulting in a full day of delay every 219 years, roughly twelve cycles.

Now the Draconic (Eclipse) calendar results from the alignment of the moon in a lunar node about every 346.62 days, a reality that is completely independent of the solar and lunar cycles.

 255 draconic months (lunar nodes) = 6,939.11 days. 20.02 eclipse years. The less than half a day difference every 19 years with the lunar and solar calendars results in eclipses often occurring on the same day of the year every 19 years for up to century.

300/44 Cycle Similar eclipses 300 years minus 44 days apart. My favorite example is the twin 300 year pairs on America and the Mideast.

391 Year Grattan Guinness Cycle: 391 years is the nearly perfect alignment of solar, lunar, and draconic yearly calendars. It is the only place after the number 19 where the three calendars match dramatically.

391 solar years equals 4836.005 lunar months and 5248.002 Draconic months.

That is, 391 solar years equals precisely 403 lunar years and precisely 412 eclipse years.

At the 391-year juncture, the epact of lunar years is 12 and the epact of draconic years is 21.

(Thanks to Chris Grant of Brigham Young University for precise modern calculations).

There is reason to suspect the 391-year pattern may bear evidence in history, but this work can only breeze by this prospect. Here's one way to look at it. Call it the "Babylon to Babylon" approach.

Seven Impulses of the 391 Year Cosmic Clock in Western History

1. 747 BC (Feb 26) the first truly agreed upon date in the history of western civilization. The Crowning of King Nabonassar of Babylon.
2. 356 BC Birth of Alexander the Great, Greek Conqueror and founder of Greco-Roman Empire.
3. 35 AD Jesus executed sometime between 27 and 34 AD. Near height of Roman Empire. Beginning of the Christian Church.
4. 426 AD Sack of Rome by Visigoths 410 AD. Decline of Roman Empire and dissolution in 476AD. Beginning of Papacy. Augustine publishes City of God. Beginning of "Dark Ages"
5. 817 AD Holy Roman Empire begins in Europe 800 AD as Charlemagne is crowned by Pope Leo III. Expansion of Papal power. Muslim Golden Age. Middle of "Dark Ages."
6. 1208 AD Roman Church persecution of "heretics" and "infidels" begins in 1209 with the Albigensian Crusades. The Inquisition authorizes torture in 1252 in what becomes a centuries long campaign against Jews and heretics. Spain and Portugal's colonial efforts, papal power extends across the Atlantic. Between disease and war perhaps 150 million native Americans and Europeans died.The Maximum Expression of the Dark Ages.
7. 1599 AD Reformation. British begin colonizing North America May 14 1607. The last Empire is born. The period of the maximum expression of war, slavery, conquest, freedom and creativity.
8. 1990 AD (Aug 2) Iraq invades Kuwait. Empire circuitously returns to Babylon. The American Empire sends troops, preparing for war against Baghdad, which sits atop the ruins of Babylon. After defeating ruler Saddam Hussein (Saddam means "one who confronts" in Arabic). This is exactly the completion of seven 391-year cycles. On Jan 14 1991, war breaks out and ground troops actually enter Iraq on Feb 26, the 2738th anniversary of the first King of Babylon. US Coalition forces first occupied the ruins of Babylon in April 2003.

This is a rough sketch of a reality that perhaps resonates in these long cycles. Taken from the 70 AD temple destruction, five periods of 391 years lead us to 2025.

There are other Cycles and Time Periods involving Religious Calendars, particularly Hebrew.

Septiform Reality: 7 as basis for cycle. A week. Sabbath day.
The ancient laws of 3 and 7.
7 year Sabbath year: sabbath of land, indentured servitude was typically 7 years.
11 Year: 136 lunar months matches up within a day or so to eleven years.
30 Year: 371 lunar months matches closely.
33 year Messianic Cycle: 33 years epact equals full lunar year. 33 solar years equals 34 lunar years. Jesus and Alexander the Great die at 33.
40 years: Time of purification. 40 years for Moses in exile. 40 years wandering in wilderness. King David had 40 year reign. Jesus had 40 days fasting and temptation. "Yet 40 Days and Nineveh will be overthrown", Jonah cried. 40 days were spent by the resurrected Christ on earth before his ascension.
49 years: Jubilee 7 X 7 49 years is full cycle.
50 years is Jubilee year.

70 years: Exile of 70 years for Jews in Babylon. Biblical typical lifespan of man. David died at 70. Temple destroyed in 70 AD.

99 years: 3 messianic cycles.

100 years: classic century mark

120 years: Bible. Moses age at death. Time given as warning for Noah.

400 years: period of captivity in Egypt, rough approximation of slavery in the Americas. First slaves taken from natives in 1492, first Africans brough it in 1526.

490 years Prophetic period of Book of Daniel....Temple to Christ

1000 years. Millenium.

1260 days (years)—mentioned in Daniel...”Time, Times, and half a time” 3 and ½ x 360 days or years, depending whose interpreting. reign of antichrist archetype/literal. The numbers 1335 and 2300 are mentioned in the same Book.

2106 years. Zodiac Age. The click of one constellation to the next on the zodiacal wheel of 25,800 years takes 2106 years for the sun of the vernal equinox to rise on the next constellation . It is said we have been in the age of Pisces since 68 BC. If there was strict science, this puts the arrival of the Age of Aquarius at 2038, but there is no clear boundary between agreed upon spots in the constellations and much debate. We are in the midst of Vernal Equinox eclipses on 2015 and 2034.

2520 Years. A full season: This is 7 x 360 years.

The 500-year historic impulse.

There are many different repeating spans of time which could be tickings of the Cosmic Clock.

For instance, eclipses can be separated by :

177-day intervals. 355-day lunar year intervals.

1211-day (7 eclipse seasons), 2422 day (7 eclipse year) Hepton intervals.

7 year, 7 month, 7 day Octon Personality eclipses.

19 eclipse year Saros personality repetitions.

19-year Metonic repetition of same moon phase.

19, 38, 57, 65, 84, 103 and 122 Draco-Metonic year repeating eclipses on the same date.

300 years minus 44-day cycle of similar shape and geographical eclipses.

And the 391 Year Perfect cycle. There is also a 391 solar year alignment of solar years, draconic and anomalistic months. This is probably the most precise return of the cosmic clock as it relates to our conception of calendar time.

391 Years= 4836.005 lunar months and 5248 draconic (eclipse months)

The actual movement of the sun in relation to human solar calendar time was slightly off until Pope Gregory in 1582 corrected the issues. 391 years is the most perfect reappearance of the cosmic positioning of the moon and sun in relation to earth. Below are two eclipses 391 and 6 ½ hours apart. They are very close to the same shape and latitude.

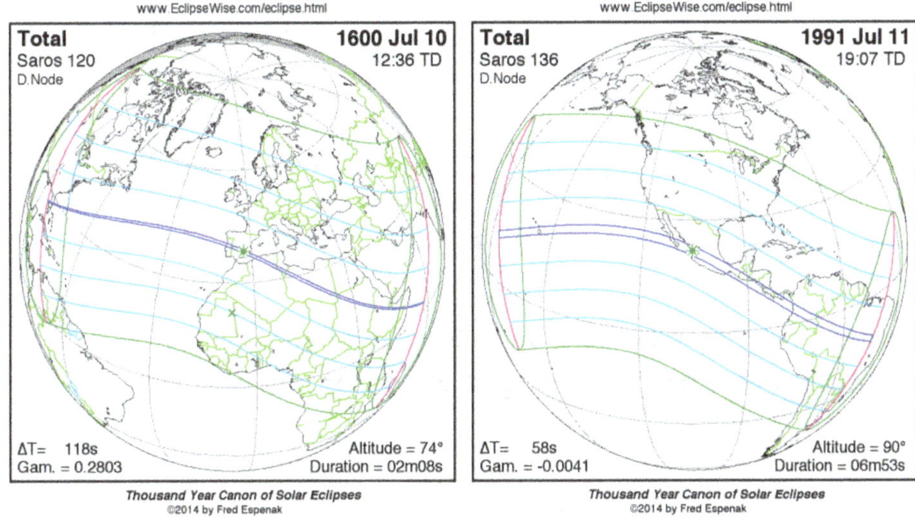

Courtesy of NASA

There is another 500-year apparent impulse where solar or lunar eclipses return again to occur on essentially the same calendar dates from 2000 BC to 1000 AD. This repetitive clicking means all three cosmic calendars have reset on an anniversary. Major religious figures and movements seem to pop up on that cyclical, approximately 500-year impulse. The average appears to be about a 491-year period before the Gregorian Calendar:

Anyway, here are nine points in time creating eight distinct periods because eclipses happen on the same days in the same order. If seen as an octave, it would be seven notes finishing on the eighth, the repeating "do" note-as in do, re, mi, fa, so, la ti, do. The time frames are rough:

1. (do)2,000 BC Abraham/ Egypt/ Babylonian beginnings
2. (re) 1500 BC Moses/ Peak of Egypt/ Indian Vedas/ Zoroaster
3. (me)1000 BC King David/Solomon/ Decline of Egypt
4. (Fa) 500 BC Buddha/ Confucius/ Pythagoras/ Persian-Greek Empires/ Jewish Exile and Return
5. (so)1 AD Jesus and Apostles/ Jewish Temple destroyed 70 AD/ Roman Empire Ascendant
6. (la) 500 AD Fall of Rome/ Islam /Beginning of European nations
7. (ti) 1000 AD Eastern Orthodox Christianity/ Holy Roman Empire/ Early Russia
8. (do) 1500 AD Martin Luther/ Protestantism/ Colonization of New World/ Slave Trade

We arrive then at 2000, which as we know, just happened a few years ago.

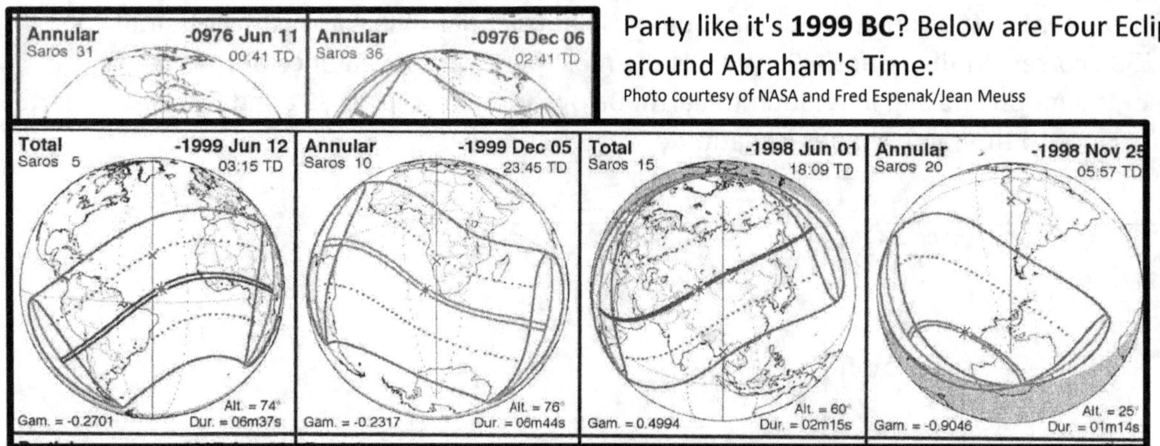

Party like it's **1999 BC**? Below are Four Eclipses from around Abraham's Time:

Photo courtesy of NASA and Fred Espenak/Jean Meuss

Note dates carefully**: June 12, Dec 05, Jun 1 and Nov 25 (then look below)**

Now Look Below at Four Eclipses from Moses' Time: ca. **1476 BC ; 521 years later**

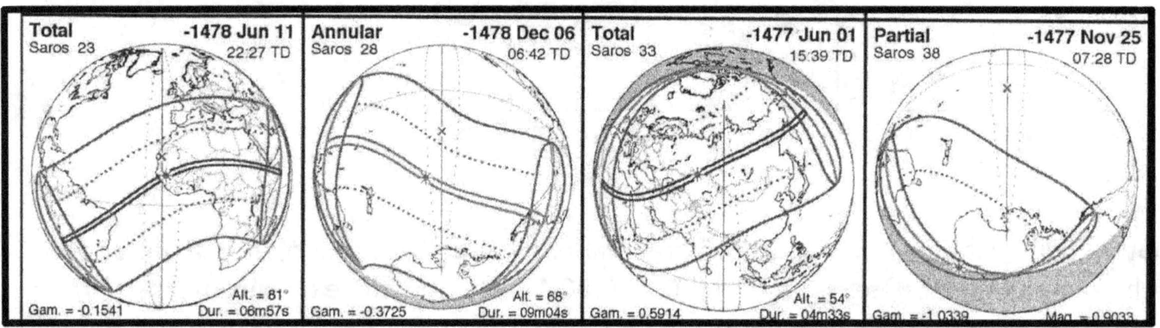

Photo courtest of NASA and Fred Espenak/Jean Meuss

The first eclipse is one day earlier, the second is one day later, the last two are the same. Even the eclipse geography is amazingly similar between the four 1999 BC and the four 1476 BC eclipses. This is the perceived passage of time along the solar calendar, with the Julian leap year.

courtesy of NASA

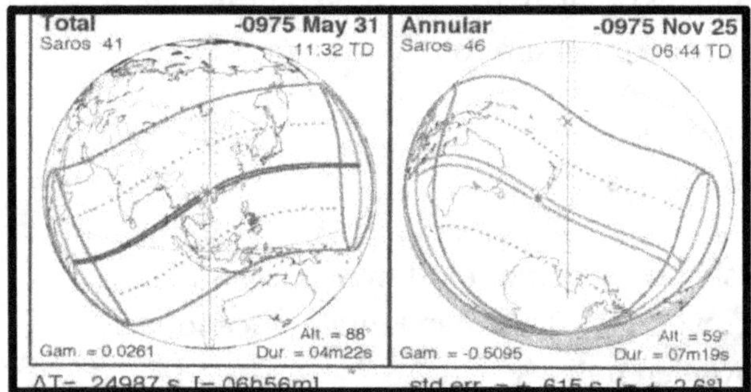

Here are essentially **Same dates: Eclipses of time of King David and Solomon (975 BC)**
497 years after the Moses Metonic, eclipses from the Reign of King David around 1010 BC to 970 BC. Solomon built the Temple in Jerusalem around 970 BC. Dates of **Nov 25** are the same as the 1476 BC. What was **Jun 1** has become one day earlier here at **May 31**. The last three eclipses here appear in very approximate geographical regions of the globe as the last three eclipses in 1476 BC and 1999 BC.

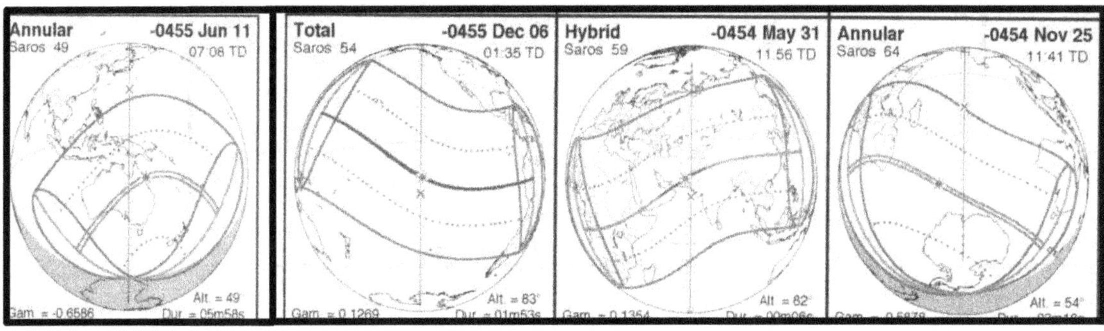

courtesy of NASA

455 BC

Above: Now 520 years after the 975 BC eclipses and we have a snapshot around the time of Buddha/ Confucius/ Pythagoras and Babylonian and Persian Empires. 455 BC is often also the date given as the order from King Artaxerxes for Nehemiah to rebuild Jerusalem. We find the same dates **-June 11, Dec 06, May 31/June 1, Nov 25**. and similar shape of eclipses in very similar geographies!

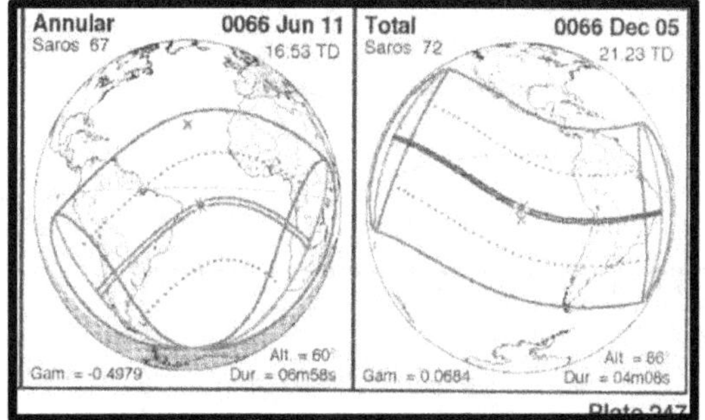

66 AD Death of the Apostles/ Destruction of Jewish Temple (left) Jump 520 years from the Buddha Metonic. We are 33 years after death of Jesus, four years before the Jewish temple is destroyed in 70 AD. Paul and nearly every other Apostle martyred before 67 AD. There's the same dates we've been seeing, **June 11 and Dec 05,** and almost same geography...see below.

587 AD—Muhammad (below) 587 AD 521 years after 66 AD- From Around Muhammad's 18th Birthday . **Jun 11/12 and Dec 05** and maintaining similarities in eclipse shapes as well.

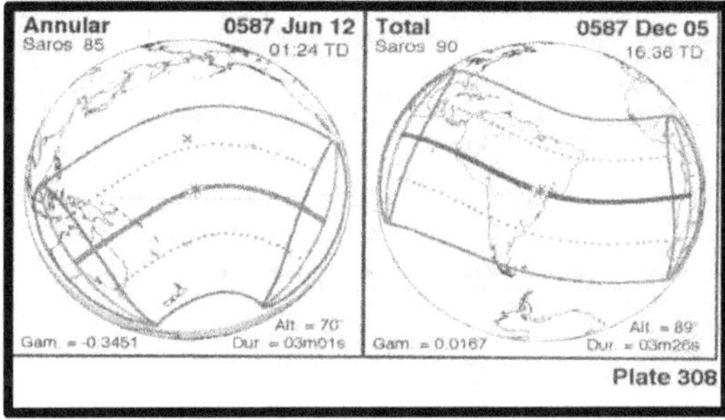

The Julian Calendar seriously departs from Solar Reality around the time of Muhammad. The Metonics of around **1054 AD and 1500 Ad** appear skewed because of the incredibly drifting solar calendar which was finally adjusted by the Gregorian Calendar Change in 1582.

Below is our Current Time: **2029**...notice similar months and days

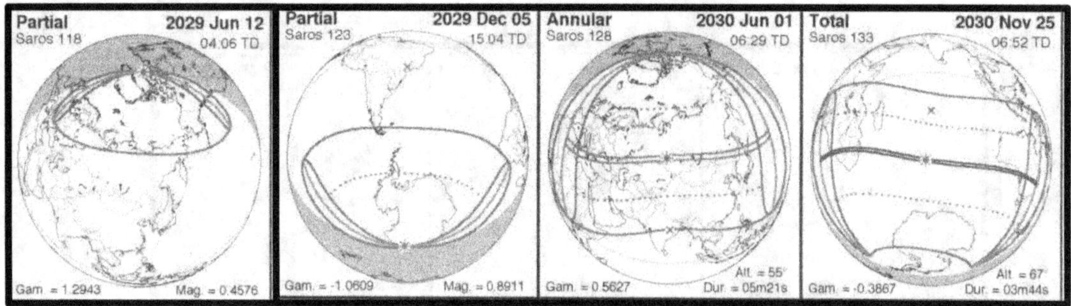

455 BC

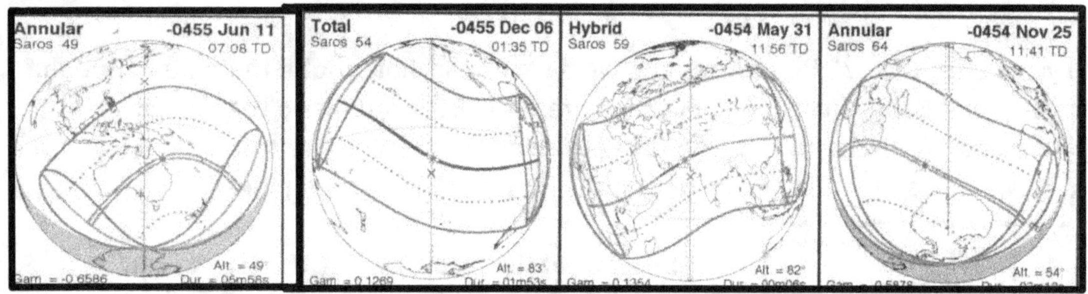

976 BC (extremely similar geography to eclipses above)

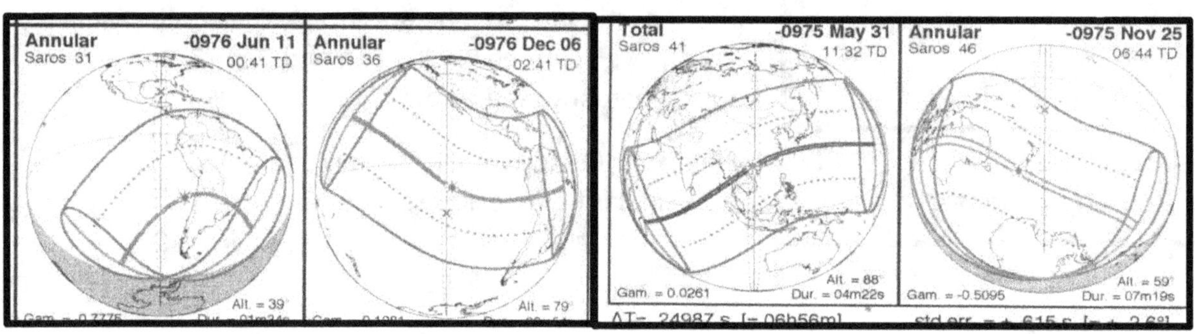

Moses' Time: ca. **1476 BC**

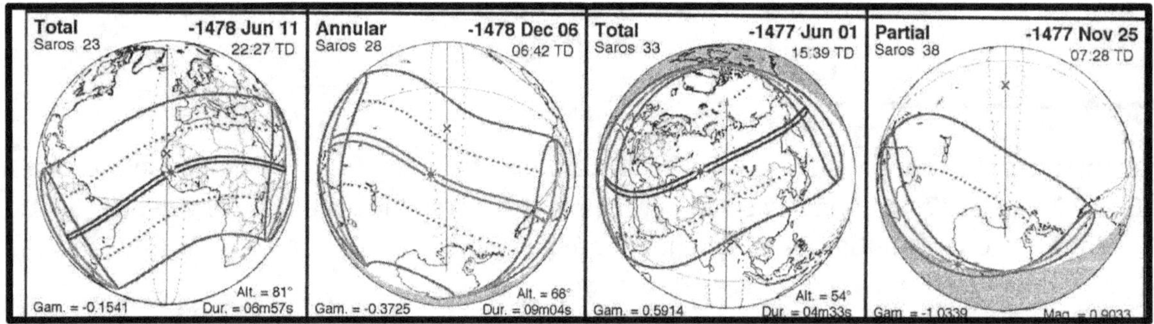

Photo courtesy of NASA and Fred Espenak/Jean Meuss

It's a Cosmic Clock, my friends. It's around 500 years.

There is obviously a truer cosmic beat at 19 years and at 391 years but this 500-year reality is difficult to deny.

Chapter 18

Lincoln and Kennedy in Eclipses
Peculiarities in the Metonic Cycle
Involving the Martyred Presidents

Peculiarities in the Metonic Cycle
Involving the Martyred Presidents

There are only two presidents with closely paired solar eclipses landing on both their birth and death days, Abraham Lincoln and John F. Kennedy. These paired eclipses also occurred during their lifetimes. They also had other peculiar solar eclipses occur during and just around their lifetimes.

There are many well-documented synchronicities between the lives and deaths of Kennedy and Lincoln. I repeat a few of these common knowledge peculiarities on the following page. Afterwards, I present what is possibly the first close look at solar eclipse patterns surrounding their lives and deaths.

These eclipse patterns are not ominous in the commonly understood sense. Sensitive consideration might reveal hints of truth much as one might receive by listening to a piece of music. This is not a subject that can be "summed up", and it carries challenges even to describe.

Having solar eclipses appear multiple times on both one's birthday and death day during your lifetime is not "common". It does happen sometimes, and that fact does not necessarily mean the people with that distinction are all great or super special. A rough attempt to determine birthday/death day eclipse peculiarity was made in the last section.

Clearly Lincoln and Kennedy are peculiar characters, both in their importance in American History, and by nature of the circumstances surrounding their murders.

I present this research with the deepest respect for the memories of these two Americans. I have come to believe that their example of sacrifice was marked in the Cosmic Timekeeping of God.

Kennedy and Lincoln Common Knowledge Synchronicities

The following list appeared perhaps as early as 1965 in the Scientific American column of mathematician Martin Gardner (using the narrative voice of his pseudonym, Dr. Matrix). It appeared in book form in 1967. The list of coincidences had already been printed in the Aug 10 1964 issue of Newsweek and Aug 21 1964 Time magazine based on an earlier printout Gardner had personally distributed to friends and associates. As with nearly everything Gardner published, they are carefully researched facts, not opinions. Since that time, as many as 200 different other coincidences have been asserted by various sources. Some are more interesting (and true) than others. Oddly, several "news" outlets (including Wikipedia, Snopes and USA Today) have seen fit to call the Kennedy/Lincoln coincidence phenomenon an "urban legend" and even run "fact-checks", typically repeating minor incidental false claims by misinformed chroniclers of the synchronicity.

Fifteen claims of synchronicity are beyond dispute.

1) Lincoln was elected President in 1860. Kennedy in 1960.
2) Lincoln was elected to Congress in 1846. Kennedy in 1946.
3) Both were deeply involved in rights for African-Americans.
4) Both were assassinated on Friday, seated beside their wife.
5) Each had lost a son while in office.
6) Both were shot in the head from behind.
7) Lincoln was killed in Ford's Theater. Kennedy was killed in a Lincoln (made by Ford)
8) Both men were succeeded by southerners named Johnson.
9) Andrew Johnson was born in 1808. Lyndon Johnson was born in 1908.
10) The first name of Lincoln's private secretary was John (John Nicolay). The last name of Kennedy's private secretary was Lincoln. (Evelyn Lincoln)
11) Kennedy and Lincoln were assassinated by Southerners who were killed before coming to trial.
12) Booth shot Lincoln in a theater and fled to a barn. Oswald is said to have shot Kennedy from a warehouse and fled to a theater (*the movie "War is Hell" was playing*).
13) The names Lincoln and Kennedy each have 7 letters.
14) The names Andrew Johnson and Lyndon Johnson each have 13 letters.
15) The names John Wilkes Booth and Lee Harvey Oswald each have 15 letters.

In the article, Gardner finishes the coincidences listed by his alter-ego, Dr. Matrix with this quote from the Old Testament Apocrypha *The Wisdom of Solomon*:

"Thou (God) hast ordered all things in measure and number" -*Wisdom of Solomon* Chapter 11 V:20

Gardner ends his Chapter with this last unusual piece of history:

"The first public suggestion that Lincoln be the Republican candidate for president is believed to be a letter written Nov 6 1858 and published in the Cincinnati Gazette. The writer, Israel Green (a druggist in Findlay, Ohio) proposed a ticket of Lincoln for President and John Kennedy for Vice-President. The proposed Kennedy was John Pendleton Kennedy, a prominent author and politician, who had been Millard Fillmore's Secretary of the Navy. Greens letter can be found in The Magazine of American History Vol 29 1893 page 282." (Gardner ca. 1965 reprinted in *Amazing Numbers of Dr Matrix*)

Kennedy and Lincoln Eclipses

Kennedy and Lincoln's birth and death day eclipses are different than the two other Presidents, Jefferson and Harrison, previously mentioned for having birthday/death day eclipses during their lifetimes. The difference is that Kennedy and Lincoln's birth and death day live near each other in the eclipse calendar. The exact anniversary dates are closely embedded within the Metonic cycle nearby each other and repeat in the same position around every 350 to 500 years or so.
The Metonic Cycle is the repeating and ever-changing link between the solar, the lunar and the eclipse year. Don't worry if you don't have it all figured out. Who does?

Eclipse year=346.62 days...sun and moon returning to the general same spot (node) at same time. Eclipse season= 173 days, but complete eclipses usually happen 177 or 178 days apart when linked inside a single year. That is one half a lunar year.

Lincoln's birth and death dates (Feb 12 and Apr. 14/15) appear 3 eclipse years (346.6 x3) apart in the Metonic Calendar, or 1033 days. This means the position of moon and sun have returned to the exact same node (place in the sky) at the end of that 1033 days.

Kennedy's "semester series" eclipses are a continual side by side Metonic couplet 177 days apart, or one eclipse season plus four days. They are at opposite nodes, one north and one south. Kennedy's birth and death days just happened to be essentially one eclipse season apart. I am not sure who else has that distinction. Certainly it happens, but it's a hard thing to just google.

There are 7 eclipses on exact birth/death days immediately around the life of Abraham Lincoln. This is assisted by the fact that Lincoln died over the course of nine hours spanning two days.

We have 5 eclipses on exact birth or death days during the life of John F. Kennedy, but a total of eleven within his extraordinary side by side Metonic in the 20th century.

The maximum expression of eclipse associations in Presidents (or any public figures I have looked at, for that matter), appears to occur with Abraham Lincoln and John F. Kennedy. These two men, in turn, were maximum expressions of the American Presidency. Maximum leadership in a nation which became the maximum imperial/cultural power of the modern era.

First Overview Look at Lincoln/Kennedy Dates before we go on.

I include eclipses which border the exact birthday/death days here, or shortly after their death. There are reasons to do this, based on orbital, historical and leap year reality. Remember what George van Den Beergh said: "A day is 48 hours!" First we need to establish facts in the basic peculiarity.

Abraham Lincoln was Born Feb 12, 1809--- Died April 14/15 1865. Lincoln was shot late on April 14th and pronounced dead 9 hours later on the morning of April 15th 1865

Birthday solar eclipses on Feb 12 1812/ Feb 12 1831/ Feb 12 1850/
Death Day solar eclipses on April 14 1809/ April 14 1828/April 15 1847/ April 15 1866

John F. Kennedy Born May 29, 1917---Died November 22, 1963

List of all JFK eclipses in 20th Century:

Birthday: May 28th 1900/ May 29 1919/ May 29 1938/ May 30 1946/ May 30 1965/ May 30 1984
Death Day: Nov 22 1900/ Nov 22 1919/ Nov 21 1938/ Nov 23 1946/ Nov. 22/23 1965/ Nov 22 1984

Again, out of 46 Presidents- only 4 have eclipses during their lives on both their birth and death days. There is a mildly interesting decreasing age progression pattern in these four Presidents.

Jefferson was 81 years old when he died.
Harrison was 68 years old when he died.
Lincoln was 56 years old when he died.
Kennedy was 46 years old when he died.
If you keep going with non-presidential deaths of closely associated leaders:
Bobby Kennedy was killed at 42.
Martin Luther King at 39.
JFK Jr. at 38.

Of all the Presidents and celebrities we looked at, only JFK died essentially on an ecliptic birthday. He lived for 16,978 days, essentially 49 draconic years.

To recap, we only found four Presidents of the US who had complete solar eclipses on birth and death days. Three completes on exact birth/death days in the life of Thomas Jefferson. There were two complete eclipses on exact birth/death days in the life of William Henry Harrison.

ABRAHAM LINCOLN- BORN FEB 12 1809 / DIED APRIL 14/15 1865

His birth and death dates appear in reverse order in solar eclipses in the first three years of Lincoln's Life. Exactly three eclipse years (1033 days) apart, they remain linked in such a way throughout time.

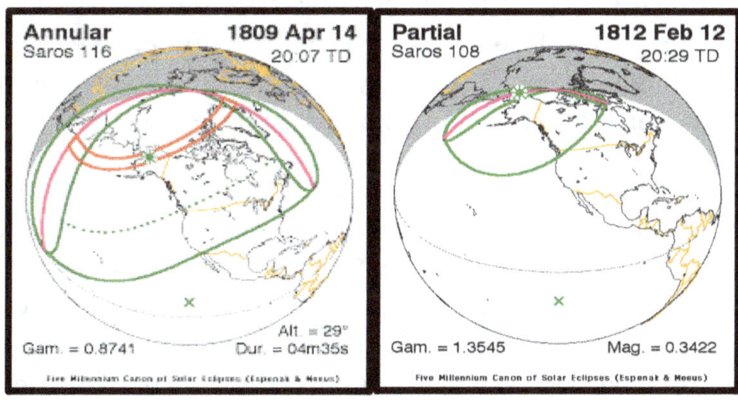

Courtesy of NASA—Espenak and Meuss

On the left, the Lincoln Assassination Premonition Eclipse two months after Lincoln's Birth and 56 years before Assassination. On the right, solar eclipse on Abraham Lincoln's third birthday.

Apr 14 1809 Ring of Fire North America (Saros 116) A ring of fire eclipse over North America occurs two months after Lincoln's birth on April 14 1809, the exact day he would be shot 56 years later. Centerline over what becomes Bethel, Alaska. Bethel means "House of God". Southern boundary of partiality through Lincoln's birthplace in Kentucky and future home in Illinois. It travels near Washington DC in a rough approximation of the Mason-Dixon line.

Lincoln is shot exactly 56 years later (on Passover) on April 14 1865.

Feb 12 1812 Partial Solar North America (Saros 108) Lincoln's birthday of Feb 12 appears in the eclipse cycle on his third birthday in 1812. It is a partial eclipse over northwest North America, maximum eclipse at almost the same time and same spot (in Alaska) as the April 14 1809 eclipse.

The Feb 12 Metonic and April 14/15 Metonic continue in profound relation until his death.

April 14 1828 and February 12 1831

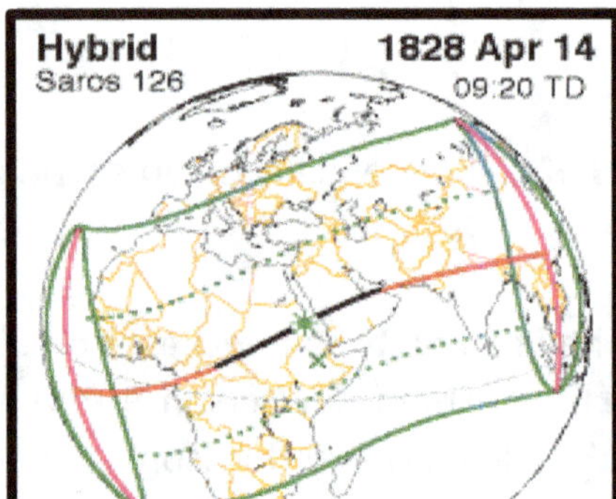

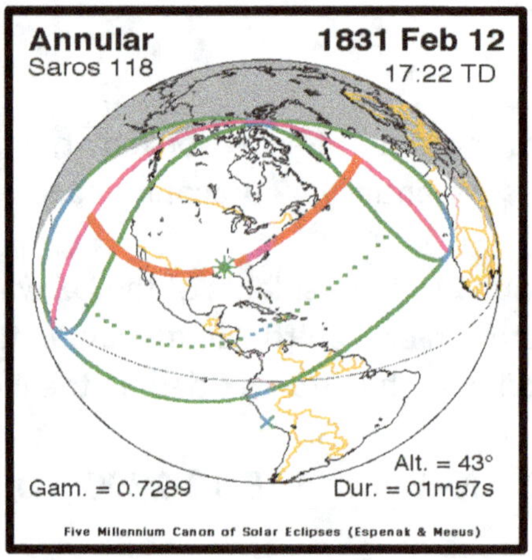

Courtesy of NASA—Espenak and Meuss

Two geographically profound eclipses arrive 19 years later on Lincoln Birthday/Deathday Metonic.

Apr 14 1828 Hybrid **Africa/Arabia/ India (Saros 126)** An amazingly similar eclipse path to the previously shown Thomas Jefferson 18th Birthday eclipse which occurred 65 years and one day before. Lincoln is now 19 years old and his assassination exactly 37 years away. Interesting eclipse travels across Africa through the continent into Arabia (92% in Mecca) and ends in the Himalaya.

Feb 12 1831 (Saros 118) Lincoln's 22nd Birthday/ Nat Turner Ring of Fire in USA Abraham Lincoln is now having his 22nd birthday on Feb 12 1831. A solar eclipse travels through the South and up the eastern coast. Annularity in Washington DC and very close to New Orleans, annularity in Atlanta and Cape Cod. Perhaps the most widely viewed solar eclipse in the USA since the nation's founding.
Only adult President in our survey in a position to view a solar eclipse on their birthday.
Virginia slave and mystic Nat Turner (nicknamed "the prophet") witnessed this eclipse and took it as a sign from God to begin planning an insurrection against slave holders which began in August. It was the bloodiest slave revolt in US history. Turner was hanged November 11 1831 in Jerusalem, Virginia.

1831 to 1837 Lincoln in New Salem, Illinois (From Abraham Lincoln Historical Society)

In 1831, Abraham Lincoln was a young man of 22 when he and a couple of companions floated the Sangamon River in a flatboat on their way to New Orleans. In mid-April, he neared New Salem, Illinois, when his boat became stranded on the nearby mill dam. A crowd gathered to watch the men work to free the boat. Under Lincoln's direction, the crew unloaded the cargo from the stern causing the flatboat to right itself. Lincoln then went ashore and borrowed an auger from Onstot's cooper shop, drilled a hole in the bow allowing the water to drain out, causing the flatboat to ease over the dam.
Denton Offutt, who had hired Lincoln to man the flatboat, was impressed with Lincoln's handling of the incident, and awarded him with the offer of a clerk position in his store. However, when Lincoln returned from New Orleans, the store was not yet open, so Lincoln took a variety of other jobs, including helping to pilot a steamboat down the Sangamon River to Beardstown on the Illinois River.

New Salem was where Lincoln lived when he did his service in the military during the Indian wars, Lincoln saw no service, but did help bury five colleagues who had been scalped near Kellogg. Lincoln became postmaster and was much beloved. On April 15, 1837 (the exact midpoint between 1809 eclipse and death in 1865), on a borrowed horse, with everything he owned in two saddlebags, Lincoln moved to Springfield, the place he would call home for the next 24 years of his life.

Last Metonic before his Death: April 15 1847 and February 12 1850

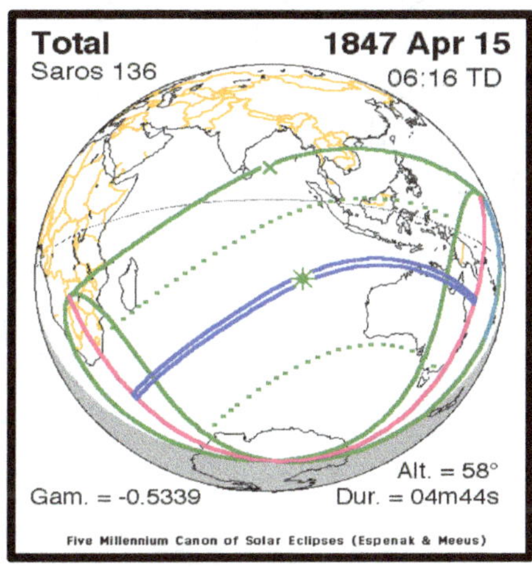

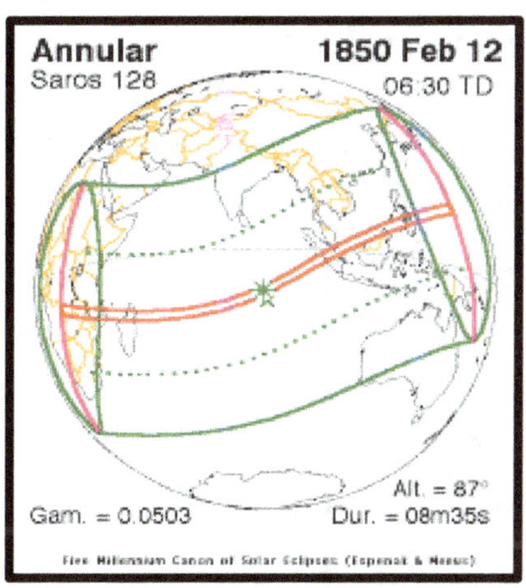

Courtesy of NASA—Espenak and Meuss

Apr 15 1847 Indian Ocean to Darwin, Australia Total (Saros 136) Eclipse sixteen years after New Salem boat incident. Exactly ten years after Lincoln moves to Springfield and 18 years exactly before his death. When this eclipse begins at sunrise over the Indian Ocean, it is still April 14th in much of America.

In 1847, Abe Lincoln is 38 years old and is serving in the US Congress. In December of 1847, he famously introduces legislation demanding to know the "spot" where hostilities with Mexico originated, in an effort to determine the government's excuse for the ongoing Mexican War, which Lincoln opposed as an imperial land grab.

 The eclipse itself takes aim at Darwin Harbor, Australia- eight years after it is named such in 1839 by a ship captain after Biologist Charles Darwin, who famously shared an exact birthday with Lincoln.

Feb 12 1850-Ring of Fire Last Birthday eclipse Africa to Philippines (Saros 128)

Lincoln and Darwin's birthday appears for the last time as an eclipse during their lifetimes on their mutual 41st birthdays. Feb 12 will not be the date of a solar eclipse again until the year 2203. 1850 was the year of the *Compromise of 1850,* which temporarily avoided war by allowing expansion of slavery west into Utah and New Mexico and admitting California as a free state.
Saros 128 also brings three modern eclipses of note:
May 10 1994 USA Ring of Fire. Annularity includes Lincoln home of Springfield, Illinois exactly 136.00 eclipse years after Lincoln's assassination and 156 solar years after birth of Lincoln's assassin John Wilkes Booth on May 10 1838.
May 20 2012 USA Ring of Fire 18 years later, Saros 128 travels from sunrise at Hanoi, Vietnam to end at sunset over Bush Family home of Midland, Texas.
June 1st 2030 Athens/Constantinople Eclipse.

Eclipses around Lincoln's Assassination

There was a partial lunar eclipse on April 11 1865, three days before Lincoln's murder, and two days after surrender of the Confederate Army at Appomattox. It was not visible in America.

Abraham Lincoln is shot in the back of the head while seated next to his wife at Ford's Theater in Washington DC on Good Friday, the evening of Passover, April 14 1865, exactly 56 years after the eclipse Metonic began on 4-14-1809(two months and two days after he was born).
The shooting occurs at 10:14 pm on April 14th. Lincoln breathes his last at 7:22 am on April 15th.
Ten days after Lincoln's death a profound total solar eclipse began in the last stronghold of Slavery in the Western Hemisphere and crossed to the slave coast of kAfrica:

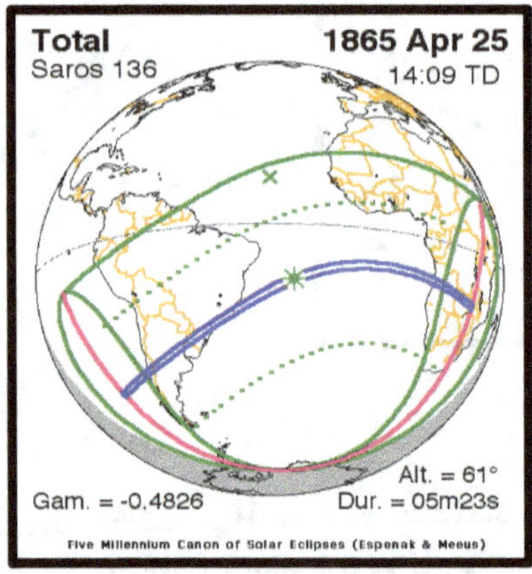

Courtesy of NASA—Espenak and Meuss

Apr 25 1865 Rio De Janeiro Eclipse South America to Africa Total (Saros 136) Eleven days after Lincolns murder and the day before John Wilkes Booth is killed in a barn after eluding capture for eleven days. Eclipse travels from South America to Africa. Goes through Brazil, where millions of African American slaves and native peoples toiled in bondage as the eclipse passed overhead. Brazil was the last nation in the western hemisphere to abolish slavery (in 1888). Totality in Rio De Janeiro, the largest slave port in the history of the world. America's great slave port Charleston, would eclipse 6 months later. Path travels to Africa with totality in Angola, source of the first slaves brought to Jamestown in 1619.

Saros 136 is paired as a semester series (½ lunar year) with Saros 141, the great ring of fire, which returns 177 days later with one of the most astonishing cross-country eclipses in US history.

Farewell Abraham Eclipse October 19 1865

188 days after his assassination, a ring of fire eclipse divides the United Sates of America. The extraordinary cross-country path begins at sunrise over Mount Saint Helens and western Washington. The path bisects the nation from Northwest to Southeast, the first eclipse to do so since its founding. The path touches 20 states and territories, unrivaled in USA history.
Annularity over Lincoln's birthplace in Kentucky and childhood home in Illinois. Path crosses modern Lincoln, Nebraska. The only west coast to east coast ring of fire eclipse in USA history. One of only five coast to coast complete solar eclipses (along with 1806, 1869, 1918 and 2017) in US history. Follows a very similar path to 2017 Great American Eclipse. Being a long eclipse of nine minutes and 27 seconds and occurring during the typically fair month of October, it would have likely been seen by more Americans than any ring of fire in US history up to that point. The centerline of eclipse goes over Confederate strongholds of Nashville and Atlanta and exits country at Charleston, South Carolina, the largest slave port in the USA. Eclipse traverses the Atlantic to end at sunset over Western Africa, centerline over Portuguese slave port of Dakar, Senegal-the westernmost city on the African continent.

Eclipse occurred on the 84th Anniversary of the victory of the American Revolution at Yorktown on Oct 19 1781. 84 years is a profound anniversary on the Metonic cycle (consisting of three 19-year periods, an 8-year lurch, then one 19 year Metonic), It's also 88.5 eclipse years or 177 eclipse seasons.

Saros 136 and Saros 141, occurring 177 days apart, provide the Western hemisphere solar eclipses of 1865. They have been inseparably paired in what is called a semester series through much of the Second Millennium to this day. These same Saros return in 1919 for the JFK birthday and deathday eclipses (May 29 1919 and November 22 1919) 54 years and 33 days later.

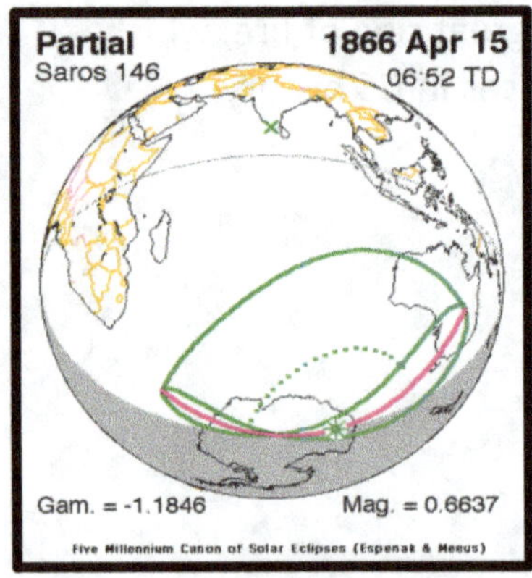

There is also an April 15th 1866 partial on the 1st anniversary of Lincoln's death. It is the last time either his exact birth or death Metonic appear for several centuries.

Courtesy of NASA—Espenak and Meuss

LINCOLN METONIC Round-Up

There were three solar eclipses on Lincoln's exact birthday of February 12 *during* his lifetime: they happened 19 years apart on 1812, 1831, and 1850. The next Feb. 12 solar eclipse is 2203 AD. The 1831 eclipse (when he turned 22) would have been visible to him, depending on weather.

In 1869 and 1888 the Metonic actually slides backward to Feb 11 for two eclipses. Then, in 1896, it lurches forward to Feb 13.

19 years later, in 1915, it's Feb 14 or Valentine's Day, where it stays for much of the 20th Century. That same Metonic has reached Feb.17 for the upcoming Feb 17 2026 Ring of Fire.

And so it goes.

There were three eclipses on Lincoln's Death Day of April 14/15 during his lifetime, and one on the first anniversary of his death. They happened 1809, 1828, 1847 and 1866. He would not have been able to see any of them. There were opportunities for Lincoln to see several solar eclipses during his life, and we'll look at that in a minute.

I have not come across any historical personage with as many exact eclipses on birth and death days as Lincoln during their lifetime, though JFK comes close.

By 1874, Lincoln's Death Day Metonic has "moved" to April 16th. The Titanic eclipse happens on April 17 1912 (three days after the sinking). The Metonic has moved all the way to April 19 by the time of the 1939 WWII Hitler Birthday Eve / Darwin Death Day Eclipse.

Now a brief look at actual eclipse paths in the USA surrounding the life of Abraham Lincoln.

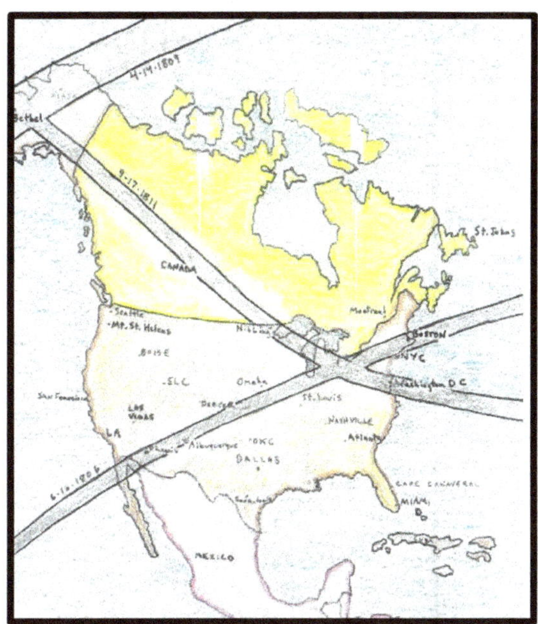

Solar eclipses in USA around Lincoln's birth: The Cleveland Cross 1806 to 1811

There were only three solar eclipses in the USA between 1783 and 1820: a cluster from 1806 to 811.

1806 Jun 16 Total Eclipse (Saros 124) The first complete solar eclipse in the US in 23 years. Almost fully Cross-country two and ½ years before Abe Lincoln's birth and 62 eclipse years exactly (within two days) before his death. The only total cross-country eclipse in US history to travel southwest to northeast coast to coast

1809 Apr 14 Ring of Fire (Saros 116) Passes over Bethel, Alaska: "House of God" in Hebrew. Bethel in The Bible is where Abraham (Abram) built an altar to God. Lincoln will be shot 56 years later on this day which occurs 2 months and 2 days after he is born on Feb 12 1809.

1811 Sep 17 Ring of Fire (Saros 141) The great Saros 141 crosses the 1809 eclipse in Bethel. Crosses Canada and bulk of Northeast US (12 states). Washington DC and Jamestown are in path. The calendar midpoint of the 1806/ 1811 eclipses is Feb 4 1809, a week before the President's birth on Feb 12 1809. Lincoln's home of Springfield, Illinois is in the path of 1806 eclipse. Washington DC, where he served and died, is in the 1811 Eclipse. The 1806 and 1811 eclipses cross in Cleveland, Ohio which was founded in 1796. There is then a ten-year break from solar eclipses in the USA until 1822.

Two Ring of Fires Six Months Apart 1821-1822

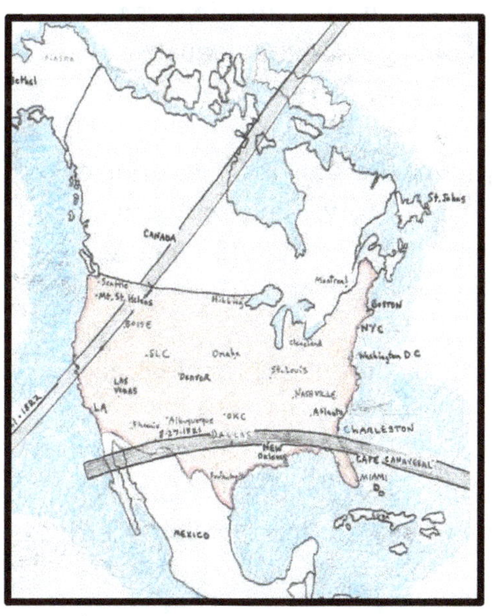

1821 Aug 27 Ring of Fire (Saros 132) An almost Cross Country Eclipse through the Deep South.

1822 Feb 21 Ring of Fire/ Hybrid (Saros 137) Occurs over mostly un-colonized far west, including Mount Shasta. These eclipses are noteworthy for the fact they occur 178 days apart, or exactly one half a lunar year. That is the shortest time between American eclipses in history- and the shortest time possible (177 to 178 days) between complete eclipses on earth. **Numeric curiosity between 2 -21-1822 eclipse and the next complete eclipse in the USA, which will occur 2-12-1831/** 2-21-1822 and 2-12-1831 eclipses occur 9 years minus 9 days apart, or essentially 8 solar years plus one lunar year apart. 2-21 and 2-12 are obviously anagrams.Also, if the full numbers of either eclipse are added, they equal the same amount:

(2-21-1822) 2+2+1+1+8+2+2 = 18 and
(2-12-1831) 2+1+2+1+8+3+1 = 18

Trail of Tears Eclipses 1831 to 1838—All three likely visible to Lincoln

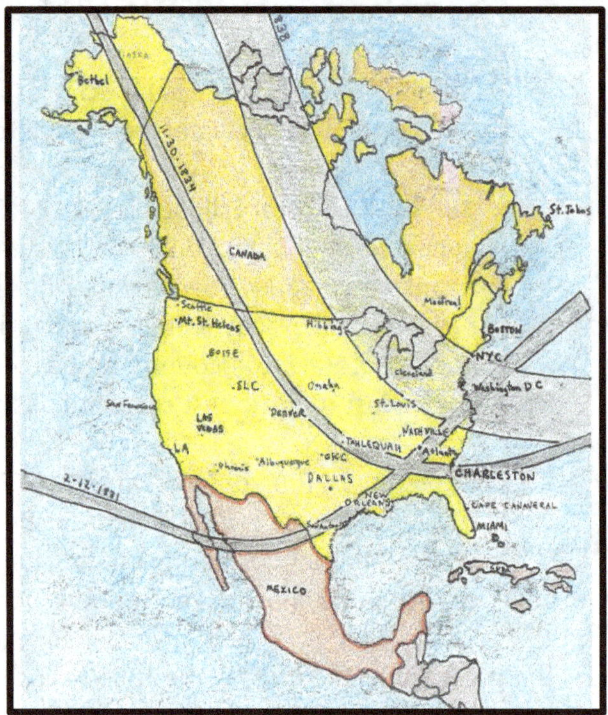

1831 Feb 12 Lincoln Birthday/ Nat Turner Ring of Fire (Saros 118) Starts in lower map. Abraham Lincoln's 22nd birthday. The eclipse that inspired Nat Turner's slave revolt. In all, 9 slave states experience the eclipse, probably the most geographically possible at any one time. The only eclipse in entire survey that occurs on someone's birthday while they are an adult in the nation they live.

1834 Nov 30 Trail of Tears Total Eclipse (Saros 120)-starts in middle of map One of the most unusual eclipse paths in history traces the Trail of Tears from Georgia to Oklahoma at the same time as the forced marches were proceeding on that trail. Atlanta, Charleston, Columbus GA. Tahlequah, Oklahoma (final destination of Cherokees) has its only eclipse besides 1694. The great Saros 120.

1838 Sep 18 Washington DC Ring of Fire (Saros 122) In the worst year of the Trail of Tears, a ring of fire across 13 northern states. This may be the widest eclipse path in US history. Many northern cities in the path: Detroit, Cleveland, Columbus OH, Pittsburgh, Philadelphia, New York City, Baltimore, Washington DC (where Lincoln would die exactly 28.001846 eclipse years later, or 9706 days)

Another cosmic masterpiece from the Eternal Artist. The eclipses actually progress chronologically from south to north balanced both geographically and in rhythm.

Between the 1831 and 1838 eclipses are exactly 7 years, 7 months and 7 days (an Octon—exactly 8 eclipse years). The 1834 eclipse is situated at the precise Metonic mid-point on the calendar (8 eclipse seasons) *and* also at the geographic midpoint of the map.
The Saros numbers are all evenly spaced as well-118, 120, and 122.

Lincoln would have been in a position to potentially witness all three of these eclipses.
After the 1838 eclipse, there is a 16-year break from complete solar eclipses in the USA, until 1854.

1854 to 1869- Civil War Eclipses

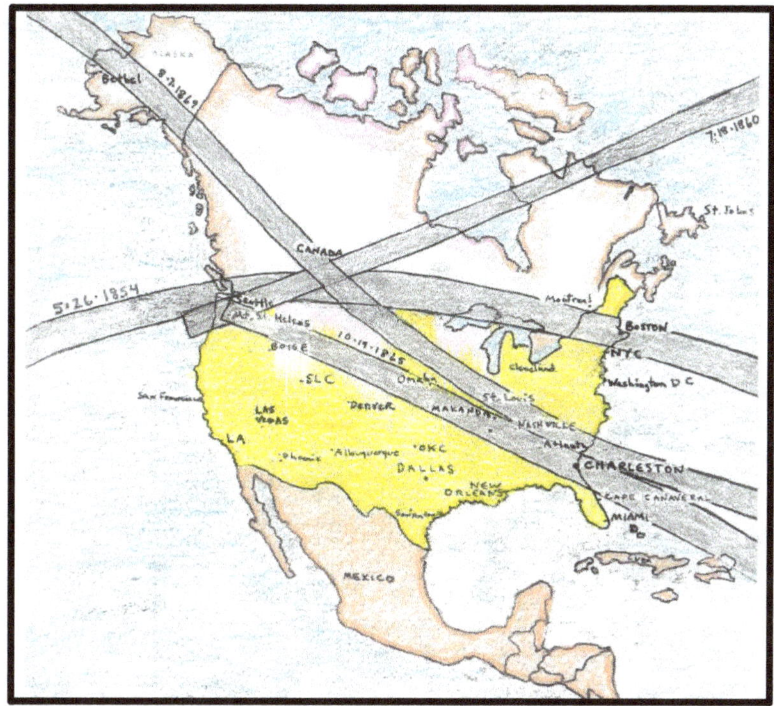

1854 to 1869

1854 May 26 Ring of Fire (Saros 135) No other eclipse has ever so perfectly straddled the northern US border. From Washington to Maine, path touches all 12 states of the northern border. The only actual metropolitan city on the path is Boston. Eclipse happens four days before passing of the Kansas-Nebraska Act on May 30, 1854, which opened up new US territories to slavery, prompting violent clashes which accelerated the race to Civil War.

1860 Jul 18 Total Eclipse (Saros 124) Three and ½ months before Lincoln is elected in November, a total eclipse over the Northwest, including Seattle, founded 9 years before. The secession of the southern states began five months later on December 20, 1860.

1865 Oct 19 Farewell Abraham Ring of Fire (Saros 141) 188 days after the assassination of Lincoln on Good Friday April 14 1865. Cross-country eclipse starts over Mount Saint Helen's at sunrise and crosses the entire USA to exit at the notorious slave port of Charleston, SC. A total of 20 states and territories were witness, the most ever in history.

1869 Aug 7 Total Eclipse (Saros 143) From Bethel, Alaska (now part of the USA after the 1867 purchase from Russia), where Abraham built an altar in the Bible and where the first assassination premonition eclipse passed over on April 14 1809, the eclipse cuts across the heart of the nation, including Lincoln's town of Springfield, Illinois. Eclipse 3 months after "The golden spike" at Promontory, Utah completed the first transcontinental Railroad.

Within 20 years of this eclipse, nearly every buffalo in America was dead and the final free native tribes of America would be largely subjugated.

Abraham Lincoln Metonic/ Jewish Calendar Oddities

Interesting Israel and Religious Dates on this map

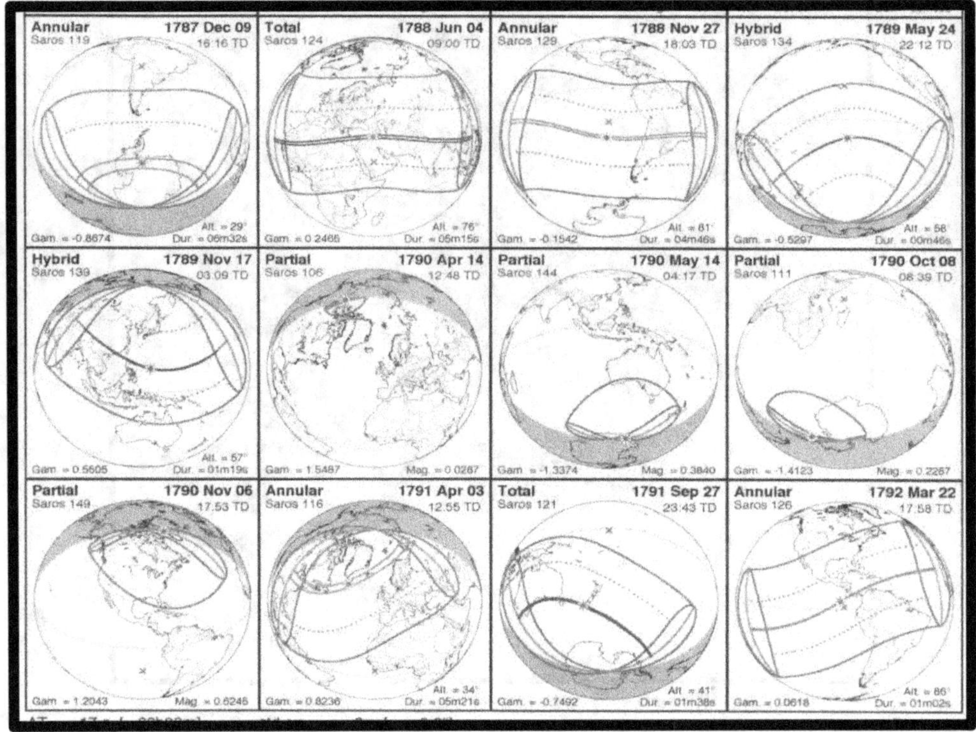

Courtesy of NASA—Espenak and Meuss

Jefferson's Birthday of **April 13** dominated the mid-April Metonic in the 18th Century. The NASA map **above** shows the first time Apr 13 moves into Lincoln Assassination Metonic, **April 14 1790**, 19 years 2 months and 2 days before Lincoln is born. Eclipse is middle row, second from left.

April 14 1790 1720 years to the day after the day Jerusalem was first surrounded by Rome in 70 AD. **Apr 14:** Date of Passover possible Jesus Crucifixion in 32 AD. Lincoln assassination day in 1865. Titanic sinking in 1912. Splitting of the atom 1932.An almost imperceptible partial, The Apr 14 1790 eclipse covered less than 3% of the sun. This is the last eclipse of Saros 106 after 1300 years.
Saros 106 began its life with a partial on January 23, 456 AD. Yes, that's right. On 1 23 456 AD.

Top row: **June 4 1788** Total eclipse through Syria and Nineveh 179 years before the last day of "Gentile" control of Jerusalem in 1967.

Top row: **May 24 1789.** Eclipse occurs 19 days after start of French Revolution.May 24 is day of Pentecost 50 days after traditional Jesus Crucifixion date of Apr 3 33 AD. That is when the Holy Spirit was first given to men.. Bob Dylan is born 152 years later.

Middle Row: **May 14 1790-** 158 years to the day before Israeli nationhood in May 14 1948. 19 years after odd "177" May 14 1771 eclipse, which happened 177 years before Israeli Independence.

Bottom Row: **April 3 1791-** 1758 years from traditional Date of Crucifixion in 33 AD

Metonic Oddities Surround Lincoln's Birth
A Trio of Future Assassinations Embedded in Eclipse Cycle

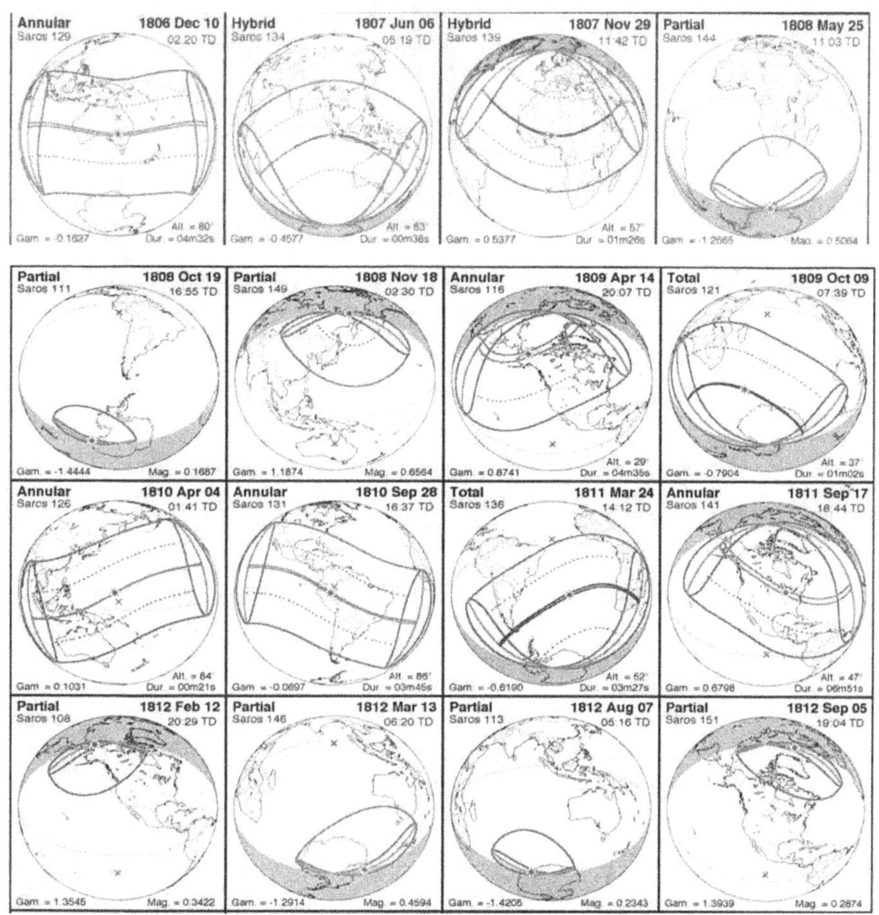

Courtesy of NASA—Espenak and Meuss

All solar eclipses from 1806 to 1812 in image. Lincoln's first death day eclipse (Apr 14) is on the second row, second from right. Lincoln's 3 year birthday eclipse (Feb 12) is on the bottom row left.

Top row is **June 6** 1807. June 6 is the date Robert Kennedy is assassinated 161 years later in 1968.

On the third row down on left is **April 4 1810,** April 4 is the date Martin Luther King will be assassinated 158 years later in 1968. April 4 1810 is 1777 years since Christ slept in the tomb by traditional dating.

The Bobby Kennedy and King dates are exactly 1033 days apart or exactly 3 draconic years. They are permanently embedded in the Metonic Cycle every 350 to 500 years. Lincoln's deathday and birthday eclipses are also 1033 days apart embedded in the Metonic Reality, exactly 3 eclipse years apart.

Aesthetic Numeric curiosity:
Double digit appearances- in day and month- in the four most famous assassinations in the history of the USA.
Lincoln's shooting has two fours: 4-14 (4-14-1865-Friday)
Martin Luther King's has two fours: 4-4 (4-4-1968-Thursday)
John F. Kennedy has two ones and two twos: 11-22 (11-22-1963-Friday)
Robert Kennedy has two sixes: 6-6 (6-6-1968-Thursday)

Darwin and Lincoln

Charles Darwin was born on the exact same day as Abraham Lincoln, Feb 12 1809.

It is surely one of the great oddities of history that two of the most famous and influential people of modern times were linked in such a way. The nature of both their fame and influence were wildly different, of course.

Darwin came to worldwide prominence in 1860, after the Nov. 1859 publication of "Origin of Species" laid out his argument for natural selection.

Lincoln also came to worldwide prominence in 1860, capturing the Presidency after laying out his determination that America could not survive half-slave and half-free.

Lincoln was martyred at 56. Darwin had a heart attack at 72.

Darwin's death occurred at 4pm April 19th, 1882, seven years and a day before Hitler was born on April 20 1889.

Darwin shared the three birthday eclipses with Lincoln. It's unclear if he saw any of them.

Darwin's death day of Apr 19 first appears as an eclipse 57 years after his death (57 is a full 3 x 19-year Metonic cycle) on April 19 1939- in the last solar eclipse before WWII. It crossed extreme northern Europe hours before Adolf Hitler's 50th birthday on April 20 1939.

JFK METONIC ECLIPSES ARE SIDE BY SIDE THROUGHOUT TIME
THE EXTRAORDINARY JFK BIRTHDAY/DEATH DAY METONIC REALITY

As Abraham Lincoln was likely the Maximum Expression of American personality in the 19th Century, so John F. Kennedy was perhaps the Maximum Expression of that American Personality in the 20th century. In turn, the American Personality was clearly the maximum expression of personality in the global eye. That American Maximum Expression likely achieved its most influential maximum expression between 1947 and 1974, with midpoint around the murder of JFK.

John F. Kennedy was born May 29 1917 and was murdered on Nov 22 1963, the latter date being one of the most important in the post-civil war history of the United States. John F. Kennedy's birthday and death day have an unusual place in the Metonic Cycle. They reside side by side during his lifetime and in perpetuity. This means May 29th Eclipses will always be followed by a Nov 22nd eclipse that same

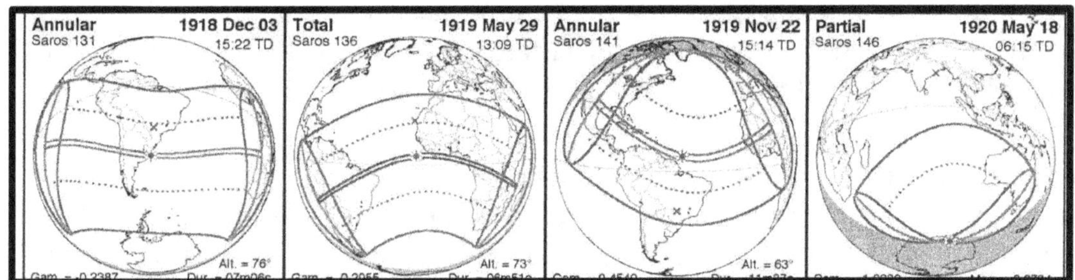

year (with a one-day possible variation on either end). I have calculated the odds at 2.6 million to one for any human, much less a President. But who knows?

Courtesy of NASA—Espenak and Meuss

Above, in the middle of the map, the Metonic of JFK birthday and death day sit side by side in 1919 when JFK is two years old. His birthday and death day are forever enshrined side by side in the cycle of eclipses 177 days apart (essentially one eclipse season or "semester series"). Since Kennedy was both born and killed on a Metonic, he was exactly 49 eclipse (Draconic) years old -minus a few days-when he was murdered. As we've noted elsewhere, Metonics are often given a one-day flexibility because of the realities of rotation, time zones, leap years and date lines.

Summary of JFK Metonic Eclipses in 20th Century/ JFK born May 29 1917 Died Nov 22 1963

May 28 1900 and Nov 22 1900. The first eclipses of the 20th century were on the JFK Metonic.

May 29 and Nov 22 1919 (when he turned 2) perfect birth and death day alignment eclipses. Both western hemisphere. One in the USA.

May 29 and Nov 22 1938, (when he turned 21) again perfectly on dates

May 30 and Nov 23 1946 (when he turned 29) 1946 is the year he's first elected to Congress). Both partial eclipses- on the 8-year Metonic lurch.

May 30 and Nov 22/23 1965 (2 years after death) the first complete solar eclipses on earth after his assassination occur on his birthday/death day back to back Metonic.

May 30 and Nov 22 1984 (21 years after his death). After 1984, the Metonic moves on with May 31st and Nov 23 2003 eclipses. By 2011, The Metonic has progressed to June 1st and Nov 25th.

JFK METONIC: First Solar Eclipses of the 20th Century

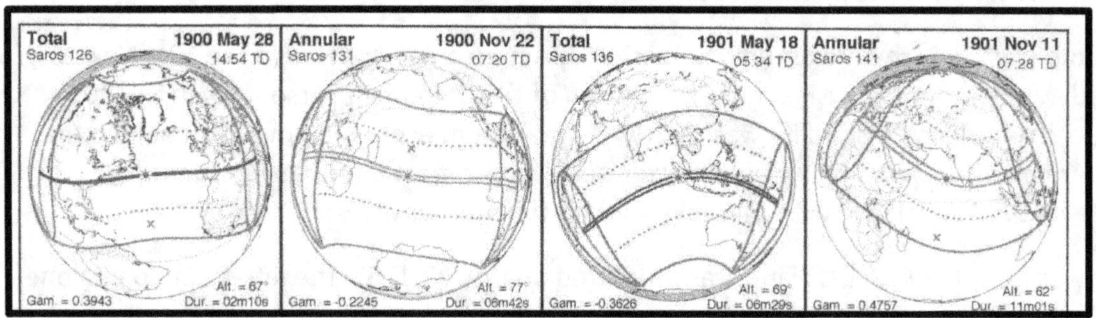

Courtesy of NASA—Espenak and Meuss

In above map, the very first eclipses of the 20th Century are on the Kennedy Metonic--May 28 1900 (one day before his birthday 17 years later) and Nov 22 1900 (exact death day 63 years later). They are followed by the Mount Saint Helens/ John Paul II Metonic of May 18 -19 years before John Paul II birth and 79 years before Helen's Eruption and Armistice Day of Nov 11th right in a row.

May 28 1900 Total (Saros 126) The first solar eclipse of the 20th Century was a total Eclipse connecting many empires- past, present and future on the JFK birthday Metonic (he would be born 17 years later on May 29 1917). Beginning in the Pacific at sunrise, the eclipse crosses Mexico and hugs the US gulf Coast after crossing Corpus Christ (98% totality). It brings totality to New Orleans, Montgomery, Mobile and Raleigh--exiting at Virginia Beach. 98% totality in Washington DC. 92 % in New York City, 90% in JFK birthplace of Brookline, Mass.

The eclipse crosses the Atlantic to bisect Portugal and Spain, the first colonizers of the New World-- entering at famous city of Porto, where the Romans had established trading operations in the centuries before Christ. Eclipse bisects Spain almost exactly 402 years after Columbus left on his third voyage to the Americas in 1498, and 400 years after Columbus was sent in chains back to Spain for trial in 1500.

The eclipse then traverses the Mediterranean and the path of totality directly crosses Algiers, Algeria, and Tripoli, Libya before ending a few miles south of the Valley of the Kings in Luxor, Egypt, which will be revisited by two eclipses in 2027 and 2034.

Nov 22 1900 (Saros 131)--Empire Connections

The second eclipse of the 20th century occurs 63 years to the day before the Murder of John Kennedy.

It begins just east of Angola and crosses one of the most infamous slave sources in Africa, Benguela, the port where the first African slaves originated who were brought to Jamestown in 1619. The eclipse then travels across the southern part of the continent, exiting through Mozambique and southern tip of Madagascar. Path passes through much of northwest Australia before dying out.

The Great 1919 Metonic Pair

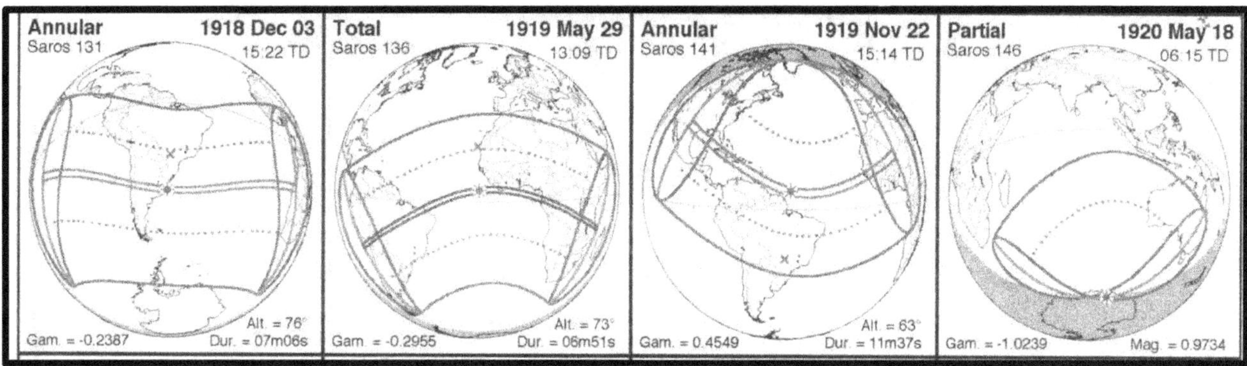

Courtesy of NASA—Espenak and Meuss

1919: JFK Birthday and Death Day in Perfect Pair

These are two of the most peculiar eclipses of all time.

May 29 1919 Einstein's Eclipse Total (Saros 136)

On John F. Kennedy's second birthday (he was born May 29 1917), Einstein's Theory of Relativity is finally proved by observation of the total solar eclipse. This is considered among the most important cosmic events in the history of science. Two teams, one in Brazil and one in Africa take simultaneous celestial light measurements to prove Einstein's theory correct. The eclipse has the longest total eclipse duration (at nearly 7 minutes) since 1416 AD. Saros Cycle 136 reappears notably in the Valley of the Kings Eclipse, on August 2 2027.

Nov 22 1919 Ring of Fire JFK Death Premonition Eclipse - LBJ, Texas and Cuba (Saros 141)

The JFK Death Premonition eclipse 177 days after the Birthday Eclipse, exactly 44 years before JFK assassination, thereby making the dates of JFK's birth and death commemorated in successive eclipses- both occurring in his home hemisphere. 30% coverage at Kennedy's home in Boston.

This tremendously long 11 minute and 37 second Ring of Fire Eclipse begins at sunrise in Texas, the state where JFK would be murdered 44 years later. Path of annularity goes over Stonewall, Texas, home to a then 10-year-old Lyndon Baines Johnson, who would replace JFK as President exactly 44 years later on the afternoon of 11-22-1963. Annularity in capitol of Austin and in Houston.

Path crosses almost entire island of Cuba west to east in their greatest eclipse of the 20th century. Cuba would be the nation which JFK's destiny would become intertwined. Path continues across Haiti, Jamaica, then travels the Atlantic to Africa, where the path reaches the countries of Senegal and Mali, both of whose presidents visited Kennedy's White House (the first African leaders to do so).

Eclipse ends after crossing ancient city of Timbuktu. Saros Cycle 141, which reappears in Spanish Ring of Fire of Jan 26 2028

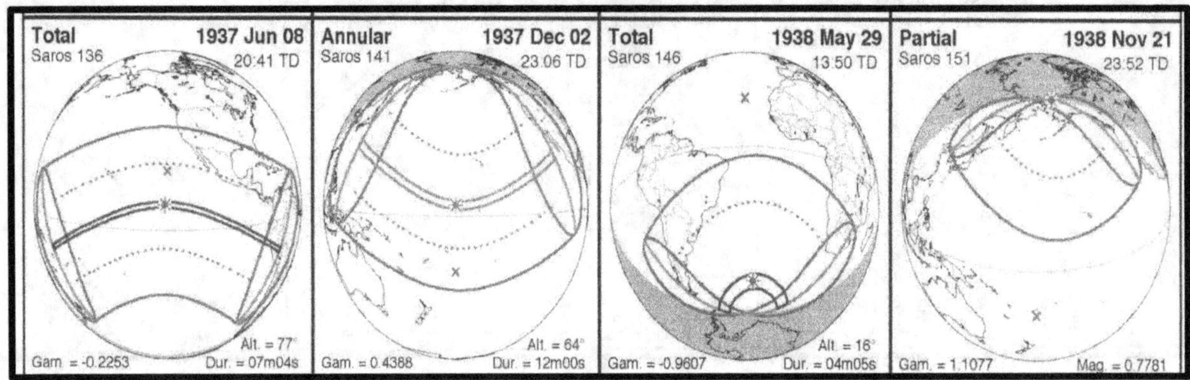

Courtesy of NASA—Espenak and Meuss

May 29 1938 (Saros 146) A solar eclipse occurs on Kennedy's 21st birthday over the site of Earnest Shackleton's Endurance Expedition 22 years after the climax of Shackleton's famed 800 mile traverse of the Antarctic Ocean and mountaineering traverse of South Georgia. (May 10 1916)

Reminder that an eclipse occurred on Abe Lincoln's 22nd birthday.

Nov 21/22 1938 (Saros 151) Death Day Metonic for a partial eclipse touching USA and USSR that says Nov. 21st on map, though straddling the date line, also was on Nov. 22 simultaneously.
JFK's father Joseph Kennedy was appointed Ambassador to Britain in 1938.

1946 The Eight Year Lurch

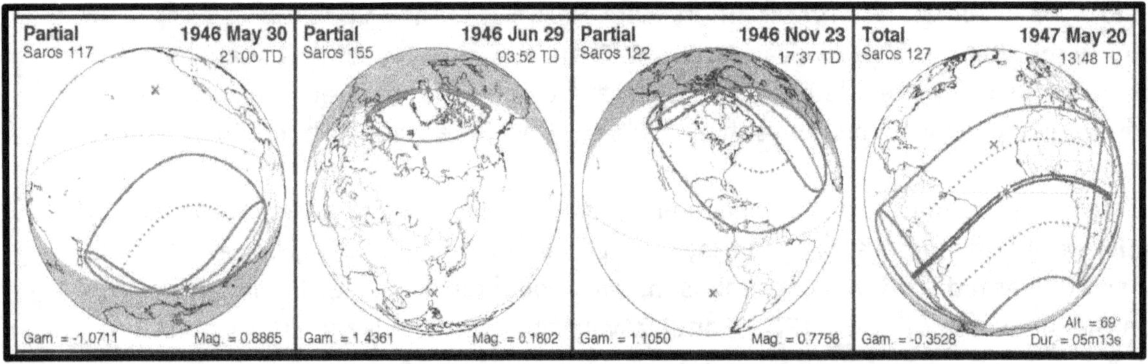

Courtesy of NASA—Espenak and Meuss

1946 is the year JFK is elected to Congress, 100 years after Lincoln is elected to Congress. In 1946, as the Metonic bends into its eight-year lurch, the days stretch out over two partial eclipses- still on the Metonic May 30 1946 and Nov 23 1946. They are separated by a partial eclipse on June 29, which barely registers (only 18% coverage of sun).

The November 23rd partial was visible in North America with 49% coverage in Massachusetts, where Kennedy won his election with 72% of the vote 18 days before on Nov 5 1946.

1960 to 1965, A Broad look at all solar Eclipses around JFK Assassination

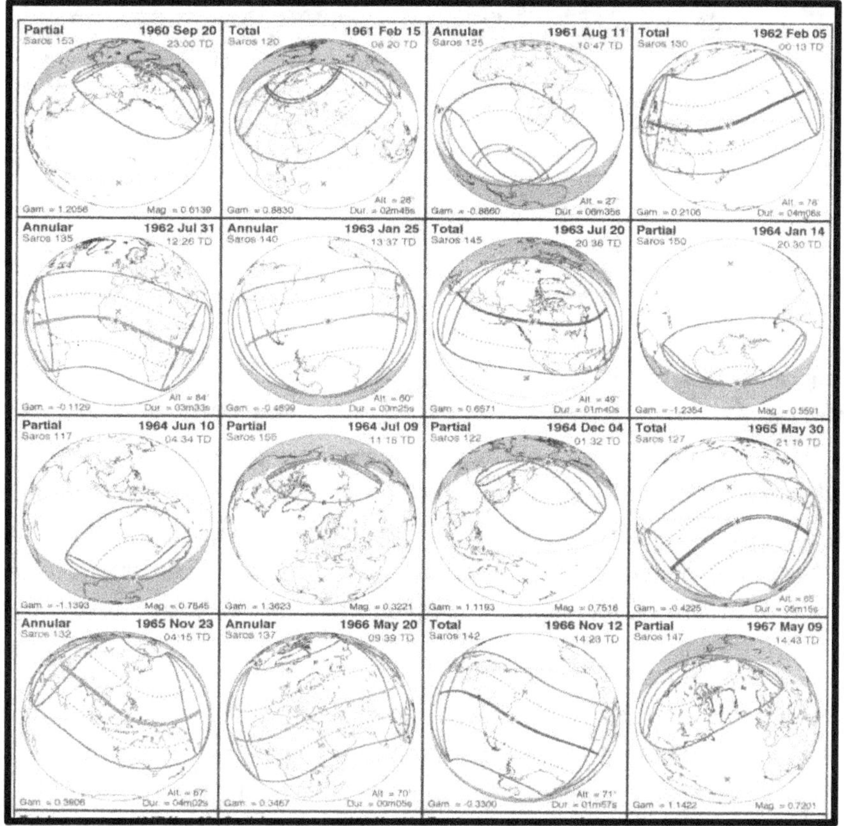

Courtesy of NASA—Espenak and Meuss

Above map of all solar eclipses on earth between Sep 1960 and May 1967. 7 have JFK associations:

1960 Sep 20 (Saros 153) Last eclipse before US Election which elected JFK is a partial over the USA.
1961 Feb 15 (Saros 120) The first eclipse after the US election goes over the Iron Curtain and USSR
1962 Feb 5 (Saros 130) Eclipse over Solomon islands where JFK's boat PT 109 went down in 1943.
1963 Jul 20 (Saros 145) The last eclipse before JFK's assassination travels from Japan to the US six years to the day before Moon Landing. Totality only in Maine. 95 % in Boston.
Jan 14/15 1964 (Saros 150) The first solar eclipse after JFK murder is a partial on Martin Luther King's 35th birthday, stretching across two days due to South Pole location.
May 30 1965 -Total (Saros 127) First complete eclipse on earth after JFK assassination arrives on his Birthday Metonic 555 days after Dallas, one day after what would have been his 48th Birthday.
Nov 22/ 23rd 1965 (Saros 132) The next eclipse on earth is 177 days later and begins while it is still Nov 22 in the USA, with a solar eclipse over war-torn Southeast Asia on the 23rd. Southeast Asia has their 5th solar eclipse in 17 years!

In the week before this eclipse 240 US servicemen were killed in Vietnam, the most ever up until that time. 80 to 90% coverage in Solomon Islands, site of the sinking of Boat PT 109.

The first two complete solar eclipses on earth after JFK's assassination occur essentially on the birth and death day of the subject of that assassination.

1984 - LAST JFK METONIC

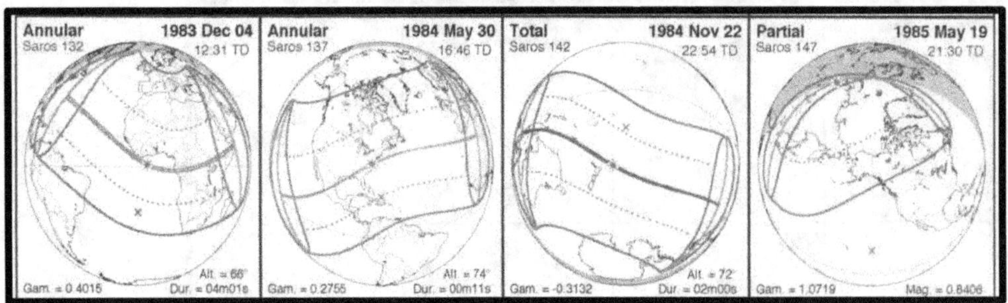

Courtesy of NASA—Espenak and Meuss

May 30 1984 USA Ring of Fire (Saros 137) JFK Metonic Birthday

99.74% maximum in Virginia. 95% coverage in Washington DC. The white House of Reagan/Bush was darkened. JFK would have been celebrating his 67th Birthday hours before if he had not been murdered. True annularity in Atlanta, MLK's home and in New Orleans, birthplace of Lee Harvey Oswald and site of the JFK conspiracy trial. 90% in JFK's birthplace of Brookline Massachusetts, 80% in Dallas Texas, where he was murdered. 93% in New York City. The eclipse crosses the ocean and enters Africa over Casablanca (white house) and Marrakesh (Land of God) --ends in desert of the Sahara.

Nov 22 1984 JFK Death Metonic Total (Saros 142)

177 days later. Total solar eclipse appearing in the same year as the JFK birthday Metonic for the last time, the 21st anniversary of the Kennedy Assassination is cosmically acknowledged. 2 and 1/2 weeks after the Reagan/Bush ticket sweeps to re-election in a landslide. These Metonic linked 5-30 and 11-22 eclipses are the only two solar eclipses of 1984. The only nation in the 11-22 path is Papua New Guinea, with totality in the capital of Port Moresby. Site of PT- 109 Kolombangara Island in totality.

Courtesy of National Archives

PT-109 And John Kennedy

One of the most famous war stories in American history involves the heroics of young Lieutenant John F. Kennedy, who miraculously helped save his crew after their small boat was rammed by a Japanese destroyer on August 3 1943 in the Solomon islands of New Guinea, inside the eclipse path of the Nov 22 1984 solar eclipse 21 years to the day after his death. The incredible story of JFK's physical and mental heroism was enshrined in both a book and a movie, and Kennedy's indisputable bravery helped propel him to the Presidency 20 years later.

November 22 1984 was Thanksgiving Day.

A powerful Nor'easter Storm began impacting the US Thanksgiving Day Nov 22 1984, beginning between Florida and Cuba. Presidential elections were 16 days prior. They brought one of the largest landslides in US history, with the Reagan/Bush ticket defeating Walter Mondale/ Geraldine Ferraro 525 electoral votes to 13, even winning JFK's home state of Massachusetts with 51.22 percent.
The number one song in America was "Wake me up Before you Go Go" by Wham. 1984 would be the last year JFK Birthday and Death Day Metonic would appear in the same year for several centuries They had begun in 1900, with the first two solar eclipses of the 20th Century. He would have been 67 in 1984. Both brother Robert Kennedy (Nov. 20) and son John F. Kennedy Jr. (Nov. 25) were born in bookends around that same November Metonic where he was martyred .

Kennedy Metonic Appears around every 520 years. Here it is in 1044 AD

Yep, there's those days (May 29 and Nov 22) in the middle of the row again, except...in 1044 AD!

All images on page courtesy of NASA Espenak and Meuss

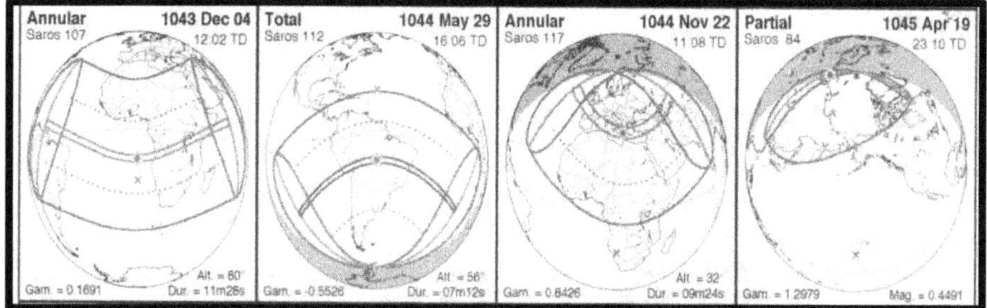

Here below is the Metonic in 1565 AD...first two in the row

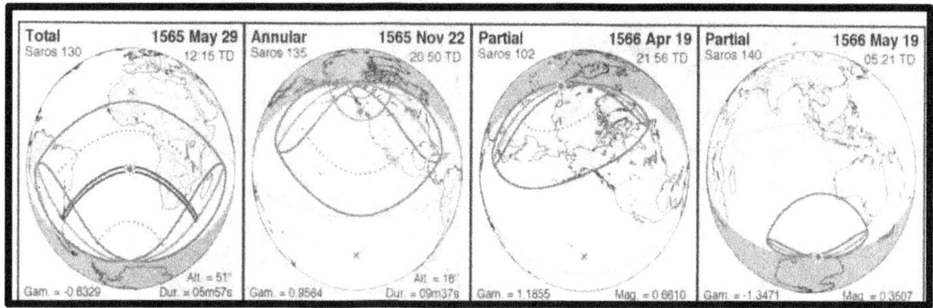

Look! Same days and **Similar eclipses in the Metonic 1496 years apart between 504 AD and 1900 AD**

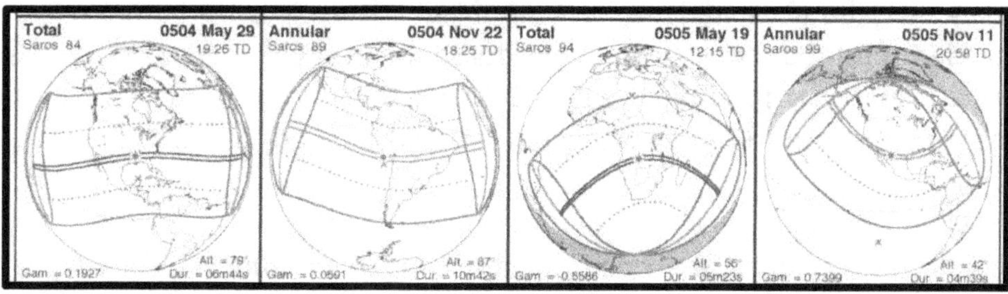

compare to set below:

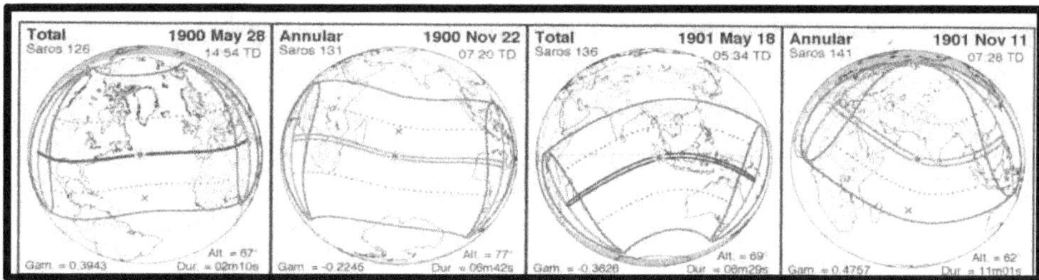

And way back when in 1626 BC:

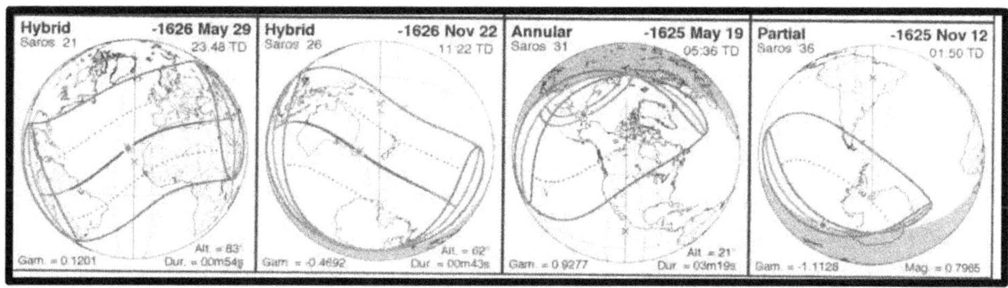

John F. Kennedy, CS Lewis and Aldous Huxley all died on November 22 1963

Three of the most interesting minds of western civilization all died within hours of each other on November 22 1963. Not only Kennedy but also:

CS Lewis (middle). Lewis was the greatest and most widely read Christian thinker of modern times, with many books of extraordinary clarity, humor and depth on the subject of the Christian faith. He was also the author of the Chronicles of Narnia series, widely considered the best children's book series of all time, with more than 100 million copies sold. He died 5:30 pm Greenwich Time Nov 22 1963 in Cambridge, England, one true hour before the shooting of President Kennedy at 12:30 pm CST in Dallas.

Aldous Huxley (right). Huxley was perhaps the most accurate futurist of the 20th Century, and, along with Lewis, one of the intellectual giants of 20th Century Britain. From a renowned family of Evolutionists and Atheists, he became a kind of scientific mystic. He wrote nearly 50 books, including the highly influential *Doors of Perception,* but is best known for his 1932 novel, *Brave New World,* which envisioned a future society drugged and pleasured into passive submission. He died in California at around 5:30 pm Pacific Time, seven hours after JFK and eight hours after Lewis.

Overview of all Solar Eclipses in USA 1900 to 1984

Above: USA to Egypt May 28 1900
(Kennedy's Birthday Metonic)

Above: Temple of Sun/ Cape Kennedy Jun 28 1908
(site of Moon Launch/ Cape Canaveral in path)

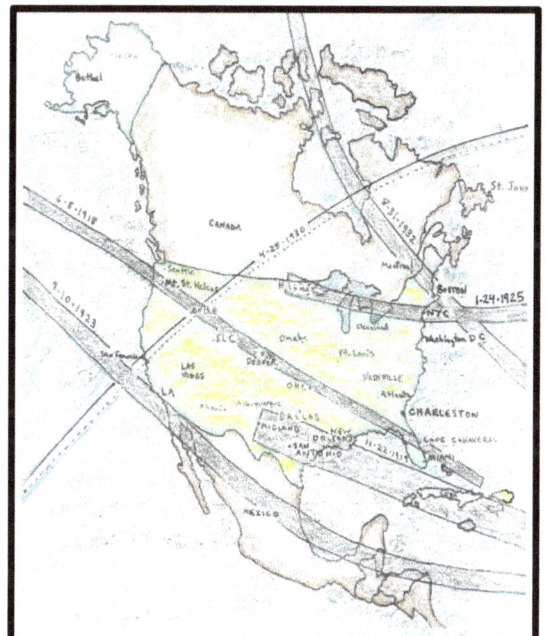

Above: Eclipse Swarm 1918 to 1932

Boston next to path of 1925 and 1932 eclipses.
1918 Cross Country eclipse Cape Kennedy.
11-22-1919 JFK Death Premonition

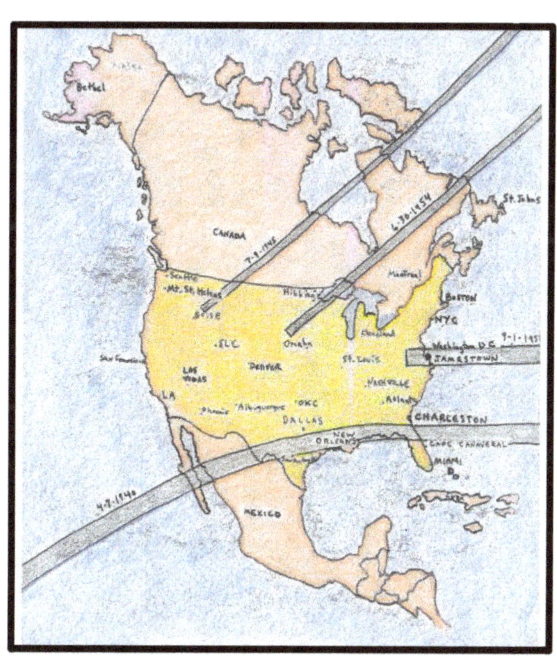

Above: 1940 to 1954 Emanations
Includes 7-9-1945 A Bomb USA to USSR
1951 Jamestown to Angola

Oct 2 1959 Boston Single City (Saros 143) Total Eclipse begins just west of Boston and gives that city a total solar eclipse for the first time since 1806 (it won't have another until 2200) just as its most famous modern son is campaigning for the Presidency of the United States. Kennedy gives a speech that night on the campaign trail in Rochester, New York. It is the only major city in America to witness the eclipse. Kennedy would be elected President the following year on Nov. 8th.

Jul 20 1963 Japan to USA Total (Saros 145) 97 % at Bush Family Compound in Kennebunkport Maine four months before JFK murder.

Mar 7 1970 USA Eastern Seaboard (Saros 139) Total Eclipse. the southeastern seaboard had mostly cloudy weather though Florida remained fair. 93% totality at Cape Canaveral 230 days after men landed on the moon. Totality at Charleston South Carolina, the greatest Slave Port in North America..

95 % totality in Washington D.C., 96% in New York and 97% in Boston. Totality in Cape Cod, home of the Kennedy's

At Left : 1979 and 1984 Eclipses

Feb 26 1979 Mount Saint Helens/Microsoft

May 30 1984 JFK 67th Birthday – Washington, Atlanta, New Orleans

Chapter 19

Jesus and The
2000th Anniversary Timelines

Jesus Timelines

We are living in the 2000th Anniversary of the Crucifixion and Resurrection of Jesus. It so happens that the Cosmic Clock is also aligned in the same location as when he died and rose again.

It is believed that Jesus of Nazareth was born somewhere between 6 BC and 1 AD. There is no year 0 in our calendar, which moves directly from 1 BC to 1 AD. The traditional dating of the Birth of Jesus (the dating that establishes our calendar) would be Christmas 1 BC. What would be the odds that a solar eclipse happened on that date? Well, one did. And it so happens that lunar eclipses happened on two possible dates of his crucifixion. We will get to that, and much more, in this chapter.

There is plenty of debate on the year of Jesus' birth, and the year of His Crucifixion. I don't have any truly new opinions on the subject but will present eclipse realities surrounding the time of Jesus and our current era. I think the eclipse realities point to certain symmetries.

Historically, given evidence both within the Bible and from historical sources, the Crucifixion and Resurrection of Jesus Christ occurred between 27 and 34 AD. This covers the time frame established by Roman historians as the governorship of Judea by Pontius Pilate, who oversaw the execution of Jesus. It's stated in the Scripture that John the Baptist (who baptized Jesus) began his own ministry in the 15th year of the reign of Tiberius, which would place the beginning of John's Ministry around 27 to 30 AD. Jesus's Ministry is said to have lasted about three years. These dates further narrow the field of possibilities. Also, Jesus was killed essentially at Passover, and this also affects calculations.

Regardless of the strong opinions, for nearly 2,000 years, most scholars agree that the Crucifixion of Jesus Christ occurred between 27 to 34 AD. Many emphatically assert one or the other of the different dates and you can read their reasoning elsewhere. Some claim earlier dates. The traditional date, which our calendar is based on, is also the scholarly favorite. That date is Friday April 3 33 AD. The date of April 14 32 AD also has strong ecliptic associations.

Obviously, the 2,000th solar anniversary of all potential dates rapidly approaches.
The 2,000[th] Eclipse Year Anniversary already occurred sometime between 1925 and 1932.
The 2,000[th] Lunar Year Anniversary occurred between 1967 and 1974.

Here are all seven credible possible dates of the Crucifixion, using necessary historical, calendar and astronomical information, as presented by hundreds of scholars over the centuries.
Apr 10 27 AD
Apr 28 28 AD
Mar 23 29 AD
Apr 7 30 AD
March 28 or Apr 25 31 AD
Apr 14 32 AD-Passover total lunar eclipse on other side of world from Jerusalem
Apr 3 33 AD -The traditional date and scholarly favorite. Passover Partial lunar eclipse in Jerusalem.
Apr 23 34 AD -Isaac Newton's calculation

We will look at eclipses and cycles during the life of Jesus, as well as around the 2,000th anniversary of the Crucifixion.

We will look at the similarities of the current 19-year Equinox Metonic (2015 to 2034) and the Metonic that encompassed the adult life and death of Jesus Christ (14 AD to 33 AD). It becomes clear that we live in the same cosmic alignment of sun and moon in relation to earth as He did.

Again, this is astronomy, not astrology. The anniversary marker I use is the solar year and calendar date we are all familiar with. Jesus lived towards the beginning of the Julian Calendar, before it got out of alignment. Our current Gregorian matches the early Julian astronomically.

But first, for those interested in the eclipse year anniversaries of the birth and death of Jesus.

Eclipse Year and Lunar Year Anniversaries

Again, an eclipse year is 346.62007 days. The eclipse (draconic) year is when the sun and moon return to about the same place (node) in the sky, or ecliptic plane.

There is almost exactly 20 eclipse years in 19 solar years. The difference is just one week.
2,000 eclipse years equals almost exactly 1898 solar years.
The traditional 2,000[th] Eclipse Year anniversary of Jesus Birth would be around 1892 to 1899.
The 2000th Eclipse Year anniversary of the Crucifixion/Resurrection somewhere between the years 1925 and 1932. If we accept the traditional dating of April 3 33 AD, then then 2000th Eclipse Year anniversary of the Crucifixion occurred on Sunday April 12 1931.

The atom was first split at Cambridge University in England on April 14 1932, close to the 2000th eclipse year anniversary of Newton's calculation of the Crucifixion at Apr 23 34 AD.

The 2000th Eclipse Year Anniversary of the final destruction of the Jewish Temple in Aug 10 70 AD occurred Aug 21 1968. As it happens, the USSR invaded Czechoslovakia that day.

The Lunar Year Anniversary of the Birth of Jesus would be between 1934 and 1941.
The Lunar Year Anniversary of the Crucifixion between 1967 and 1974.

Eclipses during Jesus Life

One of the first things that come to mind is this: were there interesting solar eclipses during the birth, life and death of Jesus? If so, being as how nearly all people in history viewed eclipse as messages, might this have affected the mindset of the people of the Middle East, regarding how they might have anticipated Messiah?

No complete eclipse path passed directly over Israel during Jesus's life. However, several provided distinct deep partial eclipses, and others would have been visible to neighboring nations who would have brought word of them. The Roman Empire of the day was filled with travelers, as Rome had established a network of roads and regulated safe shipping. It seems likely that word of eclipses were spreading. As we've considered already, eclipses were understood to be omens by all cultures, and the cosmic stage was set for Messiah.

Darkness at Noon?

One of the most common questions about Jesus and eclipses concerns the Crucifixion, because of the "darkness" that accompanies it which is mentioned in the Gospels. Three of the four Gospels - Matthew, Mark and Luke — mention the sky becoming dark as Jesus hung on the cross.
"It was now about noon, and darkness came over the whole land until three, because the sun's light failed," Luke 23:44.

It can be definitively stated that this Crucifixion darkness was not a solar eclipse, at least as we know them to occur. Jesus was crucified at full moon, on Passover. Solar eclipses can only occur at the new moon. The darkness that accompanied the Crucifixion of Jesus was either supernatural in origin or the result of an atmospheric reality like a dust storm or smoke from a forest fire.

That being said, as far as lunar eclipses go, there were lunar eclipses on two Passovers in a row, 32 AD and 33 AD. A total lunar eclipse occurred on April 14 32 AD Passover but was on the other side of the world from Jerusalem. There was a partial lunar eclipse at moonrise on Passover April 3 33 AD that was visible from Jerusalem.

Interestingly, if Jesus either died in either 32 or 33 AD, which most scholars say he did, then he died the day of a Passover lunar eclipse, though only the 33 AD was visible from Jerusalem.

These lunar eclipses are interesting as far as "signs" go, but clearly also can't be taken as providing the "darkness" mentioned in the Gospel accounts.

Jesus Birth and 2000th anniversary of Birth in two Christmas Solar Eclipses?

Oddly, there was actually a Dec 25/26 solar eclipse on Jesus' traditional Birthday in 1 BC.
This historical fact was probably first discovered in the late 1800's by Theodore Oppolzer as he created his Eclipse Canon, though he did not say much about it. These modern maps are from NASA. The Jesus Birth Eclipse 1 BC is on the left, the Christmas 2000 Eclipse is on the right.

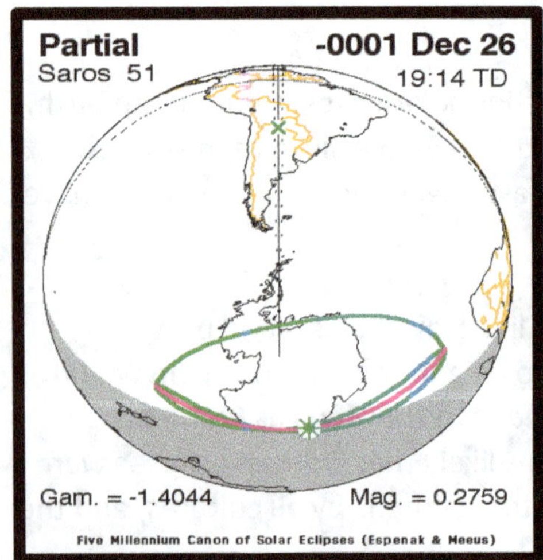

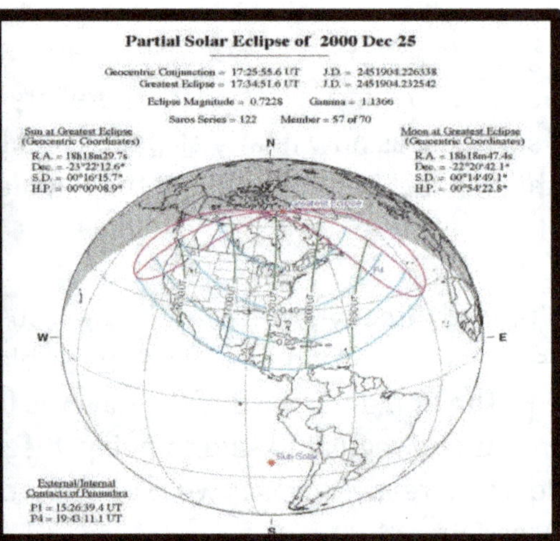

maps courtesy of NASA –Espenak/Meuss

The 01 BC Map says Dec 26, but south pole eclipses that time of year usually hover between two days, It's probably both a Dec 25 and 26th eclipse on Jesus traditional birth year. And if its a few hours off one way or another, so what? Clearly no one saw this eclipse. It was a very minor partial with only penguins around, but it did signal a total lunar eclipse that would be coming to Israel two weeks later on Jan 10, 1 AD (yes that's 1-10-1). That would be the first eclipse of any kind of the first Millennium. Exactly 2000 years later, as you see in the map on the right, there was a Christmas deep partial eclipse over North America (Saros 122). This was the very last solar eclipse of the second Millennium. 44.44% coverage in New York City. 40 % in Washington DC.

I find nowhere else that this interesting Christmas solar eclipse correlation has been made in print. Indeed, very little was actually made of the 2,000[th] Anniversary of the birth of Jesus at all. Attention seemed focused on Y2K in 1999 and I can't remember what we were all doing in December 2000.

If traditional dating is accepted, the dating that provides our calendar, then the 2,000[th] anniversary of Jesus' Birth occurred in 1999 or 2000. There was a tremendous solar eclipse on Aug 11 1999, which is discussed in detail at the World Eclipse section. Among other places, it crossed India, Pakistan, Nineveh (Iraq), Turkey, and Europe (totality at Hitler's Birthplace). There was 74 % coverage in Jerusalem/Bethlehem. It was the last complete solar eclipse of the Millennium.

Four partial solar eclipses occurred in 2000 (the most possible), including the Christmas USA eclipse we just mentioned. July 2000 in a very rare event saw two partial solar eclipses on the 1st and 31st. They sandwiched a mighty solar storm, the Bastille Day Event of July 14 2000.

There were two Total Lunar Eclipses in 2000, both visible from Jerusalem.

Concerning Tribulation Timelines:

Interestingly, there are supposedly 49 distinct uses of the number seven in the Book of Revelation, not counting repetitive uses in the same verse.

There is a seven-year end-times tribulation period in Revelation, with a half of a seven year period (3 and ½ years) also emphasized, just as it is in the Old Testament Book of Daniel.

Clearly the Hepton Cycle creates powerful seven-year and 3 and ½ year periods, but there are other seven year periods as well. For instance, the Spanish Judgment eclipse of Aug 12 2026 and the March 20 2034 Great Civilization eclipse are an Octon, which is 7 years, 7 months and 7 days apart.

There are many who believe in a Pre-Tribulation Rapture and many who do not. I believe the eclipse could give comfort to either camp and I don't see any indication in the eclipses of one theological direction or another. I also know better than to step into the Rapture/Tribulation debate. People get fired up! I believe in the return of Jesus Christ in His full glory when He's ready.

The timing is unclear to me, but I still enjoy the conversations around it.

Cosmic Circumstances Surrounding Jesus Life and Death:

We'll now shift our focus to complete solar eclipses potentially visible from the Middle East around the life of Jesus. We only portray eclipses where more than 50 % of the sun was likely to be covered in occupied Palestine where Jesus was raised. First, eclipses around his birth:

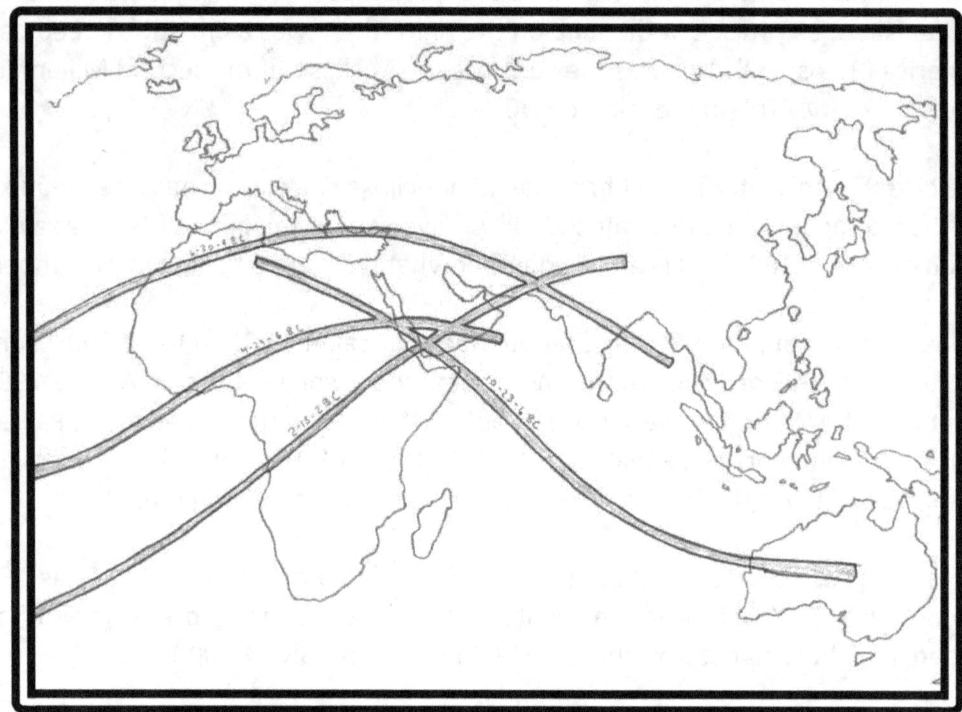

Hepton Eclipse Swarm around birth of Jesus 9 BC to 2 BC.

9 BC Jun 30 Total Saturday (Saros 75) Three years before the earliest possible birth of Jesus, this eclipse would have been well-known in the Roman Empire and Holy Land, with 80% to 90% plus coverage in Palestine. Totality in modern Tripoli, Greek Island of Crete ("birthplace of Zeus"), southern Asia Minor, Lebanon, Syria just north of Damascus. Centerline goes over Babylon (modern Baghdad, Iraq), ancient Persian capitals of Susa, Pasagardes and Persepolis. Path crosses modern Karachi Pakistan and bisects India. Eclipse lasting 6 Minutes and 14 seconds. Virtually everyone would have known about this eclipse, it having crossed populated regions in dry summer months.

6 BC Apr 29 Ring of Fire Thursday (Saros 67) Africa, crossing northern ancient Nubia just south of modern Egypt border, annularity path comes very near Mecca, ends just after crossing Arabia.

6 BC Oct 23 Total Saturday (Saros 72) 177 days later, total eclipse begins in Libya, crosses Egypt, intersects last eclipse in Nubia, and heads out into the Indian Ocean to die in the heart of Australia.

2BC Feb 15 Total Saturday (Saros 69) Seven Eclipse years exactly after the 9 BC eclipse, another Saturday Hepton from Africa to Arabia. Eclipse actually begins in South America directly over modern Rio De Janeiro, crosses Atlantic and Africa to exit after perfectly bisecting Ethiopia. Crosses southern Arabia and dying just before the Himalayas in northern India.

3 of these occur on Saturday. The people of Jesus time would have experienced three Sabbath Day partial solar eclipses, two of them likely very noticeable.

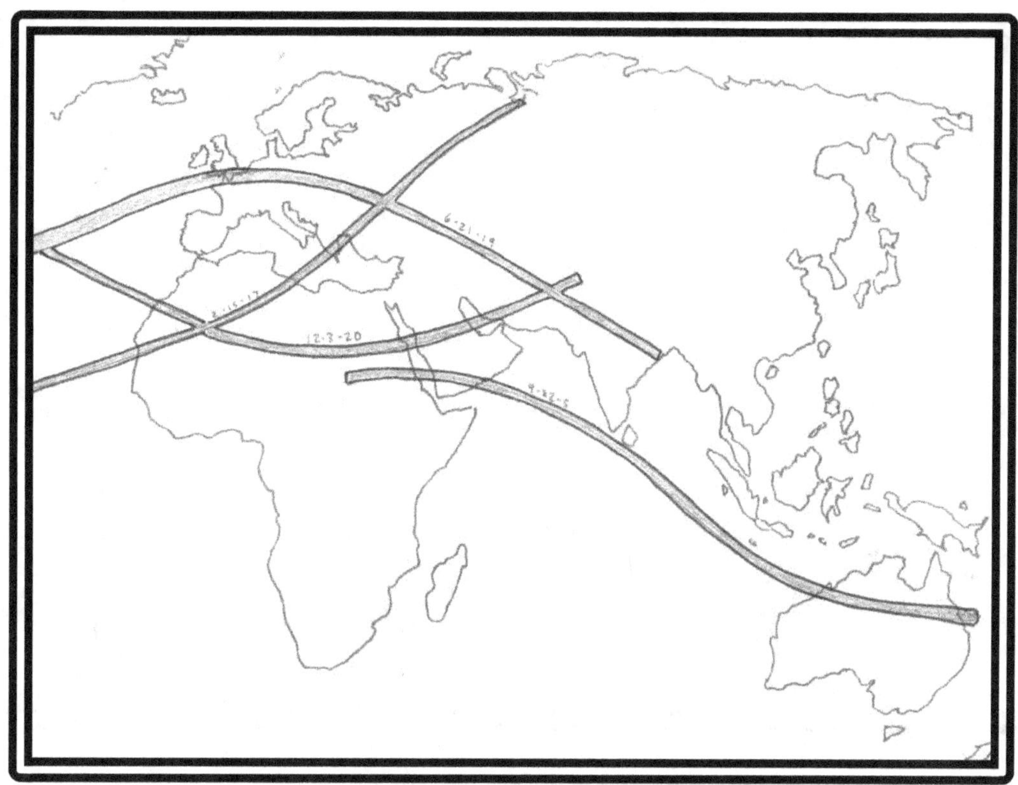

Jesus Youth to Early Adulthood 5 AD to 20 AD

5 AD Sep 22 Ring of Fire Thursday (Saros 73) This ring of fire would likely have 60% or less coverage of sun in Palestine, but probably have been noticeable. Jesus would be anywhere from 6 to 11 years old. Also notable that it occurs on equinox. Path from modern Sudan to Australia.

17 AD Feb 15 Hybrid Wednesday (Saros 79) First complete eclipse in region in 12 years. Up to 70% coverage where Jesus lived. He would be a young man anywhere from 18 to 23 years old. Path travels North Africa, Libya, across Greece. Crosses modern Ukraine and Russia. Totality in modern Odessa. Saros 79 returns to Greece in AD 71 on the Spring Equinox after the Jewish Temple is destroyed.

19 AD Jun 21 Total Friday (Saros 66) Summer Solstice. A tremendous path that crosses 20 modern countries. Perhaps around 50 or 60 percent coverage in Holy Land. Modern England, Germany, Russia, Iran etc… in path. Ends at sunset at eastern edge of Indian subcontinent over ancient city of Bhubaneswar or *"Lord of the Three Worlds"*, referring to Shiva.

20 Dec 3 AD Hybrid Thursday (Saros 81) This eclipse had good coverage in Palestine, perhaps 70%. Path travels across Arabia, splitting the distance between Medina and Mecca, into Persia.

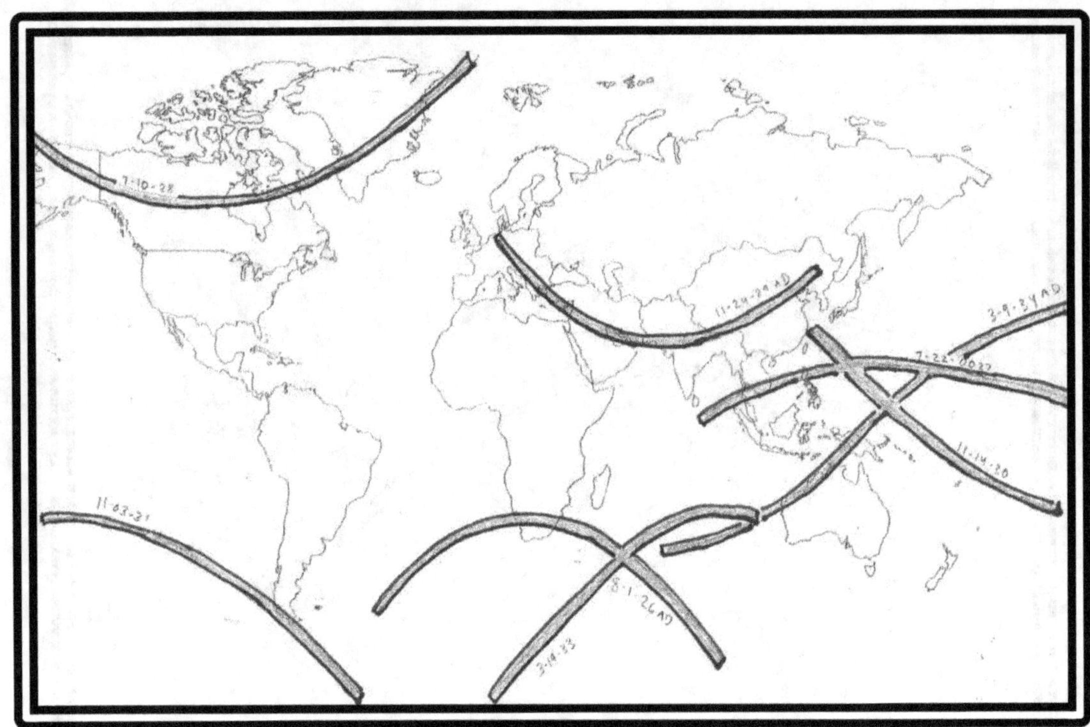

Jesus Ministry and Death 26 to 34 AD

Seen above, eight total eclipses from 26 AD to the spring of 34 AD. Though a couple of cool visual patterns appear, none of these eclipses were visible in Palestine except 29 AD Nov 24 Total eclipse. We see an apparent Aleph and Tav in the southern hemisphere.

29 AD Nov 24 Total Jesus Ministry Eclipse Saturday (Saros 62) This would have been the most impressive eclipse in the Holy Land in many decades, perhaps even surpassing the 9 BC eclipse in percentage of coverage in Jerusalem, which saw maybe 90% eclipse. Path of Totality begins in Europe - crosses Berlin and site of Auschwitz in Poland. Path over Syria, Jordan, Iraq and Arabia. This would have been a noticeable Sabbath day event, and if traditional dating is accepted, occurred around the time John the Baptist's and Jesus' ministry began.

Bottom Middle of map shows **26 AD Aug 1 and 33 AD Mar 19** Hepton Cross with apex in lower Indian Ocean south of Africa. They are each 2001 years (plus one day) away from the Hepton Cross over Egypt which occurs Aug 2 2027 and Mar 20 2034.

33 AD Mar 19 Total Solar Eclipse Friday

This is the solar eclipse that occurs closest to Passover in the years of Jesus Ministry. It would not have been visible from any populated area. It touches no land except a tiny portion of Antarctica. It appears in the map as part of the Hepton Cross over the southern Indian Ocean. Occurring the day before Spring Equinox, it is partnered with the famous partial Blood Moon Eclipse that rose fourteen days later on April 3 33 AD, the traditional date of Jesus Crucifixion. If that traditional date is accepted, this solar eclipse occurred two weeks before the Crucifixion, on the last new moon of Jesus's life.

The Mar 19 33 AD eclipse is 2000 years minus 11 days from the Bering Strait Eclipse of Mar 30 2033 and 2001 years minus one day from the Great Civilization Eclipse of Mar 20 2034.

All Total Lunar Eclipses during Life of Jesus

Lunar eclipses were seen as omens for the Jewish people. Here are all total lunar eclipses visible in the Holy Land during the time frame of Jesus' life. (There were none in 6 BC or 5 BC)

04 BC Mar 23
04 BC Sep 15
01 AD Jan 10
03 AD May 4
03 AD Oct 29
07 AD Feb 20
07 AD Aug 17
10 AD Dec 10
11 AD Jun 4

Then: Blood Moon Tetrads of 13 and 14 AD same month and day as 2013 and 2014 :

13 AD Apr 14 Passover—weak partial not visible from Jerusalem- corresponds to 2014 Blood Moon
13 AD Oct 7 Tabernacles—weak partial not visible from Jerusalem corresponds to 2015 Blood Moon
14 AD Apr 4 (Blood Moon Passover)—Total: visible from Holy land compare to 2015 Blood Moon
14 AD Sep 27 (Blood Moon Tabernacles)Total: visible from Holy land -compare to 2015 Blood Moon

(Now continuing total lunars visible from Israel)
18 AD Jul 16
21 AD May 15
21 AD Nov 8
22 AD May 4
25 AD Mar 3
28 AD Dec 20
29 AD Jun 14 (this Total Lunar Eclipse occurs 5 months before Great Solar Eclipse of Nov 24 29 AD. There seems a high potential these eclipses taken together could have been considered important signs. They accompany traditional dates of the ministry of Jesus and John the Baptist)

More Blood Moon Tetrads—same day and month and holiday as 2033 and 2034 Tetrads:
32 AD Apr 14 Total not visible from Jerusalem- Blood Moon Passover
32 AD Oct 7 Total barely visible from Jerusalem- Blood Moon Tabernacles
33 AD Apr 3 partial at moonrise visible from Jerusalem Passover-traditional Crucifixion date
33 AD Sep 27 Strong partial/ almost total visible from Jerusalem on Feast of Tabernacles

There were 18 total lunar eclipses visible from Jerusalem between 6 BC and 34 AD. All Blood Moons on Passover and Feast of Tabernacles are exactly 2001 years to the day from our current Blood Moons.

The Passover/Feast of Tabernacles Blood Moon Metonic occurred once more in 51 AD on Apr 15 and Oct 8 with two partial eclipses. After 51 AD the lunar eclipse Metonic moved on.

We now compare eclipse dates to find we are in the same Metonic cycle, with the same hands pointed to the same numbers on the same cosmic clock, as the one that occurred at the Crucifixion of Jesus Christ. Eclipses occur on essentially the same days 2001 years apart.
First the Lunar evidence:

Blood Moon Tetrads: This Metonic Return is distinguished by lunar eclipses on both Passover and Feast of Tabernacles, two important Jewish holidays linked to the full moon.

Two lunar eclipses of 13 AD (Apr 14 and Oct 7) mirrored in two eclipses of 2014 (Apr 14 and Oct 7).
Two lunar eclipses of 14 AD (Apr 04 and Sep 27) mirrored in two eclipses of 2015 (Apr 04 and Sep 27).
And then, 19 years later:
Two lunar eclipses of 32 AD (Apr 14 and Oct 7) mirrored in two eclipses of 2033 (Apr 14 and Oct 7).
Two lunar eclipses of 33 AD (Apr 3 and Sep 27) mirrored in two eclipses of 2034 (Apr 3 and Sep 27).

These dates are on Passover and Feast of Tabernacles, the High Holy Days of Judaism.

We live in the same cosmic alignment of sun and moon in the same Roman Calendar Time that occurred 2001 years ago. The 2,000 year anniversary of the Crucifixion and Resurrection of Jesus Christ is clearly a big enough deal that the Creator of all things marked it to resonate between our time and his.

Solar Eclipses on the Spring Equinox. The traditional crucifixion date is Apr 3 33 AD.   There was a Spring Equinox (Eve) Eclipse of Mar 19/20 33 AD.
2001 years later we have an eclipse on Spring Equinox Mar 20 2034.
Both Equinox solar eclipses 2001 years apart are accompanied by Lunar Tetrads on High Holy Days.

Apocalypse Then? Preachers announced Lunar Tetrads which announced Solar Equinox Metonic

In 2014, several apocalyptic preachers and Rabbis made news by commenting on the so-called "Blood Moon Tetrads", sets of four total lunar eclipses occurring on successive years (2014 and 2015) precisely on the Jewish High Holy Days of Passover and Feast of Tabernacles.

Only four such Tetrads occurred in the last 600 years:

- **1493 and 1494** Blood Moon High Holy Day Total Tetrad occurred following expulsion of the Jews from Spain in 1492 and "discovery" of the New World by Columbus.

- **1949 and 1950** Blood Moon High Holy Day Total Tetrad occurred following 1948 declaration of Jewish state. The first eclipse happened April 12/13, one month after hostilities ceased.

- **1967 and 1968** Blood Moon Total Tetrad occurred around the 1967 "Six Day War," where Israel took city of Jerusalem for the first time in almost exactly 2000 eclipse years. The first eclipse happened April 24, or about six weeks before the war began.

- **2014 and 2015** Blood Moon Total Tetrads on High Holy Days

The so-called "Blood Moon " Tetrad of 2014 and 2015 brought successive total lunar eclipses on the high holy days of Judaism. Passover (April 15) and Feast of Tabernacles (October 8) in 2014 and again Passover (April 4) and Feast of Tabernacles (Sept 28) in 2015. Nestled between these events was an

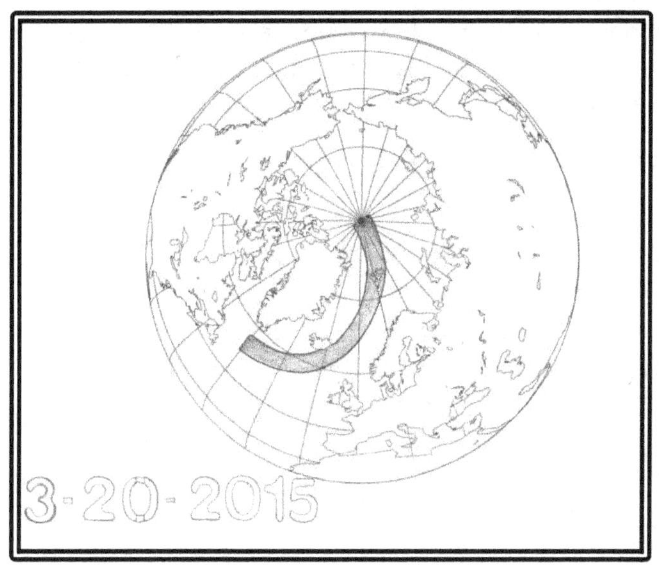

extraordinary **Spring Equinox Mar 20 2015 Solar Eclipse (left)** which ended directly at the North Pole. It is unclear whether any eclipse in history has ever done such a thing.

2032 and 2033 has a Total Tetrad, with four total lunar eclipses. 2032 is not on holy days. 2033 is. **2033 and 2034** is a High Holy Day Tetrad, and eclipses are total in 2033 but not in year 2034.

Why is the 2,000th Anniversary of Jesus a big deal?

The 2,000 anniversary of the Crucifixion/Resurrection is important for three reasons. Upon reflection, there appear symmetries in History and especially in Biblical history. There is a common instinctive urge towards the recognition of anniversary. The importance of symmetry and anniversary are both physically reflected in eclipse cycles. Eclipses appear to reach a maximum expression of symmetry and peculiarity around the current time.

"Dearly beloved, do not be ignorant of this one thing: that one day is as a thousand years with the Lord, and a thousand years as one day."
> 2 Peter 3:8

"A thousand years in your sight are like a day that has just gone by."
> Psalm 90-4

Millennial symmetry in eclipses provide a symbol of Jesus' death, crucifixion, and resurrection. Jesus apparently slept in the ground two days and rose early on the third. Two thousand years of his absence ends with a rising Son. The idea of a thousand-year reign of Jesus (Millennium) is definitively presented in The Book of Revelation, the last book of the New Testament.

'Then I saw an angel descending from heaven, holding in his hand the key to the abyss and a huge chain. He seized the dragon—the ancient serpent, who is the devil and Satan—and tied him up for a thousand years. The angel then threw him into the abyss and locked and sealed it so that he could not deceive the nations until the one thousand years were finished... I also saw the souls of those who had been beheaded because of the testimony about Jesus and because of the word of God. These had not worshiped the beast or his image and had refused to receive his mark on their forehead or hand. They came to life and reigned with Christ for a thousand years.' -Rev. 20: 1-5

Signs of the Times:
The final beast system has arrived and is pushing free will to its ultimate limit with technological tyranny. This is happening far faster than anyone supposed. There can be no story without free will.

Do we live in the same Cosmic Arrangement as Jesus? Well, Lunar, solar and eclipse calendars are all lined up the same. Proof? Below are four 19-year periods in history with almost identical matching Eclipse dates. All 42 of our current 2014 to 2034 solar eclipse dates are listed on left.
The Jesus Metonic of 14 to 35 AD is beside it directly on right.
18 days are exact matches. 12 are one day ahead. 10 are one day behind and two dates have no match. I cannot imagine a closer match for such a flexible phenomenon as a single calendar day day separated by 2000 years.
Here are four moments in history when solar eclipses lined up.

__2015 to 2034__	__14 to 33 AD__	__554 to 573 AD__	__488 to 469 BC__
2015 Mar 20	0014 Mar 19	0553 Mar 19	-468 Mar 19
2015 Sep 13	0014 Sep 13	0554 Sep 12	-469 Sep 12
2016 Mar 9	0015 Mar 9	0555 Mar 09	-487 Mar 09
2016 Sep 01	0015 Sep 2	0555 Sep 01	-487 Sep 01
2017 Feb 26	0016 Feb 26	0556 Feb 26	-486 Feb 26
2017 Aug 21	0016 Aug 21	0556 Aug 21	-486 Aug 21
2018 Feb 15	0017 Feb 15	0565 Feb 16	-485 Feb 15
2018 Aug 11	0017 Aug 10	0565 Aug 11	-477 Aug 12
2019 Jan 06	0018 Jan 06	0558 Jan 05	-484 Jan 06
2019 Jul 02	0018 Jul 01	0558 Jul 01	-484 Jul 01
2019 Dec 26	0018 Dec 26	0558 Dec 25	-484 Dec 25
2020 June 21	0019 Jun 21	0559 June 21	-483 Jun 21
2020 Dec 14	0019 Dec 15/14	0559 Dec 14	-483 Dec 15/14
2021 Jun 10	0020 Jun 10	0568 Jun 11	-482 Jun 10
2021 Dec 4	0020 Dec 03	0568 Dec 04	-482 Dec 04
2022 Apr 30		0561 Apr 30	-481 Apr 30
2022 Oct 25	0021 Oct 24	0561 Oct 24	
2023 Apr 20	0022 Apr 19	0562 Apr 19	-480 Apr 19
2023 Oct 14	0022 Oct 14	0562 Oct 14	-480 Oct 14
2024 April 08	0023 Apr 09	0563 Apr 08	-479 Apr 9/8
2024 Oct 02	0023 Oct 03	0563 Oct 3	-479 Oct 02
2025 Mar 29	0024 Mar 28	0572 Mar 29	
2025 Sep 21	0024 Sep 21	0564 Sep 21	
2026 Feb 17	0025 Feb 16	0565 Feb 16	-477 Feb 17
2026 Aug 12	0025 Aug 12	0565 Aug 11	-477 Aug 12
2027 Feb 06	0026 Feb 05	0566 Feb 06	-476 Feb 06
2027 Aug 02	0026 Aug 01	0566 Aug 01	-476 Aug 0
2028 Jan 26	0027 Jan 27	0567 Jan 26	-475 Jan 25
2028 Jul 22	0027 Jul 22	0567 Jul 22	-475 Jul 22
2029 Jan 14	0029 Jan 15	0568 Jan 15	-474 Jan 14
2029 Jun 12		0568 Jun 11	
2029 Jul 11	0028 Jul 10		-474 Jul 11
2030 Jun 01	0029 Jun 01	0569 May 31	-473 Jun 01
2030 Nov 25	0029 Nov 24	0569 Nov 24	-473 Nov 25
2031 May 21	0030 May 21	0570 May 20	-472 May 20
2031 Nov 14	0020 Nov 14	0570 Nov 13	-472 Nov 14
2032 May 9	0031 May 10	0571 May 09	-471 May 09
2032 Nov 03	0031 Nov 03	0571 Nov 3	-471 Nov 3
2033 Mar 30	0032 Mar 29	0572 Mar 29	-478 Mar 29
2033 Sep 23	0032 Sep 23	0572 Sep 23	-470 Sep 23
2034 Mar 20	0033 Mar 19	0573 Mar 19	-469 Mar 20

The Metonic Clock returns to the same alignment about every 400 to 500 years. As the Gregorian Calendar "straightened out" the solar year by correcting leap years more accurately, it appears a "truer" clock has been revealed at 391 but there is the reality of orbits etc… Even calendar time is at the mercy of perception.

You can get a feel for this repetition by following a certain date. How about Christmas?

The Christmas Metonic

The larger Metonic clock ticked around to certain dates about (loosely) every 400 to 500 years. The clock shortened to around 300 to 400 with the advent of the Gregorian Calendar.
Perception, decree, reality. We currently live towards the end of a "Christmas Metonic", or a Metonic Cycle where the Dec 25 date appears on some eclipses. I use exact dates of Dec 25 Universal Time (time/date at Greenwich England) for this study but allow for some Eclipses that cross the date line.

I do include partial eclipses in this study, but only once do I include an eclipse that says Dec 26 on its official NASA map. That is the 1 BC eclipse, which almost certainly ran over both days. The idea of a "day" is a very fluid thing around the south pole in the winter. In fact, Van Den Bergh says, "a day is 48 hours" and that's as real as anything.

There are 25 Christmas Eclipses between about 1500 BC and the end of our study in 2057, a time frame of nearly 3500 years. I give a short history round-up that is obviously not comprehensive, just a bare minimum of "western civ" to orient ourselves in history.

I. Decline of Egyptian Empire/ Rise of Israel
Christmas Eclipses occurred on: 1108 BC, 1089 BC, 1024 BC.

Ramses the 11th died around 1070 BC, the last Pharaoh of the New Kingdom, signaling the beginning of the permanent decline of the Egyptian Empire. King David traditionally born 1040 BC/ died 970 BC. The time elapsing without Christmas Eclipses from 1024 BC to 549 BC is 475 years.

II. Return of Jews from Babylon/ Persia / Greek Empire Christmas Metonic
Christmas eclipses occurred on: 549 BC, 503 BC, 484 BC

Babylonian exile formally ended in 538 BC, when Cyrus the Great of Persia gave permission for the Jews to return from their exile in Babylon. Greece defeats Persia at Battle of Marathon in 490 BC. Buddha came around 500 BC. Beginning of Greek Philosophy.
Time without Christmas Eclipses after 484 BC to 66 BC is 418 years.

III. Roman Empire/Jesus Millennium Metonic:
Christmas eclipses on: 66 BC, 47 BC, 01 BC, 37 AD, 56AD

Rome controls the known world. Julius Caesar installed the Julian Calendar in 46 BC. Important for our consideration: I include Dec 26 on 01 BC in Antarctica, which was likely to have been on both Christmas and the day after (such long days are common at the south pole in around the solstice).

Time between Christmas Eclipses of 56 AD and 558 AD is 502 years.

IV. Transformation of Roman Empire / Rise of Islam:
Christmas eclipses on: 558, 577, 596 AD.

Fall of Rome. Prophet Muhammad born around 570 AD. Legendary King Arthur defeats Saxons in England ca. 500 AD. Beginnings of Political structures in France, Germany Spain, etc...

Time between Christmas Eclipses of 596 AD and 1060 AD is 464 years.

VI. European Empire Ascendance:
Christmas eclipses on: 1060 and 1098 AD.

Norman conquest of England in 1066. Peak of Holy Roman Empire in 11th Century. European Crusades take Jerusalem in 1098. Vikings reach North America (ca. 1000). Peak of Kievan Rus, the precursor to modern Russia. Split between Eastern and Western Christian Churches (1054). The Christmas Eclipse of 12-25-1098 notably crosses southern Italy and bisects Greece.

Time between Christmas Eclipses of 1098 and 1562 is 464 years, identical time frame as last period.

VII. World Empire Ascendance:

Christmas Eclipses on: 1562, 1581, 1582, 1628, 1666.
Colonization of much of the world by European empires. The use of mass industrial conquest and slavery. We have to take a Detour: Look at the dates! Why is this sixth empire such a strange Metonic? Wait a second! Eclipses on 1581 **and** 1582? Time has changed!

Focus: The Gregorian Calendar Change

The Gregorian Calendar was instituted by Pope Gregory in 1582, to replace the Julian Calendar, which had fallen behind the actual Equinoxes. The Gregorian "fixed" the Calendar by jumping ahead 11 days from October 4 to Oct 15 1582, a span very close to a perfect Metonic week (which is 11.33 days).

It so happened that the second to last eclipse before this change was on Christmas, Dec 25 1581.
In normal years we might expect to have a Dec 14 eclipse in 1582 one lunar year later. But because of the 11 day leap, instead there is a second Christmas, on Dec 25 1582.
Thus, somewhat amazingly, the only eclipse date to appear two calendar years in a row in human history was Christmas. It happened in 1581 and 1582.
What's funnier is that the second of these Christmas Eclipses, in 1582, actually crossed tiny Christmas Island in Indonesia. No kidding.

Time between the Christmas Eclipses of 1666 and the next Christmas Eclipse of 1935 is only 269 years. a far shorter span compared to all previous seemingly because of the new Calendar reset in 1582.

Gregorian Calendar Change
Two Christmas Eclipses in a Row..first in 1581 (second from right)

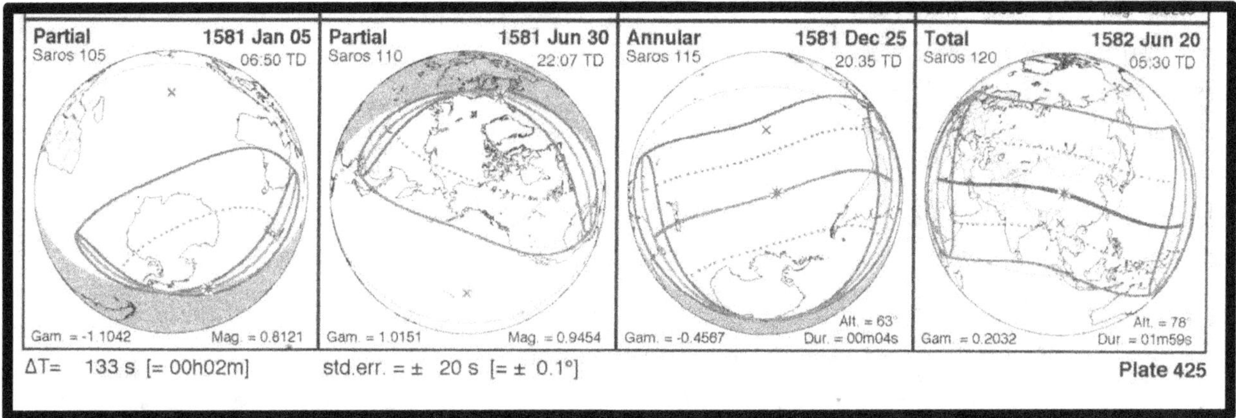

(Maps courtesy of NASA/ Fred Espenak and Jean Meuss.)

The next year, Gregorian Calendar jumps 11 days...so we have: on left)

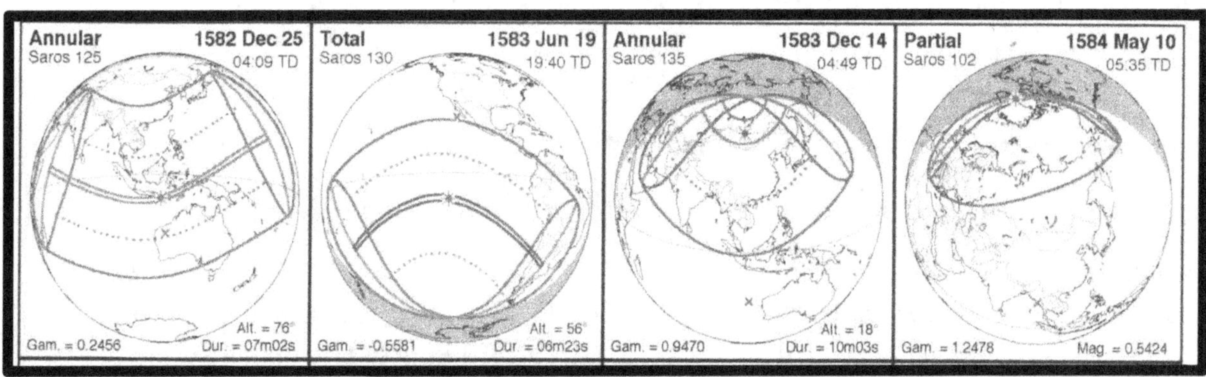

(Maps courtesy of NASA/ Fred Espenak and Jean Meuss.)

The one above in 1582 happens over Christmas Island. Fun!

VII. The Final World Empire:

This is the world we live in. Look around. Christmas Eclipses rolling around again.

Christmas Eclipses on:
Dec 25 1935 (Ring of Fire Antarctica)
Dec 25 1954 (Ring of Fire South Africa)
Dec 25 2000: Officially, this was the last eclipse of the second millennium, occurring on Christmas as a partial over North America—23% at Mount Saint Helens. 44.44% at NYC, 40% at Washington DC, 23.00% at Dallas, 47% at Kennebunkport, Maine,
Dec 25/26 2038: Starts over Australia and New Zealand on Dec 26, then crosses the International date line to end on Christmas Dec 25 in the south Pacific.
Dec 25/ 26 2057: Antarctica beginning on Christmas and ending on December 26th.

Metonic Weirdness:

So, the spacing between periods of Christmas Eclipses is:
475 years between 1024 and 549 BC.
418 years between 484 and 66 BC .
502 years between 56 AD and 558 AD.
464 years between 596 AD and 1060 AD.
464 years between 1098 and 1562.
269 years between 1666 and 1935 (due to Gregorian calendar adoption?)
250 years between 2057 and 2307.

There seemed to be a fairly stable Metonic pattern of spaces of about 400 to 500 (average about 470) years between clusters of Christmas Eclipses until the calendar adoption.

Then comes the new calendar in 1582, which apparently decreases the Metonic expanse. Only 269 years separate Christmas Eclipses between 1666 and 1935. Some of the shortening is just the Metonic stretching around date lines and Polar time zones.
How long is the Cosmic Clock taking to truly turn all the way around? Possiblly 391 years. That's how long it takes for the Draco-Metonic Cycles to reconnect after the 19-year alignment.

It's unclear how many people on earth have any idea how to consider this fundamental phenomenon.

Now, let's take a look at the Time of Jesus compared to Ours

Roman Empire Millennium Metonic: **Modern World Empire Metonic:**

Xmas eclipses on:
Dec 25 66 BC (partial then 19 year lapse)   Dec 25 1935 (Ring of fire then 19 year)
Dec 25 47 BC (ring of fire then 46 years)   Dec 25 1954 Ring of Fire then 46 years)
Dec 25/26 1 BC (partial -then 19 years)   Dec 25 2000: (partial the 19 years)
Dec 26 18 AD (ring of fire then 19 years)   Dec 26 2019 (Ring of fire then 19 years)
Dec 25 37 AD (ring of Fire then 19 years)   Dec 25/26 2038: (ring of Fire then 19 years)

Dec 25 56 AD (Total)   Dec 25/ 26 2057: (Total)

The short eclipse summary above is essentially proof that we live in the same sun/moon calendar alignment as occurred around the time of Jesus Christ.

66 BC to 56 AD Covers total span of 122 years, the maximum for any recurring date on the eclipse Metonic. From 1935 to 2057 is span of 122 years, the maximum for any recurring date on Metonic.

Summary: There is an obvious symmetry to the Christmas eclipses around the birth and death of Jesus and around the 2,000[th] anniversary of that event which we are living through. Each of these cycles provides a total of 5 eclipses over the course of 122 years. The middle (3rd) eclipse in both groupings is partial. Both of these partials occur at the turn of a millennium.

The eclipses are also spaced identical distances apart in years...19, 46, 19, 19.19

Except for the first eclipse in each group, even the *type* of eclipses are the same.
66BC to 56 AD— partial, annular, partial, annular, annular, total.
1935 to 2057 AD-annular, annular, partial, annular, annular, total.
The Christmas eclipse of 37 AD and the Christmas Eclipse of 1954 both crossed South Africa.

If the Crucifixion was 33 AD, then it occurred 99 years after 66 BC, the first Christmas eclipse in the Jesus Eclipse cluster. 99 years from the first 20th Century Christmas eclipse (1934) brings you to 2033. If you believe Jesus died at 33 years of age in 33 AD, then the year he was born in 1 BC had a solar eclipse on Christmas. If we begin Western Imperial History at around 1500 BC, we live in a seventh empire as determined by approximate 500 Cosmic Year time frames, defined by repetitive intersections of solar, lunar and draconic calendars. We appear to be living in a cosmic mirror of the time around the Life of Jesus Christ. But it's not quite a mirror. It's more like a clock, or a story.

Again, four of our most important days of human civilization are shared between the Jesus Metonic and our current day. Spring Equinox, Fall Equinox, Summer Solstice and Christmas.

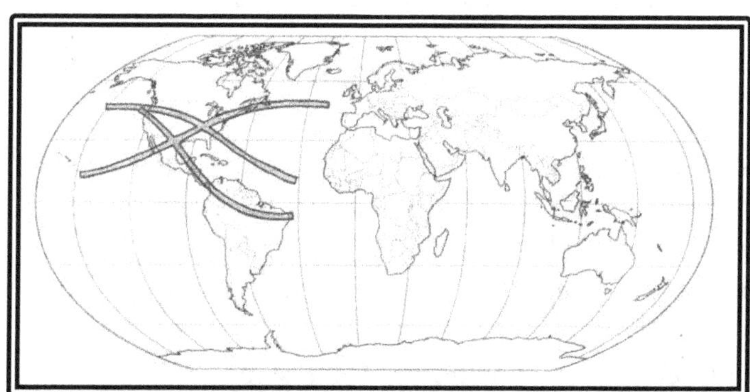

The 2000th Jesus Anniversary Eclipses Here in Our Modern World

First, another quick look (left) at the American Aleph of 2017 to 2024, which is centered on North America and the USA. It is 2000 years removed from Jesus early adulthood.

Below are all total solar eclipses from 2026 to 2034 (no ring of fires pictured)

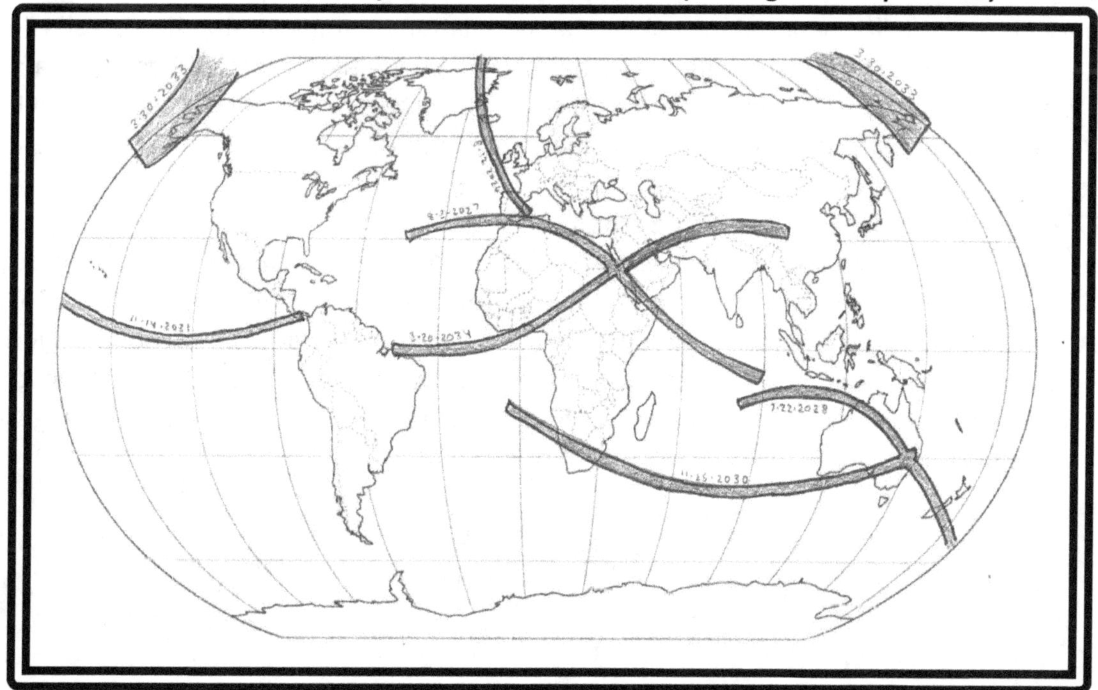

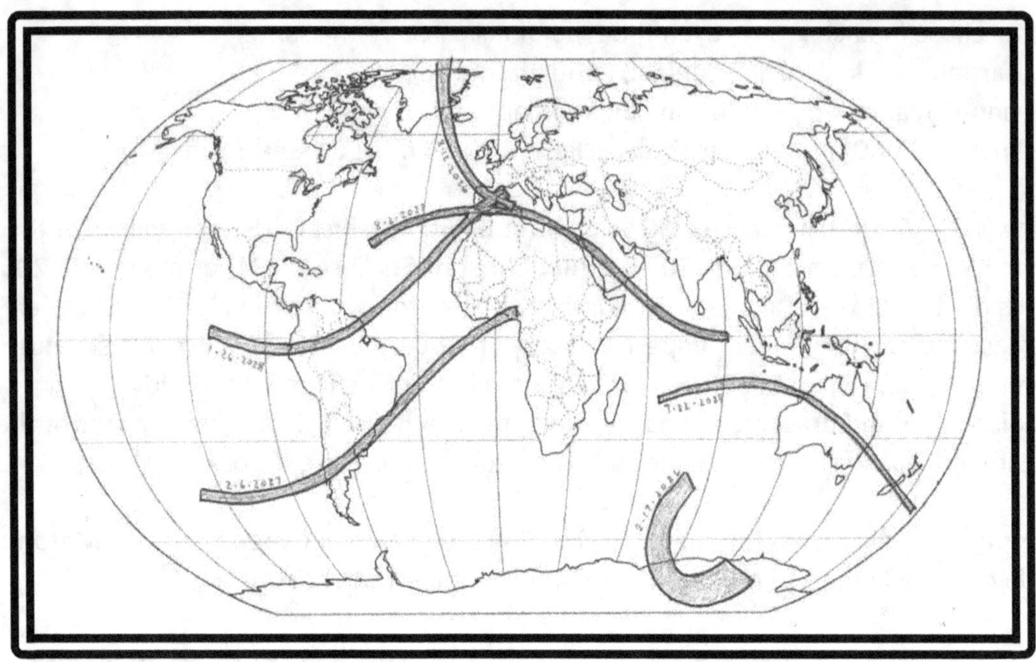

Solar Eclipses 2026 to 2028—The 2000th Anniversary of 26 to 28 AD

Few argue for a 26 AD Crucifixion dating. But 26 AD is a good place to start, as it was established by historic authorities as the beginning of Governorship of Judea by Pontius Pilate. Also, as there are no solar eclipses in 2025, the clean break from the 2024 USA Eclipse allows for coherent assessment. Some argue for a Thursday Apr 10 27 AD Crucifixion Date, and some hold to a 28 AD event.

Aug 12 2026 Spain/Reykjavik Total (Saros 126) After 112 years without a Total eclipse, this is Spain's first of 3 complete eclipses in 18 months, two of them totals. 500 years after Spain began the African Slave Trade to the New World in 1526. Eclipse occurs 128 years to the day of the peace treaty that ended the Spanish-American War on Aug 12 1898, basically ending the 600 year Spanish Empire. The extraordinary path of totality avoids almost all human populations, curling around the British Isles— 91% coverage in London, 92% in Paris. The only other nation to experience totality besides Spain is Iceland, and amazingly that occurs only on the bare sliver of western coast that includes its capital of Reykjavik (whose metro area contains 60% of Iceland's population).

Feb 6 2027 Slave Coast Eclipse Ring of Fire (Saros 131)
Less than 6 months after Spanish Judgment of Aug 12 2026, Ring of Fire Eclipse. Judgment Eclipse. Again, still 500th Anniversary Year of the importation of African slaves to South America in 1526, the eclipse path traces backwards the maritime routes followed from slave traders from South America to the "Slave Coast" of Africa -Benin, Ghana and Nigeria--ending over notorious port of Porto-Novo in Benin, one of the centers of the Slave Trade. Annularity just offshore of Rio De Janeiro, the largest slave port in the history of the world.

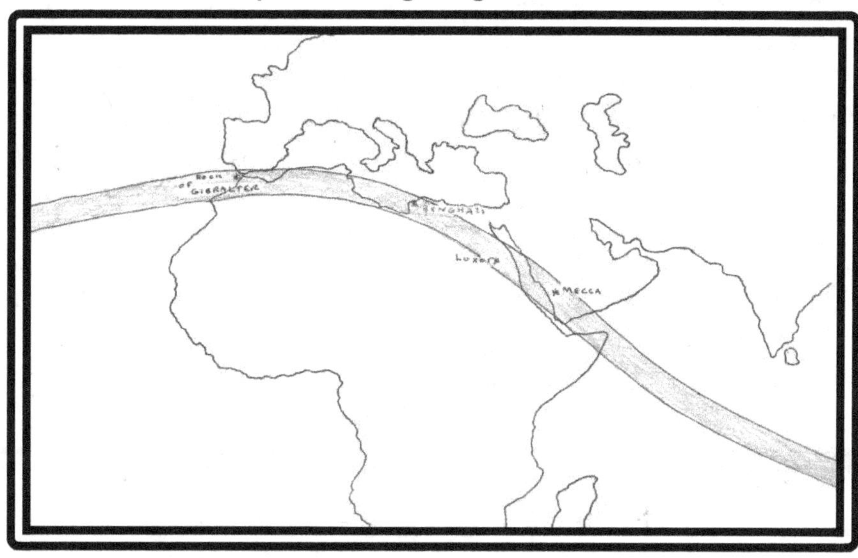

August 2nd 2027 Valley of the Kings Total Eclipse (Saros 136)

Hepton Cycle Monday Eclipse. 355 days (Exactly one lunar year) after Spanish Total Eclipse.
1211 days after USA second line of Cross.
3633 days (exactly 10.5 eclipse years) from Great American Eclipse of Aug 21 2017.
2422 days (7 eclipse years) from its own coming Cross, the Spring Equinox Eclipse of Mar 20 2034.

Straddling the straits of Gibraltar-Europe experiences totality simultaneously with Africa. Totality in Spain, Rock of Gibraltar, Tangier, Tripoli. Path over Egypt along the Nile bringing maximum eclipse to the Valley of the Kings in Luxor, the metropolis of the sun gods, the necropolis of the Pharaohs. Crosses into Arabia--including totality in Mecca, exiting into Gulf of Oman through Yemen, and passing over the Horn of Africa before ceasing in the southernmost Maldives in the Indian ocean.
83% coverage in Jerusalem. If you were going to bet on a judgment eclipse...

This will be the longest duration total eclipse for the rest of the 21st Century at 6 minutes 22 seconds. The great Saros Cycle 136, whose last appearance was Wuhan Totality of July 22 2009. Centerpoint and maximum at the Valley of the Kings, the spiritual heart of the mystery religions.

Jan 26 2028 Amazon to Spain Ring of Fire (Saros 141) Pictured Previous Page

Amazon, Spain and Portugal. Spain's 3rd eclipse in 18 months. Across the Amazon of South America, including Brazil. Eclipse crosses Atlantic, entering Portugal at Sagres ("holy"). Annularity just south of Lisbon. Like 2026, path ends just beyond border of Spain, at border of France. Occurs on "Australia Day", Australia's national holiday. See the next eclipse in series 177 days later.

Jul 22 2028 Australia/New Zealand Total (Saros 146) Pictured Previous Page

177 days after Australia Day Eclipse, Australia is bisected with totality in Sydney and Botany Bay, where the first prisoners were brought to colonize the continent 240 years and 177 days earlier, in 1788. One lunar year from the Mideast Eclipse of 2027. 19 solar years exactly after Wuhan Total eclipse of Jul 22 2009. Also crosses southern island of New Zealand, exiting over city of Dunedin.

2029-Only PARTIALS AND LUNAR ECLIPSES

Some have suggested a **March 23 29 AD** Crucifixion/Resurrection, including H. Grattan Guiness though today this is among the least popular dates. 2029 brings the 2000th Anniversary of the tremendous Nov 24 29 AD Total solar eclipse, which would have been seen throughout the Holy Land (90% and more) and the Roman Empire.

2029 only has Partial Solar eclipses:
January 14th 2029 (the eve of what would have been Martin Luther King's 100th birthday-Jan 15 1929)--visible only in North America—also June 12 2029, July 11 2029 and December 5 2029

Then comes 2030

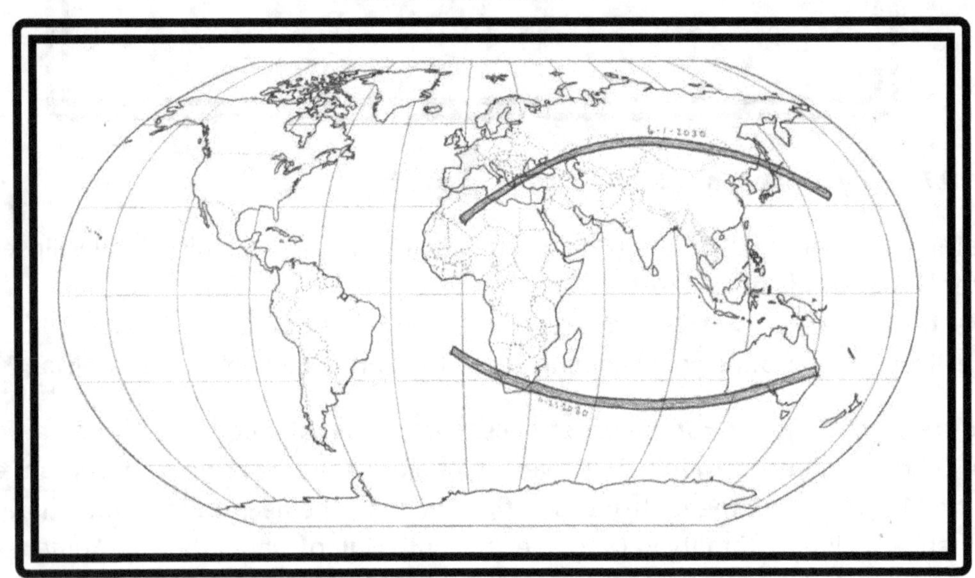

2030--TWO GREAT HEPTON ECLIPSES

2000th anniversary of possible Crucifixion to Friday, April 7th 30 AD, among the most popular dates for the Crucifixion. Two great eclipses form a symmetrical cup in the north and south of the eastern hemisphere. Both are Hepton Eclipses, one from each of the two series.

Jun 1 2030 Athens/Constantinople Ring of Fire

Ring of Fire. An extraordinary path beginning in Algeria and ending in Japan. Great cities along the path include Tripoli, Athens, Istanbul (Constantinople), Sevastopol, Volgograd and Sapporo. Ring of Fire. Judgment Eclipse. Another extraordinary path beginning in Algeria and ending in Japan. 63% in Jerusalem. 32% in Wuhan.

Nov 25 2030 South Africa to Australia--British Empire Empire

Total Solar Eclipse. Hepton Cycle Monday Eclipse. 1211 days from Aug 2 2027 Mideast Eclipse. 2422 days from USA 4-8-2024 eclipse. Creating an extraordinary cup of empire (see all land masses to north of eclipse). Almost all open ocean yet crossing South Africa and completing Tav over Australia. Totality in Durban, South Africa. reaching almost to Brisbane and the Gold Coast in Australia.

2031 and 2032

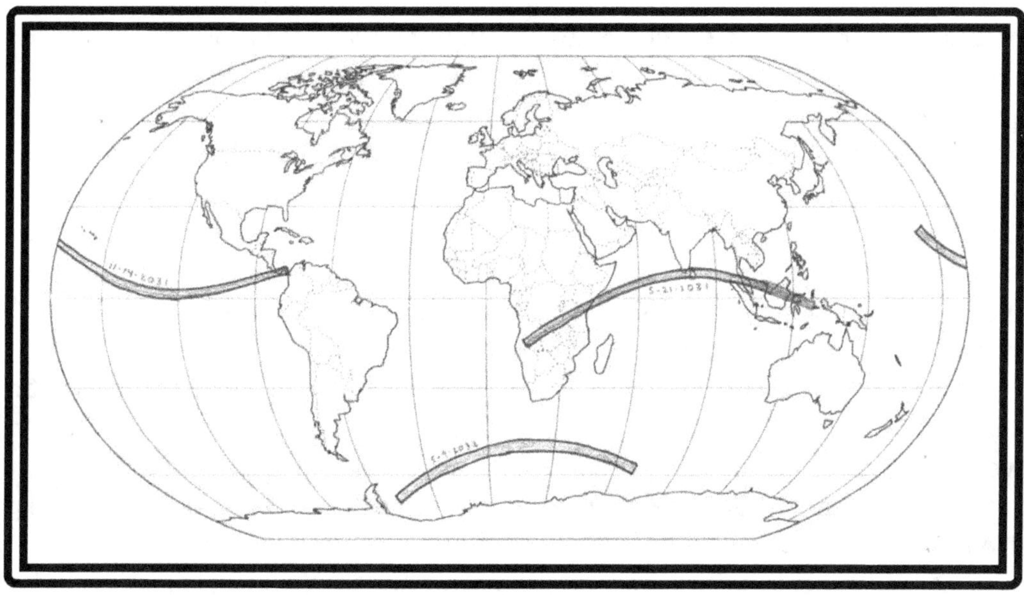

(Above) 2031 and 2032

 Seventh Day Adventists and others hold to the date of an Apr 25 31 AD Crucifixion.

May 21 2031—Africa to Malay/Dar Es Salaam Ring of Fire (Saros 138)

Across the middle of Africa, from Angola, exiting over Dar Es Saalam (place of peace) in Tanzania. Crosses Indian Ocean and extreme southern India, then Sri Lanka, again tracing 2004 tsunami path backwards towards Indonesia. Totality in Ernakulum ("the abode of Lord Shiva") and Jaffna, Sri Lanka. Just north of Kuala Lampur. Bisecting Malaysia before ending in Indonesian Islands.

Nov 14 2031- Isthmus of the Continents of Western Hemisphere.(Saros 143)

An open ocean eclipse that touches virtually no land ending with totality at sunset over the most extraordinary land bridge on planet earth, the Isthmus of Panama. The Isthmus separates North and South America by a thin strip only 50 miles wide. Incredibly, the next total eclipse (3-30-2033) will straddle the Bering Strait.

Two total eclipses in a row connect hemispheres of earth while touching very little land.

2032---ONE RING OF FIRE and Two Blood Moons Visible from Jerusalem

Based on a possible April 14 32 AD Crucifixion date. Many argue for the 32 AD Crucifixion based on math from the Bible's Book of Daniel. There was a total lunar eclipse on Passover April 14 32 AD not visible from Jerusalem. **Blood Moons In 2032:** there are two Total Blood Moons (April 25 and October 18). They don't land on the High Holy Days, but they are both visible from Jerusalem. Combined with the April 14 2033 Eclipse, that makes three straight Total Lunar Eclipses in a single year and a half that are visible from Jerusalem.

April 14 in History:

A Total Lunar Eclipse occurred at Passover Apr 14 32 AD, but it was not visible from Jerusalem. A Total Lunar eclipse occurs exactly 2,001 years later April 14 in 2033, and it is visible from Jerusalem.

A look at dates in relation to the Passover 32 AD Lunar Eclipse and Possible Crucifixion

1) The atom was first split in Cambridge, England on Apr 14 1932, exactly 1900 solar years after the eclipse.
2) Passover 32 AD Eclipse occurred 1833 years before Lincoln shot on Apr 14 1865.
3) Passover 32 AD Eclipse occurred exactly 1890 years before Titanic hits iceberg.
4) Passover 32 AD Eclipse occurred exactly 2040 (2040.0236) eclipse years before Martin Luther King killed on Apr 4 1968.

(These four events are four of the most important events in modern history, metaphorically.)

5) Passover 32 AD is exactly 2000 years minus 11 days from Apr 25 2032 Total lunar eclipse over Jerusalem.
6) Passover 32 AD is exactly 2001 years to the day from Apr 14 2033 total lunar eclipse on Passover in Jerusalem rising at sunset.

There is one complete solar eclipse in 2032, pictured on previous page. It is:

May 9 2032 Ring of Fire (Saros 148) A ring of fire eclipse with a path that reaches no land lurks deep in the southern hemisphere in the Antarctic Ocean.

ALSO IN 2032

Partial Solar Eclipse **11-3-2032 (Saros 153)** A partial solar eclipse in November visible from Asia. No coverage in Jerusalem.

Transit of Mercury: There is also a transit of the planet Mercury across the face of the Sun on November 12/13 2032 This will be the first Mercury transit since Nov 11 2019, just as pandemic began in China (101 years after Armistice Day 1918).13 years is about the longest time between Mercury Transits that can occur. Transits of Mercury occur about 13 times a century.

Again, there will be:
Total Lunar Eclipse –April 25, 2032—rises at sunset in Jerusalem.
Total Lunar eclipse- October 18, 2032—rises in later evening in Jerusalem.

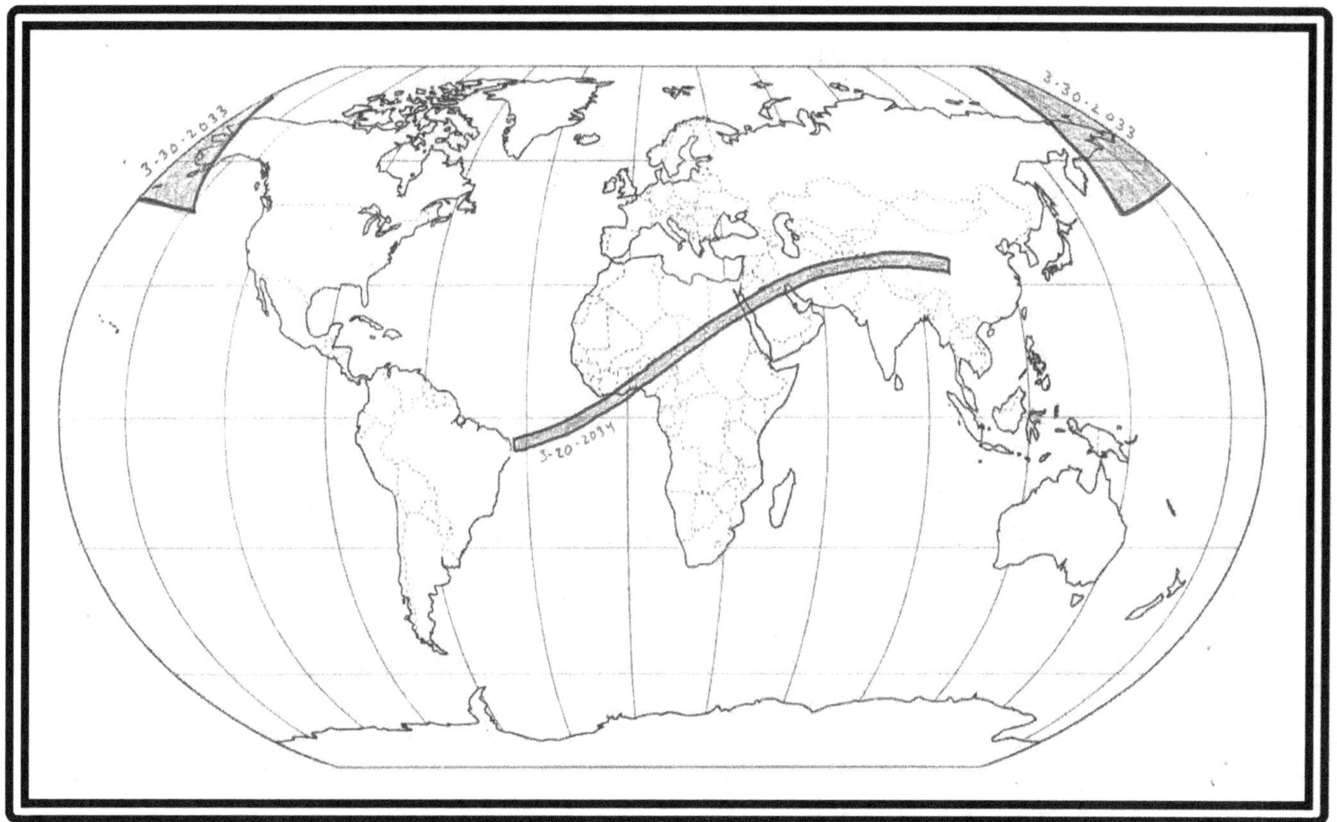

2033 and 2034 Total Eclipses

Two total eclipses occur on earth between late 2031 and early 2035, making them unique in their isolation. They are also the only total eclipses which occur around the most likely 2,000[th] Anniversary of the death and resurrection of Jesus Christ. They are pictured above.

The March 30 2033 Total Eclipse is the only one in our entire survey like it. It covers the Bering Strait, straddles both hemispheres and the International Date Line with an eclipse path maybe a couple of hundred miles wide.

The March 20 2034 Total eclipse happens exactly one lunar year later, the second half of the Monday Hepton Cross whose peculiarity began this book. Maximum Eclipse of this eclipse occurs in a desert region over north-central Chad in Africa. The distance from this maximum point to the Bering Strait is remarkably similar in both directions. If you travel northwest, it is roughly 7800 miles (12,600 km) to the Bering Strait. If you travel Northeast, it is roughly 7100 miles to the Bering Strait.

In other words, the illustration above is just as symmetrical as it looks.

Remember that 2032 and 2033 are also home to the Lunar Blood Moon Tetrads.

March 30th 2033—Great Bering Strait Eclipse (Saros 120)

Four days before the traditional and most widely accepted 2,000th Anniversary of the death of Jesus Christ (April 3 2033), eclipse perfectly crosses the Bering Strait, giving totality to both hemispheres at the same time. It occurs at the only place this could possibly happen, with both Russia and the USA having a total eclipse simultaneously. No other nation will experience the path of totality.

The Bering Strait is credited with once having a land bridge, which allowed for the population of the Americas during the last ice age. Few towns are in the eclipse path, which is sparsely populated. Significantly, those towns include Provideniya (pop. 2,000) in Russia, which means "Providence" and Nome (pop.3600), Alaska in USA, one of the centers of the great Alaska Gold Rush of 1897 to 1907. No one is sure what the name "Nome" means.

SAROS 120

2033 is the last complete eclipse from the great Saros 120, which began the 19-year Metonic with a North Pole Eclipse mentioned before on Mar 20 2015 and which ends with the next eclipse.

List of some Great Saros 120 Eclipses with USA Relations:

1. Apr 26 1492 On Isthmus of Panama. Last Eclipse Before America discovered.
2. Jul 10 1600 America to Spain First Total of 17th Century (Cape Canaveral)
3. Jul 21 1618 Cross Country USA
4. Oct 27 1780 Revolution USA
5. Nov 30 1834 Trail of Tears USA
6. Jan 1 1889 Far West Buffalo New Years Eclipse USA
7. Jan 25 1925 NYC Roaring Twenties USA
8. Jul 20 1944 Iran to Southeast Asia -first eclipse of great Southeast Asia Swarm
9. Feb 15 1961 Iron Curtain –Europe/Russia
10. Feb 26 1979 Mount Saint Helens/ Microsoft USA
11. Mar 9 1997 USSR Judgment
12. Mar 20 2015 North Pole
13. Mar 30 2033 USA/Russia and Date Line linked at Bering Strait. Numeric oddity of 3-30-33.

Then Saros 120 ends its run of complete eclipses that began on Aug 11 1059 AD.

The Bering Strait Eclipse is one lunar year before the Equinox Eclipse of Mar 20 2034 that ends the 19 year Equinox Metonic, bringing to a close all reasonable possible 2000 year anniversaries of the Crucifixion/Resurrection. The only other solar eclipse in 2033 is a partial on Sep 23.

See if you can keep this straight: Sir Isaac Newton's calculation for the Crucifixion was Apr 23 34 AD. The great solar eclipse of 2034 occurs on Mar 20. Isaac Newton died on Mar 20 (1727). Newton was born on Christmas Day. But that was before the calendars changed in England. Now if you google it, the network often says he was born on January 4. Because It can say anything It wants to. Newton was probably the most famous Scientist/ Bible Scholar of all time.

Regardless, the 2,000th anniversary year of 34 AD is home to a great celestial phenomenon.

March 20 2034--The Great Civilization Eclipse (Saros 130)

Hepton Cycle Monday Eclipse. Second half of the Mideast Tav or Cross.
1211 days from Nov 25 2030 British Empire Eclipse.
2422 days from Mideast Aug 2 2027 eclipse.
3633 days from North America Apr 08 2024 eclipse
One lunar year from Bering Strait Eclipse.

Eclipse travels from the Coast of South America across Africa to Egypt, scraping Muhammad's tomb in Medina (almost totality) as it crosses Arabia, onward through Iran, Afghanistan, Pakistan (totality in Islamabad), India, Tibet and ending in China. Jerusalem will see 74% coverage of the sun.

This Spring Equinox Eclipse occurs 19 years after the North Pole Spring Equinox eclipse. The Spring Equinox is on or around New Year's Day for Hindu, Persian, Muslim, Zoroastrian and Bahai Calendars. The Equinox or the new or full moon nearest it was once New Year's Day for many if not virtually all ancient cultures of the Northern Hemisphere.

2033 and 2034---IMPRESSIVE PECULIARITIES

April 3rd 33 AD is the traditionally accepted date of the Crucifixion, when there was a lunar eclipse visible at moonrise in Jerusalem. As shown earlier, Total Lunar eclipses re-occur on the high holy days of Judaism in April 14th and October 8th 2033, Passover and Feast of Tabernacles, 2000 years later. Passover 2033 falls on April 14. That is 2001 years after the April 14 32 AD possible crucifixion date and 15 days after a Total Solar Eclipse on 3-30-2033 that connects both hemispheres simultaneously.

Blood Moons 2034

2034 sees the last of the Blood Moon Tetrads, with lunar eclipses on the High Holy days of Judaism completing the 19-year Metonic Reality begun with the 2014/2015 Tetrads.

A penumbral lunar eclipse arrives on the night of Passover, April 3 2034, beginning "visibility" at 7:53 pm in Jerusalem, though the shadow will be so light that it won't be noticeable. April 3 2034 is the exact 2001st anniversary of the Crucifixion of Jesus Christ, if you accept traditional dating.
This is two weeks after the Equinox Eclipse that creates a Cross over Egypt.

There will also be a partial lunar eclipse on Feast of Tabernacles, Sep 28 2034, visible from Jerusalem.

And that's all for this work. Thank you for joining me. May God bless you all.

"There will be signs in the sun, moon and stars. On the earth, nations will be in anguish and perplexity at the roaring and tossing of the sea. People will faint from terror, apprehensive of what is coming on the world, for the heavenly bodies will be shaken. At that time they will see the Son of Man coming in a cloud with power and great glory. When these things begin to take place, stand up and lift up your heads, because your redemption is drawing near "

-The words of Jesus *Luke 21:25-28.*

Sand Sheff is a musical entertainer and author. Born in Arizona and raised in Oklahoma, he currently lives in the Four Corners region of the American Southwest. In 2017, he published Eclipse Miracle, a children's book about eclipses. His latest work, Eclipse Witness, was three years in the making.

Some praise for Sand Sheff's last book *Eclipse Miracle: The Sun is The Same Size as the Moon in the Sky*

"A wonderful book by Sand Sheff, with great scientific and spiritual content for all ages and backgrounds." Dan McGlaun Eclipse 2024.org

"There aren't very many children's books about intelligent design. This is a welcome addition."
Dr. Guillermo Gonzalez author of The Privileged Planet

"It has to be one of the most beautifully presented stories for children! It contains science for beginners, geography, creation and God lessons, and how to overcome fear…"
Dr. Katina Michael, editor-in-chief IEEE Technology and Science magazine

www.ingramcontent.com/pod-product-compliance
Lightning Source LLC
Chambersburg PA
CBHW081141130726
47996CB00009B/2928